INORGANIC
THERMOGRAVIMETRIC
ANALYSIS

by

CLÉMENT DUVAL

Directeur de Recherche au C.N.R.S.
Directeur du Laboratoire de Recherches Micro-analytiques (E.N.S.C., Paris)
Président du Groupe de Chimie analytique

SECOND AND REVISED EDITION

Translated from the French manuscript

by

RALPH E. OESPER PH. D.

Professor Emeritus of Analytical Chemistry, University of Cincinnati

ELSEVIER PUBLISHING COMPANY
AMSTERDAM—LONDON—NEW YORK

1963

SOLE DISTRIBUTORS FOR THE UNITED STATES AND CANADA

AMERICAN ELSEVIER PUBLISHING COMPANY, INC.

52 VANDERBILT AVENUE, NEW YORK 17, N.Y.

SOLE DISTRIBUTORS FOR GREAT BRITAIN

ELSEVIER PUBLISHING COMPANY LIMITED

12B, RIPPLESIDE COMMERCIAL ESTATE,

RIPPLE ROAD, BARKING, ESSEX

LIBRARY OF CONGRESS CATALOG CARD NUMBER 63-8403

WITH 80 ILLUSTRATIONS AND 13 TABLES

Preface

"Il y aura toujours des Bernard Palissy pour brûler leurs meubles."   G. URBAIN

Late in 1952, I wrote in the Postscript of the first edition: "Now that the initial impulse has been given, my sincerest wish is to see a second edition of this modest work supplemented by the researches of my fellow-workers in all parts of the world."

The first edition has been out of print for three years or more and I am happy to report that my wish has been fulfilled beyond my wildest hopes. By 1st January, 1961, there were 52 models of thermobalances and 10 of these were available from dealers throughout the world. I have read 2200 papers in periodicals from all quarters of the globe or, in more precise terms, I have regularly consulted the following through the given dates, *i.e.* through 1960.

Bulletin signalétique du C.N.R.S.,	No. 11–12, December 1960
Chemical Abstracts,	No. 22, 25th November 1960
Analytical Abstracts,	No. 12, December 1960
Analytica Chimica Acta,	No. 6, December 1960
Analytical Chemistry,	No. 12, December 1960
Journal of Inorganic and Nuclear Chemistry,	No. 12, December 1960
Bulletin de la Société chimique de France,	No. 12, December 1960
Mikrochimica Acta,	No. 6, December 1960

For practical reasons, the present edition will take into account the papers published up to January 1961. Those appearing after this arbitrary cut-off date will be held for the third edition. In some instances, we have been able to see only abstracts; the originals were not available in France and the authors did not respond, in most of these cases, to our requests for reprints.

Up to the present it has been analytical chemistry which has profited most from the recent progress of thermogravimetry, and this is the reason the word *analysis* has been retained in the title, but the reader will discover that this edition deals also with kinetic studies, investigations of catalysis, reactions in the solid state, and studies of the behavior of standard materials for titrant solutions, and also of various substances which are not employed in analysis. It seems appropriate to recall here the circumstances and train of thought which led me, from 1946 on, to devote my attention to the work on precipitates that had been carried out by the Japanese school.

A glance at the current reviews of analytical chemistry, or simply at the semimonthly *"Analytical Chemistry"* section of *Chemical Abstracts* will show that gravimetry, at least in inorganic chemistry, is being superseded by methods which seemingly are more rapid, such as titrimetry, colorimetry, polarography, coulometry, spectrography. These branches of analytical chemistry have made wonderful progress in the past 20 years and it may well be asked whether the funnel, the filter paper, the crucible and the balance with its knife-edges and discontinuous weighings, heritages of a long tradition, are not destined to disappear soon, and give way to other more modern devices capable of equal precision.

However, the balance is indispensable to the carrying out of the methods other than gravimetry, save perhaps coulometry, and it is still necessary when determining the faraday.

In fact I have always wondered why authors are so prolific with details specifying the conditions of precipitation (volume of liquid to be determined, maximum and minimum concentration, volume and titre of the reagent, its temperature, pH of the solution, volume of buffer, the time necessary for ageing the precipitate, the porosity of the filter crucible or ashless filter paper, the preparation and volume of wash liquids), while often even in the larger classical texts, nothing is said about the subsequent treatment except the simple direction "...and ignite to constant weight..." or such vague phrases as "heat not above a dull red", "heat so that the bottom of the crucible becomes red", or "heat to the highest possible temperature", etc. This was excusable in the time of Berzelius and Rose, but should no longer be tolerated eighty years after Le Chatelier invented the thermoelectric pyrometer. This state of affairs is extremely regrettable, for, in spite of the remarkable progress which analytical chemistry has made in the fields of aqueous systems, its empirical character is preserved, as well as "those old outmoded habits" which pure chemists are so prone to mock.

While visiting Ludwigshafen on the Rhine in 1946, I was impressed by the automatic execution of organic combustion analyses by the methods that had recently been published by W. Zimmermann. From that time on I was never free of the idea of evolving rapid automatic

inorganic gravimetric analyses which would require no preliminary separations and which could be recorded graphically as is done for infra-red
spectra. Happily the Chevenard thermobalance had progressed beyond
the experimental stage, but there remained a vast amount of work to be
done before the new methodology was adapted to routine analyses that
could be done by a technician.

With the aid of 17 collaborators, we have prepared and pyrolyzed
around 1200 precipitates which have been proposed for use in inorganic gravimetric analysis, sifting the worthwhile from the poor methods, and then judging them from the standpoint of their adaptability to
automatic gravimetry. We dealt particularly with those quantitative procedures which require a heating below 200°, with the result that some of
these procedures were completed in only 12 minutes. Furthermore, the
curves obtained on graph paper were such that the operator was no longer
exposed to the danger of making an incorrect weighing, and with such
permanent records available he could postpone, if he wished, his consideration of the findings to a convenient time long after the analysis
was made.

We have greatly appreciated the comments of some of our colleagues,
especially those whose methods and findings I have criticized or improved.
In this connection, I will cite only F. E. Beamish and W. A. E. McBryde,
who in their contribution to C. L. Wilson and D. W. Wilson's *Comprehensive Analytical Chemistry* (Elsevier Publishing Company, 1960)
wrote: "By far the greatest contributions made toward the appreciation
of appropriate heating and drying temperatures for precipitates have
been made by Duval who, through a long series of publications, has
recorded thermograms to guide the heating practice for most of the
gravimetric methods."

We have also been much interested in the many criticisms of our
publications. Some of these (less than 10) are justified. As to the others,
we will attempt in Chapter 5 to give the reasons for the lack of agreement
by pointing out the variables to which a thermolysis curve is subject.
It must be remembered that thermogravimetry is still in its infancy
and is still cutting its teeth, and despite the latest advances in electronics,
I have not yet been able to find any substance which gives the same
decomposition temperatures on five commercial thermobalances of
different makes, even though *all* of the heating conditions apparently
were alike.

In my opinion, this lack of agreement does not constitute too serious
a defect with respect to automatic gravimetry, since I accomplish the

weighing–*i.e.* the measure of a length–in the middle of a horizontal if it exists.

By itself, thermogravimetry is not capable of resolving even in approximate fashion all of the problems presented to it, and even if the trace contains a horizontal indicative of constant weight. This is why thermogravimetry is used in conjunction with conventional chemical methods, such as differential thermal analysis, X-ray analysis, and other procedures involving magnetic properties, and infra-red especially, etc.

The reader will discover that this second edition does not conform to the true sense of the term, *i.e.* it is not merely a complement to the first edition. Our objective in the previous edition was to gather into the form of a monograph our own many papers which were scattered among the various issues of *Analytica Chimica Acta* together with a reproduction of all of the curves. Our present objective is to present also the work of other investigators. It is no longer feasible to reproduce around 5000 curves, of which some are so small (often reduced 16 times) that their interesting details are no longer visible. Accordingly, in Chapter 4 we have shown the types of curves most frequently encountered. Actually, it is not the *form of the curve* which interests the worker, since it changes from apparatus to apparatus, but rather whether the curve includes horizontals and at what temperatures, and he is also desirous of learning the nature of the compound or compounds related to these horizontals.

I have still another wish to record before this manuscript is turned over to the publisher. I sincerely hope that the third edition will not register the debasement of thermogravimetry. If certain innovations such as the derivative curve by Erdey appear to be quite successful and point to the things to come, to us it will seem entirely useless now in 1961 to trace out certain curves point by point after we have replaced one of the pans of a knife-edge balance by the wire supporting the crucible descending into the muffle furnace set up on a laboratory table, and all this merely to dehydrate some preserves or ham.

Alas, why must this preface end on a mournful note? Since the first edition was issued, we have lost three world-renowned scientists, three pioneers in this field, whose long careers were filled with useful accomplishments. Honda died in 1954, and Guichard and Chevenard departed this life in 1960. A likeness and short biographical sketch of each are properly included in this volume. Although their researches were along different lines, they all attained eminence. Without their accomplishments, this book could not have seen the light of day.

Paris, 1st June, 1961   CLÉMENT DUVAL

Contents

PART TWO – THE THERMOLYSIS CURVES

CONTENTS

PART ONE

The Thermobalances

We have described in detail only those thermobalances which are of historical interest and those which are commercially available at present.
All of the temperatures given in this book are in degrees C.

CHAPTER 1

Brief Historical Review

1. Definitions

The thermobalance is an apparatus which can be used to detect and, if required, record photographically or graphically the changes in the mass of a substance being heated or cooled, as a function of the temperature or the time or those of a substance maintained isothermally, as a function of time.

The curve or trace produced is known as a *thermogram*, or *thermolysis curve* or *pyrolysis curve* or *thermo-weighing curve*.

The ensemble of the techniques which may be accomplished by the kinetic or dynamic method constitutes *thermogravimetry*, which must be clearly distinguished from thermal analysis.

2. The precursors *

Efforts to learn the changes in mass of a heated substance go back many years, the objective being either to follow the course of a dehydration, an oxidation, or especially to discover the temperature at which a substance begins to decompose, a datum of importance for metallurgical operations. This is why through all of the 19th and in the early years of the 20th century we find investigators constructing loss of weight curves point by point as a function of the temperature, the weighings being made with an ordinary balance after the specimen had cooled. This technique was employed for instance by Truchot[1] who heated pyrites and pyrrhotite as early as 1907 using a method resembling the present thermogravimetry. In 1912, Urbain and Boulanger[2] constructed an electromagnetic compensation balance to follow the efflorescence of

---

* After the manuscript of this book was completed, I have found that with Riesenfeld[118] Nernst has equipped, his quartz microbalance with an electric furnace and that Brill[119] has registered the first thermolysis curve using calcium carbonate and operating up to 1200° by continuous heating.

References p. 22

hydrated salts. I saw this apparatus in 1922 when I began to work in Urbain's laboratory. The trials apparently were not encouraging because Urbain said to me one day when I told him that I wished to use this apparatus: "The results depend too much on the heating conditions." I do not know what became of this remarkable and unique device, which contained the germ of the idea of the present thermobalances.

3. The first thermobalance

The word *thermobalance* was introduced into the science by Honda[3] in 1915. He designed and put into use the first apparatus worthy of this name; it will be described presently.

K. Honda (1870-1954).

Dr. Kotara Honda was born in the Aichi Prefecture on 23rd February, 1870. After graduating from the University of Tokyo, he continued his scientific training there and was awarded the Doctor of Science degree in 1903. From February 1907 to February 1911, he studied in Germany, France and England, and on his return to Japan was appointed to a professorship at Tohoku University. He was made Director of the Iron and Steel Institute in May 1919, and was elected to the Imperial Japanese Academy in 1921. His scientific accomplishments can be seen in the eight books which he authored and in 188 contributions to the *Science Reports* issued by Tohoku University. These achievements brought him the Grand Cordon of the Rising Sun, conferred by the Japanese government in 1954. He died at the age of 84 on 12th February, 1954.

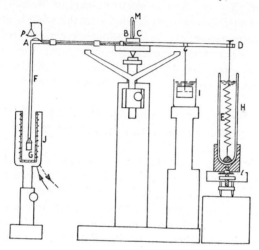

Fig. 1. Diagram of the Honda thermobalance.

The Honda thermobalance consists of a balance with a quartz beam (Fig. 1). A porcelain tube F, which goes down into the electrically heated furnace J is attached to one arm and carries the porcelain or magnesia dish G. The other arm has a fine steel spring E immersed in oil contained in a Dewar flask H; I is a second damper. The adjustment is relatively easy: after placing a centigram weight (necessary for calibration) on the small pan p, the position of the Dewar vessel is adjusted so that the scale-reading seen through the eye-piece is zero.

The original Honda paper[3] contains thermolysis curves of manganous sulphate, $MnSO_4 \cdot 4H_2O$, of gypsum, $CaSO_4 \cdot 2H_2O$ and chromic anhydride CrO_3.

Saito[4] came into Honda's laboratory in 1920 and began by modifying the thermobalance. In a remarkable piece of thesis research, dealing especially with natural sulphides and oxides, he constructed 175 thermolysis curves, including 27 for pyrites. In these curves he–prior to Vallet–showed the influence of the rate of heating, the grain size, the speed of the air current, the composition of the surrounding atmosphere (hydrogen, carbon dioxide, sulphur dioxide).

Later, Shibata (who in the meanwhile had worked in Urbain's laboratory) and Fukushima[5] rendered the Honda balance self-recording by converting it into a null-type balance. Going back to the idea of Urbain and Boulanger, they attached an electromagnet to the end of the beam; the current is controlled in such fashion that the scale is always seen at zero. The determination of the variations in mass is reduced to the registration of a continuous current. In their paper, these workers give the thermolysis curves of Mohr's salt (ferrous ammonium sulphate), cadmium sulphate, cupric sulphate, barium chloride, and calcium carbonate.

Afterwards, Somiya[6] in his turn modified the Honda thermobalance so that it was possible to use it with the furnace at 1300°. He reported on an analysis of lime.

Subsequently, the Japanese scientists undertook a general investigation of the precipitates employed in gravimetry–a study which I was destined to generalize. They chose to study families (oxinates, thionalidates, etc.) rather than groups of anions and cations; their findings are given in Part 2 of this book. We note particularly the names: Noshida[7], Ishibashi[8], Umemura[9], Kiba and Ikeda[10], Ishii[11], Kitajima[12], Takeno[13], etc. In addition, Kobayashi[14] published a general summary on gravimetric analysis by means of the thermobalance; he has given a list of the temperatures appropriate for about 300 precipitates starting from 27 notes published in Japan from 1925 to 1940. This study was very useful to me later on. Because of it I was able to persuade Chevenard to modify his thermobalance to render it suitable for analytical studies.

4. The Guichard thermobalance

Starting in 1923, Guichard, assisted by his students, began a series of studies directed principally to the realization of a linear elevation of temperature with respect to time[15-21].

Marcel Guichard was born at Charenton (Seine) on 17th December, 1873. He was successively apothecary, assistant at the Institut de Chimie de Paris, private assistant to Moissan, and secretary of the *Traité de chimie minérale* which was edited by Moissan. He became professor of inorganic chemistry at the Sorbonne and, at the close of his career, a chair of chemical analyses and measurements was created for him personally. He was an outstanding teacher, illustrating his lectures with masterly demonstration experiments, and initiating his students into the benefits of reading the great publications in their original form. One of his papers, issued prior to World War I, dealt with the determination of the atomic weight of iodine; it has always been regarded as a model of its kind. He died at Cachan (Seine) on 7th February 1960.

M. Guichard (1873-1960).

Guichard's pioneering character in this field is well summed up by his declaration[18]: "The method of studying chemical systems through their changes in weight when subjected to temperatures varied in regular fashion may be widely employed, provided certain precautions are taken. It will be usefully applied to certain operations of quantitative analysis."

However it must be pointed out that in none of his papers does Guichard cite Honda or the latter's students. It is equally true that Guichard never claimed to be the discoverer of thermogravimetry. He improved this technique, brought it to a very high point, and studied it with much critical sense. His work deserves high acclaim but none the less I believe that some of his former coworkers do his memory a great

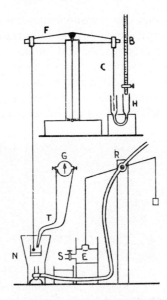

Fig. 2. Diagram of the original Guichard thermobalance.

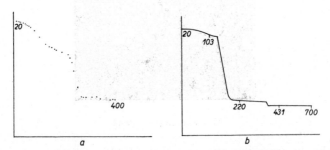

Fig. 3. Dehydration of disodium hydrogen phosphate
(a) according to Guichard, (b) according to Duval.

wrong by proclaiming "that with his hands he built the first thermobalance."

In the original Guichard thermobalance (Fig. 2) a regular increase in
the temperature was ensured by a gas burner fed via a constant level
device R consisting of a valve with attached float. The water E which
supports the float runs slowly and regularly through a sensitive valve S,
thus causing slow and gradual opening of the valve R. The balance is

compensated hydrostatically. The amounts of oil poured in are exactly proportional to the variations in the weights found if the balance is maintained at zero throughout. A loss in weight of 100 mg corresponds to an addition of 9 ml oil. Convection currents begin to make themselves felt above 400°. The sensitivity and reproducibility up to that point are better than 0.5 mg. The historic example reproduced in Fig. 3a relates to the dehydration of disodium hydrogen phosphate, the curve having been made point by point. Fig. 3b shows the pyrolysis curve of the same salt determined 25 years later with a continuous recording apparatus. The compounds with 12 and 7 H_2O, the anhydrous phosphate, and the pyrophosphate can be clearly seen in both figures. We thus find, said Guichard[16] in the study of the rate of dehydration with regularly increasing temperature, a new method capable of establishing rapidly the state of the water in hydrated substances. Definite hydrates manifest themselves in the curves by level stretches which are more or less distinct; adsorbed water gives no indication of this kind.

Later, Guichard substituted an electric furnace for the gas heating and replaced air by other gases, chosen especially for the experiments in mind, and which escape via the space between the furnace and the rod supporting the heated crucible[17].

In the meantime, McBain and Baer[22] constructed the thermobalance, with a quartz thread in a vacuum, to follow experiments dealing with sorption. The principle was to be used again later by Loriers[23], Gulbransen[24], and Murthy, Bharadwaj and Mallya[25].

In 1936, Vallet[26,27] made a noteworthy and systematic investigation into the effect of the rate of heating, the shape of the crucible, the rate of evolution and the composition of the gaseous decomposition products, and the initial state of the substance being studied (particularly cupric sulphate). In his graduation diploma of higher studies[28] he dealt with ammonium chloroplatinate, the reduction of chloroplatinic acid with hydrogen and oxidation of the resulting divided platinum. Vallet obtained a linear variation in the temperature by means of a liquid rheostat and an accurately calculated cam; Fig. 4 gives a clear picture of the mechanism[27]. He also used a mercury regulator placed in an auxiliary oven. He was able to obtain a strictly linear rate of heating in which the rise from 20° to 600° took three days.

Around the same time, Schreiber[29] set his furnace so as to give a rise of 5° per hour in order to isolate certain hydrates. However, Vallet pointed out that a decrease in the heating rate is not always favorable for the appearance of intermediate compounds. Schreiber studied the

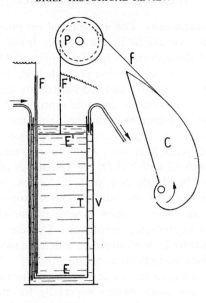

Fig. 4. The device used by Vallet to ensure a linear rate of heating.

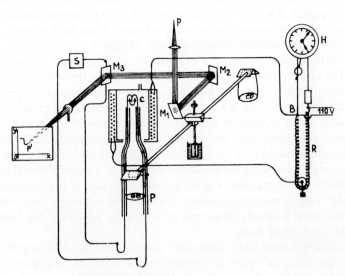

Fig. 5. Dubois apparatus; the first set-up enabling the making of
a continuous photographic record.

dehydration of dithionates in particular. In his thesis, Selva[30] also used a linear rate of heating to isolate hydrates.

As early as 1931, Cochet proposed an interesting application of the Guichard thermobalance for industrial purposes. He suggested[31] that the capture of nitrogen by calcium carbide to yield cyanamide be followed in this way.

Rigollet[32] introduced the idea of inverting the electric furnace–thus placing the opening at the bottom–to eliminate the convection currents and to allow working in different gases whose densities are less than unity at temperatures above 100°.

This idea of a bell-like furnace was revived some years later by Chevenard and his associates.

Before going further, it is necessary to explain the setup developed in Guichard's laboratory by Dubois[33]. It is shown in Fig. 5. The compound under study is placed in a crucible C which is suspended from an alumel-chromel couple attached indirectly to the beam of a balance (sensitivity 2 mg, glycerol damper). A vertical, cylindrical pencil of light falls on the mirror M_1, fixed to the prolongation of the central knife-edge at an angle of 45°. The light is reflected from M_1 to M_2 and then to M_3, and finally the image of P, strikes at P′ a photographic paper or plate. The temperature of the furnace is raised at a uniform rate by means of the following simple device. A 12-A helical resistance R is connected to the winding-up chain of a pendulum clock H, while the furnace (10 Ω resistance) is shunted across one of the poles of the electricity supply at B and the axle of a metal pulley which rolls along the resistance R. With 7 A initially in the resistance and 3 A in the furnace, the temperature can be raised uniformly from 30° to 400° in 4 hours.

The thermocouple is placed in the substance being heated and causes rotation of the mirror M_3 of a galvanometer whose sensitivity can be adjusted by means of the resistance S. On the paper XOY, the abscissa OX represents the time, and the ordinate OY the variation in weight. Dubois tried to define the individuality of the various manganese oxides by means of this apparatus.

In particular, he heated the compounds: MnO_2, $MnSO_4 \cdot 5H_2O$, $Mn(NO_3)_2$, $MnCl_2$, MnC_2O_4, $MnCO_3$. It seems entirely reasonable to inquire why such a fine accomplishment was not developed further and what was the reason for its not being commercialized even in the face of the progress made by the workers in Guichard's laboratory in the field of thermogravimetry.

5. The Longchambon thermobalance

Longchambon[34], working with a similar assembly, provided the lower half of the furnace (open at the top) with a coil through which a slow current of air saturated with water vapour could be introduced at a given temperature. It was possible, by changing the length of the pendulum, to vary the rate of heating within very wide limits, namely 0–1100° in from 4 hours to 4 days. Longchambon used the familiar device in which the beam of light is interrupted at equal intervals of time by a toothed disc for the automatic control of the variation of the fusion temperature

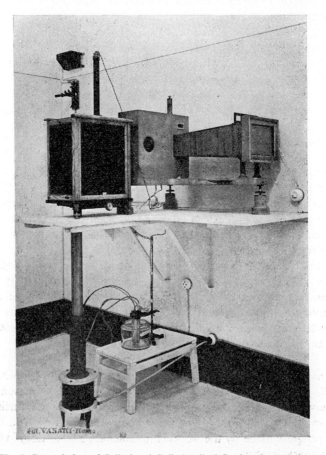

Fig. 6. General view of Collari and Galimberti's deflection thermobalance.

as a function of the time. Instead of a continuous curve, there resulted a curve made up of dots or dashes and this was adequate to permit the reconstruction of the complete curve unambiguously. Longchambon investigated the dehydration of kaolinite, muscovite (in order to evaluate the effect of the initial physical state), sepiolites (Meerschaums) and palygorskites, as well as the dissociation of calcite, dolomite, and giobertite.

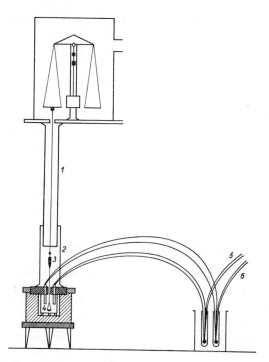

Fig. 7. Diagram of Collari and Galimberti's thermobalance. (1) Fixed protection tube for platinum wire supporting the crucible. (2) Gliding tube to protect the same. (3) Tensor weight of hanging wire. (4) Platinum cylinder. (5) Thermo-electric couple to Le Chatelier-Broniewski galvanometer. (6) Thermo-electric couple to control galvanometer.

It was likewise about this same time that Skramovsky, Forster and Hüttig[35] succeeded in obtaining a photographic record on a rotating cylinder, with the aid of a motor whose speed could be controlled, by reflection of a ray of light on a mirror carried at the end of the beam of a knife-edge balance. Deformations in the recorded curve doubtless were due to the difficulty of synchronizing the speed of rotation of the cylinder with the mechanism raising the temperature linearly.

Shortly thereafter, Collari and Galimberti[36] linked a 1/10-mg Sartorius analytical balance (damped by vaseline oil) with a Le Chatelier-Broniewski differential galvanometer. A loss of one-tenth of a mg in the weight of the sample corresponded to a displacement of about 2 mm of a luminous point on the photographic plate.

Fig. 7 shows (in elevation) the diagram of the balance, and Fig. 6 is a photograph of the assembly with which these workers obtained mass–temperature curves of a magnesite (99.03% $MgCO_3$), $SrBr_2 \cdot 6H_2O$, and artificial magnesium carbonate resulting from the dehydration of the trihydrate.

Finally, from the period preceding World War II, we can cite the papers by Müller and Garman[37] and by Gill[38] who combined a calcining furnace with an analytical balance, the thermocouple being suspended into the charge and supported by the balance. The apparatus was used especially for studying the behaviour of minerals. Orosco[39], in turn, gave a new form to the thermobalance but unfortunately we have been unable to obtain a diagram of this instrument. In various instances he showed that the degree of granulation (down to 1 mm) is without effect on the curve.

6. The Chevenard thermobalance

In 1936, after stainless steels had just been developed, the laboratory of the Société Fourchambault, Decazeville and Commentry at Imphy (Nièvre) began a series of studies to determine the increase of weight in these materials when heated. This is why the arm of the earliest apparatus carried a stirrup intended for the support of the specimen wires bent into the form of hairpins. Chevenard desired above all to preserve the photographic shell of the dilatometer; in addition since he wished to make the determinations over a period of several days in a place where cars of ore and coal were passing by, he decided to adopt the bifunicular suspension of the Kelvin electrodynamometer for the beam. Also he carefully explored the Rigollet and Dubois idea of the inverted furnace to avoid convection currents. Because of World War II, publication was delayed for 6 years and the apparatus was exhibited to the Société chimique de France in collaboration with Waché and de la Tullaye[40].

Pierre Chevenard was born on 31st December, 1888 at Thizy (Rhône).

He studied at the Ecole des Mines de Saint-Etienne (Loire) and graduated first in his class in 1910. All of his scientific career was spent at Imphy (Nièvre) as Director of Metallurgical Researches and as professor in the Ecole de Saint-Etienne. He created special steels for the catalysis equipment of Georges Claude and for the metrological apparatus of C. E. Guillaume. Chevenard constructed around 150 scientific devices, the best known being the dilatometers, thermaomgnetometers, micro-

P. Chevenard (1888-1960).

machines for measuring creepage, etc. He was a member of the Académie des Sciences, president of numerous scientific bodies, Commander of the Legion of Honour, Knight of the Order of Leopold, and recipient of the Transenster medal. After a lengthy illness, he died at Fontenay-aux-Roses (Seine) on 14th August, 1960. He was a man of extreme simplicity, always ready to be helpful, and extremely affable. His pronounced modesty made him appear timid at times. My associates and I are greatly indebted to him for having so generously equipped the laboratory.

The photographically recording thermobalance was put on the market in 1945 but the heating was not a linear function of time. The program involving the use of this device was not established until 1947 (Société Adamel). Because of my insistence, the photographic paper was replaced by a pen (1948); the unique prototype with a horizontal cylinder always functions in the laboratory; it led to the model with vertical paper, interchangeable with the photographic housing (1953), and then to the model with a vertical cylinder and thermobalance functioning in vacuo (1959). The Chevenard photographic thermobalance was subsequently modified by Gordon and Campbell[41] in such fashion as to render the registration visible electronically (use of an Atcotran transformer). In 1947, I started the publication in the *Analytica Chimica Acta*[42] of a series of analytical studies inspired by those of the Japanese chemists. In these reports I focused attention on automatic gravimetric analysis[43], and then on the study of solid state reactions[44], analytical standard materials[45], etc.*

Gravimetric determinations without separation have been undertaken in numerous cases by Unohara[46].

7. The progress in thermobalances

No less than fifty thermobalances had been described by 1961, and ten of these are commercially available. Some of them will be discussed in detail; as to the others, their particular features will be pointed out here.

As early as 1947, Jouin[47] had constructed an interesting balance with the furnace placed below the apparatus insulated with cork and asbestos. The heating was linear (1000° in 3.5 hours), the recording was photographic, and a displacement of 50 mm corresponded to 100 mg. These features resulted in an apparatus suitable for studying clays, lignites, charcoal, Tunisian pyrites, and esparto grass. The proof had now been provided: a beam balance could be made recording in a laboratory free of vibrations.

Burriel-Marti and Barcia-Goyanes[48] likewise replaced one of the pans of a semi-microbalance by a platinum wire descending into a furnace placed on a table; a Sunvic regulator assured an approximately linear heating (Fig. 8).

At the U.S. Bureau of Standards, Mauer[49] likewise modified an or-

---

* My 17 collaborators in these various researches were: Jean Besson, Pierre Champ, Monique De Clercq, Annie Dautel, Marly Doan, Thérèse Dupuis, Raymond Duval, Pierre Fauconnier, Yvette Marin, Josette Morandat, André Morette, Simonne Panchout, Simonne Peltier, Janine Statchtchenko, Suzanne Tribalat, Nguyen Dat Xuong, Colette Wadier.

dinary analytical balance by introducing a solenoid beneath one of the pans. This apparatus was able to record mass–time curves and thermal analysis curves simultaneously.

Bartlett and Williams[50] and Blazek[51] also employed electromagnetic compensation.

Chain balances have been widely used. The Testut Company has given special attention to an arrangement devised by Carton and then studied by Jacqué at the Ecole Polytechnique (see below).

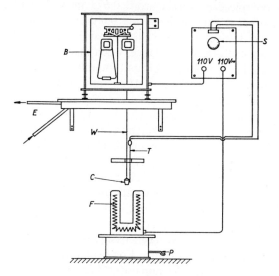

Fig. 8. Diagram of Burriel-Marti and Barcia-Goyanes thermobalance. B: semi-microbalance; E: water; W: platinum wire; F: furnace; P: pedal for furnace; C: crucible; T: thermocouple; S: Sunvic regulator.

Monk[52] provided a chain balance with small motors; it is inclosed in an exhausted bell jar and functions up to 700° under 0.01 μ. He has published degassing curves of aluminium, carbon, and Teflon. The apparatus of Hyatt, Cutler and Wadsworth[53] is also of the chain and programmed type.

Among the torsion thermobalances, we will cite those of Wendlandt[54], Cueilleron and Hartmanshenn[55], Iddings[56], Gulbransen[24], Izvekov[57], Oshima and Fukuda[58], Murthy, Bharadwaj and Mallya[25]. In the Iddings thermobalance, a perfectly insulated quartz wire serves as the basic element. The movement of the beam actuates a servo-mechanism controlling its position. The current passing through a spool placed in the electrical system is proportional to the weight of the sample. The recording of the difference of potential at the terminals of the spool allows an immediate trace of the mass as a function of the time or the temperature.

Barret[59] constructed a microthermobalance based on the McBain balance.

The remarkable microthermobalance of Gulbransen includes a framework made from rods of fused quartz 1/8th of an inch in diameter and a quartz beam (piece of stirring rod) 0.673 inch. The tungsten wires supporting the beam and the loads are 0.001 inch in diameter and are attached by fused silver chloride. The sensitivity is $0.3 \cdot 10^{-6}$ g with a sample weighing 684 mg. This balance, when inclosed in a glass tube and fitted with an auxiliary device, was found suitable for studying the kinetics of film formation from $-90°$ to $+1000°$ and under pressures ranging from 10^{-5} mm to 1 atm.

Among the differential instruments, we will note first of all that of De Keyser[60], placed on the market by the Sartorius concern; it is described in Chapter 2. It has been modified by Lambert[61] and also by Waters[62], Erdey, Paulik and Paulik[63]. Splitek[64] combined differential thermal analysis with thermogravimetry (112 references) and this was also done by Papailhau[65], Herold[66], Markowitz and Boryta[67]. The Société Adamel, at the request of a large French firm, constructed an apparatus of this kind in 1948; I believe this is the sole example.

Certain thermobalances have been designed to function over a wide range of temperature and pressure. For example, Rabatin and Card[68] work up to 40 atm. The Cordes apparatus[69] functions in a vacuum; that of Linseis[70] can be used up to $1550°$ at 0.2–0.4 mg pressure, and that of Stonhill[71] is designed for operations in a dangerous atmosphere. The reaction chamber is made of Inconel. A detector for water-soluble gases (HF, NH_3, etc.) is included. The electromagnetic Gregg balance[72], which functions between $-200°$ and $+1200°$, will reveal a change of 0.1 mg in 25 mg. Peters and Wiedmann[73] work up to $1200°$ with their balance; with a maximum load of 10 g the sensitivity is 0.08 mg.

Bastian and Colombié[74] describe and illustrate a thermobalance which serves for corrosion tests on metals and alloys in a number of atmospheres: carbon dioxide, hydrogen, hydrogen sulphide. The pressures range from $5 \cdot 10^{-1}$ torr to 30 kg/cm². The specimen is placed at the upper end of a silica rod inclosed in an Fe–Ni–Cr tube encircled by an electric furnace whose maximum heating rate is $600°$ per hour. The maximum permissible temperature at the highest pressure is $700°$. The furnace is well insulated from the moving parts of the balance. The latter includes a horizontal needle with bifilar tungsten suspension, damping and counterpoise, concave mirror whose movements are recorded optically, or electronically amplified on a screen. The loading and unloading under pressure are accomplished electromagnetically. The sensitivity is close to 0.1 mg, and 1 mg corresponds to a displacement of 3 mm of the spot on the scale.

Mauret[75] has described, under the name of thermobalance, a very simple apparatus heated by infra-red rays, which may be used up to 200° and which is very suitable for obtaining dehydration curves.

We wish to mention also the D.A.M. apparatus and the C.N.R.S. apparatus[76], the Eyraud electronic thermobalance[77] and that of Stanton which are on the market, that of Brefort[78], that of Scholten, Smit and Wijnen[79], of Lambert[80] for sedimentation, of Banks[81], of Groot and Troutner[82], of Kinjyo and Iwata[83] in which the crucible is supported by the couple, of Formanek and Bauer[84], of Satava[85], of Kalinine and Kuznetsov[86], of Palei, Sentyurin and Sklyarenko[87], of Teetsel, Munroe, Williamson, Abbott and Stoneking[88].

Among the more recent techniques meriting attention here are: fractional thermogravimetric analysis by Waters[89], derivative analysis by Erdey, Paulik and Paulik[63] which has led to the apparatus called the Derivatograph, the study of proportional systems by continuous gravimetry by Lowe[90], and inverse stathmography by Skramovsky[91].

I also think it proper not to forget all those who, after the event, rediscovered thermogravimetry, for instance Zagorski[92], Gregg and Winsor[93], Spinedi[94], Rogers, Yasuda and Zinn[95] by reporting some new observations in this field.

Subsequent to the publication of studies dealing with thermogravimetry, various textbooks of analytical chemistry have mentioned the new techniques in their sections dealing with Gravimetry. Two important mixed congresses have been held on differential thermal analysis and thermo-weighing, namely at Kazan (U.S.S.R.) 19th to 23rd June, 1953 and at Dallas, Texas (U.S.A.) in 1956 under the chairmanship of Saul Gordon. Under the title *Automatic and Recording Balances*, Gordon and Campbell[96] published an important summary containing 165 references in which are discussed the thermobalances now on the market, along with their descriptions, characteristics, and performances. Kacnel'son and Edel'shtejn[97] have also published a review of the various balances and microbalances with automatic registration.

I wish also to call attention to various lectures given at conferences and published in various reviews: The use of the thermobalance in analytical chemistry by Claisse, East and Abesque[98], La thermobalance en chimie minérale by Portevin[99], La gravimétrie automatique (in Turkish) by Orhan[100] with 54 references, Sur la thermogravimétrie et la gravimétrie automatique by van Tongeren[101], Revue sur la thermogravimétrie, with 25 references by Barcia-Goyanes[102], Review, theory, equipment, with 20 references by Wilson[103], Thermogravimetric techniques by Campbell, Gordon and Smith[104]. A Thermogravimetric balance for student experiments was described in 1960 by Delhez[105].

Aside from many symposia, I personally have lectured at a number of public meetings. Almost all of these lectures subsequently appeared in print:

At: Utrecht (Netherlands) (1948)[106].

Paris, Maison de la Chimie (1949)[107].

Graz (Austria) (1950), precipitates described in the text by Hecht and Donau[108].

Clermont-Ferrand: Association française pour l'Avancement des Sciences (1949).

Toulouse, Association française pour l'Avancement des Sciences (1950).

Paris, Maison de la Chimie: Automation in analytical chemistry (1950)[109].

New York City, Diamond Jubilee of the American Chemical Society: Continuous weighing in analytical chemistry, with 111 references (1951)[110].

New York City, 12th Congress of the International Union of Pure and Applied Chemistry: Continuous weighing in inorganic chemistry (1951).

New York City, American Society of Microchemistry: Applications of the thermobalance to microchemistry (1951).

Delft (Netherlands), Fundamental ideas of thermogravimetry (1951).

Gembloux (Belgium), Continuous weighing in analytical chemistry (1951).

Liège (Belgium), idem (1951).

Ghent (Belgium), idem (1951).

Paris, Maison de la Chimie: Recent progress in automatic gravimetry (1951)[111].

Madrid (Spain), Thermogravimetry and its applications (1952)[112].

Frankfurt am Main (Germany), New method of continuous weighing (1952)[113].

Paris, Maison de la Chimie, Microanalysis of alkali metals (1952)[114].

Geneva and Lausanne (Switzerland), New advances in thermogravimetry (1953).

Paris, G.A.M.S., Recent progress in thermogravimetry (1955)[115] with demonstration.

Paris, Faculty of Pharmacy, Analytical applications of thermogravimetry (1957)[116] with demonstration.

Washington, D.C. (U.S.A.), Bureau of Standards, Thermogravimetry and its applications (1958).

Baton Rouge (U.S.A.), Louisiana State University: Thermogravimetry and its applications (1958); 2 lectures with demonstrations.

Schenectady, N.Y. (U.S.A.), General Electric Company: Thermogravimetry and its applications (1958).

Philadelphia, Penn. (U.S.A.), Pennsylvania Symposium: Thermogravimetry and its applications (1958).

Birmingham (England), Midlands Society: Continuous weighing and its applications (1958).

Nottingham (England), Boots Drug Company: Continuous weighing and its applications (1958).

Erlangen University (Germany): New advances in thermogravimetry (1958).

Montpellier (France), Faculty of Pharmacy: Recent progress in thermogravimetry.

Paris, C.E.G.O.S.: Recent progress in microanalysis (1960).

Budapest (Hungary), International Union of Pure and Applied Chemistry: Study of solid state reactions by thermogravimetry and infra-red spectrometry (1961).

In addition, I have given each year two 1-hour lectures on thermogravimetry to the students of the Ecole Nationale Supérieure de chimie de Paris.

Despite of all these efforts, and in spite of the enormous amount of work throughout the world on this subject and after reading, abstracting or rejecting the 2200 papers which have been consulted in composing this second edition, I still feel it appropriate to close this first chapter with the statement by Beamish and McBryde[117]:

"Thermogravimetry is yet in its infancy and these differences of opinion expressed in analytical literature, while initially frustrating, will yield to a much better understanding of the character and mechanisms of ignitions."

References p. 22

REFERENCES

[1] P. Truchot, *Rev. chim. pure et appl.*, 10 (1907) 2.

[2] G. Urbain and C. Boulanger, *Compt. rend.*, 154 (1912) 347.

[3] K. Honda, *Sci. Repts. Tohôku Imp. Univ.*, 4 (1915) 97; *C.A.*, (1915) 2610.

[4] H. Saito, *Proc. Imp. Acad., (Tokyo)*, 2 (1926) 58.

[5] Z. Shibata and M. Fukushima. *Bull. Chem. Soc. Japan*, 3 (1928) 118.

[6] T. Somiya, *J. Soc. Chem. Ind. Japan*, 31 (1928) 217; 32 (1929) 249.

[7] I. Noshida, *J. Chem. Soc. Japan*, 48 (1927) 520.

[8] S. Ishibashi, *J. Chem. Soc. Japan*, 61 (1940) 125, 130, 133.

[9] T. Umemura, *J. Chem. Soc. Japan*, 61 (1940) 25.

[10] T. Kiba and T. Ikeda, *J. Chem. Soc. Japan*, 61 (1940) 25.

[11] K. Ishii, *J. Chem. Soc. Japan*, 52 (1931) 167, 229, 232, 461, 727, 730, 774.

[12] I. Kitajima, *J. Chem. Soc. Japan*, 53 (1932) 566; 55 (1934) 199, 201, 228.

[13] R. Takeno, *J. Chem. Soc. Japan*, 54 (1933) 741.

[14] M. Kobayashi, *Sci. Repts. Tohôku Imp. Univ.*, 29 (1940) 391.

[15] M. Guichard, *Bull. Soc. Chim. France*, 33 (1923) 258.

[16] M. Guichard, *Bull. Soc. Chim. France*, 37 (1925) 62, 251, 381.

[17] M. Guichard, *Bull. Soc. Chim. France*, 39 (1926) 113.

[18] M. Guichard, *Bull. Soc. Chim. France*, 2 (1935) 539.

[19] M. Guichard, *Compt. rend.*, 199 (1934) 133.

[20] M. Guichard, *Ann. Chim. Paris*, 9 (1938) 323.

[21] M. Guichard, *Bull. Soc. Chim. France*, 5 (1938) 675.

[22] J. W. McBain and A. M. Baer, *J. Am. Chem. Soc.*, 48 (1926) 600.

[23] J. Loriers, *Rev. Mét.*, 49 (1952) 807.

[24] E. A. Gulbransen, *Rev. Sci. Instr.*, 15 (1944) 201.

[25] V. Murthy, D. S. Bharadwaj and R. M. Mallya, *Chem. & Ind. London*, (1956) 300.

[26] P. Vallet, *Bull. Soc. Chim. France*, 3 (1936) 103; *Comp. rend.*, 198 (1934) 1860.

[27] P. Vallet, *Thesis*, Paris, 1936; *Ann. Chim.* Paris, 7 (1937) 298.

[28] P. Vallet, *Diploma of Higher Studies*, Paris, 1928, No. 355.

[29] J. Schreiber, *Thesis*, Strasbourg, 1933, No. 105; *Ann. Chim. Paris*, 1 (1934) 38.

[30] L. Selva, *Thesis*, Strasbourg, 1935, No. 111.

[31] A. Cochet, *Z. angew. Chem.*, 44 (1931) 367.

[32] C. Rigollet, *Diploma of Higher Studies, Paris*, 1934, No. 552.

[33] P. Dubois, *Thesis*, Paris, 1935; *Bull. Soc. Chim. France*, 3 (1936) 1178.

[34] H. Longchambon, *Bull. Soc. franç. Minéral.*, 59 (1936) 145.

[35] S. Skramovsky, R. Forster and G. Hüttig, *Z. physik. Chem. (Leipzig)*, B, 25 (1934)1; *J. Iron Steel Inst. (London)*, 11 (1936) 587; *Collection Czechoslov. Chem. Communs.*, 9 (1937) 302.

[36] N. Collari and L. Galimberti, *Boll. sci. fac. chim. ind. Bologna*, 18 (1940) 1.

[37] R. H. Müller and R. L. Garman, *Ind. Eng. Chem.*, 10 (1938) 436.

[38] A. F. Gill, *Can. J. Research*, 10 (1934) 703.

[39] E. Orosco, *Ministério trabalho ind. com. Inst. nac. tecnol. (Rio de Janeiro)*, 1940; *C. A.*, 35 (1941) 3485.

[40] P. Chevenard, X. Waché and R. de la Tullaye, *Bull. Soc. Chim. France*, 10 (1944)41.

[41] S. Gordon and C. Campbell, *Anal. Chem.*, 28 (1956) 124.

[42] C. Duval, *Anal. Chim. Acta*, 1 (1947) 341.

[43] C. Duval, *Compt. rend.*, 224 (1947) 1824; 226 (1948) 1076.

[44] C. Duval, *Proc. Intern. Symposium on Reactivity of Solids, Gothenburg*, 1952 Vol. 1, p. 511.

[45] C. Duval, *Anal. Chim. Acta*, 13 (1955) 32.

[46] N. Unohara, *J. Chem. Soc. Japan.*, 73 (1952) 488; 74 (1953) 329; 76 (1955) 357, 359, 615.

[47] Y. Jouin, *Chim. & Ind. (Paris)*, 58 (1947) 24.

[48] F. Burriel-Marti and C. Barcia-Goyanes, *Anales real soc. españ. fís. y quim.*, 47 (1951) 73.

[49] F. A. MAUER, *Rev. Sci. Instr.*, 25 (1954) 598.
[50] E. S. BARTLETT AND D. N. WILLIAMS, *Rev. Sci. Instr.*, 28 (1957) 819.
[51] A. BLAZEK, *Silikáty*, 1 (1957) 158; *C.A.*, (1958) 16806.
[52] G. W. MONK, *J. appl. Phys.*, 19 (1948) 485.
[53] E. P. HYATT, I. B. CUTLER AND M. E. WADSWORTH, *Am. Ceram. Soc. Bull.*, 35 (1956) 180.
[54] W. W. WENDLANDT, *Anal. Chem.*, 30 (1958) 56.
[55] J. CUEILLERON AND O. HARTMANSHENN, *Bull. Soc. Chim. France*, (1959) 164, 168, 172.
[56] F. A. IDDINGS, *Dissertation Abstr.*, 20 (1959) 867.
[57] I. V. IZVEKOV, *Trudy Krym. Filiala Akad. Nauk S.S.S.R.*, 4 (1953) 81.
[58] Y. OSHIMA AND Y. FUKUDA, *J. Soc. Chem. Ind. Japan*, Suppl. binding, 33 (1930) 251.
[59] P. BARRET, *Bull. Soc. Chim. France*, (1958) 376.
[60] W. L. DE KEYSER, *Nature*, 172 (1953) 364.
[61] A. LAMBERT, *Bull. soc. franç. céram.*, 28 (1955) 23.
[62] P. WATERS, *Nature*, 178 (1956) 324; *Coke and Gas*, 30 (1958) 341.
[63] L. ERDEY, F. PAULIK AND J. PAULIK, *Acta Chim. Acad. Sci. Hung.*, 13 (1957) 117; *C.A.*, 50 (1956) 3952.
[64] R. SPLITEK, *Hutnické Listy*, 8 (1958) 697; *C.A.*, (1958) 19261.
[65] J. PAPAILHAU, *Bull. soc. franç. Minéral.*, 82 (1959) 367.
[66] A. HEROLD, *Bull. Soc. Chim. France*, (1960) 533.
[67] M. M. MARKOWITZ AND D. A. BORYTA, *Anal. Chem.*, 32 (1960) 1588.
[68] J. G. RABATIN AND C. S. CARD, *Anal. Chem.*, 31 (1959) 1689.
[69] J. F. CORDES, *Chem. Ingr. Tech.*, 30 (1958) 342; *C.A.*, (1958) 14410.
[70] M. LINSEIS, *Keram. Z.*, 11 (1959) 54; *C.A.*, (1959) 17589.
[71] L. G. STONHILL, *J. Inorg. & Nuclear Chem.*, 10 (1959) 153.
[72] S. J. GREGG, *J. Chem. Soc.*, (1955) 1438.
[73] H. PETERS AND H. G. WIEDMANN, *Z. anorg. Chem.*, 298 (1959) 202.
[74] P. BASTIAN AND M. COLOMBIÉ, *Métaux*, 34 (1957) 447.
[75] P. MAURET, *Compt. rend.*, 240 (1955) 2151; 246 (1958) 3450; 248 (1959) 1808; *Bull. Soc. Chim. France*, (1956) 655; (1958) 1083; *Compt. rend. soc. Sci. phys. et nat. (Maroc)*, (1957) 2.
[76] ANONYMUS, *Nature (Paris)*, (1958) 238.
[77] C. EYRAUD AND I. EYRAUD, *Laboratoires*, 12 (1955) 13.
[78] J. BREFORT, *Bull. Soc. Chim. France*, (1949) 524.
[79] P. C. SCHOLTEN, W. M. SMIT AND M. D. WIJNEN, *Rec. trav. chim.*, 77 (1958) 305.
[80] A. LAMBERT, *Bull. soc. franç. céram.*, No. 45 (1959) 19.
[81] J. E. BANKS, *Anal. Chim. Acta*, 19 (1958) 331.
[82] C. GROOT AND V. H. TROUTNER, *Anal. Chem.*, 29 (1957) 835.
[83] K. KINJYO AND S. IWATA, *J. Chem. Soc. Japan*, 72 (1951) 958.
[84] Z. FORMANEK AND J. BAUER, *Silikáty*, 1 (1957) 188.
[85] V. SATAVA, *Silikáty*, 1 (1957) 188.
[86] P. D. KALININE AND A. K. KUZNETSOV, *Zhur. Fiz. Khim.*, 32 (1958) 1658.
[87] P. N. PALEI, I. G. SENTYURIN AND I. S. SKLYARENKO, *Zhur. Anal. Khim.*, 12 (1957) 329.
[88] F. M. TEETZEL, M. A. MUNROE, J. A. WILLIAMSON, A. E. ABBOTT AND D. J. STONEKING, *U.S. At. Energy Comm.*, NLCO-713 (1958); *C.A.*, (1958) 11482.
[89] A. WATERS, *Anal. Chem.*, 32 (1960) 852.
[90] R. P. LOWE, *Mech. Eng.*, 76 (1954) 653.
[91] S. SKRAMOVSKY, *Silikáty*, 3 (1959) 74; *C.A.*, (1959) 21122.
[92] Z. ZAGORSKI, *Przemysl. Chem.*, 31 (1952) 326.
[93] S. J. GREGG AND G. W. WINSOR, *Analyst*, 70 (1945) 336.
[94] P. SPINEDI, *Ricerca sci.*, 23 (1953) 2009; *C.A.*, (1954) 868.
[95] R. N. ROGERS, S. K. YASUDA AND J. ZINN, *Anal. Chem.*, 32 (1960) 672.
[96] S. GORDON AND C. CAMPBELL, *Anal. Chem.*, 32 (1960) 271R.
[97] O. G. KACNEL'SON AND A. S. EDEL'SHTEJN, *Khim. Promyohl. S.S.S.R.*, 3 (1959) 262.
[98] F. CLAISSE, F. EAST AND F. ABESQUE, *Chemical Institute of Canada Congress,* February 1953, Quebec, (Canada).

99 A. PORTEVIN, *Ind. chim. belge*, 16 (1951) 157.

100 S. ORHAN, *Folia Pharm. (Istanbul)*, 3 (1955) 84.

101 W. VAN TONGEREN, *Chem. Weekblad*, 46 (1950) 847.

102 C. BARCIA-GOYANES, *Inform. quím. anal. (Madrid)*, 9 (1955) 159.

103 C. L. WILSON, *J. Roy. Inst. Chem.*, 83 (1959) 550.

104 C. CAMPBELL, S. GORDON AND C. L. SMITH, *Anal. Chem.*, 31 (1959) 1188.

105 R. DELHEZ, *J. Chem. Educ.*, 37 (1960) 151.

106 C. DUVAL, *Anal. Chim. Acta.*, 2 (1948) 432.

107 C. DUVAL, *Chim. anal.*, 31 (1949) 173, 204.

108 C. DUVAL, *Mikrochemie*, 36 (1951) 425.

109 C. DUVAL, *Chim. anal., special number*, 33 (1951) 52.

110 C. DUVAL, *Anal. Chem.*, 23 (1951) 1271.

111 C. DUVAL, *Chim. anal.*, 34 (1952) 55.

112 C. DUVAL, *Anales edafol. y fisiol. vegetal. (Madrid)*, 12 (1953) 143.

113 C. DUVAL, *Dechema Monograph.*, 21 (1952) 150.

114 C. DUVAL, *Chim. anal.*, 34 (1952) 209.

115 C. DUVAL, *International week of Higher Studies, G.A.M.S.*, 1957, p. 125.

116 C. DUVAL, *Mises au point de chimie analytique pure et appliquée et d'analyse bromatologique*, Masson et Cie, 1958, 6th Series; pp. 19-66; 122 references.

117 F. E. BEAMISH AND W. A. E. MCBRYDE, in C. L. WILSON AND D. W. WILSON, *Comprehensive Analytical Chemistry*, Vol. IA, Elsevier Publishing Company, Amsterdam, 1959, p. 452.

118 W. NERNST and E.H. RIESENFELD, *Ber.*, 36 (1903) 2086.

119 O. BRILL, *Z. anorg. Chem.*, 45 (1905) 275.

CHAPTER 2

Deflection Type Thermobalances

1. General

Thermobalances are of two kinds: those in which the deflection of the beam under the action of the load is measured, and those of the null type in which the beam is brought back to its original position either manually or automatically as soon as the load changes. The first type will be discussed here, and the second type in Chapter 3.

According to Gordon and Campbell[1], the selection of the type of automatic recording balance to be constructed or purchased depends largely upon the application and a consideration of the physical and chemical parameters involved. Among these are the load, range of weight change, sensitivity, accuracy, speed of response, environmental conditions of atmosphere, temperature and pressure, type of recording system, and constructional complexity of the apparatus.

The classical method of recording with deflection type balances is that of measuring the balance beam deflections photographically by means of a light beam reflected from a mirror mounted on the balance beam. This technique involves recording the deflection directly on a photographic paper wrapped around a motor-driven drum.

The deflections may be measured electronically by a shutter attached to the balance beam and operating so as to intercept a beam of light impinging upon a phototube. The precision, sensitivity and the time of response is much more dependent on the electronic or electromechanical systems employed for the automatic graphic recording than on the balance itself. At present, it is difficult to obtain a precision better than $\pm$ 0.2%, even if the analytical balance used is fundamentally capable of a higher precision.

Mention was made in Chapter 1 of the thermobalances constructed by Dubois[2], Jouin[3], and Longchambon[4], Vytasil *et al.*[5] and by Orosco[6] for mineralogical specimens, and Spinedi[7] for oxidation of metals.

References p. 64

Couleru[8] constructed a simple balance whose beam had unequal arms and which was pen-recording.

Brefort[9] at the chemical laboratory C of the Sorbonne devised a recording thermobalance which heated linearly at the rates of 20°, 30°, 40°, or 60° per hour, or it could be kept at any desired temperature between 0° and 600° by means of a thermostat.

The balance (sensitive to 0.2 mg) was provided with an oil damper; its pointer carried a micrometric scale which allowed the operator to follow its movement. The regulator was a Tom apparatus from the firm of Chauvin and Arnoux. The recording device consisted of a voltaic photocell placed behind an aperture more or less concealed by a mobile shutter suspended on the pan of the balance not bearing the crucible. This assembly was illuminated by a beam of parallel white light with constant intensity. The cell delivered a current which was a function of the position of the shutter and therefore of the mass of the sample. A shunted galvanometer sent its spot of light on a sheet of transparent millimeter paper.

The thermobalance of Bartlett and Williams[10], which was used to follow the oxidation of metals, has one end of the beam of an analytical balance fastened to the base with a strain gauge wire so that the resistance of the wire increases as the sample suspended from the other end of the beam undergoes an increase in weight. This strain wire is part of a Wheatstone bridge circuit from which a signal proportional to the weight of the sample is fed into a potentiometric recorder.

2. The Chevenard thermobalance

This was the first instrument of the kind to be put on the market in several models[11]. It was originally developed for the study of the dry corrosion of metals:

"As has been known for some considerable time, particularly from the work of Portevin and Chaudron and their pupils, unoxidizable alloys owe their resistance not to chemical indifference, a privilege of the noble metals, but to the protection afforded by an adhering and quasi-impermeable continuous film of oxide. It is therefore important to study the formation of this film and to establish the origin of its protecting qualities. The thermobalance now to be described was evolved to this end".

The photographically recording instrument will be described first and then the instrument in which a pen is used.

In the first case, it is necessary to distinguish between heating at the rate of the furnace itself and heating at a predetermined rate. The two types of instrument enable substances to be heated in air, nitrogen, hydrogen, etc., and at ordinary pressures. The use of pressures other than atmospheric is still under investigation.

Details of the assembly and adjustment of the apparatus are to be found in the brochures published by the makers.*

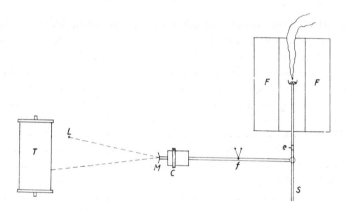

Fig. 9. Diagram of the Chevenard thermobalance.

Photographic registration

The duralumin beam f (Fig. 9), 23 cm long, is supported by two sets of tungsten wire (0.05 mm diameter). At one end it carries a perfectly vertical silica rod S which is well ballasted. The substance being studied or determined is contained in a crucible (or on a plate, or suspended from a small frame if the sample is a metal wire) which is placed at the top of the silica rod. The other end of the beam is fitted with a counter-poise C and carries a mirror M which catches the well-diaphragmed incident ray originating from a small motor car light bulb. The reflected ray strikes a photographic paper (24 × 30 cm) carefully wrapped round a cylinder T which is made to revolve at a uniform rate (generally one revolution in 10 minutes, 3 hours, 24 hours, or 8 days) by a clockwork or synchronous motor. The temperatures are measured to within 1° by means of a platinum-rhodium/platinum couple whose hot junction,

* Société Adamel, 6 Passage Louis-Philippe, Paris 11.

housed in a silica sheath, reaches to the level of the substance being heated.

The calibration in weights is made by placing an overload of 50 mg, in the form of an aluminium wire analogous to a microbalance weight, on a small platform e fixed to the silica rod. On the recordings, a gain or loss of 50 mg corresponds to *ca.* 25 mm on the paper after drying.

The balance itself

Fig. 10 is a photograph of the base of the thermobalance with its protecting parts removed for the sake of clarity; the names and functions of the various parts are indicated in the legend. The apparatus is, of course, placed on a table or foundation to protect it from air currents and vibration. The figure shows the method of suspending the silica rod by means of the tungsten wire S (diameter 0.05 mm), like the wires replacing the central knife-edge. Verticality of the rod which supports the crucible is ensured first by the levelling screws C, then by the oil dampers Cu_1 and Cu_2, and finally by the metal eccentric D. (In recent models this eccentric is replaced by four screws set at right angles to each other.) The support Bu limits the movement of the furnace and makes certain that the substance to be heated is always in the centre of the furnace and in the same position relative to the sheath of the hot junction of the thermocouple.

After placing the crucible containing the substance to be pyrolysed in the carrier ring of the silica rod (Fig. 11 a, b, or c), the position of the beam is adjusted by means of the counterpoise Cp. If the substance is going to lose weight, the mirror M must be adjusted to the highest possible position; when a gain in weight is expected or when carrying out an automatic determination it must be as low as possible. If, for reasons of economy or convenience, it is desired to record two or three curves on the same paper, the position of the beam must be adjusted before each recording, so that the starting point of the curves is known with certainty, by observing the initial position of the spot on the frosted glass of the photographic film holder. It must be ascertained before each experiment that the silica rod does not rub against the wall of the furnace or on the cover D_1 (Fig. 12) which protects the oil dampers from dust.

It is usually recommended to arrest the beam with the set screw V each time after adjusting the position of the counterpoise. However, the writer must confess that he has long neglected to take this precaution and that even so he has never had the unpleasant experience of seeing the suspension wires break. The beam normally carries an opaque disc

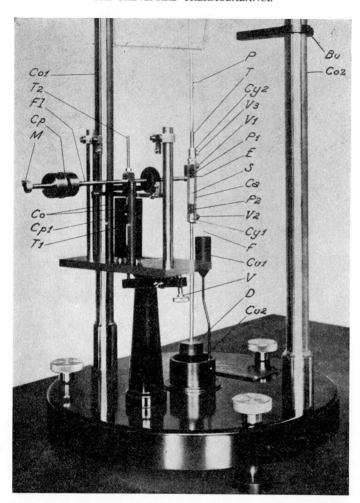

Fig. 10. Photograph of the base of the thermobalance (actual diameter 30 cm). Bu: support for the furnace; Co_1, Co_2: columns supporting the furnace; Cu_1, Cu_2: cups containing the two dampers and some oil; D: eccentric permitting vertical adjustment of the silica rod; V: set screw of the balance; Cy_1, Cy_2: adjusting screw of the tungsten wire (0.05 mm in diameter) permitting articulation between the beam and the rod; V_1, V_2: heads of Cy_1, Cy_2; F: rod supporting the upper damper; S: tungsten wire; Ca: framework supporting the rod and supported by the wire S; P_1, P_2: clamps holding the wire S; E: end of the beam; V_1: screw holding the collar of the silica rod; T: brass sleeve of the silica rod; P: platform carrying weights for the calibration; Fl: beam; M: mirror; Cp: counterpoise; T_1, T_2: adjusting rods; Co: column supporting the balance; Cp_1: movable member serving for transporting the balance. (The two tungsten wires supporting the beam are not visible.)
(By courtesy of Adamel.)

References p. 64

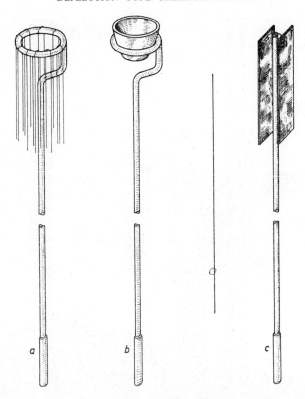

Fig. 11. Diagram of silica rods.

which prevents access of light to the black-walled casing which encloses the balance itself. Fig. 12 is a general view of the original type of thermobalance with the bell-shaped furnace which moves up and down between metal guides, and the photographic housing containing the revolving cylinder; a silica rod and crucible can be seen on the table. The minimum length of space needed for this type of installation is 2 m. The following auxiliary equipment is not shown: a 110/8-V transformer, a 0–7A ammeter, a 0–110 V voltmeter (optional), the relay (permitting conversion of the furnace to a thermostat; see p. 32), the automatic rheostat, and the temperature regulator.

The electric furnace

The furnace is cylindrical with a central tube, made of silica, closed

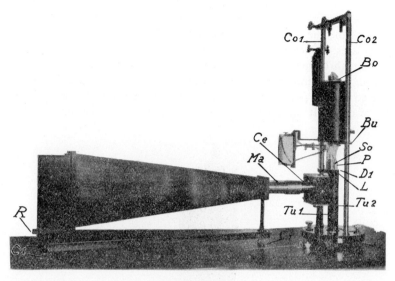

Fig. 12. Photograph of the balance and the casing (actual length 1.32 m). C, C_1, C_2: levelling screws; R: metal platform supporting the photographic chassis; Tu_1, Tu_2: metal sheaths protecting the damping device, the extremity of the beam, and the silica rod; So: cold source; Co_1, Co_2: vertical metal guides for the furnace; Bo: cover of non-corrosive alloy carrying the sheath of the hot junction; Bu: shoulder check for the furnace; P: silica rod supporting the crucible (another silica rod and a porcelain crucible are on the table); D_1: metal cover protecting the oil of the damping device against dust; L: metal tongue closing the slot of D; Ce: flange protecting the beam; Ma: metal sleeve permitting adjustment of the balance and the casing. (By courtesy of Adamel.)

by an almost unoxidizable metal cap which carries the sheath protecting the hot junction of the thermocouple. In order to annul the magnetic field created by the electric current, the winding of the heating element is bifilar; it consists of the well-known nickel-chromium alloy RNC 3.

The furnace is balanced by a cast-iron counterpoise; the height of the shoulder check is so adjusted that when the furnace is lowered it comes to a stop with the substance to be heated in its centre. It functions via a 110–220 V rheostat.

The rheostat

The twin-slide, vertical rheostat needs only a small base. Its windings, which can be used in series or in parallel, are made of a difficultly oxidizable nickel- and chromium-rich resistance wire with a high resistivity (alloy RNC 1) and are wound round two insulating tubes open at both ends. A passage is thus formed which ensures adequate cooling, thereby allowing the rheostat to be placed near the apparatus whose functioning it regulates. The cross-section of the wire used to wind the rheostat is not constant throughout its length. Therefore, the wire is previously heated at a high temperature and in a suitable atmosphere, so that it becomes covered with a poorly conducting, but continuous and very adhering, film of oxide. It can then be wound on the tube with the turns touching

Fig. 13. Rheostat with clockwork motor.

each other and without any fear of short-circuiting between the turns. Some of the oxide is scrapped off the wire to allow contact with the brushes of the slide; these brushes consist of small graphite prisms which last almost permanently.

It is a very tedious business to adjust the slides by hand. They are therefore generally connected up to a three-speed clockwork motor (Fig. 13), the speeds of which can be readily interchanged by means of a system of gears similar to that in motor cars. Corresponding to these three speeds, there are three heating curves, which for one particular furnace are shown in Fig. 14. Starting with a current strength of 2 A the most rapid speed causes the temperature to rise to 1000° in *ca.* 145 min, at the second rate *ca.* 195 min are necessary (which is exactly the time taken by the recording cylinder to make one complete revolution), while at the slowest rate it takes *ca.* 195 min to reach a temperature of 530°.

Since the clockwork motor may be stopped at any desired moment, the temperature curve of the furnace can be varied enormously; certain

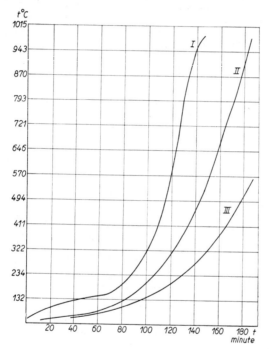

Fig. 14. The three heating curves of a furnace controlled by a rheostat of the type illustrated in Fig. 13.

trajects deviate little from a straight line. A device giving a strictly linear rate of heating is mentioned further on (p. 37).

The upper limit of the furnaces generally used is 1050° and it is not advisable to exceed this temperature. For the curves recorded in this book it has only occasionally been necessary to go beyond 1000°; and in any case, temperatures as high as this are quite useless for the automatic method of determination.

References p. 64

The thermostat

The thermostat is of the expanding wire type and is illustrated in Fig. 15. In order to control the temperature properly, such an apparatus must be able to counteract very rapidly the fluctuations in temperature due to changes in the heating current. It must be placed very close to the wall of the furnace; it must have a very low heat capacity; and it must be sensitive to the slightest changes in temperature.

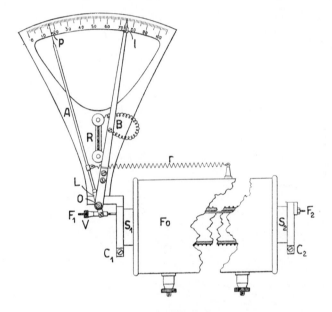

Fig. 15. The thermostat.

The wire F_1F_2, secured at F_2 to the invar collar C_2, is attached at F_1 to the small lever L which can turn about the axis O. A silver contact P and a needle which moves over the dial are fixed to the free end of the arm A (joined to the lever L). The dial and the lever support are secured to the silica tube S_1S_2 of the furnace by the invar collar C_1. The wire is held taught by the spring r. When the temperature rises, the needle moves to the right in the case of Fig. 15 (but upwards when the furnace is in position in the thermobalance) until the contact P touches a small piece of silver foil l fixed to the arm B. This completes an electric circuit and via a relay diminishes the heating current or cuts it out completely; the temperature then falls, but the contact is hardly broken before the relay causes the heating current to return to its original strength; the temperature rises again; and so on. The resistance R is included to suppress sparking when the contact is broken and the screw V enables the zero to be accurately adjusted.

With this apparatus it has been possible to maintain the temperature of the furnace at 175° ± 0.1° for 3 hours, with a concomitant variation in room temperature between 16° and 16.4°. The current strengths in amperes necessary to maintain constant temperature by means of a thermostat of the type shown in Fig. 15 and the corresponding deflections of the pointer A in mm are recorded in Table 1 at intervals of 100°.

TABLE 1

CURRENT STRENGTHS FOR MAINTAINING CONSTANT TEMPERATURE

Temperature (°C)	Deflection (mm)	Current strength (A)
100	10.0	1.4
200	23.0	2.1
300	38.0	2.8
400	54.6	3.5
500	71.7	4.1
600	90.0	4.5
700	109.0	5.1
800	130.0	5.7
900	151.8	6.3
1000	174.0	7.0

It is necessary to calibrate the scale for each use of the furnace, because the total expansion of the wire F_1F_2 depends not only on the temperature in the centre of the furnace but also on the conditions at the ends. Different readings are obtained when the furnace is closed at one end and at both ends. The dial therefore cannot be graduated in temperatures and carries a simple angular scale.

The lighting arrangement

The point source of light used for tracing the curve on the photographic paper is obtained by suitably diaphragming the beam from an 8–12 V motor car lamp (Fig. 16) and is projected via a condenser on to the concave mirror situated at the end of the balance beam. The diaphragm consists of a metal plate pierced with a small hole, diameter 0.1 mm.

During the adjustment, to facilitate fixing the position of the light spot on the frosted glass, a luminous spot (5 mm in diameter) is substituted for it and the diaphragm is altered.

A narrow slit over the point opening in the diaphragm is covered by a screen. When an electric current passes through the electromagnet attached to the diaphragm the slit is uncovered and a dash, several mm

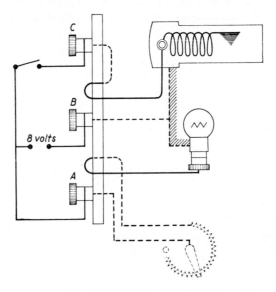

Fig. 16. The lighting arrangement.

long, is traced on the photo-sensitive paper below the curve. In this way the temperatures on the final recording can be fixed by reference marks. The operator passes a current through the electromagnet for 5 seconds each time the needle of the thermocouple millivoltmeter passes a pre-determined division on the scale. The frequency of these points is a function of the slope of the curve and it is important to determine them frequently below 200° since it is especially at low temperatures that losses in weight are found (elimination of water, ammonia, etc.). The calibration of the thermocouple can then easily be related to the curve and interpolation carried out where necessary. Clearly, if the heating curve of the furnace is completely known, an automatic arrangement can take the place of the operator. A very simple device has been constructed in which a brass cylinder is made to revolve at the rate of one revolution per hour by means of a synchronous motor. It has rows of notches which meet a fixed contact every two, five, or fifteen minutes. A weight–time or weight–temperature curve can be obtained as desired.

The furnace with a predetermined temperature cycle

The expanding wire F, which was discussed in connection with the thermostat (Fig. 15) and which is fixed to the invar collar of the furnace, is attached to the small lever which carries the arm A. The latter has at its end a flexible strip 1 ending in a small cylinder c. The wire is kept taught by the spring r of fixed weight, while the contact P fixed to the flexible strip 1 normally touches the flattened end of the rod t which is joined to the terminal B_2 of the relay R (Fig. 17).

The contact between P and t is very sensitive and causes attraction of

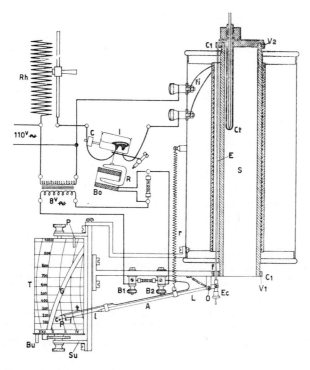

Fig. 17. General diagram of the installation with a predetermined heating cycle.

the magnetic part of the electromagnet Bo, is whereupon the mercury of the circuit breaker l closes the heating circuit and the temperature of the furnace rises. The wire F then expands and c moves until checked by the stop G; this causes the contact P to separate from the rod t; the current in Bo is therefore cut off; the circuit breaker l rocks under the action of the counterpoise C; and cuts out the heating current. The furnace cools, and so on.

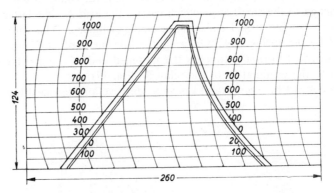

Fig. 18. An unrolled guide for predetermined heating cycles.

The nature of the stop G still has to be explained. It consists of a triple thickness of strong cardboard glued to a strip of paper wrapped round a small drum T which is rotated by clockwork (one revolution per 12 hours or per 24 hours). It is shown unrolled in Fig. 18. The form of this guide can be chosen according to the purpose in mind, but the simplest idea is to make it rectilinear when unfolded (helical when on the drum), *so that the heating will be linear and so that the weight–temperature and weight–time curve can be obtained simultaneously.* By using the guide illustrated in Fig. 18 the temperature of the furnace will rise to 1000° at the rate of 100° per hour, will be maintained at 1000° for 1 hour, and will then decrease at the rate of 150° per hour.

The operator can have at his disposal a whole series of interchangeable guides which can be fixed to the drum T in a matter of seconds. As far as analytical operations are concerned, it is best to make the rise to 1000° in $2\frac{1}{2}$ hours.

If only the range 350–700° is of interest then in order to gain time the drum T can be arrested; using the maximum current and without the guide G, the temperature is allowed to rise unhindered as far as 350° (this takes 7-10 minutes); the drum is then released and the point c comes into contact with the guide.

Naturally, if the drum is stationary, the furnace will remain constant at the temperature indicated on the guide—moreover, a temperature which can be reached at any desired rate of heating. If there happens to be no

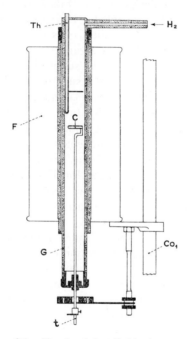

Fig. 19. Diagram of the silica sheath installed in the centre of the furnace.

relationship between the temperature as indicated by the thermocouple galvanometer and the temperatures marked on the sheet of paper wound round the drum T, the zero is adjusted accurately by means of the set screw Ec at the end of the wire F_1F_2.

This short description shows the very great ease with which *predetermined heating cycles*, as put into effect by Chevenard and Joumier at the Aciéries d'Imphy (Nièvre), can be manipulated.

References p. 64

Operation in a controlled atmosphere

Experiments at different temperatures, but under ordinary pressures, have recently been carried out in hydrogen, argon, nitrogen, and carbon dioxide instead of in air. The assembly needs to be adjusted rather carefully; however, this new technique naturally leads to results of extreme interest.

The gas to be used is introduced at the top of the vertical silica tube G which fits the centre of the furnace F (Fig. 19); t is a silica rod which ends in a ring carrying the crucible C; Th is a cavity for housing the hot junction of the thermocouple, while Co_1 is one of the bars supporting the furnace and to it is attached the arrangement by which the silica tube is sealed. The hydrogen, for example, supplied from a cylinder through a reducing valve, is passed through a sulphuric acid bubble counter and then through a small auxiliary furnace containing copper turnings maintained at 450°; next, it goes through a tube filled with silica-gel impregnated with cobalt chloride, through another bubble counter, and finally into the silica tube G where its movement is checked by a horizontal baffle placed above the substance being heated.

After purging the tube G with gas for 45 minutes, the supply of hydrogen is adjusted to one bubble per second, for example. The gas escaping at the bottom of the tube carries along with it any reduction products formed.

The following striking example demonstrates the possibilities of this technique. Fig. 20a represents the pyrolysis curve of silver chloride heated freely in air–it does not begin to dissociate until 600°. In hydrogen, however, silver is already present at 450°; the bend in the curve indicates the formation of a mixture AgCl+Ag, not of a "subchloride" (Fig. 20b).

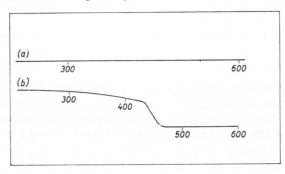

Fig. 20. Pyrolysis curves of silver chloride heated (a) in air, (b) in hydrogen.

3. Automatic pen-recording thermobalance

Fig. 21 is a general view of the apparatus (not available commercially), while Fig. 22 shows the recording system with the cover removed. As this apparatus is the prototype, we cannot enter into details of its construction. However, its essential feature is that the beam is maintained in a state of constant oscillation, thus replacing by a dynamic position of equilibrium the static position of equilibrium which is partly determined by frictional forces.

Fig. 21. Photograph of the new pen-recording thermobalance.

Instead of the mirror there is a feeler which via a special relay starts the current in a servo-motor; this last transmits its to-and-fro movement to an endless screw and in turn to an inked pen which traces the pyrolysis curve on a cylinder which completes one revolution every 3 or 24 hours (Fig. 23).

The switches from left to right at the base of the apparatus have the following functions: to release the beam, to start the heating current in the furnace, to start the servo-motor, to start the recording cylinder

Fig. 22. The same with cover removed.

revolving, and to control the movement of the pen. The heating cycle of the furnace is predetermined; when working in hydrogen, a silica sheath can be introduced into its centre, as indicated in the preceding section. A circuit breaker (not visible on the photographs) enables the temperatures to be marked while the operator is following the deflections of the galvanometer connected to the hot junction. The curves traced are as delicate as those obtained by the photographic procedure.

Experience has shown that one operator can follow the recordings being made simultaneously by three thermobalances.

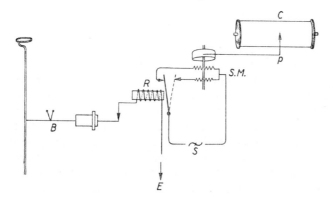

Fig. 23. Diagram of the pen-recording device.

Fig. 24. Close-up of Chevenard thermobalance showing electronic conversion. (Authors' courtesy.)

The performance of a pen-recording Chevenard thermobalance has been studied by Simons, Newkirk and Aliferis[12] (see Chapter 5).

The inability to observe the curves during the course of the reaction under study has led Gordon and Campbell[13] to an electronic conversion of the Chevenard thermobalance. A suitable transformer to convert balance-beam displacements into electrical signals has to be accurately linear, light-weight, and frictionless in operation, also sensitive to very small deflections, relatively inexpensive, and involve no more than a minimum modification of the balance. The device (Alcotran) finally selected was a linear variable differential transformer, used in conjunction with a demodulator, which satisfactorily met all of these requirements.

Newkirk[14] used a clear quartz tube and a mirror, and worked in a controlled atmosphere.

Fig. 24 shows a close-up of a thermobalance with electronic conversion.

After about a thousand runs with the balance fitted for pen recordings as pictured in Fig. 21, the Société Adamel put on the market an assembly combining the latter with the photographic balance; it is shown in Fig. 25. The balance proper, *i.e.* the right side of the figure remains as it was; the small concave mirror M of Fig. 10 is unscrewed and replaced by the silver feeler of Figs. 22 and 23. The inscriber pen has no special features; it still functions mechanically but is so designed that it adjusts itself to the same height as the photographic housing of Fig. 9, and does so in several minutes. Accordingly, the operator can have at his disposal both methods of recording and still enjoy the economy of needing only one furnace and one balance.

In the new electronically-recording balances, which may be designated as IA, IB, IC, and which function at 220 V, the maximum load is 20 g, and the sensitivity is such that 1 mg corresponds to from 0.5 to 2 mm on the graph. Under especially favorable operating conditions, it is possible for 1 mg to correspond to 5 mm.

The thermobalance itself is suspended by thin strips or wires and is fitted with a magnetic damping device. The furnace is heated by a wound resistance element made of platinum-rhodium; it has a double wall, and is cooled by circulating water. It is provided with a fritted alumina cover, and a safety valve for the water comprising an electromagnetic cut-off for the entering water and a valve for the exiting water. The furnace can be brought to 1500°. The curve is recorded on a drum whose diameter is 93 mm; the paper is 300 mm × 250 mm. In model IA, the drum is driven by a two-speed motor, which will give one revolution in 3 hours or in 24 hours. In the models IB and IC, this rotation is additionally

Fig. 25. Adamel thermobalance with pen and recording vertical chart.

controlled by a MECI potentiometer pyrometer especially adapted for this purpose. The mass–time curve is recorded in the IA model; either the mass–time or the mass–temperature curve with the IB model according to the operator's desire, while the IC model records both curves simultaneously by means of two styli (Figs. 26 and 27).

The principle on which this instrument functions is simple. A beam of

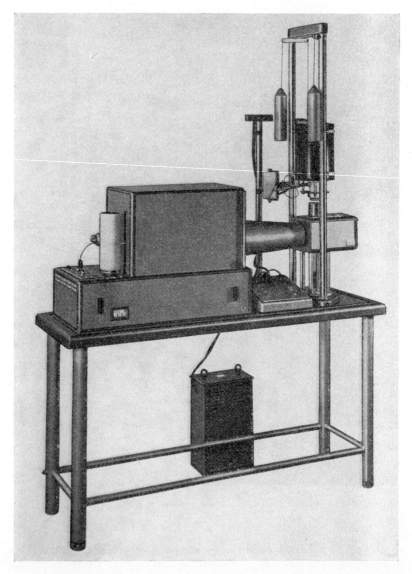

Fig. 26. General view of Adamel thermobalance 1A, 1B, or 1C.

light produced by the lamp of the recorder is concentrated by a lens and projected on the mirror of the set-up and then reflected by the mirror so that it falls on the vertical path traversed by a photoresistance cell.

Fig. 27. The same with cover removed.

As soon as this cell is slightly illuminated, a motor causes it to descend until it meets the spot positioned below it. A current strong enough to activate a sensitive relay immediately passes through it and the direction

References p. 64

in which the motor is turning reverses, and the cell again rises until it is too weakly illuminated which causes it to descend again.

However, because of the rapid displacement of the slide, the cell is obliged to rise to a certain height before provoking the reversal of its movement, and this results in the disadvantage of producing too thick a line if the pen is attached to the slider. Therefore the recorder is provided with a device for thinning the line. This device is composed essentially of two carriers; one, bearing the cell, oscillates continuously with an amplitude which is kept rigorously constant; the other, bearing the recording pen, does not move under the action of the first unless a phenomenon intervenes. The interplay of the two carriers is regulated by a stop screw. If the spot is stationary, it produces alternate ascents and descents of the first carrier with constant amplitude and, if the pen is attached to this carrier, it will produce a line whose thickness corresponds to the vertical displacements of the cell. Since the play between the two carriers is equal to the stroke of the pen attached to the second carrier, the latter will not move so long as the spot itself remains immobile; it will rise or fall only when the spot changes its position. The curve recorded in this fashion will have only the thickness of the line traced by the non-oscillating pen, even though it follows the slightest movement of the spot.

The models IA, IB, IC may become IIA, IIB, IIC if they serve in experiments under greatly reduced pressures or in trials conducted in a gas other than air. The thermobalance proper is fitted with a tight chamber enclosing the entire mechanism; this chamber is provided with openings for adjusting the secondary pump and control gauge. A liquid nitrogen trap serving as baffle extends partly under the thermobalance and partly under the secondary exhausting pump. This trap not only improves the vacuum but also prevents the entrance of impurities into the pump. Finally, there is a device designed by C.R.C. or control chrono-relay which serves for the complete protection of the furnace when the thermobalance is used in a tight enclosure.

4. The Stanton thermobalance

This recording thermobalance* is shown in Fig. 28. It is a deflection type instrument in which a capacitance-follower plate located above the balance beam is used to measure the movement electromechanically and

---

* Stanton Instruments Ltd., 119 Oxford Street, London W 1.

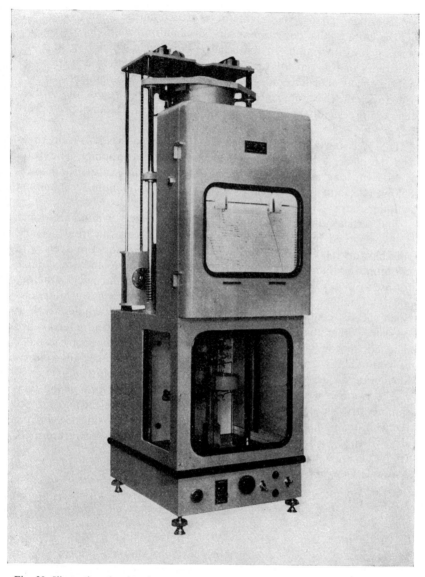

Fig. 28. Illustration showing the compactness of design of the Stanton thermo-recording balance.

References p. 64

to record the change in weight for a predetermined beam movement. At the close of this deflection a servomechanism brings electric weight-loading into operation, so that changes in weight up to 1 g at 1 mg sensitivity or 0.1 g at 0.1 mg sensitivity may be followed and recorded automatically. Dual recorders and a temperature programmer are provided with the thermobalance to obtain the change in weight and furnace temperature simultaneously as a function of time on curvilinear chart paper.

The precision air-damped analytical balance has special knife edges and the planes are made of synthetic optically flat sapphire. The beam is 12.5 cm long and made of hard brass; it is standardized and interchangeable, and is fitted with precision spun aluminium air damping cylinders.

The vertical tube type furnace with grooved silica or alumina former and top closed by 4-inch deep alumina plug is gage nichrome wound suitable for 1000°; its internal bore is 5 cm. Alternative models with platinum/rhodium bifilar wound furnaces are available and are suitable for use to 1400°. Both models have been designed to take inner refractory sheaths (mullite) so that samples may be heated in a gas atmosphere. Such sheaths cannot be sealed at the base and this precludes work at pressures other than 1 atm. Bifilar windings prevent field effects when magnetic or conductive materials are heated. The thermocouple is made of platinum–13% rhodium. A simple programmer system gives nearly linear rate of heating at 350° per hour or 250° per hour.

The recorder-controller contains an electronic twin-pen giving continuous strip side by side records of both change in weight and furnace temperature with time. The chart of high grade, low shrinkage paper shows twin 5-inch rectilinear tracks scaled mass and temperature in °C. Its speeds are 3, 6, and 12 inch per hour.

A side view of the thermobalance is shown in Fig. 29.

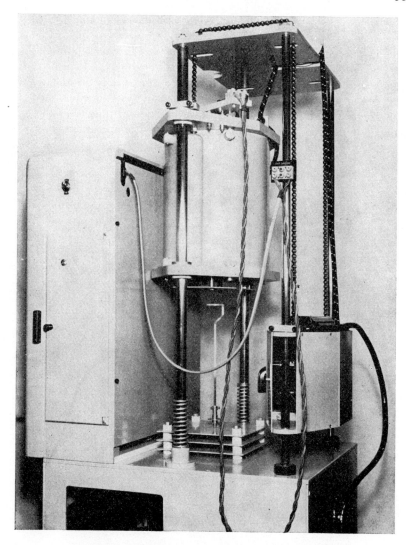

Fig. 29. Side view of the thermo-recording balance showing recorder, furnace with counter-balancing transformer and silica riser rod holding a crucible.

5. The Brabender thermobalance

This instrument (Fig. 30) was developed in the chemical laboratories of the Steinkohlenbergbauverein at Essen, (West Germany).* In it the stylus attached to the end of the balance beam is automatically pressed against the pressure-sensitive chart paper at regular intervals of a few seconds, thus recording the change in weight as a function of time. This thermobalance is designed for ambient and controlled temperatures with 1 g

Fig. 30. Brabender recording deflection thermobalance.

---

* C. W. Brabender Instruments Inc., 50 East Wesley Street, South Hackensack, N. J. (U.S.A.).

weight loss range measured on the 17.5-cm strip chart. It is supplied with a 1200° furnace and a pyrometric programming controller having interchangeable templates for various heating cycles.

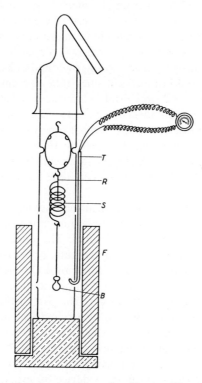

Fig. 31. Diagram of a torsion thermobalance after Murthy, Bharadwaj and Mallya. F: furnace; B: bucket; S: spring; R: reference rod; T: thermocouple.

6. Torsion thermobalances

These instruments were inspired by the McBain setup. The assemblies designed by Gulbransen and Iddings have been noted in Chapter 1. That of Izvekov[15] is obtained by attaching a mirror to a tungsten wire spring balance in such fashion that a pencil of light rays is reflected by the mirror and made to fall on photographic paper wound on a rotating cylinder.

The instrument devised by Murthy, Bharadwaj and Mallya[16] is likewise very simple. It consists essentially of a McBain quartz fiber spring balance (Fig. 31). The sensitiveness of the springs ranges from 8–10 cm per g weight. Hooke's law is well followed in this region. The load ordinarily employed for thermogravimetric work is about 100–300 mg. By making use of a quartz reference rod, the elongation of the spring is measured with the aid of a travelling microscope (smallest reading = 0.002 cm). The substance being studied is placed in quartz or platinum receptacles suspended from the spring by long hooks. The spring and the container are housed in a vertical silica tube (2.5 cm diameter) which is closed at its lower end and firmly clamped in position. Apertures are provided at the closed end of the tube to allow the vapours to escape. The silica tube is inserted in the vertical tube furnace which is closed at the bottom. The furnace can be heated to 1000° and can be regulated to a rise of 3–5° per minute. A chromel-alumel thermocouple inserted close to the container serves for reading the temperature of the material inside of the furnace.

The Loriers balance[17] was designed for studying the changes in weight which result when such metals as cerium are heated in oxygen. The inclosure consists of two vertical tubes. The upper tube P (Fig. 32) is made of Pyrex and can be moved vertically by means of a toothed support S_1 and so made to fit into the lower tube Q which is made of quartz. The lower tube rests on a spring r whose contraction assures that the joint R between the tubes stays tight. The part of the quartz tube which contains the sample can be heated by a cylindrical electric furnace which moves independently of the tube by means of a second toothed support S_2. The open end of this tube is narrow and passes through the spring r and extends through the base supporting the apparatus, which is provided with an opening for this purpose. A device for evacuating the tubes may be fitted at this end x.

The upper tube contains a spring Re made of elinvar (Fe–Mn–Cr–Ni alloy). This very sensitive spring consists of about 20 spirals (15 mm radius) of 0.25 mm wire. It is suspended from a solid hook of a joint connecting the tube with an inlet cock r_1 for gas. A small Plexiglas scale g is hung below the spring; it has a micrometric graduation down to one-tenth of a millimeter. The movements of the graduation are read with reference to the horizontal wire of the cross wires illuminated by the bulb L of a telescope (magnification 25 ×). This mobile telescope which may be moved along a vertical support T can be fixed opposite the micrometer. The sample s is suspended from the beam by a long hook made of Pyrex or quartz. Constrictions e e′ prevent any heating

of the Pyrex tube or the beam. A small disk d damps the oscillations of
the spring. The furnace is equipped with a thermocouple Th.

The thermobalance described by Campbell and Gordon[1] at the Pitts-
burgh Conference on Analytical Chemistry in March 1957 is a vacuum
apparatus using quartz or metal springs and a differential transformer as
transducer. See also Hooley[18] and Stephenson, Smith and Trantham[19].

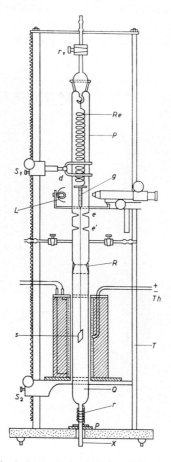

Fig. 32. Diagram of Loriers' thermobalance for studying oxidation under a controlled
atmosphere. Re: elinvar spring; Th: thermocouple.

7. The Thermo-Grav

This instrument* has been described by Müller[20] and developed by
Gordon and Campbell. It records thermogravimetric curves of samples
in vacuo or in controlled atmospheres at temperatures up to $1000°$. It
can record changes in weight as a function of temperature; it may be
programmed for a selected heating rate, and also for changes in weight
as a function of time at a constant temperature.

A general view of the heating system is shown in Fig. 33 and a schematic

Fig. 33. Thermo-Grav balance. General view.

---

* Manufactured by the American Instrument Co., 8030 Georgia Avenue, Silver Springs,
Md. (U.S.A.), Bulletin No. 2304.

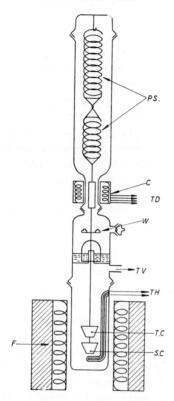

Fig. 34. Diagram of the Thermo-Grav. PS: precision springs; F: furnace; TH: thermocouple; SC: sample crucible; TC: tare crucible; TV: to vacuum pump; W: weight calibrator; C: coil; TD: to demodulator and recorder.

representation is given in Fig. 34. The instrument consists primarily of a highly precise spring balance enclosed in a glass vessel and it can be operated at controlled temperatures under a vacuum or a controlled gaseous atmosphere, at or below ambient pressure. Spring deflections, proportional to changes in sample weight, are converted into electrical signals by movement of the core in a linear differential transformer C. After passing through the amplifier and demodulator, these signals are presented as weight changes on the Y-axis of an XY recorder. The X-axis of the recorder receives an input signal which corresponds either to the furnace temperature or to the elapsed time.

References p. 64

The suspension system carries a weight calibrator pan, an oil-dashpot, a damper, and at the bottom fused quartz rings to hold the sample crucible SC and the tared crucible TC. The furnaces are designed to allow consecutive determinations without delay due to cooling-down time. The furnaces can be rotated and elevated into position with ease and precision. The lower quartz jacket is readily removed and is fitted with a lateral port for rapid evacuation and subsequent purging or filling with any desired gaseous atmosphere. The system is shown in Fig. 34. The furnace control and programmer can be set for any one of 16 temperature spans with a choice of seven rates of temperature rise: 5°, 10°, 20°, 50°, 100°, 200°, 500° per hour; or for maintaining the furnace at constant temperatures. The recorder adjustment permits X-axis spans (full scale deflection) for temperature increments of 200°, 500°, and 1000°.

The thermocouple in series with a fixed resistor generates a current proportional to its temperature. The voltage across the potentiometer is maintained constant by means of a Zener diode. Since the potentiometer is driven at a fixed rate, its output voltage will be proportional to time. A resistor in series with the arm of this potentiometer can be varied by means of a switch, and the output is therefore a function of the switch position. It is calibrated in °C per minute and the potentiometer arm is calibrated in minutes.

A third input is a small sweep current. This is driven a 7 r.p. m. and provides a small modulating signal to the amplifier. Thus the furnace is switched on and off seven times a minute, the length of "on" time being determined by the error signal between the thermocouple and the potentiometer outputs. A proportional control is used to adjust the stability of the system and to prevent temperature cycling. This is similar to the gain control in any servo loop.

A blocking diode prevents the relay from being energized when the summing junction is positive (indicating too hot) but allows it to be energized and apply power to the furnace when the summing junction is negative. A non-linear converter is required in order to read the temperature of the sample in digital form and to be able to drive the recorders linearly with respect to temperature, because a thermocouple is not sufficiently linear for the purpose. The converter is a standard servo loop with a non-linear feedback potentiometer, whose output matches the thermocouple. Linear potentiometers with proper taps and shunts are used. Thus, the shaft rotation is a linear function, not of voltage, but of temperature. The readout dial counter is coupled to it to drive the recorders.

8. The Wendlandt thermobalance

Wendlandt[21] has adapted a torsion-wire type instrument (capacity 0–120 mg) made by Verenigde Draadfabrieken, Nijmegen (Netherlands). The smallest scale division was 0.2 mg; hence, weighing could be read to 0.1 mg. The recording drum consisted of an aluminium cylinder 3 inch in diameter and 10.5 inch in length connected to the torsion wire shaft of the balance. The rotation of the drum was controlled by a r.p.m. reversible synchronous motor. However, since this speed caused to much overshoot, the shaft speed was reduced with a 4 to 1 reducing gear (Fig. 35).

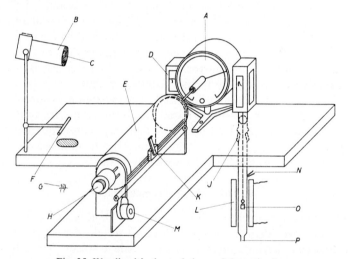

Fig. 35. Wendlandt's thermobalance. Schematic view.
A: torsion balance, B: light source, C: slit and lens, D: beam mirror, E: recording drum, F: mirror, G: photocells, H: drum motor, J: combustion tube joint, K: pen carriage, L: furnace, M: pen drive motor, N: thermocouple, O: platinum sample pan P: exhaust gas connection,

The recording pen was a size 000 Leroy lettering pen suitably mounted on a sliding carriage. The latter was drawn across the slide bar with a 1 r.p.m. synchronous motor, which was provided with a friction clutch so that the pen carriage could be manually reset to the starting position.

The furnace is constructed by first winding 15 feet of No. 22 gage Nichrome alloy V resistance wire (1.01 Ω/ft.) into a coil 1/8 inch in diameter. This is then wound onto an asbestos-covered Vycor glass tube, 2.5 cm in diameter and about 25 cm long, at about 0.25 inch spacings.

The completed windings were adequately covered with asbestos insulation.

The sample is contained in a platinum pan, 1 cm in diameter and 0.5 cm high, suspended in the furnace by a platinum wire connected to the balance beam. The temperature rise of the furnace is controlled by gradually increasing the input voltage by means of a 6-revolutions-per-day synchronous motor connected to the shaft of a Powerstat. The motor-driven Powerstat input voltage is controlled by means of another Powerstat. With an input voltage of 60 V and the motor-driven Powerstat started at 20 V, the temperature rise of the furnace is approximately linear at 5.4° per minute.

The optical system contains a plane mirror, 0.25 inch in diameter, attached to the weight hook of the balance beam. A beam of light from a 50-W lamp is focused on this mirror by means of a lens and an adjustable slit. The light reflected from the mirror is focused by a lens mounted on the balance case and then reflected downward onto the two photocells by means of another plane mirror. The photocells are enclosed in a light-tight cabinet to prevent interference from stray light.

During a typical decomposition run, a slow stream of air is passed through the furnace by means of a water aspirator connected to the lower part of the furnace. A blank run shows that there is no apparent weight change (less than 0.1 mg) for a temperature rise from ambient to 850°. A simpler apparatus has been developed by Banks[22].

9. Differential thermobalances

An interesting instrument has been devised by De Keyser[23].* It is shown in Fig. 36. Two identical samples of equal weight are suspended into identical furnaces from opposite ends of a balance beam. These furnaces are heated at the same linear rate but one is kept approximately 4° cooler than the other. Since the reactions involving changes in weight in the two samples occur consecutively, there is a weight differential whose magnitude is proportional to the heating rate. Beam deflections corresponding to these weight differentials are recorded photographically by means of a mirror attached to the beam so as to deflect a light onto the photosensitive paper wrapped around a drum. The resulting curve approximates the derivative of the weight change plotted as a function of temperature, the area under this curve being the total change in weight of the sample. Ranges of weight change of 5, 10, and 20 mg can be obtained for sample weights within the capacity of an analytical balance.

---

* Manufactured and offered for sale by the Sartorius Werke, Göttingen (West Germany)

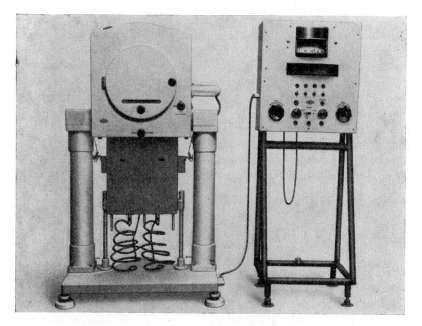

Fig. 36. Sartorius-De Keyser differential thermobalance.

The technique of differential thermogravimetric analysis has proved valuable because the curves are the derivative of primary weight change, and therefore closely resemble the complementary differential thermal analysis curves with which they are compared in many thermoanalytical investigations. Another differential thermobalance was developed by Erdey, Paulik and Paulik[24, 25]. Their apparatus, which is known as the Derivatograph, can be used for running differential thermal analysis, thermogravimetry, and derivative thermogravimetry simultaneously.*

The Derivatograph is shown schematically in Fig. 37. As may be seen, the porcelain tube (3) of this differential balance (a modified automatic balance with air dampers) is provided with a cavity in which a thermocouple (4) has been threaded in such fashion that its hot junction (5) extends into the crucible containing the sample being studied. A crucible (16) of the same size and shape is attached to the end of another porcelain tube (17) for the inert comparison material. This tube contains another thermocouple (18) and is fixed in the base of the balance (19). The terminals of the thermo-elements are mounted in opposition; the circuit includes

---

* This instrument is sold by Orion G.V.E.M., 15 Lágmányosi, Budapest (Hungary).

a millivoltmeter (6) calibrated in degrees, a galvanometer (8), a precision resistance, and an interrupter (15).

In a crucible (5) of the type shown in Fig. 37, the thermojunction is directly in contact with the sample being studied, but the experience of these workers has proved that a crucible having the form (20) may be used. In the latter instance, the junction is outside of the crucible and measures the temperature of the air inclosed between the walls of the crucible and the porcelain tube (3). This arrangement has the advantage that melted and liquid samples can be investigated.

Fig. 38 shows how the automatic recording is accomplished. A thin plate, provided with a vertical aperture (22) is attached to the pointer of

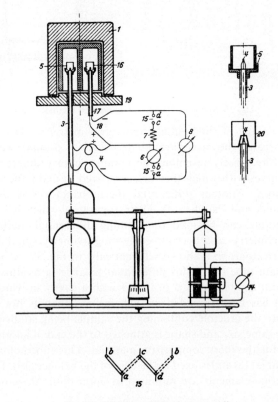

Fig. 37. Diagram of "Derivatograph".

the balance (21) and is illuminated by light from the incandescent bulb (23). The light falls on a recording drum (25) driven by a synchronous motor. The image of the aperture yields a curve TG (mass–time or temperature) on the sensitive paper. A curve of this kind is shown in Fig. 68 in Chapter 4. In addition, the galvanometers (6, 8, 14) indicating the temperature of the sample T, the difference of the weights in the two crucibles DTA, and the temperature difference DTG are installed on the other side of the drum in such manner that their spots yield the three aforesaid curves on the other half of the paper. Fig. 68 shows thermograms obtained in this manner with a sample of clay.

At the same time and in addition to these measurements, four lighting bulbs (29, 30, 31, 32) are placed at the four corners of the opening of the recording drum (Fig. 38). It is then possible to provide the thermogram with a time scale by inserting a release, analogous to the kind used in a photographic apparatus, and controlled by a clockwork movement.

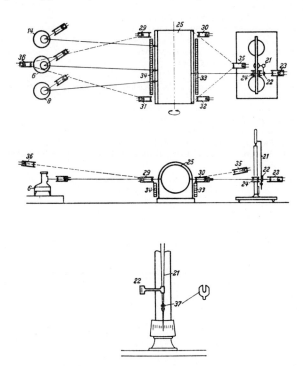

Fig. 38. Automatic recording of temperature, thermal curve, differential thermal analysis curve, and derivative thermal curve.

Furthermore, a scale of weights and a temperature scale are inscribed on the sensitive paper, either before or after the measurement. Templates (33) and (36) which bear calibrated apertures are used for this purpose.

REFERENCES

[1] S. GORDON AND C. CAMPBELL, *Anal. Chem.*, 32 (1960) 271R.
[2] P. DUBOIS, *Thesis*, Paris, 1935; *Bull. Soc. chim. France*, 3 (1936) 1178.
[3] Y. JOUIN, *Chim. & Ind. (Paris)*, 58 (1947) 24.
[4] H. LONGCHAMBON, *Bull. Soc. franç. minéral.*, 59 (1936) 145.
[5] V. VYTASIL, *Silikáty*, 2 (1958) 285.
[6] E. OROSCO, *Ministério trabalho ind. ecom., Inst. nac. tecnocol. Rio de Janeiro*, 1940; *C.A.*, 35 (1941) 3485.
[7] P. SPINEDI, *Ricerca Sci.*, 23 (1953) 2009; *C.A.*, (1954) 868.
[8] A. COULERU, *Rev. prod. chim.*, 40 (1937) 611.
[9] J. BREFORT, *Bull. soc. chim. France*, (1949) 524.
[10] E. S. BARTLETT AND D. N. WILLIAMS, *Rev. Sci. Instr.*, 28 (1957) 919.
[11] P. CHEVENARD, X. WACHÉ AND R. DE LA TULLAYE, *Bull. Soc. Chim. France*, [5] 10 (1944) 41.
[12] E. L. SIMONS, A. E. NEWKIRK AND I. ALIFERIS, *General Electric Co.*, Report No. 56 RL-1522, May 1956.
[13] S. GORDON AND C. CAMPBELL, *Anal. Chem.*, 29 (1957) 298.
[14] A. E. NEWKIRK, *Anal. Chem.*, 30 (1958) 162.
[15] I. V. IZVEKOV, *Trudy Krym. Filiala Akad. Nauk. S.S.S.R.*, 4 (1953) 81.
[16] V. MURTHY, D. S. BHARADWAJ AND R. M. MALLYA, *Chem. & Ind. (London)*, (1956) 300.
[17] J. LORIERS, *Rev. Met.*, 49 (1952) 807.
[18] J. G. HOOLEY, *Can. J. Chem.*, 35 (1957) 374.
[19] J. L. STEPHENSON, G. W. SMITH AND H. V. TRANTHAM, *Rev. Sci. Instr.*, 28 (1957) 380.
[20] R. H. MÜLLER, *Anal. Chem.*, 22 (1960) 77A.
[21] W. W. WENDLANDT, *Anal. Chem.*, 30 (1958) 56.
[22] J. E. BANKS, *Anal. Chim. Acta.*, 19 (1958) 331.
[23] W. L. DE KEYSER, *Nature*, 172 (1953) 364.
[24] F. PAULIK, J. PAULIK AND L. ERDEY, *Z. anal. Chem.*, 160 (1958) 241.
[25] L. ERDEY, F. PAULIK AND J. PAULIK, *Nature*, 174 (1954) 885; *Acta Chim. Acad. Sci. Hung.*, 10 (1956) 61.

CHAPTER 3

Null Type Thermobalances

1. General

In a balance of the null type, a sensitive element detects the slightest deflection of the beam, which is horizontal in an automatic balance, or vertical in the electromagnetic suspension type. An adjusting force, which may be mechanical or electrical, is applied to the beam through the action of servo-hookups and returns it to its null position. This adjusting force, proportional to the change in mass of the heated substance, is recorded either directly or by means of an electro-mechanical transducer. In many instances, a type of recording balance which functions in the cold is subsequently remodeled into a thermobalance.

The null balances are the most numerous and the first to come to mind since from earliest times it has been the custom to place weights in a discontinuous fashion on one of the pans so that the pointer or the reference mark is brought in front of a division selected in advance. But we are going to deal here with a beam which tends to be deflected continually.

The earliest "cold" balance devised by Angström[1] consisted of an analytical balance with a magnet suspended from the beam into a solenoid. As soon as the substance underwent a change in weight, the current created by induction in the coil was manually adjusted in such wise as to leave the beam at the null position. This current was proportional to the weights of the sample and was read on the scale of a reflecting galvanometer. A similar apparatus enclosed in a glass case for working under reduced pressure and in a controlled atmosphere was devised by Mikulinskii and Gel'd[2].

If the adjusting force is to be applied *mechanically*, weights may be added to or removed from one pan. Honda[3] used this method in his first thermobalance. Alternatively, a rider may be shifted along the beam or its extension, or the progressive effect of a helical spring may be applied, or a chain may be lengthened or shortened, or a liquid (water, oil) may be added or removed in such manner as to alter the Archimedes buoyancy, as in the first Guichard balance[4], or a hydraulic pressure may be applied in accord with Pascal's principle.

When the adjusting force is applied *electromagnetically*, this may be accomplished either by changing the position of a mild iron armature in a coil, or of a magnet in a balance, or of one coil in another. The necessary current may be obtained directly from the null detection circuit or by means of a servo-potentiometer.

In the very original type devised by Waters[5], the adjusting force may be said to be *electrochemical* since it arises from the coulometric solution of a metal or its deposition on the electrode suspended from the beam.

Other methods include: changing the capacity of a condenser, altering the power of a differential transformer as function of the displacement of the armature, or changing the flow of nuclear radiation. The photoelectric detector utilizes the change in intensity of a light source acting on a phototube. It requires a light source, an intermediate diaphragm or a mirror, and either a simple or a double cathode phototube. The shift in the position of the diaphragm attached to the beam intercepts the light ray in such fashion as to increase or decrease the intensity of the light acting on the phototube in accord with the changes in weights. It is also feasible to mount a mirror on the beam so as to reflect the light onto a double cathode tube, the light passing alternately from one photocathode to the other.

As to the recording, there is a choice, just as in the case of the balances in which the beam is deflected. The choice may be made from among:

(a) Source of light–mirror–photographic paper.

(b) Pen connected to the slider of a potentiometer, or to the drum of a chain balance, or to the end of the beam, or to the feeler.

(c) Electronic setup–current passing into the coil of an electromagnet, current produced in a differential transformer, in a photoelectric cell, in a variable conductance transducer, charging a condenser, etc.

2. The Urbain balance

This balance, which was mentioned in Chapter 1, was designed for studies of the efflorescence of salts[6]. Urbain stated: "I have found a satisfactory solution for the problem at hand: (1) by replacing the stamped weights with the attraction exerted by a solenoid on a magnetized needle suspended vertically to one of the ends of the balance beam, (2) by placing the pan hung from the other end of the beam in the centre of an electric resistance furnace whose temperature may be adjusted at will and measured at any time by a sensitive couple, (3) by making the enclosure tight and sturdy enough to support a vacuum, (4) by introducing into the balance inclosure reagents capable of absorbing the gases as soon as they are produced. The balance being constantly brought back to zero by adjusting the intensity of the current which is passing through the solenoid, the magnetized needle always remains in the same position with respect to the solenoid and the attraction is proportional to the intensity of the current. The use of a shunt suitable for the terminals of the galvanometer placed in the circuit of the solenoid makes it possible at each instant to read the weights of the substance directly from the scale. The balance, whose triangular glass beam is 8 cm long, can support 100 mg. The three knife edges consist of three tightly stretched platinum wires 0.01 mm in diameter. The sensitivity is 100 mg for the load given above." This was written in 1912.

3. The Testut thermobalance

The Testut thermobalance, devised by Carton,* and shown in Fig. 39, employs a chain balance. The earliest application of the chain balance was made by Müller and Garman[7]. The beam is kept in constant os-cillation on both sides of its position of static equilibrium; the static equilibrium position of the beam in the definition which involves frictional forces is replaced by the position of dynamic equilibrium in which the frictional forces are continuously compensated by the input of an auxiliary servo-motor.

The dynamic equilibrium is secured by means of a platinum contact placed on the upper beam (Fig. 40), which can make contact with a wire of the same metal attached to the framework of the balance. This contact governs a relay which in turn directs the movement of the servo-motor in one direction or the other. The chain winds and unwinds on a spool firmly attached to another spool which supports the recording stylus. The upper beam transmits its oscillatory motion to the lower beam; consequently the two beams are in a state of continual oscillation and they mutually transmit their respective indications. The lower beam is situated within the tight case of the balance; it is provided at one end with a counterpoise and a permanent magnet whose poles alternate with respect to those of the magnet of the upper beam. The other end supports

---

* Placed on the market by Etablissements Testut, 8 Rue Popincourt, Paris 11.

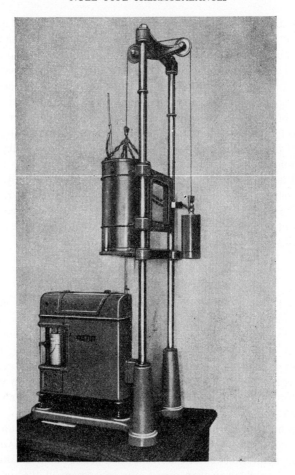

Fig. 39. Photograph of Testut thermobalance.

the crucible during the experiment; it is suspended by a stainless steel wire. In certain models (such as shown in the photograph) the furnace is placed above the balance; in others it is below.

The recording is made on a cylinder (150 mm high and 140 mm in circumference) fitted with millimeter chart paper. The apparatus has a range of 50 g; the sensitivity is such that 2 mg are equivalent to 1 mm, and the total variation along the ordinate corresponds to 300 mg.

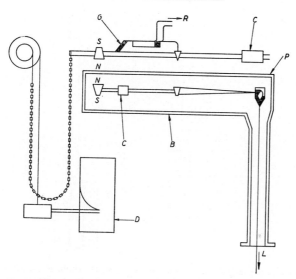

Fig. 40. Testut null type thermobalance. G: gold knife edge; C: counterweight; B: brass casing; P: Plexiglas lid; R: reverser; L: laboratory tube; D: recording drum.

4. The Waters thermobalance

This apparatus[5] is a null type differential thermobalance with coulometric restoring force and photoelectric null detector. It is a quite unique recording differential thermobalance for the investigation of coals (Fig. 41).

An ordinary analytical balance is employed designed in such fashion that the sample may be balanced by means of a silver electrode suspended from one of the ends of the beam and immersed in a silver coulometer.

Changes in the sample weight are compensated by corresponding changes in weight of the silver electrode due to a control current passing through the cell. The necessary voltage and current control are effected by means of a photocell null detector and magnetic amplifier with an output rectifier. The voltage drop across a precision resistor in the output circuit makes it possible to record a portion of the controlling voltage or current as a measure of the differential weight change. Integration of this function by electromechanical means also makes it possible to secure an accurate record of the cumulative weight change at the same time on a multipoint recorder.

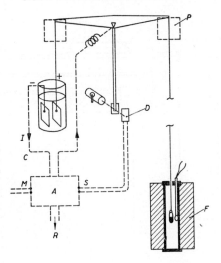

Fig. 41. Waters' thermobalance. P: air-damping pot; D: photocell detector; F: furnace; C: control-current; A: amplifier; M: main supply; S: signal input; R: to recorder.

5. The Groot and Troutner thermobalance

These workers[8] have made use of the electromagnetic principle in their null type recording thermobalance, which is equipped with a light source-shutter-photocell null detector. The current required to maintain the null position was recorded as a function of time equivalent to the linear heating rate of the furnace.

The furnace is located beneath the balance. The sensitivity is 0.3 mg and the accuracy is within ± 0.3, ± 0.4, and ± 0.8 in the 50-, 100-, and 200- mg ranges, respectively. 10 minutes is required to record a full range weight change. This thermobalance was designed to follow weight losses only. With the equipment used, the temperature rose from 20° to 600° in 14 hours and 20 minutes; the recorder chart speed was 2 inches per hour. Accordingly, 1 inch on the chart corresponded to 19.8°. The electric furnace (not described) would reach 1200° as the upper limit.

Blažek[9] developed another thermobalance of this type; it can also be used to obtain differential thermal analysis curves and for gas analysis. It employs a solenoid-iron core electromagnetic weight compensator with a photoelectric null detector with electromagnetic feedback.

Fig. 42. Mauer recording null-thermobalance.

6. The Mauer thermobalance

Brown et al.[10] combined an analytical balance with a recording potentio-meter. The current produced by the double cathode cell when the balance is deflected varies the current supplied to the damping coil in such manner as to bring the beam back to the null position through the action of a

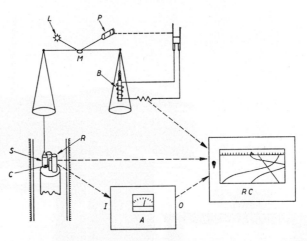

Fig. 43. System for simultaneous recording of weight change and thermal analysis curve. S: sample; R: reference; I: input; C: control thermocouple; RC: recorder; A: DC amplifier; O: output; M: mirror; B: solenoid; L: light source; P: phototube.

coil attached to the magnet suspended from the beam. Mauer[11] returned to this design (Fig. 42).*

The null detection system consists of a light source, a beam-mounted mirror, and a dual phototube which controls the current through the restoring coil. This current is a linear function of the change in weight, and is recorded by measuring the voltage drop across a precision resistor in series with the solenoid. The standard balance can be used for recording a basic full-scale change in weight of 100 mg. An incremental weight loading device has been added to increase the range of weight change to several grams.

The system for the simultaneous recording of weight change curves and thermal analysis curves is shown in Fig. 43.

The furnace is placed on a table. It has an alundum core and is insulated with magnesia. The winding consists of 62 feet of 20 gauge wire (80% platinum, 20% rhodium). The attainable temperature is about 1300°, and the heating rate is 12.5° per minute when a 135-V source is employed.

---

* Available from the Niagara Electron Laboratories, Andover, N.J. (U.S.A.).

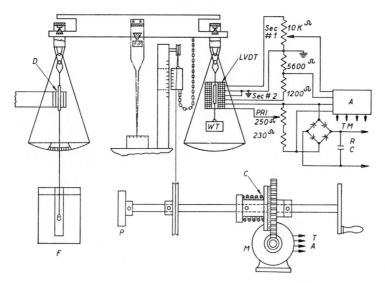

Fig. 44. Garn's thermobalance. F: furnace; D: damper; P: potentiometer; M: motor;
C: clutch; WT: weight tare; A: amplifier; RC: recording circuit; TA: to amplifier; TM:
to motor; LVTD Model 060 SL, Schaevitz Engineering, Camden, N.J.

7. The Garn thermobalance

This automatic recording balance[12] was designed and constructed for use
in thermogravimetric and related studies. It is an electronically controlled
null-point instrument employing a linear variable differential transformer
as the sensing agent (Fig. 44).

The potentiometer turns from 340° to 350° when the position of the
chain goes from 0 to 100 mg. Fig. 45 shows the recording circuit of
the balance.

8. The Eyraud thermobalance

Fig. 47 gives a general view of this remarkable continuous recording
instrument.* See particularly[13-16].

The mobile part (Fig. 46) consists of a beam of an ordinary analytical
balance (1), load limit 200 g, sensitivity better than 0.1 mg, made of

* It may be purchased from the Société pour la diffusion d'appareils de mesure et de
contrôle (D.A.M.), 6 Avenue Sidoine Apollinaire, Lyons 5, (Rhône) (France).

References p. 77

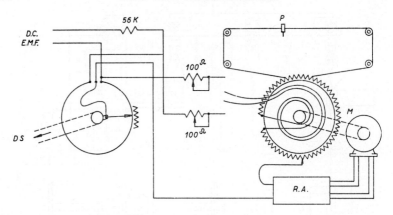

Fig. 45. Recording circuit for balance. P: pen; M: motor; RA: recorder amplifier; DS: to drive shaft.

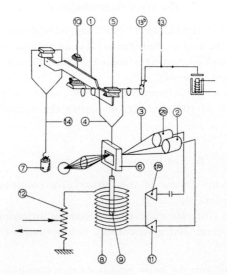

Fig. 46. Diagram of Ugine-Eyraud thermobalance.

polished aluminium alloy and fitted with two agate knife-edges. It is housed, along with its suspensions (4) and (14) inside a vacuum chamber whose polished inside walls facilitate rapid removal of gases. The position of the mobile parts is registered by means of a photoelectric cell (2). The exciting light ray (3) is directed in a plane perpendicular to that of the balance beam to reduce the effect of parasitic oscillations of the suspension (4) around the axis of the corresponding knife-edge (5). For the same reason, the mobile interception screen (6) is placed as close as possible to the knife-edge. The changes in weight of the sample (7) are automatically compensated by the magnetic force developed by a coil (8) on a small permanent magnet (9). The strength of the electric current gives a measure of the weights even though the inclination of the beam is not a variable independent of the equilibrium forces[16, 17].

The necessary and sufficient conditions for achieving this effect are: (a) one of the extremities of the magnet must be in a field which is practically uniform, the other end being in a region of the field that is practically null, and (b) the mobile parts must be adjusted once for all to the indifferent mechanical equilibrium by setting a counterpoise (10).

Fig. 47 gives an over-all view of the Eyraud thermobalance.

The operating mechanism (11) and (11b) has two functions: (a) to produce the equilibrium current which flows through the coil (8) by amplification of the illumination current from the photocell (2), and (b) to deliver a damping current in the coil (8) proportional to the speed of the mobile parts, by amplification and correction of the phase of the variation of the illumination current of the photocell (2 bis). The potential difference at the terminals of the potentiometer (12) may be applied to a millivoltmeter, or to a manual potentiometer, or preferably to a recording potentiometer. A transistor manipulator (13) distributes at will four tared riders (13 bis) (10, 20, 40, 80 mg) which make sixteen combinations possible.

The electric furnace (not described) goes up to 1000°. It is provided with a water jacket and is fastened at three points which makes for easy centering.

9. Various other thermobalances

Hyatt, Cutler and Wadsworth[18] have developed a thermobalance for ceramic investigations. It employs a chain balance connected through a servo-mechanism with the system: light source-mirror-double cathode

Fig. 47. Photograph of Ugine-Eyraud B-60 thermobalance .

cell as zero detector, and a precision potentiometer which is in relation with the motor of the chain.

Kinjyo and Iwata[19] have developed a relatively simple thermobalance using a vacuum thermocouple in series with an electromagnetic restoring force coil and a battery, as a transducer for recording weight changes as a function of the manually regulated restoring force currents.

The *Linseis thermobalance*[20] employs a photoelectric null detector to operate a servo-motor-driven chain along with the pen of a strip-chart recorder with full-scale ranges of 0.25 to 4 grams. The sample is suspended from the balance beam into a vertically mounted furnace. The latter reaches 1550° and is fitted with a temperature programming controller. The apparatus can operate as a sedimentation balance. It is available from German dealers.

The *Netzsch recording null thermobalance* is also on the market.* It consists of a short beam symmetrical knife-edge balance system with a photoelectric null detector which drives a servomotor-operated chain to maintain the balance in equilibrium and to produce a signal proportional to the change in weight. A full-scale range of 100 mg is recorded on a 5-inch chart, simultaneously with the temperature, up to 1500°, for loads up to 100 g. Linear heating rates of 1° to 10° per minute can be obtained.

REFERENCES

[1] K. Angström, *Ofers. Kongl. Vitenskaps. Akad. Förh.*, (1895) 643.
[2] A. S. Mikulinskii and P. V. Gel'd, *Zavodskaya Lab.*, 9 (1940) 921.
[3] K. Honda, *Sci. Repts. Tohôku Imp. Univ.*, 4 (1915) 97.
[4] M. Guichard, *Bull. Soc. chim. France*, 37 (1925) 258.
[5] P. L. Waters, *Nature*, 178 (1956) 324; *J. Sci. Instr.*, 35 (1958) 41; *Coke and Gas*, 20 (1958) 252.
[6] G. Urbain, *Compt. rend.*, 154 (1912) 347.
[7] R. H. Müller and R. L. Garman, *Ind. Eng. Chem., Anal. Ed.*, 10 (1938) 436.
[8] C. Groot and V. H. Troutner, *Anal. Chem.*, 29 (1957) 835.
[9] A. Blažek, *Silikáty*, 1 (1957) 158.
[10] F. E. Brown, T. C. Loomis, R. C. Peabody and J. D. Woods. *Proc. Iowa Acad. Sci.*, 59 (1953) 159.
[11] F. A. Mauer, *Rev. Sci. Instr.*, 25 (1954) 598; *Instrumentation*, 7 (1955) 36.
[12] P. D. Garn, *Anal. Chem.*, 29 (1957) 839.
[13] I. Eyraud, *J. chim. phys.*, 47 (1950) 104.
[14] C. Eyraud, *Compt. rend.*, 238 (1954) 1511.
[15] C. Eyraud, *Technica*, 177 (1954) 2.
[16] C. Eyraud and I. Eyraud. *Laboratoires*, 12 (1955) 13.
[17] C. Eyraud and R. Goton, *J. phys. radium*, 14 (1953) 638.
[18] E. P. Hyatt, I. B. Cutler and M. E. Wadsworth, *Am. Ceram. Soc. Bull.*, 35 (1956) 180.
[19] K. Kinjyo and S. Iwata, *J. Chem. Soc. Japan.*, 74 (1953) 642.
[20] M. Linseis, *Keram. Z.*, 11 (1959) 54.

* Gebrüder Netzsch, Sartorius-Werke A.G., Göttingen (Germany).

CHAPTER 4

Applications of the Thermobalances

1. Introduction

As soon as it became feasible to make a continuous recording of the gains or/and losses in weight of materials being heated or cooled, there was much more hope of successfully resolving many problems in such fields as physical and analytical chemistry, metallurgy, geology, and biology. When curves have to be constructed point by point, there is always danger that a slight detail may be missed; furthermore the operator must be present and exercise constant supervision. For example, one of the earliest trials at Imphy, in which a prototype of the Chevenard balance was used in a study of the 18/8 rustproof alloys, extended over 40 days and 40 nights. However, up to the present, thermogravimetry has been used predominantly in the study of the precipitates employed in gravimetric analysis. The students trained by Honda initiated this field of study; it has been generalized by Duval.

If the thermobalance is used in association with chemical methods, differential thermal analysis, magnetism, absorption spectrography (especially infra-red), colorimetry, etc., its field of application becomes almost limitless; it provides a means of penetrating more deeply into the structure of compounds and mixtures. Solid state reactions may be readily explored in this way, likewise thermochemical reactions, or activation energies, or sedimentation in aerosols and powdered materials, or specific surfaces can be measured through the weights of absorbed gases; furthermore, such things as enthalpies, absorption phenomena, adsorption and desorption, sublimation rates, etc. are more accessible. Since the beginning of the century, numerous laboratory operations have been conducted with automatic registration, but only at room temperatures, but now this technique can be extended to around 100°.

The great flexibility of the modern thermobalances makes it possible to operate at any heating rate, or by cooling, isothermal heating, in controlled atmospheres, or by dehydration, rehydration, removal or combination with heavy water, or operations from the temperature of liquid air up to 1500°, or at pressures ranging from near that of a cathodic vacuum up to 20 atmospheres.

We wish to stress emphatically that thermogravimetry by itself cannot resolve all of the problems of structure; in fact, on one hand it is possible to obtain a horizontal indicating constant weight from impure materials, and on the other hand, a single compound may isomerize along a horizontal, at constant temperature, with the result that the structure is not the same at the two ends of the horizontal. Perhaps it was facts of this kind which have led the inventors during these last years to employ the furnace in the same manner to construct the pyrolysis curve and the differential thermal analysis curve.

There is no need to stress the importance which electronics has acquired in the operation of the modern assemblies, especially in the case of the null type thermobalances.

It seems fitting to close this introduction by a quotation from Guichard[1] whose studies were noted in Chapter 1. "This method (thermogravimetry) should be checked occasionally by very prolonged experiments at constant temperature, but even in this case it shortens the trials because it indicates the temperature regions where it seems appropriate to make such experiments."

2. Study of precipitates used in analysis

In the first edition of this book, all the curves recorded for these precipitates were reproduced (reduced to $1/4$ size). This practice will not be followed in the present text. Actually, these curves may be collected into several well characterized groups which will be defined below. The essential thing for the analyst to know is not so much the form of the curve, which varies considerably with the quantity of water retained by the precipitate, and with the rate at which the air passes through the furnace, but rather he should know whether this precipitate does or does not yield a line which is perfectly horizontal or slightly inclined, and if so the approximate temperatures between which this line is obtained. He must also realize that these temperatures are influenced by the weight of the sample and by the rate at which its temperature is raised.

Case 1

A horizontal is obtained beginning at ordinary temperatures and con-

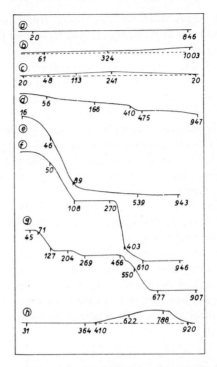

Fig. 48. Pyrolysis curves for: (a) silver thallium arsenate, (b) tellurium or cadmium, (c) nitron nitrate, (d) iron potassium lithium periodate, (e) aluminium hydroxide, (f) cobalt anthranilate, (g) copper selenate, (h) precipitated palladium.

tinuing as far as 1000°. Instances are lead sulphate $PbSO_4$, silver thallium arsenate $TlAg_2AsO_4$, etc. (See Fig. 48a).

Case 2

A straight line is obtained here also, but it is inclined (at least in air or oxygen). A line of this kind is obtained invariably when precipitated tellurium is heated, or metals such as copper, cadmium, etc. which have been deposited electrolytically (See Fig. 48b). The analyst thus knows that such materials should be weighed in the cold after they have been dried in a desiccator. The oxidation is irreversible in the instances just cited, but in other cases the curve or rather the ensemble of the two straight lines comes back to the starting level if the product is allowed to cool from a given initial temperature and by using the same rate during the descent as during the ascent. Pertinent examples include precipitated gold, nitron nitrate (See Fig. 48c).

Case 3

The curve descends continuously without any stops regardless of the heating rate or the quantity of material involved. Therefore it is impossible to tell at what moment a constant weight is obtained and the precipitate does not merit consideration for gravimetric analysis. A case in point is the triple iron potassium lithium periodate (Fig. 48d). Consequently, the author's attempts[2] to use this material, which supposedly has the formula $KLiFeIO_6$, have been without success. This is unfortunate since the factor for lithium is very favourable and the salt can be precipitated in the presence of the other alkali metal ions, provided they are not present in too great excess. Actually, the salt (which is used for titrimetry and in spot test analysis) is not suitable for gravimetric purposes because its composition is not constant and furthermore its pyrolysis curve shows a continuous descent. The excess of potassium which it would need to contain in order to exist gives a variable weight in the residue above 947°.

Case 4

The curve descends until a horizontal is obtained. This is the case most frequently encountered with the hydrates of aluminium oxide, and the hydrous oxides of beryllium, gallium, thorium, etc. The horizontal obviously corresponds to the anhydrous oxide: Al_2O_3, BeO, Ga_2O_3, ThO_2, etc. Accordingly, the curve (Fig. 48e) represents the pyrolysis of a hydrated aluminium oxide precipitated by means of potassium cyanate in accord with the study by Ripan[3]. The trace shows two breaks; the first, around 100° obviously reflecting the end of the loss of the moisture, and the other in the neighbourhood of 510°, indicating the end of the evolution of the combined water. The horizontal of the oxide begins above this temperature. It is very important to emphasize that when dealing with curves of this kind, namely those which reach their horizontals very slowly, there is danger that the observed start of the horizontal may be several hundred degrees too low if the operator uses a balance of low sensitivity and if the inscribed line is too thick. This is why we urge that samples be taken along the descent and that the disappearance of the bands due to water be checked by infra-red absorption.

It may also happen that at the start the curve may begin horizontally and that the loss in weight does not appear until the temperature is between 60° and 100°. It may then be said that the material is not moist nor does it have any water of absorption but only so-called water of crystallization. If, after the temperature has gone above 100°, the material continues to lose water, there is great likelihood that the water makes up part of the constitution of the material, as in chromium alum according to Harmelin[4]. It may also happen that, despite all appearances, the material really contains no water as such but the latter is produced at the expense of OH groups concealed in the compound as is true of the "monohydrates" of inorganic sulphates which Lendormy[5] suggests should be written MSO_5H_2 rather than in the conventional manner: $MSO_4 \cdot H_2O$. These two examples clearly indicate that thermogravimetry by itself cannot resolve these important findings.

Case 5

The curve includes two parallel horizontals. This case is observed with numerous salts or organic inner complexes. The liquids with which the precipitates were washed (alcohol, ether, etc.) come off first, then the horizontal of the pure precipitate appears; a downward course of the line corresponds to the decomposition of the sample, an evolution of gaseous products, and then the second horizontal corresponds to the weight of the residue: oxide, carbonate, ash, unless the crucible is completely empty after sublimation and/or combustion. The curve of cobalt anthranilate (Fig. 48f), precipitated in accord with the procedure given by Wenger, Cimerman and Corbaz[6], is a pertinent instance. They recommend that the precipitate be dried at 120–130°. The curve given by this anthranilate indicates the loss of moisture up to 108°; then a horizontal begins and extends to 290°. (The loss suffered by 201.5 mg between the extremities of this horizontal was less than 0.1 mg.) Accordingly, the temperatures recommended by these workers are correct. The destruction of the organic matter is indicated by the change in direction of the curve at 403°. Above 609°, we find the equally good horizontal which pertains to the oxide Co_3O_4. A very simple calculation, which is based on weights of this latter oxide, shows that the weights of the anthranilate, as measured on the paper in mm, agree within at least 0.1%. This is one of the better gravimetric methods.

Case 6

The curve has several horizontals parallel to each other. This is a rather rare occurrence since one of the materials corresponding to one of the horizontals often begins to decompose before it is completely formed. This happens notably with the various hydrates of sulphates of the magnesian series, unless the heating is conducted very slowly as was done by Fruchart and Michel[7] in the case of the heptahydrate $NiSO_4 \cdot 7H_2O$ who raised the temperature 0.6° per minute. In this way, it is possible to distinguish the horizontals of the hexahydrate, tetrahydrate, dihydrate, monohydrate, and finally the oxide NiO.

The curve shown in Fig. 48g relates to the pentahydrate $CuSeO_4 \cdot 5H_2O$, of which 250 mg was pyrolyzed at the rate of 300° per hour, in an "aluminite" (porcelain) crucible, and passed through a 100-mesh sieve. It may be seen that this salt is stable up to 60°, that it progressively loses 4 molecules of water from 60° to 140°, and that the monohydrated salt gives a horizontal from 140° to 204°. At this point the last molecule of water begins to come off, namely the one which actually was bound to the selenium in the form $CuSeO_5H_2$. The horizontal of the anhydrous compound extends thereafter up to around 450–460°. Finally, the loss by sublimation of the selenious anhydride and of oxygen leads to CuO, which likewise leads to a horizontal starting at 640°. On the down slope, below 500°, there is a break in the curve and this corresponds to a basic selenate just as in the case of copper sulphate at a higher temperature, where a basic sulphate is produced as described in Part 2 (Chapter Copper).

Case 7

When the heating is conducted in the open air, it is not too exceptional to observe ascents in the curves following descents. This happens with the sulphites which are converted into sulphates, or with selenites which go over into selenates, or with metals which oxidize, or basic oxides which take up carbon dioxide. It is similarly possible, if the furnace temperature is high enough, say up to 1200°, to find a decomposition of the compound or compounds formed during the ascending period of the curve. However, we are not dealing here with portions of a straight line as in the case of Fig. 48c shown above. Fig. 48h refers to a grey-black deposit of palladium formed after Brunck[8] by passing a stream of carbon monoxide into a solution of palladium(II) chloride. The metal is stable in the air up to

384°. The oxidation becomes distinctly evident around 410° and the resulting PdO, which is formed quantitatively, yields a flattened maximum situated between 788° and 830°. Then this oxide breaks down and the initial weight of metal is attained above 920°. The metal, when recycled, gives the same curve when reheated under the same conditions.

3. Automatic gravimetric analysis

General considerations

In analogy with the present-day tendency in organic chemistry, it is very desirable to be able to determine any required ion within a matter of 15 minutes or so with the precision usually attainable in gravimetric procedures and independently of the skill of the operator. The ideal solution would, of course, be the possibility of being able to determine simultaneously two or three different ions without having to carry out a preliminary separation.

However, the problem is rather different from that in organic chemistry where thousands of compounds, whose purity is assured by physical means, can be determined by the *same* method. In the inorganic field, two different ions generally require different methods of determination. Mostly, it is mixtures which are dealt with; and the separations–often imperfect–take up more time than the determination itself. Hence the idea of determining mixed compounds by utilizing the different horizontals afforded by the pyrolysis curve.

It was quickly realized that the problem could only be resolved after a tremendous amount of preliminary work:

(1) Construction of an apparatus able to record the useful parts of the curves mechanically or electronically in full view of the operator. This part of the problem has largely been solved.

(2) Registration of all the precipitates recorded in the literature (up to 15th July 1961 there were 1200) in order to establish whether:
   (a) the corresponding compounds acquire constant weight, and if so,
   (b) from what temperature or within what approximate range.

(3) Selection of the known gravimetric methods which satisfy all the following requirements:
   (a) quantitative and immediate precipitation,
   (b) immediate filtration with no need for the precipitate to age,
   (c) immediate drying,
   (d) production of a constant weight horizontal at the lowest possible temperature.

In addition, the price of the reagent and the choice of analytical factor, which should of course be the lowest possible, can be taken into account.

There is also the question whether for routine determinations it is more advantageous to buy a thermobalance (approximately the price of a good microbalance), to allow for its depreciation, and to pay the wages of a (non-specialized) laboratory technician who is capable of changing the substance to be determined every 15 or 20 minutes and of presenting the results in a permanent form at once or 5 years later (it is impossible to make mistakes in handling the weight); or whether it is better to pay the salary of a graduate chemist who with inexpensive equipment carries out evaporations, separations, precipitations, filtrations through paper, and ignitions of paper, who heats crucibles at ill-defined temperatures (function of the quality of the gas supplied on the particular day) for indeterminate periods of time which are generally too long, who weighs, reheats, reweighs, and risks making mistakes in the weight, and who, having done all this, has no evidence of any kind as to whether in the event of the determination being incorrect the mistake has been voluntary or involuntary. In agreement with Professor Charlot, we believe that the role of the modern graduate chemist is to create and not to perform the same analytical operations more or less with distaste throughout his whole career.

The conditions mentioned previously call for some discussion: inorganic hydroxides and sulphides should be eliminated completely, not only because of the time they take to age, not only because of their colloidal nature and the impurities they adsorb, but also because they must be heated to far too high a temperature in order to obtain constant weight as the oxide–a form with an unfavourable analytical factor.

A wide use of modern organic reagents is essential: oxine, neocupferron, nioxime, anthranilic acid, etc. furnish precipitates of complexes which dry sometimes within a few minutes and which can be weighed without having to be converted to the metal oxide. For mixed determinations a reagent must obviously be chosen which leads to horizontals far enough removed from each other. With some exceptions, it has been possible to find at least one method for the automatic determination of each ion. See "Suggested Methods for Automatic Inorganic Analysis" (p. 147).

Not all the methods acceptable in ordinary gravimetric analysis are necessarily suited to the new technique. Thus, if a precipitate takes up oxygen on heating and loses it again on cooling (e.g. nitron nitrate), the resultant oblique levels cannot, as a rule, be utilized.

References p. 124

The means of filtration

The maximum weight which the silica rod can take is 10 g. However, especially in automatic determinations where there is a fairly rapid initial loss of wash liquid, it is preferable not to exceed 5 g, so as not to disturb the adjustment of the apparatus. This weight seriously limits the choice of apparatus for the filtration. The question therefore arose whether in gravimetric analysis it is necessary to go on using so-called "ashless" filter papers and whether the laborious procedure of their combustion, considered by some as "the supreme art of the analytical chemist", must

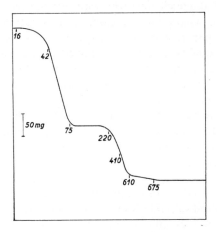

Fig. 49. How a filter paper burns.

also be preserved. A further point of interest was to learn the temperature at which combustion of a filter paper is complete under the usual conditions of heating in a crucible. Experiments with the thermobalance give identical results with papers of different degrees of fineness, *i.e.* of different textures and rates of filtration, whether dry or saturated with water, whether packed or not with kaolin (a substance which should not change weight at the temperatures assigned). A remarkable fact is that the same curve is recorded for absorbent cotton, glucose, flour, and potato starch.

Fig. 49 relates to a 12-cm filter paper (Durieux red label) saturated with water[9]. The drop as far as 75° indicates escape of water and is followed by a level, more or less horizontal, extending to 180°, which pertains to the weight of the dry paper, now slightly yellowed at the edges. Decomposition then sets in and takes place at two different rates. The water of constitution corresponding to the simplified formula for paper $C(H_2O)$ is rapidly eliminated up to 410°; at this temperature the form of the paper is intact, but it consists almost entirely of carbon (amorphous, according to the Debye-Scherrer spectrum). Between 410° and 675° the carbon burns quantitatively–rapidly up to 610° and then more slowly. At 675° the residue is perfectly white and weighs 0.3 mg.

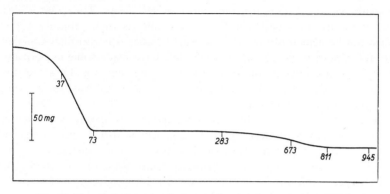

Fig. 50. The pyrolysis curve of asbestos showing the anomaly at 283°C.

Leaving aside the reducing action of carbon and carbon monoxide on many precipitates, this important result makes filtration through paper quite out of the question for the automatic method of determination. Complete combustion of the filter paper is too slow and takes place at much too high a temperature to enable the operator to carry out a complete determination in less than half an hour.

We have therefore thought of using Gooch crucibles with an asbestos mat. On heating a precipitate of ammonium phosphomolybdate in one, it is observed that the curve always shows a bend at 283°. The same molybdophosphate heated in a porcelain or glass crucible affords a horizontal which is continuous between 155° and 410°. Therefore, the anomaly can only be due to the use of the asbestos, and whether dry or moist, new or used, or washed with hydrochloric acid, it still has a pyrolysis curve analogous to that in Fig. 50. Up to 60°, the adsorbed water is driven off; the weight then remains constant as far as 283°; and above this temperature there is a gradual loss of water which is only complete at 879°. The following figures are taken from the original recording and illustrate this evolution of water for aninitial dry weight of asbestos of 170.1 mg:

Temp. (°C)	341	406	471	539	607	673	743	811	879
Loss (mg)	1.50	1.95	2.00	2.60	3.20	4.25	7.50	13.95	15.50

If the asbestos so heated is then moistened an da new Gooch crucible is made with it, it gives *the same loss-in-weight curve*. The cause is clear and if the use of asbestos in Gooch crucibles is to be continued they must not be heated above 283° (Duval[11]). Silica (quartz) filter crucibles do not change in weight up to 1035°. We have made some in porosities 2 and 3 with a capacity of 10 ml and weighing only 3 g. They are only used in special cases, because they are difficult to wash and, being extremely thin, very fragile.

As far as porcelain crucibles are concerned, we have only experimented with A2 crucibles of Berlin manufacture which also do not change in weight below 1035°. However, the first time they are heated there is a steady increase in weight from ordinary temperatures up to 1035°; the gain (oxygen) amounts to *ca.* 4 mg for a crucible weighing 4 g. The manufacturers have been unable to find a satisfactory explanation for this phenomenon, but its occurrence confirms the elementary precaution of always igniting a new crucible before use. New crucibles made of ordinary porcelain do not exhibit this behaviour.

Gooch-Neubauer platinum crucibles have a mat of porous platinum obtained by pyrolysis of ammonium chloroplatinate. This mat remains constant in weight from 407° to 538°; then the platinum oxidizes, superficially at least, and there is a maximum in the weight at 607° (see Chapter Ammonium); on further heating the oxide dissociates so that at 811° the original weight is regained exactly. The maximum gain may amount to as much as 5 mg per 123 mg platinum. Such crucibles must be used above 811°, but it may be wondered whether they are really necessary. Their role is limited to work with fluorides which, as will be seen in Part Two, we propose to determine as lead chlorofluoride or triphenyltin fluoride, without decomposition of the complex formed. When the filtration stage is reached, there are no longer, or there should no longer be, any free hydrofluoride ions; hence glass or silica can be used without fear of attack.

If a G2, G3, or G4 sintered-glass crucible is heated, it remains constant in weight up to 510–512°, after the moisture has been driven off. Above this temperature, there is always an increase in weight and the filter plate becomes unusable.

Glass crucibles can thus be used at temperatures up to 500°, but from the present point of view they have two serious faults: (a) they are difficult to wash, especially after filtering compounds like the internal complexes of cobalt, or hexacyanoferrates(II), etc.; the washing often takes more time and trouble than the determination itself; (b) 10 ml crucibles are too heavy for the carrier ring of the thermobalance, since for the capacities required, a weight between 8 and 12 g must be counted on. The rim at the bottom only serves for the manufacture and represents so much dead weight in the sequel. In addition, these filter crucibles are expensive and after being used about twenty times their porosity alters.

Fig. 51. Glass crucible with glass silk.

We therefore decided to reject them and to go back to Gooch crucibles with a glass mat. On going over the problem, it was quickly realized that filtration at the pump through commercial glass wool is a capricious procedure; more often than not, channels and cavities form which result in too rapid a filtration. The firm Saint-Gobain therefore put samples of carded glass silk at our disposal.

We have constructed crucibles of thin Pyrex glass, weighing 3 g, which fit the carrier ring of the thermobalance. About 20 holes are pierced in the bottom with a tungsten point. The crucibles have the following dimensions: lower diameter 20 mm, upper diameter 25 mm, height 30 mm. Circles of fat-free carded glass silk, the thickness depending on the fineness of the precipitate, are cut out with a 25 mm cork-borer and carefully placed in the bottom of the crucible. They are moistened slightly, sucked dry at the pump, and further dried at a convenient temperature (Fig. 51). Strictly, the mat of glass silk should only be used once, especially in view of its insignificant price, but it has been found that 4 or 5 determinations of the same substance can be carried out with one mat.

In this way we have solved the problems of lightness, cleaning (which is completely eliminated), price, and filtration (of a precipitate of any degree of fineness).

Apart from hydrofluoric acid which, of course, immediately destroys the glass silk, daily experiments on the variation in weight with different wash liquids have so far shown that the only reagent causing alteration of the weight in the cold is normal sodium hydroxide. When passed repeatedly through the mat there is an increase in weight, but is must be admitted that this is a quite exceptional wash liquid.

As far as the behaviour towards heat is concerned, glass silk produces the same pyrolysis curve as does fritted glass, *i.e.* a constant weight level up to 510°; after that the mat becomes heavier (no doubt due to fixation of oxygen), then shrinks, sinters together, and becomes unusable.

Summarizing, for automatic determinations the crucibles just discussed are applicable in 99 cases out of 100 and they are mostly used below 200°, sometimes even below 100°. Fig. 52 shows the appearance of some fibers of glass silk as seen through the electron microscope.

Fig. 52. Fibers of glass silk as seen through the electron microscope.

Principle of the automatic method of determination

The example chosen for discussion is the precipitation of copper by benzoinoxime according to the well-known method of Feigl[12]. As indicated in Fig. 53, after losing wash liquid the complex $CuC_{14}H_{11}O_2N$ is stable from 60° to 143°. The ordinate of the corresponding horizontal is

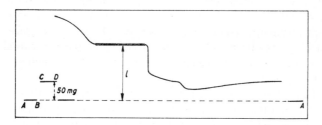

Fig. 53. Pyrolysis curve of Feigl's copper benzoinoxime complex.

measured and compared with that given by the calibration (Duval[13]).

Let us suppose that we have a well-precipitated deposit of copper-benzoinoxime complex which corresponds to the copper to be determined. A filter crucible (+ mat of glass silk) is first brought to 120° and then placed alone in the carrier ring of the silica rod.

(a) The spot is arranged to strike the photographic paper (or the pen the squared paper) as low down as possible and to draw the horizontal line AB at the right-hand side of the paper in order to define properly the X-axis. This gives the weight of the empty crucible, but not the absolute weight (for which there is no need). It takes 3 minutes for the photographic recording and only 1 second for the pen recording, if the cylinder is turned by hand in one or the other direction.

(b) A 50 mg calibration weight is placed on the platform fixed to the silica rod (e in Fig. 9) for 3 minutes (or one second for the direct recording); the spot or the pen then describes the horizontal CD. Actually, this calibration does not have to be repeated for every determination; thus the distance between AB and CD, 25.5 mm, has, in the case of one of our thermobalances, remained constant during four years' operation.

(c) The calibration weight is removed and the horizontal part EF is now recorded; this check is purely optional. (fig. 54)

(d) Next, the beam of the balance is arrested, the crucible is removed, and the precipitate is rapidly filtered into it at the pump by the usual technique. The amount of wet precipitate added should be such that ultimately the crucible does not contain more than 450–480 mg if a 24 × 30 cm paper is being used.

(e) The furnace is switched on, the needle of the thermostat is set to 100°–a point which corresponds approximately to the middle of the horizontal given by the Feigl complex–and the lamp is switched on again at an opportune moment so that the spot can trace the line GH (Fig. 54) or merely a small portion of the horizontal. When recording with a pen, a glance is enough to show whether the horizontal is reached.

(f) It then suffices to measure the distance between the horizontal AB–EF and the level GH with the help of a scale graduated in half-millimeters on glass and to compare it with the distance AB–CD which, it will be remembered, represents 50 mg. The millimeter paper can also be graduated directly in mg.

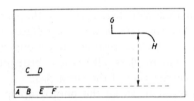

Fig. 54. Principle of the automatic method of determination.

Example of a determination: calculated 176.34 mg; found 176.00 mg.

Automatic gravimetric analysis in its simplest form thus amounts to the measurement of two distances and the results are recorded on a document which can be consulted at leisure and which is obtained independently of the skill of the operator.

The filtration is, of course, carried out while the horizontal of the previous determination is being recorded. In the case of cadmium and uranium, it has been possible to carry out a precipitation, filtration, washing, drying, and recording in 12 minutes, the furnace remaining throughout at the same temperature. It will now be understood how one and the same operator can perform up to 35 determinations of the same substance in the course of a day's work and why, in order to be able to operate rapidly, it is necessary to choose a suitable precipitating reagent yielding an immediate precipitate which is very little wetted (the uranium-oxine complex is outstanding in this respect) and which can be brought to constant weight below 200°.

Binary mixtures

A practical case is the determination of a mixture of calcium and magnesium. It is well known that the calcium oxalate precipitate carries down an unknown amount of magnesium oxalate when the magnesium is present in great excess.

In Fig. 55 the pyrolysis curves of the two oxalates are traced one above the other. Curve (a) is already known. The other curve (b) is much simpler

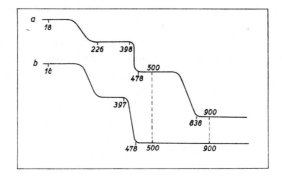

Fig. 55. Pyrolysis curves of oxalates: (a) calcium oxalate, (b) magnesium oxalate.

because it has no horizontal relative to magnesium carbonate; in other words, when the anhydrous oxalate, which is stable from 233° to 397°, decomposes, it loses carbon monoxide and carbon dioxide simultaneously and almost instantaneously. Magnesium oxide appears at 480° and at 500°, for example, there is a mixture of $CaCO_3$ and MgO. At 900° the mixture is composed of CaO and MgO. Therefore, by recording the levels at 500° and 900° corresponding to these two mixtures and comparing their heights above the calibration line two first-degree equations can be set up with the unknowns x and y as the weights of calcium and magnesium and with m and n as the known weights of the mixtures present at 500° and 900°:

$$\frac{100\ x}{40} + \frac{40.32\ y}{24.32} = m; \quad \frac{56\ x}{40} + \frac{40.32\ y}{24.32} = n; \text{ whence } x = \frac{m-n}{1.1}$$

By way of example, a synthetic mixture consisting of 0.1541 g calcium oxalate monohydrate and 0.0453 g magnesium oxalate dihydrate gave the result $x = 0.0427$ g instead of the calculated amount 0.0422 g.

Completely automatic determination

Silver nitrate is the most stable nitrate known and it can be melted without decomposition; this is the basis of a recognized method for separating silver from other metals. Starting from this fact it is possible to perform a combined determination without any separation or weighing, even at the beginning. Let us take the simple case of a silver-copper alloy (Peltier and Duval[14]).

Experiments carried out on four alloys containing only copper and silver (tested spectrographically) afforded the following titres: 996.5, 890.5, 839.2, and 796.5 $\pm$ 0.1, by the Gay-Lussac volumetric method.

Fig. 56a is the pyrolysis curve of dry, crystalline silver nitrate. A horizontal extends to 342° and is followed by a descent as far as 473°. At this temperature decomposition sets in abruptly and nitrous fumes are expelled up to 608°. Next, there is a much slower descent from 608° to 811° (decomposition of silver nitrite), which is not observed when copper oxide is present; no doubt the latter catalyses the decomposition. Above 811° the weight is again constant, pure silver being present.

Repetition of the same operation with copper nitrate hexahydrate, leads to a quite different curve (Fig. 56b). Water and nitrogen oxides are driven off up to 58°. From 148° to 200° there is a horizontal, the existence zone of a new compound which has been analysed and which corresponds to a basic nitrate $Cu(NO_3)_2 \cdot 2Cu(OH)_2$. Then, between 200° and 607° this compound decomposes–vigorously from 252°, subsequently more slowly. The residue is copper(II) oxide CuO which only becomes constant in weight at 944° (later experiments have shown that it does not decompose below 1024°).

If now an intimate mixture of copper and silver nitrates is placed in the crucible, the curve recorded (Fig. 56c) is a resultant of the two preceding ones. The horizontal relative to the basic copper nitrate is, nevertheless, well marked. From 280° to 400° there is a residue keeping constant weight ($AgNO_3$ + CuO), while above 529° a mixture of Ag + CuO is present. These results suggest the following method of operation:

(a) An empty porcelain crucible of the ordinary type, previously ignited at 800° and unweighed, is placed in the carrier ring of the silica rod.

(b) The paper is calibrated with the three parallel lines AB, CD, and EF, as already explained (Fig. 54). An unweighed sample of the alloy is then placed in the crucible and a new line parallel to AB at once gives the weight of the alloy, *i.e.* of the sum Ag + Cu.

(c) The balance is arrested. A few drops of nitric acid are added to the alloy in the crucible which is set on a sand-bath at 80°. The reaction, the dissolution in water, and the evaporation to dryness can all be carried out in less than a quarter of an hour.

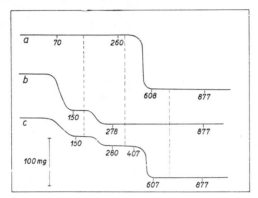

Fig. 56. Pyrolysis curves of nitrates: (a) silver nitrate, (b) copper nitrate, (c) silver nitrate + copper nitrate.

(d) The crucible is replaced on the ring and a curve analogous to Fig. 56c is recorded by the spot (or pen). The difference between the heights of the last two levels immediately furnishes the amount of NO_3 combined with the silver, from which the weight of this silver can be obtained and, consequently, that of the copper by difference. The result can also be obtained by setting up two first-order equations with two unknowns, as indicated on p. 93.

The entire operation (which is exceptionally long owing to the necessity of heating to a fairly high temperature) takes 85–90 minutes. The error (negative) in the weights of the copper and silver is fairly constant and does not amount to more than 1 part in 300; however, calculation of the error shows that it is difficult to improve upon this with the photographically recording Chevenard thermobalance in its present form. Table 2 gives some results compared with those obtained by Gay-Lussac's method on the same samples.

In the course of this determination the following operations have been abolished: (a) weighing the sample on an ordinary balance; (b) dissolution in water after decomposition of the copper nitrate; (c) collection of the

TABLE 2

ANALYSIS OF COPPER-SILVER ALLOY

Weight analysed (mg)	Thermobalance (mg)	Gay-Lussac's method (mg)
140.50	124.55	125.11
182.30	144.88	145.20
214.10	170.03	170.53
133.33	111.40	111.89

copper oxide; (d) gravimetric determination of the silver as chloride, which requires two weighings and sometimes even three; (e) dissolution of the copper oxide in sulphuric acid; and (f) electrolysis and two weighings.

Among the numerous results obtained from that time throughout the world, let us mention a set of researches by Unohara[15] dealing with mixed determinations: $Ca + Mg$; $Ba + Mg$; $Sr + Mg$; $Ca + Sr + Mg$ as the oxalates; $Ba + Mg$, $Ca + Sr + Ba$, $Bi + Zn + Cd + Pb$ as the carbonates; $Fe + Ni$, $Fe + Zn$, $Fe + Be$ as the sulphates and so on.

We hope that the reader in comparing some of the curves indicated in the second part will be able to evolve new automatic methods of determination which will avoid the necessity of a preliminary separation of the constituents.

4. Construction of a system of isotherms

As was pointed out earlier, it is possible at any time to use the furnace as a thermostat. This is why the thermobalance rendered such excellent service in reviewing the studies by Pavelka and Zucchelli[16] *i.e.* in tracing the isotherms relating to 300°, 400°, 500°, 600°, and 700° with respect to molybdic anhydride (obtained from ammonium molybdate). Each trial lasted at least 2 hours. It is easy to see in the loss of weight–time diagram, for all these temperatures (Fig. 57 a, b, c, d, e) with the ordinary precision of analysis, some straight lines parallel to the abscissae. The error calculated from the thickness of the lines traced on the photographic paper used justifies the conclusion that a relative loss of weight of 1/500th would be easily observable. The ordinary weighing errors do not enter in here because the crucible is never taken out of the furnace. No variation in weight was observed even for the isotherm at 600°, where Pavelka and Zucchelli reported a maximum loss of 0.435% at the end of 2 hours (corresponding to a deflection of 1 mm on the photographic paper).

If the isotherms of molybdic anhydride MoO_3 shown in Fig. 57 are examined carefully, it is most important to note, with respect to chemical analysis, that their parallelism to the time axis is attained more

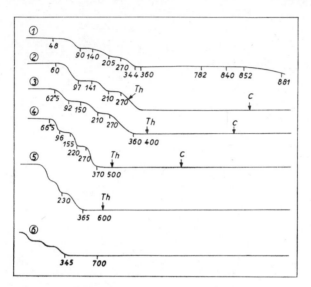

Fig. 57. Isotherms of ammonium molybdate. Th: placed in thermostat; C: attainment of constant weight.

quickly the higher the temperature (points C). This is apparent in the following tabulation:

Temp. (°C)	300	400	500	600
Time (min)	90	60	45	0

For example, a sample weighing 371 mg, which to have reached constant weight from 360° on, would have to lose 0.6 mg more to come to its theoretical weight. The losses in weight observed by these workers may have been due to this phenomenon, which certainly is not appreciable on an ordinary pan balance[17].

[5. Should a precipitate be dried or ignited?

This is a question which has already consumed much ink and paper because so many workers have wished, according to their temperaments, to give a categorical answer. Some, such as Spacu and his coworkers, propose that the precipitate be washed with ether and then dried in a desiccator under reduced pressure. Others, such as Winkler, prefer to use the temperature provided by their ovens (in this case 132°) on all occasions, while others, *e.g.* Carnot, Treadwell, etc. apparently were

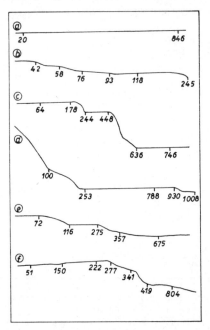

Fig. 58. Thermolysis curves for: (a) thallium silver arsenate, (b) calcium picrolonate, (c) mercurous iodate, (d) zinc hydroxide (carbonated), (e) antimony sulphide (with thiocyanate), (f) cadmium anthranilate.

unable to finish a determination to their satisfaction without applying the heat of a Meker burner or blast lamp.

Continuous recording of the weight makes it possible for everyone to reach agreement and avoid ambiguity. For each precipitate there exists a drying or desiccation zone, whose location is furnished by this author's curves, and which must be respected if correct results are desired. This viewpoint is supported by the following six examples.

Example 1

According to Spacu and Dima[18], arsenates can be determined by precipitating $TlAg_2AsO_4$, which is weighed after being dried at room temperature in vacuo. The curve or rather the straight line, which is shown in Fig. 58a, shows a constant weight from 20° to 846°. In one run, the authors[19] found initial and final weights of 360.73 mg and 360.05 mg, respectively. Hence any temperature between 20° and 846° may be selected.

Example 2

Fig. 58b represents the decomposition of calcium picrolonate Ca $(C_{10}H_7O_5N_4)_2 \cdot 7H_2O$ (not $8H_2O$). Ordinarily this salt is dried in a current of dust-free air. Sometimes the method cannot be used in the tropics; in fact, the curve reveals that the decomposition of calcium picrolonate[20] begins at slightly above 30°. Thorium picrolonate is incomparably more stable.

Example 3

Spacu and Spacu[21] recommend that mercurous iodide be weighed after it has been dried in vacuo at room temperature, whereas Gentry and Sherrington[22] advocate heating for an hour at 140°. Fig. 58c shows that this salt is stable up to 175° since it yields a perfect horizontal up to that point. The subsequent decomposition occurs in four stages up to 642°, where the crucible is completely empty. The portion of the curve between 230° and 449° corresponds to mercuric iodide, which sublimes and dissociates[23].

Example 4

When a zinc salt is treated with sodium carbonate in contact with air, a more or less carbonated hydroxide results. During the dehydration there is a sudden change in curvature at about 100° (Fig. 58d); then the loss of water becomes slower. At 200° we are dealing with a relatively stable basic carbonate, which does not release the whole of its carbon dioxide below 1000°. Consequently, the crucible may be returned to the furnace as often as the operator may desire, at 950°, for instance. Though the weight will always be constant, the result will be fallacious[24].

Example 5

The author[25] has shown that the familiar Treadwell hot-air bath and its modifications, which are commonly used to bring antimony sulphide into a suitable weighing form, should not be used because of the complications it introduces into this determination. In fact, after the loss of water and sulfur, the curve (Fig. 58e) has a horizontal between 176° and 275°, which agrees rigorously with the composition Sb_2S_3 and the existence of a homogeneous black product. The author advises that the sulphide be dried in an electric furnace for 10 minutes at 176°, and he especially urges the use of ammonium thiocyanate rather than hydrogen sulphide for the precipitation.

Example 6

The curve given by cadmium anthranilate (Fig. 58f) rises slowly up to 222°, where decomposition starts[26]. The gain in weight is slight, as may be seen from the following data which refer to 118.25 mg of the dry anthranilate:

Temp. (°C)	51	87	150	222
Loss (mg)	0.6	0.9	1.8	2.0

However, in view of the ease with which the leaflets of cadmium anthranilate dry, especially after they have been washed with ethanol, it seems preferable to keep the compound in a desiccator below 40° before weighing.

The question which heads this section can therefore not be answered categorically. Whenever a chemist attempts to find the optimum drying temperature for a given precipitate by means of trials alone, he is in much the same situation as a blind man who has lost his cane.

6. The use of the thermobalance for discovering methods of separation

The separation of gallium and iron is one of the most difficult problems of analytical chemistry. Iron is always found in commercial gallium when the specimens are examined spectrographically or tested with α, α'-dipyridyl. The thermobalance likewise reveals this contamination; numerous curves produced with gallium hydroxide show a rise (Fig. 59B) at high temperatures, notably from 800° on. The rise signals the gain of oxygen involved in the restoration of the iron to the Fe_2O_3 state. This finding has made it possible to compare the efficacies of the various methods of separation that have been proposed. All of them are unsatisfactory[27]. On the other hand, treatment with sodium sulphite (or bisulphite) is the only known procedure that yields a perfectly horizontal section in the gallium hydroxide curve (Fig. 59A). The finding proves that this reagent leaves the iron in solution (provided that no more than 1% is present initially). A curve analogous to that of Fig. 59B is obtained if a trace of

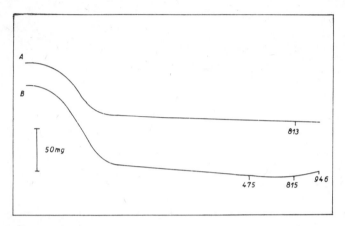

Fig. 59. Pyrolysis curves of gallium hydroxide, precipitated (B) with aniline, (A) with sodium sulphite.

iron is added to the gallium oxide resulting from the sulphite precipitation.

Less than 1 p.p.m. iron is left in the gallium which originally contained 1% of iron, and furthermore this separation by the sulphite method is achieved in a single operation and by means of a reagent which is readily available.

7. The thermobalance in gasometry

When the gas being determined is the sole product evolved on heating a solid, the thermobalance obviously can record the loss with no need for using a gas collecting tube, whose manipulation at temperatures in the vicinity of 1000° is not always a simple matter. A sample of gypsum ($CaSO_4 \cdot 2H_2O$) gave the data:

Water (moisture)	0.3%	SO_4	59.9%
Water (combined)	20.2%	CO_3	1.5%
Calcium	23.4%	$Al_2O_3 + Fe_2O_3$	traces
Silica	0.1%		

References p. 124

The pyrolysis curve of this material (Fig. 60b) is compared with that of pure calcium sulphate (Fig. 60a), namely a specimen of gypsum which contains no calcite ($CaCO_3$). The decomposition of the calcium carbonate becomes evident around 820°; the loss of 4 mg of carbon dioxide measured on the graph corresponds to 1.69% in terms of CO_3. (The classical volumetric method gave 1.5%.)

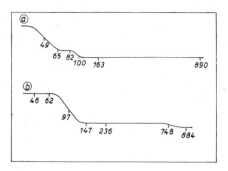

Fig. 60. Thermolysis curves for: (a) pure calcium sulphate, (b) gypsum or plaster of Paris.

8. Check of the atomic weight of carbon

In reality, Fig. 61 is the right-hand part of the Fig. 55a and relates to the thermolysis of calcium oxalate. It is known that one molecule of carbon

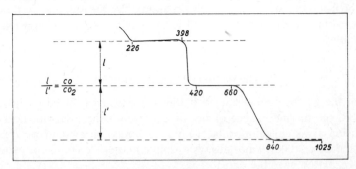

Fig. 61. Check of the atomic weight of carbon.

monoxide is given off between 400° and 420° and one molecule of carbon dioxide between 600° and 840°. Therefore,

$$\frac{CO}{CO_2} = \frac{l}{l'} \quad \text{or} \quad \frac{C+16}{C+32} = \frac{l}{l'}$$

from which it is then possible to calculate the value of C.

Six student trials, in which they prepared individual specimens of calcium oxalate, gave the following values for the atomic weight of carbon:

$$12.00 \quad 12.06 \quad 12.08 \quad 12.00 \quad 12.00 \quad 12.00$$

Accordingly this very simple experiment has given a rather good value as compared with the international value, and with no need of applying lengthy corrections or conducting extensive purifications of the gas. However, our objective was rather to prove that the thermobalance can be used to check the purity of the materials employed in the determination of atomic weights.

9. Study of sublimations

The author is convinced that the measurements of vapour tension should become less precise as the temperatures go up, assuming that the pronounced divergences observed by the various workers are due to their inability to perceive the instant at which the sublimation becomes appreciable to our senses. This applies especially to chlorides. Some examples will be given which are selected from those encountered in the course of our researches.

(i) Mercurous chloride

The weight of this salt remains constant up to 130°. Progressive heating (in an open crucible and at atmospheric pressure) at the rate of 100° per hour has resulted in the following losses of weight relative to an initial weight of 417.6 mg of dry calomel (Fig. 62a):

Temp. (°C)	130	167	283	410
Loss (mg)	0	5.8	184.0	366.6

(ii) Lead chloride

Fig. 62b shows that this salt begins to lose weight around 528°; up to this temperature there is a slight gain in weight probably due to the superficial formation of the oxychloride. The losses measured (as in the previous case) on the recording refer to 328 mg of lead chloride:

Temp. (°C)   528   675   788   826   859   915   928   946
Loss (mg)   0   4   30   51   80   180   220   292

The residue at 946° consists of lead chloride contaminated with a small amount of oxide, which apparently is due to the onset of decomposition.

(iii) Gallium oxinate

Various workers have suggested that gallium oxinate be weighed after drying between 110° and 150°. An examination of the curve (Fig. 62c) shows clearly that this complex compound loses weight from ordinary

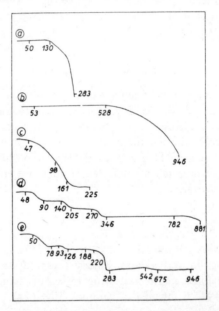

Fig. 62. Thermolysis curves for: (a) mercurous chloride, (b) lead chloride, (c) gallium oxinate, (d) dry ammonium paramolybdate; (e) uranium oxalate.

temperatures on. No horizontal appears below 180° and at this temperature the crucible is already full of carbon. The compound is manifestly sublimable and the region 110–150° corresponds precisely to the highest rate of decomposition. Therefore, the author opposes the adoption of this method for determining gallium.

(iv) Molybdic anhydride

This compound has long engaged our attention because various workers, in accord with its pronounced tendency to sublime, are much less in agreement as to the temperature at which it begins to sublime. We have seen figures ranging from 300° to 700°. The isotherms traced above (Fig. 57) already indicate that there is no danger of loss below 700°. It obviously was of interest to trace the course of the curve at still higher temperatures and this was done by Dupuis[17]. Starting at 782°, molybdic anhydride sublimed, slowly at first, and gave the following weight losses from an initial weight of 290.82 mg of dry paramolybdate (Fig. 62d):

Temp. (°C)	782	827	840	852	866	881
Loss (mg)	0	0.98	2.94	4.90	8.82	17.64

This was a very important piece of information because molybdic anhydride serves not only for the determination of molybdenum but also of phosphorus in fertilizers in the form of ammonium phosphomolybdate, in the determination of silica in the form of silicomolybdate oxine, hexamethylenetetramine, etc., and of germanium, arsenic, and many alkaloids by indirect methods. Hence, in conclusion, there is no danger of loss below 782° provided the molybdic anhydride is present alone.

10. A strange oxidation phenomenon

Under some circumstances we have found that the heating curve of certain oxidants (nitrates, chlorates, bromates, iodates, chromates, stannates, sulphamates, etc.) rises more than was to be expected from the simple variation due to the upward thrust of the heated air, even in the case of a poor adjustment of the balance. This initial gain is due to oxygen; it does not occur in nitrogen; sometimes it is reversible, in other cases it is irreversible. In the former case there is a chance that the weighing will be good but automatic determination becomes impossible. Accordingly, when a compound has to deliver oxygen it begins by taking a certain amount of this element from the air to form a peroxide or an unstable per-salt, which probably is capable of starting a chain reaction. For instance, anhydrous lead chromate is stable up to 904° but from 673° on it takes up oxygen and its curve ascends. In the chapter dealing with gold (p. 596) it will be pointed out that when this metal is precipitated by certain reagents such as oxalic acid, it may gain as much as 4 mg per 239 mg around 980°. Similarly, in the neighbourhood of 256°, which is close to the decomposition temperature of nitron nitrate, the latter undergoes a significant gain in weight (see Fig. 48c).

It is logical to discuss here various oxides of uranium which the author has been able to present on the same graph (Fig. 62e) despite the slight variation in the oxygen content which accompanies the passage from one oxide to the other[28].

This curve shows the pyrolysis of the pale green uranium oxalate, $U(C_2O_4)_2 \cdot 6H_2O$, which first of all loses 4 molecules of water between 50° and 78°. Then the dihydrate is present from 78° to 93° as shown by the short horizontal extending between these temperatures. The anhydrous oxalate is present from 126° to 158°, after which it starts to decompose, ordinarily losing carbon monoxide and carbon dioxide, notably at 250°, while the residue takes up some oxygen. Actually, it is surprising to find that the residual oxide does not have the anticipated composition UO_2; instead it corresponds to U_3O_8. This latter oxide is not within its stability zone; it combines with the quantity of oxygen necessary to produce UO_3 quantitatively at 542°. Finally, the latter reverts to U_3O_8 above 700°, and at 942° the oxide UO_2 starts to appear. These successive oxidations and reductions may be quantitatively verified on photographic paper with about 100 mg of oxide.

This example will serve to demonstrate the precision which may be attained with the Chevenard thermobalance.

11. New forms of gravimetric determination

With the collaboration of the 17 persons named on p. 16, the author has investigated the pyrolysis curves of approximately 1200 precipitates that have analytical interest. Even though, as a result of this critical examination, he suggests that only about 250 of these methods be retained, he has also recommended the revival of certain procedures which have been forgotten because their proposers did not specify the correct drying temperatures or because they destroyed the precipitate without learning that it could be weighed advantageously. In reality, a systematic choice cannot be made except on the basis of continuous weighing. Up to the present, the author has suggested 90 new methods.

It should be noted, for the future, that if a worker finds a new gravimetric procedure, a simple trial requiring about three hours–and supplying conclusive evidence–may save weeks of work. Two things should be taken into consideration:

(a) If the curve yielded by the precipitate does not contain a horizontal either on continuous heating or when subjected to isothermal treatment, the method is without value; it is useless to continue with it, even though the precipitation is complete, and the precipitate does not pass through the filter, etc. Sometimes, the precipitate can be dissolved in a suitable solvent and the determination then finished either by means of another gravimetric method or by a titration or a colorimetric measurement.

(b) If there is at least one horizontal, all of the dryings or desiccations of the precipitate must be carried out at a temperature situated between the two ends of this horizontal, with the proviso, of course, that we are dealing with a pure compound of a definite formula. It has sometimes happened, that we have recommended mixtures, such as $SiO_2.12MoO_3$ or

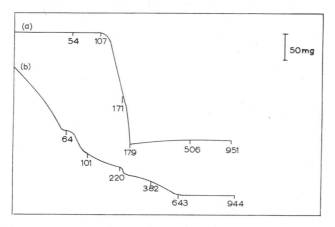

Fig. 63. Pyrolysis curves for: (a) iron cupferronate, (b) titanium cupferronate.

$CoO + 3CsNO_3$. In general, the curves have shown us that in present laboratory practice, the analysts are prone always to overheat the precipitates and to prolong the heating period unduly.

With these facts as a background, we have proposed a number of new determinations, including the following; the details will be found in Part Two, under the proper headings:
Determination of:
iron and of copper as cupferronates and neocupferronates,
gold or platinum as thiophenolates,
zirconium and hafnium as tetramandelates,
chromium as mercurous chromate,
gallium as camphorate,
thorium as sebacate, etc.

It is readily seen that the gravimetric (conversion) factor becomes smaller if the precipitate is not decomposed so as to reach the oxide as the final weighing form. Fig. 63a shows the thermolysis curve of iron cupferronate, and Fig. 63b is the thermolysis curve of titanium cupferronate. The former is clearly stable up to 100°, whereas the latter decomposes from even ordinary temperatures on.

12. Studies in pure inorganic chemistry

As we have pointed out from 1947 on, our general study was largely restricted to determining the fate of gravimetric precipitates when subjected to the action of heat. By chance we have been able to go into details regarding several points pertaining to inorganic chemistry not connected with analysis but they serve to clear up certain purely inorganic reactions. This line of investigation has been pursued even more since we have taken up the study of standards with Wadier and Servigne[28]. This latter series of studies was made necessary for the preparation of standard solutions used in our determinations, and for preparing buffer solutions, etc. Our method of approach is especially applicable to soluble materials. In this way we have learned what happens to most of the sodium salts: carbonate, bicarbonate, phosphate, tetraborate, acetate, tungstate, oxalate, thiosulphate, and also Mohr's salt, ferric sulphate, ferric ammonium alum, etc.

Along another line, we frequently have heated compounds containing deuterium oxide to compare their behaviour with that of the corresponding hydrated compounds. We have run comparisons on gypsum and heavy plaster of Paris, deuterogöthite and deuterolepidocrocite, chromium sulphates, as well as heavy bromates, iodates, sulphites, selenates.

The following are among the particular points which have claimed our attention:

(1) Rammelsberg reaction

The conversion of an alkaline earth iodate into a corresponding paraperiodate is familiarly known as the Rammelsberg reaction. For instance:

$$5 \; Ca(IO_3)_2 \rightarrow Ca_5(IO_6)_2 + 4 \; I_2 + 9 \; O_2$$

We, for the first time, have determined precisely the temperature at which the reaction occurs: calcium salt, between 550° and 887°, very rapid between 680° and 750° (Fig. 64a); strontium salt, between 600° and

748°, very rapid between 650° and 700°; barium salt, between 476° and 720°, very rapid at 610°; the other iodates which have been heated do not appear to yield paraperiodates.

(ii) Silver chromate

This salt is used for the determination of chromium or silver. We have found that between 812° and 945° each molecule loses exactly one molecule of oxygen and leaves a mixture of silver and its chromite:

$$2 \, Ag_2CrO_4 \rightarrow 2 \, O_2 + 2 \, Ag + Ag_2Cr_2O_4$$

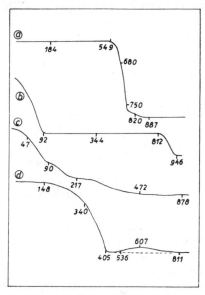

Fig. 64. Pyrolysis curves for: (a) calcium iodate, (b) silver chromate, (c) uranium "peroxide", (d) platinum oxidation.

a reaction that apparently has not been reported previously. The silver dissolves readily in nitric acid, leaving the green chromite which is not attacked by acids (Fig. 64b).

(iii) Uranium "peroxide"

The yellow precipitate obtained by treating uranyl nitrate solution with 12 volume hydrogen peroxide is usually written $UO_4 \cdot 2H_2O$. This formula is correct from the standpoint of percentage composition but it is incorrect when written in the form just given .Fig. 64c shows that up to 90° the imbibed water is given off rapidly and that the compound with the above formula is observed only at this temperature. It loses precisely one molecule of hydrogen peroxide in the interval 90–180°, and it is only then that the residue corresponds to the formula $UO_3 \cdot H_2O$ (uranic acid). The latter slowly loses 1 H_2O and then between 560° and 672° yields the horizontal level relating to the oxide UO_3. These findings indicate that the formula should be written $UO_3 \cdot H_2O_2 \cdot H_2O$. This point of view excludes the idea of the octavalency of the uranium (which is now accepted by very few authorities). The above finding is confirmed by the infra-red absorption spectrum between 6 and 15 μ. In this region there should be a triple degenerated band due to the UO_4 group. However, this is not the case; only the strong band at 920 cm^{-1} is found and it is characteristic of the UO_3 group.

(iv) Oxidation of platinum sponge

If ammonium chloroplatinate, $(NH_4)_2$ [$PtCl_6$], is heated, the decomposition is complete at 407° and a constant weight of platinum is obtained up to 538°. Then the metal takes up oxygen (the phenomenon does not take place in nitrogen or hydrogen) and some of the oxide PtO is formed. The latter is characterized by its X-ray spectrum. However, the oxidation is not complete under atmospheric pressure. The very flat maximum is reached at 607° (Fig. 64d); then the oxide decomposes and the initial weight is regained at 811°. These changes in weight (which may amount to as much as 5 mg per 123 mg of metal) have perhaps not always been taken into account when determining the molecular weights of amines. It is likewise necessary to take note of these changes when conducting reactions in a Gooch-Neubauer filtering crucible.

(v) The structure of inorganic bases

The thermobalance has been of much value in this line of investigation, which has been conducted along with a spectrophotometric study in the infra-red region. It has been possible to place these materials into two main categories: the true bases which contain OH groups: NaOH, $Ca(OH)_2$, $Cd(OH)_2$, $Mg(OH)_2$, $Ni(OH)_2$, and the hydrous oxides, which lose only the water which is in juxtaposition to the oxide ($Fe_2O_3 \cdot nH_2O$, $ZrO_2 \cdot nH_2O$, $ThO_2 \cdot nH_2O$). The two types of hydroxide may be obtained, depending on the precipitant employed. Thus Cabannes-Ott[29] has prepared both $Cu(OH)_2$ and $CuO \cdot H_2O$. Frequently, the natural product, found as a mineral, has a formula containing OH groups; for instance, göthite and lepidocrocite should be written $FeO(OH)$. On the other hand, natural magnesia (brucite) has the same structure as the artificial product.

Along the same line of thought, two kinds of rust may be distinguished: those of the $Fe_2O_3 \cdot nH_2O$ type, the others of the $FeO(OH)$ type, according to the nature of the water from which they originated[30].

13. Discovery of new compounds

In many cases, a thermolysis curve includes a horizontal stretch which corresponds to a mixture or to a pure compound which has not previously been reported in the literature. Its molecular weight may be obtained by means of a simple calculation, either starting from the formula of the initial material, or, more often, from that of the residue. Then, since the extreme temperatures of the horizontal are known, it is possible to heat another sample just to a temperature indicated, for instance, by the position of the horizontal. Pursuant to the work of Jolibois[31], who in this way characterized the uranium oxide U_3O_7, we have discovered 82 new compounds up to the present. Some of these are:

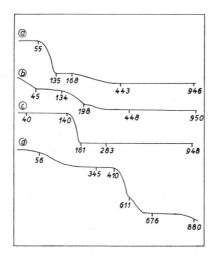

Fig. 65. Pyrolysis curves for: (a) boric acid, (b) metavanadic acid, (c) reduced silver, (d) cerium iodate.

(i) Metaboric acid

If orthoboric acid is heated progressively, it retains the composition H_3BO_3 up to 55° (Fig. 65a). No part of the dehydration curve gives any indication of the existence of pyroboric acid $H_2B_4O_7$. On the contrary, a short horizontal, which extends from 135° to 168°, corresponds to metaboric acid HBO_2, and this finding suggests a simple method for isolating the latter in a pure state. The molecular weight indicated by the graph is 43.8 (calculated 43.82). Above 168° this new acid slowly loses water and the excellent horizontal corresponding to boric anhydride B_2O_3 begins at 443°.

(ii) Metavanadic acid

If slightly moistened ammonium metavanadate is heated, a horizontal is obtained between 45° and 134° (Fig. 65b) which agrees with the composition NH_4VO_3. From 134° to 198°, the loss in weight corresponds to the elimination of all of the ammonia and the residue consists of metavanadic acid HVO_3. This finding suggests a method for preparing this acid. However, the latter is not stable and it begins to lose water at 206°. The anhydride V_2O_5 is obtained quantitatively at 448°; this anhydride is then stable up to 1000° at least.

(iii) Reduced silver

It is well known that silver can be precipitated from a solution of silver nitrate by electrolysis or by the reducing action of ammoniacal cuprous chloride, hypophosphorous acid, cadmium, aluminium, hydroxylamine, vitamin C, etc. However, if we choose the classic example of the silvering of glass, i.e. the reduction by formaldehyde and 15% ammonia, it is quite surprising to find that metallic silver does not deposit in the cold. The black deposit yields a curve (Fig. 65c) which shows a sudden descent between 140° and 160° where a lively decomposition sets in. The constant weight for silver starts only when the temperature has reached 500°. The starting material is therefore not silver. We have turned our attention to the initial precipitate; it is known that the NH groups are specific for the detection of silver. Under the experimental conditions used by us, NH_3 combines with two molecules of formaldehyde to give dihydroxymethylamine, $HOCH_2-NH-CH_2OH$. The replacement of the central hydrogen atom by silver leads to the precipitation of $HOCH_2-$

NAg–CH$_2$OH, which is commonly called reduced silver. Its calculated and found molecular weight is 184.

(iv) Cerium iodate

It had been shown previously that cerium can be precipitated by potassium iodate as $2Ce(IO_3)_4 \cdot KIO_3 \cdot 8H_2O$, which may be weighed after drying with ether. Actually, the thermolysis curve (Fig. 65d) shows a sharp drop between 410° and 650°, corresponding to a loss of oxygen and iodine. Between 650° and 746°, we have a mixture of potassium iodide and cerium metaperiodate $Ce(IO_4)_4$, the latter being a new salt. The residue at 880° contains some cerium peroxide. Even though this precipitate may not be used for the gravimetric determination, its pyrolysis curve is full of promise and opens the way for new researches in inorganic chemistry.

14. Correction of errors in analytical chemistry

While examining our findings some years ago, Fritz Feigl uttered the quip: "The thermobalance is capable of distinguishing between the good and the incompetent chemists." The present author has no desire to start a debate, but he wishes merely to present some cases in which the chemists, discoverers of the methods in question, seem not to have exercised the proper care.

(i) Copper sulphide

The bell-shaped curve of copper sulphide is explained by the diagram (Fig. 78) in Chapter Copper.

(ii) Vanadium by luteocobaltic chloride

If a vanadate solution is treated with luteocobaltic chloride, it is claimed that the structure of the precipitate differs according to whether the milieu is acid, neutral or basic. The author has found that the same product results in neutral or alkaline surroundings, namely $[Co(NH_3)_6](VO_3)_3$, which may be used as the weighing form after it is dried between 58° and 143° (Fig. 66a). On the other hand, the formula $[Co(NH_3)_6]_4(V_6O_{17})_3$ which has been proposed for the product obtained in an acidic milieu, departs from the actuality by almost 1.7% and the corresponding curve has no horizontal in the region around 100°. More accurate figures would be obtained if it were feasible to dry the product at 127°, but the product suffers decomposition at that temperature.

(iii) Tungsten by oxine

The formula $WO_2(C_9H_6ON)_2$ which has been assigned to the precipitate resulting from a tungstate solution and oxine is not correct. The composition changes from one trial to another. Although a horizontal is obtained up to 218°, it does not correspond to a definite formula, but rather to an adsorption product of oxine by WO_3. In Fig. 66b the apparent

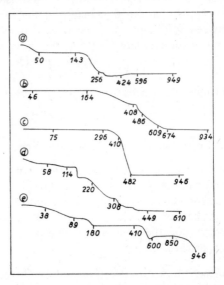

Fig. 66. Pyrolysis curves for: (a) vanadate with luteocobaltic chloride, (b) tungsten by oxine, (c) cuprous iodide, (d) germanium by cinchonine, (e) ammonium phosphomolybdate.

molecular weight according to the horizontal is 436 instead of 503.92 required by the above formula. The oxide WO_3 appears from 674° on.

(iv) Cuprous iodide

If a solution of a cupric salt is treated with sulphur dioxide, and then with potassium iodide, the resulting cuprous iodide precipitate gives the curve shown in Fig. 66c. Above 482° there is a horizontal which corresponds to the presence of cupric oxide, CuO. It therefore is quite proper to inquire into the significance of the melting point 628°, and of the boiling point 772° given in the literature for cuprous iodide, since this compound no longer exists at that temperature.

(v) Germanium by cinchonine

The author has repeated the experiments of Davies and Morgan who claim to have obtained a precipitate with the composition $H_4[Ge(Mo_{12}O_{40})](C_{19}H_{22}ON_2)_4$. Fig. 66d shows the curve corresponding to 2 mg of the metal, while another curve, not given here, corresponds to precisely 9 mg because these workers have stated that low results are obtained above 5 mg. We have found that the precipitates formed in this case do not yield reproducible curves, although a vague horizontal appears between 92° and 121°. It does not agree with the given formula, nor does it correspond to the rational formula $H_4[Ge(Mo_2O_7)_6](C_{19}H_{22}ON_2)_4$. The level which extends from 450° to 900° no longer agrees with the usual mixture $GeO_2 + 12\ MoO_3$; it contains too little molybdenum. Therefore, this procedure is not recommended for determining germanium; we prefer 5,7-dibromooxine for this purpose.

(vi) Ammonium phosphomolybdate

This salt is used at present for the determination of phosphorus, and most of the modern texts on chemical analysis prescribe the application of various correction factors for reasons which do not seem to be very clear.

We carried out the precipitation according to the directions given in the Treadwell text, which specifies that the material be weighed after prolonged heating at 170°; the formula is taken to be $(NH_4)_3PO_4 \cdot 12\ MoO_3$ after this treatment. According to Woy, gentle ignition results in the mixture $P_2O_5 \cdot 24MoO_3$. Curve 66e relates to a moist precipitate of ammonium phosphomolybdate, which is ordinarily given the formula $(NH_4)_3PO_4 \cdot 12MoO_3 \cdot 2HNO_3 \cdot H_2O$. This complex salt loses nitric acid and water up to 180°. If the recording of this same curve is recommenced, but with a precipitate which has been dried in the open air, the loss of weight up to 180° corresponds to 2 HNO_3 and not to 2 $HNO_3 + H_2O$. The difference is insignificant and has no influence whatsoever on the determination of phosphorus. A perfectly horizontal stretch extends from 180° to 410°; it corresponds closely to the formula given by Treadwell.

Beginning at 410°, two molecules of this material lose six molecules of ammonia and three molecules of water up to 540°; however, all of our recordings show that the destruction is more deep-seated; the curves undergo a distinct change of direction, and then go up again. Accordingly, it is necessary to assume a transient reduction of the molybdenum,

References p. 124

followed by a reoxidation due to the air to produce the ensemble P_2O_5. $24MoO_3$. At 540° the material in the crucible is green; it consists of a mixture of molybdenum blue and yellow phosphomolybdate. The rise in the curve is very slow. Theoretically, the horizontal is only reached between 812° and 850°; but no great error in the phosphorus follows if the ignition region is anywhere between 600° and 850°; above this temperature, the molybdic anhydride sublimes rapidly. It is now accepted that all of the possible approximate formulas for the residue may be found below 600°.

15. The thermobalance in functional organic analysis

This line of investigation is just getting under way and up till now it has dealt with the reagents which serve for characterizing aldehydes and ketones, notably 2,4-dinitrophenylhydrazine, dimedone (5,5-dimethyl-dihydroresorcin), and thiocarbohydrazide.

In general the pyrolysis curves are much simpler than the preceding, and often are reduced to horizontal straight lines up to the melting point of the compound.

16. The thermobalance in the study of solid state reactions

The thermobalance lends itself particularly well to the investigation of solid state reactions, especially if its use is combined with infra-red spectrometry as suggested by Duval and Lecomte[32]. In fact, if the reaction gives rise to a gas, it is possible to learn at what temperature the evolution begins and also where it is completed. Such reactions seldom go to completion and chemical analysis frequently is incapable of disclosing the real state of affairs. However, since 1938, we have recorded the characteristic absorption bands, notably between 6 and 15 μ, for the families of compounds containing the groups:

XO_2 (nitrites, chlorites, metaborates . . .)

XO_3 (chlorates, bromates, iodates, sulphites, selenites, nitrates, carbonates, meta-antimonates, metavanadates . . .)

X_2O_5 (pyrosulphites, pyroselenites)

XO_4 (orthosilicates, orthogermanates, sulphates, chromates, manganates, ferrates, molybdates, tungstates . . .)

X_2O_3 (thiosulphates)

X_2O_6 (dithionates)

X_2O_7 (pyrosulphates, bichromates, pyrophosphates, pyroarsenates, pyrovanadates . . .).

It is known in addition that the metal cations, when simple, are transparent in the infra-red. Furthermore, the number of permissible bands can be foreseen through application of the formula: $f = 3n - 6$, where n denotes the number of atoms in the group. For instance, the group XO_4 is capable of giving a simple vibration (v_1), a degenerated double vibration (v_2), a triple degenerated valence vibration (v_3), located at 1106 cm^{-1} in the case of SO_4, and a triply degenerated deformation vibration (v_4) at 570 cm^{-1} for the SO_4 group. If now the atomic weights of the X atoms are entered as ordinates, and the frequencies in cm^{-1} in relation to v_3 as abscissae, four curves that are practically parallel will result; the one nearest the center relates to tetravalent elements X, the second curve to pentavalent elements, the third to hexavalent elements, and the fourth to heptavalent elements.

Accordingly, if vanadic anhydride, for instance, is heated with sodium peroxide, a horizontal indicating constant weight is obtained due to a vanadate, but the thermobalance does not indicate whether it is a meta-, pyro-, or orthovanadate. This question can be resolved in a matter of 20 minutes by means of the infra-red spectrum of the solid powder, and with no need for making separations or preparing an aqueous solution, which may change the nature of the salt obtained.

Along another line, if bismuth trioxide, Bi_2O_3, is heated with oxylith, a brown powder results which is employed as oxidant in analytical chemistry. Some writers have given this product the formula of an orthobismuthate, Na_3BiO_4, while others prefer to regard it as a metabismuthate, $NaBiO_3$. It is impossible to analyze it chemically since water decomposes it along with the excess of oxylith which it retains. The infra-red spectrum indicates that the so-called bismuthate is a mixture of Bi_2O_5, excess oxylith, and sodium carbonate[33].

Among the many solid state reactions which have been reported up to the present, and which have involved the use of thermogravimetric analysis, we have selected for special note here the preparation of inorganic chromites in association with Wadier[34] (see Chapter Chromium); the study of the synthetic mixture $SiO_2 + 12 MoO_3$, and the investigation of the same mixture as produced during the pyrolysis of oxine silicomolybdate; the study of the adsorption products: potassium permanganate and barium sulphate; iodine and magnesia; the investigation of a reaction with no release of gas between strontium oxide and magnesium pyrophosphate:

$$3 SrO + Mg_2P_2O_7 \rightarrow Sr_3(PO_4)_2 + 2 MgO$$

the action of sodium peroxide on silica, which leads to an orthosilicate[32].

It was this latter reaction which Viltange[33] generalized (see Chapter Sodium) by bringing sodium peroxide into reaction with about ten inorganic oxides.

Through the combined use of thermogravimetry and X-ray spectrography, Chaudron[35] prepared various apatites (see Chapter Calcium). For instance, the reaction:

$$3 Ca_3(PO_4)_2 + CaF_2 \rightarrow [Ca_3(PO_4)_2]_3.CaF_2$$

was conducted at 600° between calcium fluoride and tricalcium phosphate.

Finally, by using the thermo-weighing curves together with differential thermal analysis, Hogan and Gordon[36] followed the solid state reaction between potassium nitrate and barium chloride:

$$2 KNO_3 + BaCl_2 \rightarrow 2 KCl + Ba(NO_3)_2$$

17. Use of differential thermal analysis

Within the last few years, a number of workers have complemented the thermogravimetric curves with differential thermal analysis curves to be in a position to interpret with more assurance the various unexpected happenings observed while making these curves. The thermogravimetric curves reveal the variation in weight, whereas those of the other type are connected with calorific phenomena: dissociation, allotropic transformation, fusion, etc., which often occur with no change in weight. It was an enticing thought for the investigators to use the same furnace to trace the two curves simultaneously; various models of dual purpose thermobalances have been noted in Chapters 1, 2, and 3.

As examples of the two curves on the same chart as obtained by Papailhau[37], we will cite the curves for a kaolinite (Fig. 67a), a manganous phosphate (Fig. 67b), and a dolomite (Fig. 67c).

(i) Kaolinite

With respect to the kaolinite, the samples weighing 300 mg were heated in an alumina crucible. They showed an initial loss of adsorbed water at around 140° (bend Ed_1 and corresponding loss of weight). Alumina, an inert substance, was used as reference material. Then the water of constitution was driven off between 500° and 650° (loss of weight and important deflection Ed_2 whose maximum amplitude is around 610°). X-rays show that the crystalline lattice of the kaolinite is destroyed during this reaction and that the anhydrous material, initially in the amorphous state, suddenly recrystallizes at an elevated temperature (exothermic reactions Ex_1 and

Ex_2 without change in weight). In accord with known observations, the X-rays reveal the formation of cristobalite, alumina, and mullite.

(ii) Hydrated manganese phosphate

In the case of artificial hydrated manganese phosphate $MnPO_4 \cdot 2H_2O$, of which 250 mg were slowly heated in a fused silica crucible, there is an

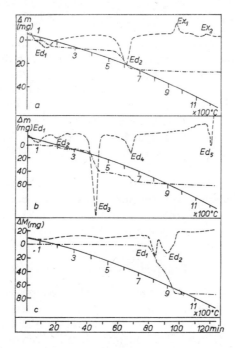

Fig. 67. Differential thermal analysis and gravimetric thermo curves for: (a) kaolinite, (b) manganous phosphate, (c) dolomite.

initial loss of water at about 100° (slight decrease in weight and reaction Ed_1). There is an additional loss of water from 170° to 240°, a more notable loss which leads to another endothermic effect Ed_2. X-rays indicate that these two reactions do not alter the crystalline lattice, which proves that the water removed is not water of constitution. A very marked endothermic reaction Ed_3 occurs from 350° to 460° accompanied by a distinct loss of weight. After this reaction has occurred, the X-rays reveal a poorly crystallized mixture of manganous orthophosphate and pyrophosphate. A new endothermic reaction Ed_4 occurs from 570° to 680° accompanied by loss of weight. At the close of this reaction, the entire sample has been converted into manganous pyrophosphate. Accordingly, the water of constitution has been eliminated in two stages designated by Ed_3 and Ed_4. Finally, around 1210°, a very distinct bend marks the fusion of the pyrophosphate.

(iii) Dolomite

The dolomite was heated in a platinum crucible; the sample weighed 165 mg. Its double decomposition is plainly revealed on the characteristic curves c. An endothermic reaction (maximum amplitude around 815°) occurs from 770° to 830° and is accompanied by a loss of weight (liberation of carbon dioxide). A new endothermic reaction appears from 830° to 970° (maximum amplitude around 900°) and a new release of carbon dioxide. X-ray studies reveal that Ed_1 corresponds to the decarbonation of the magnesium salt and Ed_2 to the loss of carbon dioxide by the calcite produced during the course of Ed_1.

These varied instances demonstrate that it would be extremely difficult to correlate accurately the various bends in the differential thermal analysis curves with corresponding irregularities in the thermal analysis curve in which the weights are recorded along with the temperatures.

18. Differential thermogravimetry

During these past few years, and especially because of the studies by De Keyser[38], and Erdey, Paulik and Paulik[39] as was pointed out in Chapter 2, there have come on the market such differential thermobalances as the Sartorius instrument and the "Derivatograph" which yield a direct recording of the curve.

The Derivatograph gives on the same sensitive paper (1) the curve T as function of time, (2) the curve TG of thermo-weighing either as function of time or as function of temperature, (3) the curve DTA for the differential thermal analysis, (4) the curve DTG of the resulting thermogravimetric analysis which greatly resembles it. An impressive example is shown in Fig. 68; it was obtained from an alumina. The inventors of the "Derivatograph" have traced analogous curves for bauxite[40], lignite, a mixture of alkaline earth oxalates, oils[41, 42].

The principle of simultaneous recordings is shown in Figs. 37 and 38.

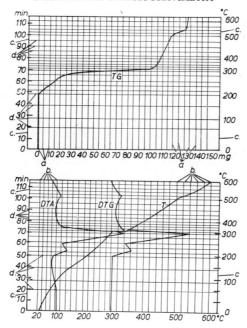

Fig. 68. Curves of an alumina as recorded with "Derivatograph".

19. Use of the thermobalance in combination with other apparatus

The important services rendered by *infra-red spectrophotometry* in the identification of the compounds suggested by the horizontals have been noted above. It is not the only auxiliary technique which may be called on. For example, Brenet and Grund[43], and Brenet, Gabano and Seigneurin[44] have employed thermomagnetism to establish the individuality of the γ- and the β-forms of manganese dioxide.

The use of *X-ray spectrography* is much more frequent. Among the numerous instances of its applications, (see Part 2), we call attention to the study by Boullé and Dominé-Bergès[45] of the conversion of mono-potassium phosphate into the pyrophosphate, and the study by Morin[46] of the acid triphosphate $K_3H_2P_3O_{10} \cdot H_2O$, of apatites by Chaudron[35], and the study by Capestan[47] of uranyl sulphamate $UO_2(NH_2\text{-}SO_3)_2 \cdot 2H_2O$, which isomerizes into the double sulphate $UO_2(NH_4)_2(SO_4)_2$.

Thermogravimetric analysis and *enthalpic analysis* have been studied jointly in the case of hydrated aluminium oxides, such as böhmite by Eyraud *et al.*[48].

Calvet[49] has built a *microcalorimeter* with 4 elements; it may be employed for thermogravimetric investigations of adsorbed materials.

20. Various studies

Vallet[50] succeeded in integrating the rate constants given by the Arrhenius law, a problem which arises notably in the interpretation of the mass–time curves obtained with the thermobalance when the rate of temperature rise is constant. This integration has been accomplished only in those instances in which there are slight elevations of temperature and for three forms of the rate constant.

Some workers have determined *activation energies* through the form of the thermolysis curves. In one of our papers with Robert[51] dealing with the dehydration of proteins previously saturated with water at 15°, we have worked either with a steady rise in temperature (50°/hour, or 100°/hour, or 200°/hour), or isothermally at 80°, 110° and 150°. Gelatin which had previously been treated with acetaldehyde yielded values of 1600 cal/mol (between 100° and 200°) and 3200 cal/mol (between 50° and 100°). These values are respectively 460 and 3500 cal/mol for untreated gelatin. Identical calculations made on the isotherms yielded approximately 600 cal/mol if one takes the logarithm of the percentage of water lost as a function of the inverse of the absolute temperature.

Cases involving cycling and even recycling have often been studied. Examples will be given with respect to boric acid[52] and germanic acid[53].

Löwe[54] employed proportional systems for continuous gravimetry and Skramovsky[55] determined the Arrhenius constants for magnesite, siderite, dialogite, ankerite and dolomite by inverse stathmometry.

21. Unexplained phenomena

During our studies we have observed certain facts, often bizarre, for which we could not offer an explanation and which required further and other investigations. Allusion to these cases will be found in the appropriate chapters in Part Two. Typical examples are the following:

(1) It is well known that Kumins suggested the excellent method in which zirconium is precipitated by mandelic acid. The corresponding curve (Fig. 69a) has a good horizontal up to 188° and it corresponds to the formula $Zr(C_6H_4-CHOH-CO_2)_4$. After the organic matter is de-

stroyed, the zirconium appears to be partially reduced. In fact, beyond 570°, the curve ascends in such manner that the constant weight for the zirconium oxide is not obtained until 950° is reached. This is not an isolated case and is observed, but in a more pronounced manner, with all of the precipitates produced by zirconium and arsenical compounds (arrhenal, atoxyl, propylarsonate, phenylarsonate, hydroxyphenylarsonate, etc.). Accordingly, we isolated[57] and analyzed the apparent mixture

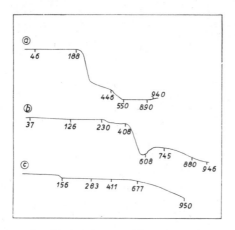

Fig. 69. Pyrolysis curves for: (a) zirconium mandelate, (b) thallium trioxide with ferrocyanide, (c) thallium trioxide with electrolysis.

formed at the minimum of the curves. Even though the composition accords with $Zr_2O_3 + ZrO_2$, the X-ray spectrum is invariably the same, namely that of pure ZrO_2.

(2) In association with Peltier[56] we have obtained the trioxide Tl_2O_3 by the ferrocyanide method and by electrolysis. The details will be found in Chapter Thallium. The curves (Fig. 69b and c) show that above 677° there are two different forms of the double oxide $3 Tl_2O_3 \cdot Tl_2O$ but their Debye-Scherrer spectra are identical.

22. Conclusion

This lengthy discussion exhibits the various aspects of thermogravimetry at the start of 1961. Analytical chemistry, inorganic chemistry, and mineralogy are among the sciences which have benefitted most from the thermobalance.

References p. 124

The prospects are that much can be hoped for from the construction of curves of these types, and from the coordinated use of differential thermal analysis and of infra-red spectrophotometry with the new technique. However, the next chapter will reveal a more pessimistic aspect; it will take up the principal sources and causes of error in the use of thermobalances and it will discuss why the findings of different workers are far from being in perfect agreement.

REFERENCES

1 M. Guichard, *Ann. Chim.*, 9 (1938) 323.
2 C. Duval, *Chim. anal.*, 31 (1949) 177.
3 R. Ripan, *Bull. Soc. stiinte Cluj.*, 3 (1927) 311.
4 M. Harmelin, *Diploma of Higher Studies, Paris*, 1958.
5 N. Lendormy, *Diploma of Higher Studies, Paris*, 1961.
6 P. E. Wenger, C. Cimerman and A. Corbaz, *Mikrochim. Acta*, 2 (1938) 314.
7 R. Fruchart and A. Michel, *Compt. rend.*, 246 (1958) 1222.
8 O. Brunck, *Z. angew. Chem.*, 25 (1912) 2479.
9 C. Duval, *Anal. Chim. Acta*, 2 (1948) 92.
10 C. Duval, *Anal. Chim. Acta*, 3 (1949) 163.
11 C. Duval, *Mikrochemie*, 35 (1950) 242.
12 F. Feigl, *Ber.*, 56 (1923) 2083.
13 C. Duval, *Compt. rend.*, 224 (1947) 1824; 226 (1948) 1276.
14 S. Peltier and C. Duval, *Compt. rend.*, 226 (1948) 1727.
15 N. Unohara, *J. Chem. Soc. Japan*, 73 (1952) 488; 74 (1953) 329; 76 (1955) 357, 359, 615.
16 F. Pavelka and A. Zucchelli, *Mikrochemie*, 31 (1943) 69.
17 T. Dupuis, *Compt. rend.*, 228 (1949) 841; *Thesis*, Paris, 1954.
18 G. Spacu and L. Dima, *Z. anal. Chem.*, 120 (1940) 317.
19 T. Dupuis and C. Duval, *Anal. Chim. Acta*, 3 (1949) 262.
20 S. Peltier and C. Duval, *Anal. Chim. Acta*, 1 (1947) 346.
21 G. Spacu and P. Spacu, *Z. anal. Chem.*, 96 (1934) 30.
22 C. H. R. Gentry and L. G. Sherrington, *Analyst*, 70 (1945) 419.
23 C. Duval and Ng. Dat Xuong, *Anal. Chim. Acta*, 5 (1951) 494.
24 M. De Clercq and C. Duval, *Anal. Chim. Acta*, 5 (1951) 282.
25 J. Morandat and C. Duval, *Anal. Chim. Acta*, 4 (1950) 498.
26 C. Duval, *Anal. Chim. Acta*, 4 (1950) 190.
27 T. Dupuis and C. Duval, *Anal. Chim. Acta*, 3 (1949) 324.
28 C. Duval, *Anal. Chim. Acta*, 13 (1955) 32, 427; 15 (1956) 223; 16 (1957) 221, 545; 20 (1959) 263; 23 (1960) 257, 541.
29 C. Cabannes-Ott, *Thesis*, Paris, 1958, *Ann. Chim.*, 5 (1960) 905.
30 J. Loisel, *Diploma of Higher Studies, Paris*, 1955.
31 P. Jolibois, *Compt. rend.*, 224 (1947) 1395.
32 C. Duval and J. Lecomte, *Proc. Intern. Symposium on Reactivity of Solids, Gothenburg*, 1952; *Compt. rend.*, 234 (1952) 2445.
33 M. Viltange, *Thesis*, Paris, 1960; *Ann. Chim.*, 5 (1960) 1037.
34 C. Duval and C. Wadier, *Proc. 3rd. Intern. Symposium on Reactivity of Solids*, Madrid, 1957, p. 265.
35 G. Chaudron, *Colloq. I.U.P.A.C., Münster*, 1954, p. 199.
36 V. D. Hogan and S. Gordon, *J. Phys. Chem.*, 64 (1960) 172.
37 J. Papailhau, *Bull. Soc. franç. Minéral.*, 82 (1959) 370.
38 W. L. De Keyser, *Nature*, 172 (1953) 364.

39 L. ERDEY, F. PAULIK AND J. PAULIK, *Nature*, 174 (1954) 885; *Acta Chim. Acad. Sci. Hung.*, 7 (1955) 27; 10 (1956) 61.
40 F. PAULIK, J. PAULIK AND L. ERDEY, *Z. anal. Chem.*, 160 (1958) 241.
41 F. PAULIK AND M. WELTNER, *Acta Chim. Acad. Sci. Hung.*, 16 (1958) 159.
42 M. WELTNER, *Acta Chim. Acad. Sci. Hung.*, 21 (1959) 1.
43 J. BRENET AND A. GRUND, *Compt. rend.*, 240 (1955) 1210.
44 J. BRENET, J. P. GABANO AND M. SEIGNEURIN, *16th Intern. Congress Pure and Appl. Chem.*, Paris, 1957. *Chimie minérale*, p. 69.,
45 A. BOULLÉ AND M. DOMINÉ-BERGÈS *Colloq. I.U.P.A.C.*, *Münster*, 1954, p. 258.
46 C. MORIN, *Thesis*, Paris, 1960.
47 M. CAPESTAN, *Thesis*, Paris, 1960; *Ann. Chim.*, 5 (1960) 222.
48 C. EYRAUD, R. GOTON, Y. TRAMBOUZE, TRAN HUU THE AND M. PRETTRE, *Compt. rend.*, 240 (1955) 862, 1082.
49 E. CALVET, *Compt. rend.*, 236 (1953) 377.
50 P. VALLET, *Compt. rend.*, 249 (1959) 823.
51 C. DUVAL AND L. ROBERT, *Compt. rend.*, 238 (1954) 282.
52 J. HALADJIAN AND J. CARPENI, *Bull. Soc. chim. France*, (1956) 1679.
53 A. PIETRI, J. HALADJIAN, G. PERINET AND G. CARPENI, *Bull. Soc. chim. France*, (1960) 1909.
54 R. P. LÖWE, *Mech. Eng.*, 76 (1954) 653.
55 S. SKRAMOVSKY, *Silikáty*, 3 (1959) 74; *C.A.*, (1959) 21122.
56 S. PELTIER AND C. DUVAL, *Anal. Chim. Acta*, 2 (1948) 211.
57 J. STACHTCHENKO AND C. DUVAL, *Anal. Chim. Acta*, 5 (1951) 410.

CHAPTER 5

Precautions to be Taken in the Use of Thermobalances

1. Introduction

It would appear logical, after having enumerated the existing thermo-balances and describing those which are commercially available in 1961, to compare their advantages and demerits, their performances, limits of error, etc. However, we are not yet ready for such a comparison. Some of the instruments are of too recent date and their users are content for the present to confine themselves to the making of recordings. Certainly there is no need for comparing these instruments merely on the basis of their sales prices.

The Chevenard thermobalances were the first to be offered for sale and have been carefully scrutinized with respect to their performance by several hundred purchasers. Consequently, since more registrations, both good and bad, have been delivered by these balances, it follows that the greater bulk of the critical information refers to them.

The temperature at which a compound decomposes or appears to begin to decompose is not a fixed point like a melting point. Ten (and doubtless as many as twelve) parameters must seemingly be taken into consideration here. This is what we propose to do now; and we shall profit here from the criticisms and comments that have been addressed to me personally.

I wish to remind the reader that our initial work on the precipitates was carried out in aluminite* crucibles, with 200–300 mg amounts of the moist precipitate, and heated with the greatest possible speed in order to simulate as much as possible the usual operational procedure of gravimetric analysis. Therefore, it is not proper to compare the decomposition of a crystal of calcite with the decomposition of the calcium carbonate obtained transiently from the breaking-down of calcium oxalate. In

---

* Aluminite is the registered name of a Limoges porcelain, whose lightness and refractory properties render it suitable for laboratory ware.

addition, it is difficult to find an analogy between the heating of lanthanum oxalate heated in a vacuum and the same salt heated under atmospheric pressure.

Only comparable things can be properly compared. We will review the principal factors operating in this field, discussing them in sufficient detail for the purpose at hand, and we shall indicate, where necessary, the remedies for the lack of success. It will then become evident that certain criticisms have not been justified.

2. The make of the thermobalance

Most of the thermobalances commercially available have a beam with a knife edge (agate, steel, etc.); only the Chevenard thermobalances employ a bifilar suspension. These latter balances are capable of yielding the same curve over a span of 25 years even though the wires have been broken and replaced from time to time. A prismatic knife edge becomes dull with use and also wears away its support; the edge becomes blunted or rounded and this may have serious consequences for registrations extending over several days, particularly in a manufacturing plant where there is much vibration. There is no assurance that it will retain the same position during the exposure. It is necessary to allow sufficient time to elapse before coming to any decision on this point, in other words, we cannot yet see it in the proper perspective.

The following factors also enter into the construction. Should the position of the furnace be inverted or not, and should it be placed above or below the balance? What should be the position of the thermoelectric couple, how fast should the air flow; what method of recording should be employed, what kind of damping devices, if any, should be included? These points will be discussed presently. In my opinion it is very difficult to compare the two types of balances, those of the type in which the beam is deflected and those of the null-deflection type, which are based on different principles and which do not meet the problems due to friction in the same manner, and also the spring balances which are governed by another law and which ordinarily do not operate in the air under atmospheric pressure, and the differential balances in which the influence of the convection currents is eliminated almost entirely.

Usually a thermobalance is standardized in the cold by placing a calibration weight on one pan or a platform; it produces either a more or less deep notch in the curve, or it causes a movement amounting to a certain number of divisions in the field of sight, or it results in an induced current. The comparison of these results appears to be difficult. The calibration is not the same for all temperatures and all loads. The curve shown in Fig. 70, which was traced by Newkirk[1] for 300 mg of calcium oxalate, shows that the movement along the four horizontals remains quite constant on the Chevenard thermobalance with pen recording, and with the temperature rising at the rate of 300° per hour, if an overload of 20 mg is added. I believe that this will hold good, all other conditions being kept the same, provided one-half of the permissible load (5 g) in the crucible is not exceeded. If the temperature–time curve is not a straight line in the range 0–1200°, and this depends on the programming arrangements of the apparatus, it is advisable to follow the example of several workers (Papailhau, Erdey, etc.) and record the heating curve on the same paper as the thermoweighing curves. A clockwork governs this program in the Chevenard-Joumier system. High precision may then be expected. (The winding of the clock should be done at regular intervals, say every Monday morning.) In most of the other assemblies, a suitable device is revolved by a synchronous motor– and the operator must be sure that it is always synchronous–, which was not always the case in France during the Second World War.

3. The thermoelectric Couple

We will single out four things in this discussion: the kind, the calibration, the position, the corrections.

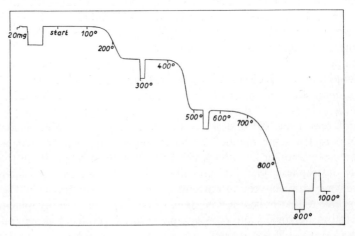

Fig. 70. Effect of a 20-mg overload along the horizontal of the pyrolysis curve of calcium oxalate.

Up till now the users of the thermobalance have used either a platinum-platinum/rhodium (10 or 20%) system or a chromium-nickel couple. The choice is determined by the cost of the apparatus and the heating range. In the former case, the calibration curve is a very open parabola up to 200°, followed by a straight line. In some thermobalances the calibration is carried out by the maker, but in most instances the operator uses the boiling points of water, aniline, the melting points of tin, lead, the boiling point of sulphur, the melting points of antimony, sodium chloride, silver, and gold. The "cold" source is not always placed in melting ice but in a metal block which follows the temperature fluctuations of the room. In order to save on platinum, this "cold" source is frequently placed near the furnace, and furthermore its temperature should be followed and noted during the heating so as to correct more or less the numbers read on the calibration curve. A correction of this kind is valid only for several degrees. It is misleading if the calibration has been made at 15° and the temperature of the room rises to 30°. Accordingly, the calibration should be made with a cold junction placed in its protective silica shield and the latter immersed half-way in the calibration material and not touching the sides of the container. In the case of gold, metal may be saved by noting the fusion temperature of a wire wound around the shield and traversed by an electric current, the whole assembly being placed in the furnace. The mechanical and electrical properties of the junction and the wires change in the course of time. Consequently, it is necessary to repeat the calibration from time to time, for example once a year, if the assembly is heated every day.

Ordinarily, the temperature of the substance itself is not determined in a thermobalance, but rather the temperature of a layer of air in the neighbourhood of the crucible. The latter temperature is of no importance in the case of automatic determinations. In some models, the hot junction comes in contact with the crucible, as is the case in the "Derivatograph". When deflection balances are employed, it is difficult to immerse the hot junction in the material contained in the crucible. This can be done in the case of the null-deflection balances, but there is danger that the platinum may be altered by the action of the products released (HF, As, etc.) or by the fused materials (KNO_3, chlorates, etc.).

We also wish to point out that if a thermostat is to be installed in connection with a programmed furnace which has previously been used for continuous heating, it is well, because of the calorific inertia of this furnace, to arrest the rise of temperature about 10° below the point which it is desired to maintain.

References p. 143

4. The nature of the substance being heated

Since 1947, we have made it a general rule to include in our published contributions the procedure employed or at least the principle underlying the method of obtaining the precipitates which we have heated. Even though the material is available in finished condition on the market, barium sulphate for example, we invariably have carried out the preparation ourselves and washed the product in accord with the directions given in the books dealing with quantitative analysis. In the case of minerals, their place of origin should be reported because of the various impurities they contain. In general, the curve obtained with a material which has been kept for some years in a cupboard or display case of a mineralogical collection will not be identical with the curve given by the same mineral recently taken from the mine. Actually, the specimens become harder in the course of time; they lose water, sulphur dioxide, arsenious anhydride, etc.; the thermal conductivity changes and the rate at which the heat of the specimen changes; generally the decomposition sets in at a higher temperature. This fact is observed all the more clearly by tracing on the same chart the thermolysis curves of a mineral and the artificial product, such as brucite and magnesia $Mg(OH)_2$. Non-superposable curves are also given by allotropic varieties; this is observed with calcite, aragonite, and vaterite. The curve is no longer the same after a precipitate has aged, when it is filtered off after 1 hour, or 24 hours, or a week. This fact is still more evident with colloidal materials and is readily accounted for in such instances. Similarly, a precipitate wetted with the wash liquids as is the case in current analytical practice does not yield the same curve as the product after it has dried to a greater or lesser extent in contact with the air. In the former case, the temperature marking the start of the horizontal of the dry compound may be recorded, which constitutes a practical problem.

If the precipitate is voluminous because of retained water, it is well to follow the advice of Newkirk[1] and employ low rates of heating. Thus if the zinc salicylaldoxime complex is subjected to a temperature rise at the rate of 380° per hour, it is impossible to find a horizontal for this complex. However, when dry, this same product yields a horizontal extending from 25° to 285° according to Rynasiewicz and Flagg[2]. A partially hydrated product (50%) shows a horizontal from 135° to 190° if the heating rate is 300° per hour. A sample, which was dry initially and then moistened to 63% gave its own horizontal, extending from 245° to 315°.

The quantity of water retained by the precipitate has an influence on the thermal conductibility of the material, and the vapour which it generates considerably alters the atmosphere in the furnace. Furthermore, all vapourization involves absorption of heat and hence has a cooling effect.

Many workers have made studies of the grain size of the materials when they were not interested in analytical problems but rather in the fate of the dry materials. The first to work along these lines was Saito[3]; he made comparative heating studies of mineral metal sulphides and arsenosulphides whose grains had been passed through sieves of various mesh sizes. Of course, his studies were guided by the functioning of pyrites roasters. The smaller the grains the lower the temperature at which the decomposition sets in.

Vallet[4] likewise was interested in the influence of the size of the grains, particularly of copper sulphate, and then, in a paper kindly written at my request, Richer and Vallet[5] compared in nitrogen and at a heating rate of 150° per hour, the threshold of decomposition of pure dry (R.P.) calcium carbonate (783°) with that of powdered calcite (802°) and of a cube of calcite weighing 352 mg (891°). This crystal retained its shape but became a little smaller, the heating being conducted in air. These workers did not give the result for precipitated still moist calcium carbonate, which provides a reason for criticism since it is rather rare for an analyst to weigh a precipitate in nitrogen and in the form of a cube.

5. The weight of the substance being heated

The weight of the substance (and therefore its volume) greatly influences the temperatures at which decomposition begins. Also, when dealing with chelates containing a metal and an organic material, which may be difficult to burn, the residual tar disappears more quickly the smaller its weight, and the level of the oxide (or final carbonate) will start at quite different temperatures. This fact is clearly shown by quinoline coordination compounds. This is why we have provided a wide margin of safety when the precipitates are heated rapidly and when the weights are between 100 and 300 mg. The heats of reaction, which may be positive or negative, alter the difference between the temperature of the sample and the furnace.

Newkirk[1] points out that the method of differential thermal analysis is possible because of the existence of this difference, and it is quite significant that high rates of heating, such as 600° per hour, are employed in this technique so as to accentuate the differential temperature.

References p. 143

Since such differential temperatures may range well above 10° at this heating rate, kinetic constants calculated from the thermogram made at such high heating rates may be unavoidably and significantly in error. When the reaction is endothermic, the effect of temperature lag and differential temperature will be additive, but when the reaction is exothermic, the effects will tend to compensate each other. That the effect of the heat of reaction can be quite large, is strikingly illustrated in the case of the oxidation of tungsten carbide in air (Newkirk[6]). As shown by the thermograms, approximately the same rate of reaction was observed with the thermobalance furnace heated to 527° and held constant,

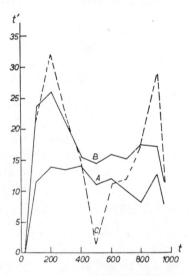

Fig. 71. Temperature lag due to sample. Decomposition of monohydrated calcium oxalate; heating rate: 600°/h. A: Crucible only; B: Crucible + 0.2 g oxalate; C: Crucible + 0.6 g oxalate.

as was observed when the furnace was heated continuously and uniformly. The effect is perhaps more clearly shown in Fig. 71 when the differential temperature between the thermocouple t and crucible t' was measured directly for the decomposition of calcium oxalate monohydrate at a high heating rate of 600° per hour for two different sample sizes. Curve A which applies to the crucible alone shows a 10–14° lag in the range 100–1000°. When a 0.2-mg sample is used, the endothermic loss of water results in a 25° lag at 200°, while the exothermic loss of carbon monoxide brings the sample nearly back to the difference observed with the crucible alone, but the lag increases again during the endothermic loss of carbon dioxide. With a 0.6-g sample, these effects are accentuated and at one point the temperature of the sample and that of the furnace are nearly the same.

A résumé of the studies by Richer and Vallet[5] will be found in Chapter Calcium; they studied the influence of the mass on the final decomposition temperature of pure R.P. pulverulent calcium carbonate and of calcite in air and in dry carbon dioxide. It should also be noted that Wendlandt, with his torsion balance which permits a maximum load of 100 mg, obtained results which frequently differed from ours with regard to the beginning of the final horizontal, for a reason which is especially accounted for by the difference in the weights of material taken.

6. The rate of heating

Most of the thermobalances are now equipped with devices allowing linear heating at speeds ranging from 0.6° to 10° per minute. However, as was just shown, the use of such high heating rates results in changes, which frequently are considerable, in the final and initial temperatures of the horizontals, and in fact certain of the horizontals may disappear or be marked only by a retrogression point or a point of inflection. The case is illustrated very well by compounds carrying much water of crystallization, such as the sulphates of the magnesium series. If it is desired to obtain one of these hydrates, the heating will therefore have to be as slow as possible, or still better, the furnace should be put on the thermostat when the expected horizontal is neared.

It was in this way that Fruchart and Michel[7] brought out the existence of the hexahydrate, tetrahydrate, dihydrate, and monohydrate of nickel sulphate by heating the heptahydrate at the rate of 0.6° per minute. In contrast, Demassieux and Malard[8] found only the horizontal of the monohydrate when they raised the temperature at the rate of 2.5° per minute.

On the other hand, Newkirk[1] demonstrated a change in the classic curve of calcium oxalate monohydrate when he reduced the heating rate from 300° to 150° per hour, namely a displacement of the horizontals toward lower temperatures. Fig. 72 shows the influence of the rate of heating on the decomposition of a sample of polystyrene in nitrogen. Sabatier[9] observed a similar effect during the dehydration of mica. When samples of a given weight were heated at rates which ranged from 12°

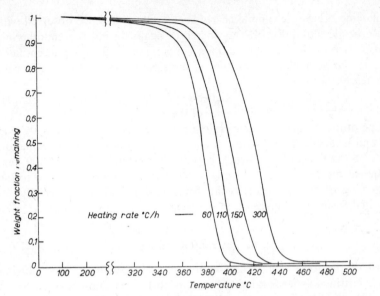

Fig. 72. Pyrolysis of polystyrene in nitrogen at different heating rates.

to 30° per hour, the threshold temperatures observed went from 735° to 815°, *i.e.* a difference of 80°.

The effect of the heating rate is very important if the thermogram is to be used for kinetic analyses.

7. The nature of the surrounding atmosphere

This topic has been studied by many workers since Saito[3] first took it up; he used hydrogen, nitrogen, oxygen, sulphur dioxide, etc. However, the substance which is being heated is not in equilibrium with the ambient atmosphere while decomposing but it perhaps is close enough to this equilibrium during the continuance of the horizontals. If the atmosphere of the furnace already contains a gas given off by the substance, an elevation of the initial and the final temperature may be expected. This was observed by Vallet[4] when he heated the pentahydrate of copper sulphate in a stream of dry air and then in a current of moist air, and likewise by Richer and Vallet[5] when they heated calcium carbonate in carbon dioxide. Accordingly, variable temperatures should be expected because even though the gases rapidly attain a temperature equilibrium, the gas which arrives cold at the bottom or at the top of the furnace must have an effect on the diffusion of the gas issuing from the crucible.

Since the vapour tension of water governs the stability of hydrates, Vallet[4] was able to make the horizontal of copper sulphate trihydrate appear or disappear at will. Similarly, Haladjian and Carpeni[10] prevented all dehydration of orthoboric acid, and they produced or prevented the formation of the horizontal of metaboric acid, etc. as they saw fit.

When Richer and Vallet[5] heated calcium carbonate in carbon dioxide and then in nitrogen, they observed that the threshold of decomposition is well defined (900°) in the first case, and rather well defined (around 500°) in nitrogen. The same thing was not found with respect to the temperature at which the decomposition stopped; in carbon dioxide this final temperature was between 914° and 1034°, whereas in nitrogen it was between 683° and 891°.

Very fortunately the differences are not this great in actual analysis, where a thin layer of the material (moist at first) is heated under atmospheric pressure. A gas or a vapour released by the substance may affect the form of the curve; in particular, the residue may be partially reduced by carbon monoxide when organic salts or chelates are heated. After this reducing gas is removed, the current of air (or oxygen) ordinarily causes the curve to ascend again. The reduction is very evident in the case of uranium (Streng's salt, see p. 208), and of cupric oxide which goes either to the metal or to cuprous oxide.

We have employed the heating of gypsum in a water vapour atmosphere, and also the heating of deutero-gypsum in an atmosphere of heavy water vapour to prepare light and heavy plasters of Paris[11].

Using a Chevenard thermobalance, and a heating rate of 300° per hour, Newkirk[1] heated two identical samples of precipitated calcium oxalate, but in one case the top of the furnace was hermetically sealed while in the other it was provided with an opening 7 mm in diameter. The two curves agreed exactly, except at the high temperature, with the horizontal of calcium oxide. The differences are still greater in the case of sodium carbonate heated in air and in carbon dioxide. At 300° per hour, the decomposition sets in at 850° in air but this threshold is not reached until 1050° in carbon dioxide (250 ml per minute). Sodium tungstate carrying 28 molecules of water yields quite different curves, at least up to 175°, depending on the degree of humidity of the ambient air. In a more recent study, Garn and Kessler[12] working in a self-produced atmosphere, investigated the thermal decomposition, in open and

References p. 143

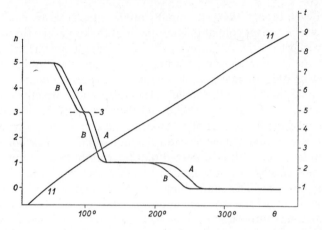

Fig. 73. Influence of dry-air flow upon dehydration of pentahydrated copper sulphate.

closed vessels, of ammonium carbonate, lead carbonate, manganese carbonate, and the dihydrate of cobalt oxalate.

8. The velocity of the gas current

This factor is of the highest importance because as we have just seen the
nature of the gas surrounding the crucible plays a principal role. Most
of the furnaces are open, and therefore they contain a mixture of air,
the gas introduced, and the gas or vapour given off by the sample on
heating. Hence it is understandable that the composition of the furnace
atmosphere changes in accord with the speed of the gaseous current,
and this results in a modification of the length and the position of the
horizontals. In one of his early experiments, Vallet[4] heated 400 mg of
identical samples of copper sulphate pentahydrate at the rate of 42°
per hour in a current of dry air, which flowed at the rate of 6.6 liters per
hour in one case and 34.0 liters per hour in the other (Fig. 73). The
appearance of the trihydrate is not influenced by the increase in the rate
of flow. The latter entails a kind of translation of the mass–temperature
curve parallel to the temperature axis in the direction of decreasing
values. This translation is less pronounced for the portion of the curve
relating to the dehydration of the pentahydrate than for the portion
relating to the trihydrate, and still less marked for this latter hydrate
than for the final portion corresponding to the dehydration of the monohydrate.

Saito[3] obtained like results, for instance when he heated manganese dioxide in hydrogen.

Allusion was made above to the ventilation of the furnace, a factor whose importance is of interest in the present connection. As a matter of fact, in the thermobalances where the furnace is open above, namely those into which the crucible is "plunged", the circulation of the gas is accomplished mostly by convection and the crucible is subjected to variable pushes and thrusts which change its actual weight and which manifest themselves by a rather regular ascent of the curve somewhat proportional to the rise in temperature. In the inverted furnaces, such as that of the Chevenard thermobalance, two or three openings are provided in the inclosing housing, and these may be closed or opened at will by the operator, who thus can vary the rate of flow of the air current. In this way, he can change the atmosphere more or less quickly and also remove the released gases at a more or less rapid rate. In our experiments, from 1936 to 1946, we have always found a perfectly horizontal straight line (in other words, one which came back to its initial position after one revolution of the cylinder) when we heated to 1050° a specimen, such as a gold or platinum crucible, which should not gain or lose weight in this interval. It was on the basis of this finding and also the entirely reliable control that we developed automatic gravimetry and noted the preliminary oxidation preceding the explosion of oxidant compounds. Since that time, various workers have published curves which are not in agreement with ours, all conditions of heating rates being respected. These workers have notably found gains and losses with specimens which we reported as having a constant weight. Struck by this lack of agreement, Simons, Newkirk and Aliferis[13] of the General Electric Company (Schenectady, N.Y.) constructed a correction curve by heating objects made of molybdenum, platinum, porcelain, etc. and they estimated that the thrust of air heated up to 1073° can cause a rise in the curve (straight line above 200°) corresponding to 5 mg in 200 mg (Fig. 74). Moreover, a correction curve presenting rather an S-shape, with a concavity between 700° and 800°, has been constructed by Mielenz, Schieltz and King[14] on the same thermobalance, the apparent gain in weight still being of the order of 5 mg in samples weighing 1.5 g up to 1000° (Fig. 75). Therefore it seemed to us that the deviations observed on heating a material maintaining a constant weight were due to the construction of the metallic housing inclosing the upper part of the furnace. Normally, this metallic cover is fitted with an opening to admit the silica sheath of the couple and it also has two vent-holes (2 mm in diameter), which

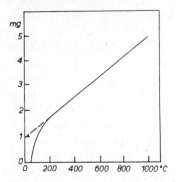

Fig. 74. General representation of apparent weight as function of temperature. (After Simons, Newkirk and Aliferis.)

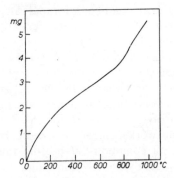

Fig. 75. Buoyancy correction for heating cycle of the thermal balance.

ordinarily are closed. In our models of the initial thermobalances, the cover was provided with openings of different diameters (used in preliminary studies) which could be closed off at will. Furthermore, these covers fitted much less closely in the upper collar of the furnace than in the recent models. The diameter of this cover moreover changes after repeated heatings over a period of about ten years.

We have provided our apparatus with metal covers with one central and six smaller openings (Fig. 76) arranged symmetrically and capable of being closed off at will. Depending on the number of openings left unstopped, it is possible to obtain with the same specimen heated in the same way, either a straight line which slopes upward (in case all of the openings are closed or opened in a disymmetrical fashion), or a horizontal straight line (in other words, returning to the starting point after one revolution of the cylinder) or a descending straight line. Therefore, it was by mere chance that our own apparatus had been provided with an adequate plug.

Later we considered the idea of providing the central opening with a progressive closing system, analogous to the iris-diaphragm of photographic apparatus, but this proved too complicated and it did not always function properly, especially when the temperature reached higher levels.

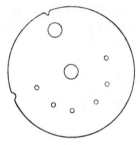

Fig. 76. The plug with six openings symmetrically placed.

Fig. 77. The plug with central opening.

We found it more simple to fit the central opening with rings of various thickness which could be adjusted like the lids of a stove or the top of a water bath (Fig. 77). The crucibles are such that the upper and the lower diameters are the same, and prior to each pyrolysis, the rings are adjusted so as to give a straight horizontal for the heated crucible, i.e. a horizontal parallel to the temperature axis[15].

Not much information is available regarding the kind of curves obtained with compounds heated under pressures above or below the atmospheric pressure. This is a very complex problem and it is changed still more if the gas or vapour released is removed or not drawn off. It seems logical to assume that by drawing off the gaseous products it will be easier to maintain the pressure constant and that the decomposition threshold should then be lowered.

It may also be expected that the influence of convection currents will be lowered by diminishing the pressure. This has been confirmed by Eyraud and Goton[17] but it has also been found that at pressures below 1 mm of mercury there arises a new source of error, which is at least of equal importance; it is due to the impact of the molecules on the sample. Eyraud has been able to remedy this in the balance which he constructed.

Some values will be given of radiometric thrusts, expressed in mg, exerted on boats about 12 mm in diameter, loaded with samples, and suspended in a quartz tube (inside diameter 17 mm) and heated in a vertical Chevenard furnace of the CT_2 type and open at both ends. The following data were recorded under a pressure of 0.1 mm Hg and at a temperature of 500°.

Location of the boat	Thrust (mg)
Upper third of the furnace	+ 2.2
Middle of the furnace	— 1.1
Lower third of the furnace	— 5.0

It has been found in addition that during the course of a continual elevation of the temperature, the boat being kept in an unchanging position, the radiometric thrust does not always change in the same direction; its value may return to zero, and then change its sign.

At 800°, and with the boat kept in a fixed position, and with the manner of heating being the same as before, the thrust attained the following values, as a function of the pressure:

Pressure (mm Hg)	5	3	1	0.1	0.001
Thrust (mg)	0.0	0.5	1.0	2.2	3.3

The results are very uncertain for bulky objects which are practically identical and which occupy the same position within the furnace. At 600° and under a pressure of 0.001 mm, the data were:

Object employed	Thrust (mg)
Quartz boat	+ 1.15
Porcelain evaporating dish	+ 2.9
Solid quartz bar	+ 1.0

Therefore the phenomenon is occasioned by the dissymmetry of the pressures exerted on the containers by the impacts of the molecules coming either from cold parts of the enclosure, or from the heated portions. An easy remedy consists in placing the boat in an intermediate enclosure which is rendered essentially isothermal by its reduced form. A small aperture is provided in the upper part to allow the suspension wires to pass through. This arrangement has been found to be very effective and satisfactory; it is necessary to use it whenever the pressure becomes less than 3 mm of mercury.

9. The nature and the shape of the crucible

A large variety of containers have been used for the material which is to be heated. These receptacles include: 3-cornered crucibles, both high and low form; cones; hemispherical dishes; square slabs, 1 cm on the side, with raised edges; crucibles whose lower part is shaped like the bottom of a bottle. These crucibles are made of various materials, ranging from platinum, gold, silver, tungsten, nickel to silica, porcelain, glass, aluminite, thoria, zirconia, and steatite. The systematic trials have not always been made with a single variable, but it may stated that there is a difference in the thresholds of decomposition in a metal crucible, which has high thermal conductivity, as opposed to crucibles made of oxides or silicates. These differences may run as high as 5–10°.

The shape of the crucible plays a dominant role even under atmospheric pressure, since the walls are heated more strongly than the center. Consequently, the thermal conductivity of the sample and the speed of diffusion of the released gases through the grains are involved. The use of a plate (slab) and a thin layer seems to provide the best solution, whereas the high-form crucible yields the worst performance in this respect. However, certain materials spatter or puff up greatly, and accordingly the use of crucibles of the high form is obligatory at times. Garn and Kessler[16] have studied these factors in detail in the case of lead carbonate and hydrated cobalt oxalate. Newkirk and Aliferis[18] have investigated the behaviour of dry sodium carbonate in crucibles made of gold, platinum, alumina, and porcelain. The influence of the kind of metal seems to be negligible.

On the other hand, the same cannot be said about the effect of covering these crucibles or leaving them open. We here again encounter a factor which was discussed previously, namely the velocity of the gas current. Everyone knows that water boils more quickly when heated in a covered casserole over an open flame. In pyrolysis, if the removal of the gas is impeded, the horizontals of the curves become longer. It is even possible to cause the appearance of horizontals. From 1936 on, we found that in the case of magnesium ammonium phosphate, there appeared to be no discontinuity between the loss of water and ammonia in an open crucible; a short horizontal or at least a break appeared as soon as the ammonia stopped coming off, provided the $MgHPO_4$ was pure, and if the same crucible was provided with a cover.

A final factor, which may not be neglected, is the attack on the crucible, *i.e.* a solid state reaction with the glaze or even the material of the crucible to produce a gas. Obviously such a reaction will result in a parasitic loss of weight. Newkirk and Aliferis[18] have furnished a striking example in the pyrolysis of sodium carbonate when a quartz crucible is employed. The attack begins at 500° in this instance.

It should be pointed out that if molybdic anhydride, MoO_3 is heated in a porcelain crucible, the bottom takes on a blue-green color. However, there is no change in weight, or more precisely, the weight does not change more than 0.1 mg in 4 g.

10. The thinness of the line

When the photographic registration is made on glazed paper, the lines, being only 0.1 mm wide, are scarcely visible. When the pen has a triangular tip made of platinum, and the paper is not very porous, and the ink contains glycerol, it is difficult to obtain a line that is less than 0.5 mm wide. The importance of this width becomes evident if calculations must be made from very small changes in weight. In certain setups, Chevenard traced the course of the curve by means of a series of points that were very close to each other and which were produced by a needle just touching the paper periodically. However, in relation to all the other factors, we consider this one to be of minor importance. Its influence especially on the sensitivity is about of the same order as that of the movements of the clock work, the play in the gears, the lag in the photoelectric cells, and the unforeseen variations in the voltage of the electric current.

11. The nature and the weighing of the residue

In most cases the final horizontal is used for calculating the molecular weights in order to learn with the aid of the ordinates of the horizontals the nature of the starting material and the intermediate products. It must be assumed that the final product has a known composition and that it does not take up water during the cooling before being weighed on an analytical balance as a control. However, many metal oxides are carbonated at the close of the operation and yield a level stretch which is quite horizontal. We have also called attention to a possible combination with the materials making up the interior walls of the crucible. Spot test analyses of the residue are therefore indicated if the calculations do not lead to a logical conclusion. It would be proper and profitable to close the first part of this text by a calculation of the relative error involved in thermo-weighings, and perhaps the requisite material for this will be at hand when the third edition is being composed. At present we have available only some partial results relating to the Chevenard thermobalance and furthermore not all of the ten factors which we have just discussed have been taken into consideration. We will do no more than cite the very laudable attempts by Simons, Newkirk and Aliferis[13] with the pen-recording apparatus, and those of Claisse, East and Abesque[19], and Mielenz, Schieltz and King[14] with the thermobalance provided with photographic registration.

REFERENCES

[1] A. E. NEWKIRK, *Anal. Chem.*, 32 (1960) 1558.
[2] J. RYNASIEWICZ AND J. F. FLAGG, *Anal. Chem.*, 26 (1954) 1506.
[3] H. SAITO, *Sci. Repts. Tôhoku Imp. Univ.*, 16 (1927) 1.
[4] P. VALLET, *Compt. rend.*, 198 (1934) 1860; *Thesis*, Paris, 1936; *Ann. Chim.*, 7 (1937) 298.
[5] A. RICHER AND P. VALLET, *Bull. Soc. chim. France*, (1953) 148; *Compt. rend.*, 249 (1959) 680
[6] A. E. NEWKIRK, *J. Am. Chem. Soc.*, 77 (1955) 4521.
[7] R. FRUCHART AND A. MICHEL, *Compt. rend.*, 246 (1958) 1222.
[8] N. DEMASSIEUX AND C. MALARD, *Compt. rend.*, 245 (1957) 1514.
[9] G. SABATIER, *J. chim. phys.*, 52 (1955) 60.
[10] J. HALADJIAN AND J. CARPENI, *Bull. soc. chim. France* (1956) 1679.
[11] C. DUVAL, J. LECOMTE AND C. PAIN, *Compt. rend.*, 237 (1953) 238.
[12] P. D. GARN AND J. E. KESSLER, *Anal. Chem.*, 32 (1960) 1563.
[13] E. L. SIMONS, A. E. NEWKIRK AND I. ALIFERIS, *General Electric Co.* (Schenectady), Report No. 56-RL-1522.
[14] R. C. MIELENZ, N. C. SCHIELTZ AND M. E. KING, *Proc. 2nd Natl. Conf. Clays, and Clay Minerals, Missouri*, 1953, Publ. No. 327 of the *Natl. Acad. Sci.*, (1954) 285.
[15] C. DUVAL, *Mikrochim. Acta*, (1958) 705.
[16] P. D. GARN AND J. E. KESSLER, *Anal. Chem.*, 32 (1960) 1900.
[17] C. EYRAUD AND R. GOTON, *Bull. Soc. chim. France*, (1953) 1009.
[18] A. E. NEWKIRK AND I. ALIFERIS. *General Electric Co.; Phys. Chem. Research*, No. 405; *Anal. Chem.*, 30 (1958) 982.
[19] F. CLAISSE, F. EAST AND F. ABESQUE, *The Use of the Thermobalance in Analytical Chemistry*, Depart. of Mines, Quebec, Canada, 1954.

APPENDIX TO PART ONE

SUGGESTED METHODS FOR AUTOMATIC INORGANIC ANALYSIS

The following table brings together the methods which we suggest for automatic inorganic analysis (plutonium and americium excepted). The precipitant is given, also the weighing form, and the limits of the horizontal stretch between which the precipitate has the formula given in the table. This precipitate is assumed to be slightly moist and taken in a thin layer; the weight is in the range 100–500 mg, ordinarily 250 mg. These limits, especially the upper limit, must be very closely observed. By way of sheer precaution, we have frequently indicated a lower temperature at which a loss of weight begins to appear. Usually, when regulating the furnace, we employ a temperature that is half-way between those of the two extremities of the horizontal. Of course, it is recommended that the determination be made by means of a pen recording. We wish to emphasize once again that our results are valid only if the crucible is not taken out of the furnace. The largest number of determinations in a single normal working day was obtained in the series determination of uranium with oxine.

Precipitating reagent	Form weighed	Temperature limits (°C)	See p.
ALUMINIUM			
Gaseous ammonia	Al_2O_3	>475	227
Bromine	Al_2O_3	>280	227
Sodium hydrogen sulphite	Al_2O_3	>412	227
Potassium cyanate	Al_2O_3	>510	227
Hydrazine	$2Al_2O_3 \cdot H_2O$	>350	227
Oxine	$Al(C_9H_6ON)_3$	102-220	234
5,7-Dibromo-oxine	$Al(C_9H_4ONBr_2)_3$	110-156	234
AMMONIUM			
Iodic acid	$NH_4H_2(IO_3)_3$	<146	186
Hydrochloroplatinic acid	$(NH_4)_2[PtCl_6]$	<181	189
ANTIMONY			
Ammonium thiocyanate	Sb_2S_3	170-290	513
Pyrogallol	$(SbOH)C_6H_3O_3$	74-140	514
Gallic acid	$(SbOH)[C_6H_2(OH)_3 \cdot CO_2]$	114-163	515
Cobaltic dichlorodiethylene- diamine dichloride	$[Coen_2Cl_2][SbCl_6]$	90-153	516
ARSENIC			
Potassium xanthate	As_2S_3	208-273	426
Hydrogen sulphide	As_2S_3	200-275	425
idem	As_2S_5	78-245	426
Silver nitrate + thallium acetate	Ag_2TlAsO_4	20-846	427
Lead nitrate	$PbHAsO_4$	81-269	428
Lead nitrate	$Pb_2As_2O_7$	320-950	428
BARIUM			
Sulphuric acid	$BaSO_4$	180 or 780-1000	530
Potassium iodate	$Ba(IO_3)_2$	320-475	530
Potassium chromate	$BaCrO_4$	<180	534
Ammonium oxalate	$BaC_2O_4 \cdot H_2O$	< 76	535
idem	BaC_2O_4	110-346	535
idem	$BaCO_3$	>476	535
BERYLLIUM			
Disodium hydrogen phosphate	$Be_2P_2O_7$	>640	173
BISMUTH			
Formaldehyde	Bi	73-150	635
Hypophosphorous acid	Bi (impure)	48-230	636
Lead	Bi	65-113	636
Ammonium carbonate	$(BiO)_2CO_3$	68-308	636
Ammonium chloride	$Bi(OH)_2Cl$	<258	637
idem	$BiOCl$	328-805	637
Potassium iodide	$BiOI$	<307	637
Sodium sulphite	$(BiO)_2SO_4$	475-946	638

Precipitating reagent	Form weighed	Temperature limits (°C)	See p.
Sulphuric acid	$Bi_2(SO_4)_3$	236-405	639
Sulphuric acid + water	$(BiO)_2SO_4$	854-962	639
Diammonium hydrogen phosphate	$BiPO_4$	379-950	640
Arsenic acid	$BiAsO_4$	47-400	640
Cobaltic diethylenediamine chloride + potassium iodide	$[Coen_2][BiI_4]_2I$	<188	641
Ammonium formate	$(BiO)_2CO_3$	45-155	641
idem	Bi_2O_3	>370	641
Thionalide	$Bi(C_{12}H_{10}ONS)_3$	45-134	644
3-Phenyl-5-thio-1,3,4-thiodiazol-2-thione	$Bi(C_8H_5N_2S_3)_3 \cdot \frac{1}{2}H_2O$	40-150	644
Phenylarsonic acid	$C_6H_6O_4AsBi$	60-300	644

BORON

Nitron fluoroborate	$C_{20}H_{16}N_4 \cdot HBF_4$	50-197	177

BROMINE

Silver nitrate	$AgBr$	70-946	432

CADMIUM

Potassium hydroxide	$Cd(OH)_2$	90-170	492
idem	CdO	370-880	492
Hydrogen sulphide	CdS	218-420	493
Sulphuric acid	$CdSO_4 \cdot H_2O$	69-120	493
idem	$CdSO_4$	320-900	493
Potassium iodide + hydrazine	$[Cd(N_2H_4)_2]I_2$	70-166	494
Disodium hydrogen phosphate	$CdNH_4PO_4 \cdot H_2O$	90-122	495
idem	$Cd_2P_2O_7$	>580	495
Ammonium molybdate	$CdMoO_4$	82-250	495
Ammonium chloride + pyridine	$[Cd(C_5H_5N)_2]Cl_2$	< 70	495
idem	$[Cd(C_5H_5N)]Cl_2$	107-177	495
idem	$CdCl_2$	270-610	495
Thiocyanate + pyridine	$[Cd(C_5H_5N)](SCN)_2$	77-101	496
Brucine + potassium bromide	$[C_{21}H_{21}O_2N_2(CH_3O)_2]_2$ $CdBr_4$	120-250	497
Reinecke's salt + thiourea	$[Cd(CNS_2H_4)_2]$ $[Cr(NH_3)_2$ $(SCN)_4]$	<167	497
Anthranilic acid	$Cd(NH_2-C_6H_4-CO_2)_2$	60-222	498
Oxine	$Cd(C_9H_6ON)_2$	280-384	499
Quinoline-8-carboxylic acid	$Cd(C_{10}H_6O_2N)_2$	90-260	499
Quinaldic acid	$Cd(C_{10}H_6O_2N)_2$	66-195	500
2-Phenyloxine	$Cd(C_6H_5-C_9H_5ON)_2$	60-335	501

CALCIUM

Ammonium oxalate	$CaC_2O_4 \cdot H_2O$	<100	281
idem	CaC_2O_4	226-400	281
idem	$CaCO_3$	420-660	281

Precipitating reagent	Form weighed	Temperature limits (°C)	See p.
Sodium sulphite	$CaSO_3$	385-410	249
Sulphuric acid	$CaSO_4$	>105	272
Iodic acid	$Ca(IO_3)_2$	160-540	271
Ammonium molybdate	$CaMoO_4$	230-1000	280
Sodium tungstate	$CaWO_4$	>400	280
Ammonium ferrocyanide	$Ca(NH_4)_2[Fe(CN)_6]$	60-280	280
Potassium ferrocyanide	$CaK_2[Fe(CN)_6]$	60-325	280
Oxine	$Ca(C_9H_6ON)_2$	90-290	283
Loretine (Ferron)	$Ca[C_9H_4NI(OH)SO_3]_2$ ·$3H_2O$	60-130	283
idem	$Ca[C_9H_4NI(OH)SO_3]_2$	165-235	283
Picrolonic acid	$Ca(C_{10}H_7O_5N_4)_2 \cdot 7H_2O$	< 40	283

CERIUM

Aqueous ammonia + hydrogen peroxide	CeO_2	450-814	542
Sulphuric acid	$Ce(SO_4)_2$	277-845	543
Oxalic acid	CeO_2	450-880	544
Oxine	$Ce(C_9H_6ON)_4$	128-233	545

CESIUM

Perchloric acid	$CsClO_4$	70-610	524
Sodium cobaltihexanitrite	$Cs_3[Co(NO_2)_6]$	at 110	525
idem	$CoO + 3CsNO_3$	200-495	526
Stannic chloride	$Cs_2[SnCl_6]$	130-344	526
Hexachloroplatinic acid	$Cs_2[PtCl_6]$	200-410	526
Sodium tetraphenylboride	$Cs[B(C_6H_5)_4]$	60-210	527

CHLORINE

Silver nitrate (from Cl⁻)	$AgCl$	70-600	253
Mercurous nitrate (from Cl⁻)	Hg_2Cl_2	<130	253
Lead nitrate (from a chlorite)	$Pb(ClO_2)_2$	< 77	254
Nitron (from a perchlorate)	$C_{20}H_{17}N_4 \cdot ClO_4$	40-232	254

CHROMIUM

Gaseous ammonia	$Cr_2O_3 \cdot 3H_2O$	440-475	302
Thiosemicarbazide	Cr_2O_3	>475	303
Potassium cyanate	Cr_2O_3	>475	303
Silver nitrate	Ag_2CrO_4	92-812	308
Mercury (I) nitrate	Hg_2CrO_4	52-256	309
Thallium (I) nitrate	Tl_2CrO_4	100-745	614
Lead nitrate	$PbCrO_4$	90-900	309
Oxine	$Cr(C_9H_6ON)_3$	70-156	310

COBALT

Yellow mercury (II) oxide	Co_3O_4	>475	337
Potassium peroxydisulphate	Co_3O_4	371-950	337

Precipitating reagent	Form weighed	Temperature limits (°C)	See p.
Sulphuric acid	$CoSO_4$	600-812	339
Potassium nitrite	$K_3[Co(NO_2)_6]$	43-160	343
Potassium cyanide	$Ag_3[Co(CN)_6]$	100-252	344
Thiocyanate + hydrazine	$[Co(N_2H_4)_2](SCN)_2$	< 80	344
Mercuric thiocyanate	$Co[Hg(SCN)_4]$	50-200	345
Potassium oxalate	Co_3O_4	>288	347
Sodium anthranilate	$Co(NH_2-C_6H_4-CO_2)_2$	108-290	347
5-Bromoanthranilic acid	$Co(NH_2-C_6H_3Br-CO_2)_2$	<194	348
3-Aminonaphthoic acid	$Co(NH_2-C_{10}H_6-CO_2)_2$	<176	348
α-Nitroso-β-naphthylamine	$Co(C_{10}H_7ON_2)_3$	80-179	350
α-Nitroso-β-naphthol (with Co $^{3+}$)	$Co(NO-C_{10}H_6O)_3$	81-140	349
β-Nitroso-α-naphthylamine	$Co(C_{10}H_7ON_2)_3$	72-218	350
Oxine	$Co(C_9H_6ON)_2$	115-295	351
2-Methyloxine	$Co(CH_3-C_9H_5ON)_2$	200-280	351
2-Phenyloxine	$Co(C_6H_5-C_9H_5ON)_2$	105-195	351
Quinoline-8-carboxylic acid	$Co(C_{10}H_6O_2N)_2$	164-336	352
	COPPER		
Hydrazine	Cu	50-97	373
Hypophosphorous acid	Cu	<194	374
Hydroxylamine	Cu_2O	<145	375
idem	CuO	350-950	376
Sodium arsenite	Cu_2O	150-288	376
Hexamine	$CuO·2H_2O$	< 90	377
Iodic acid	$Cu(IO_3)_2$	210-352	379
Sulphuric acid	$CuSO_4·H_2O$	153-250	382
idem	$CuSO_4$	303-624	382
Hydrazinium sulphate	$CuSO_4·(N_2H_5)_2SO_4$	<133	383
Alkali thiocyanate	CuCNS	<300	384
Reinecke's salt	$Cu[Cr(NH_3)_2(SCN)_4]$	<157	385
Oxalic acid	CuC_2O_4	100-270	386
Salicylaldehyde	$CuC_{14}H_{10}O_2N_2$	<105	387
Anthranilic acid	$Cu(NH_2-C_6H_4-CO_2)_2$	<225	387
5-Bromoanthranilic acid	$Cu(NH_2-C_6H_3Br-CO_2)_2$	<204	387
Quinaldinic acid	$Cu(C_{10}H_6O_2N)_2·H_2O$	<105	394
idem	$Cu(C_{10}H_6O_2N)_2$	130-273	394
5,6-Benzoquinaldinic acid	$Cu(C_{14}H_8O_2N)_2$	70-143	394
Cupferron	$Cu(C_6H_5O_2N_2)_2$	<118	388
Neocupferron	$Cu(C_{10}H_7O_2N_2)_2$	<100	389
α-Nitroso-β-naphthol	$Cu(C_{10}H_6O_2N)_2$	62-172	389
Salicylaldoxime	$Cu(C_7H_6O_2N)_2$	<150	390
Benzoinoxime	$CuC_{14}H_{11}O_2N$	60-143	390
5-Chlorosalicylhydrazide	$Cu(C_7H_6O_2N_2Cl)_2$	140-200	391
Dilituric acid	$Cu[C_4H_2O_3N_2(NO_2)_2]$	200-230	391
Oxine	$Cu(C_9H_6ON)_2$	66-270	392
5,7-Dichloro-oxine	$Cu(C_9H_4Cl_2ON)_2$	63-177	392
5,7-Dibromo-oxine	$Cu(C_9H_4Br_2ON)_2$	170-200	392

Precipitating reagent	Form weighed	Temperature limits (°C)	See p.
Quinoline-8-carboxylic acid	$Cu(C_{10}H_6O_2N)_2$	110-221	394
2-Methyloxine	$Cu(CH_3-C_9H_5NO)_2$	140-210	393
2-Phenyloxine	$Cu(C_6H_5-C_9H_5NO)_2$	90-215	393
Pyridine-2-carboxylic acid	$Cu(C_6H_4O_2N)_2$	120-300	395
Biguanide sulphate	$[Cu(C_2H_7N_5)_2]SO_4$	94-146	395
Thionalide	$Cu(C_{12}H_{10}ONS)_2 \cdot H_2O$	81-121	395
idem	$Cu(C_{12}H_{10}ONS)_2$	148-167	395
Mercaptobenzothiazole	$Cu(C_7H_4NS_2)_2$	<145	396
Potassium dichromate + pyridine	$[Cu(C_5H_5N)_4]Cr_2O_7$	< 64	398
Iodomercurate(II) + pn	$[Cupn_2]HgI_4$	<157	399
Iodomercurate (II) + phen	$[Cu(phen)_2]HgI_4$	64-90	399
Thiocyanate + isoquinoline	$[Cu(C_9H_7N)_2](SCN)_2$	<103	399

CYANATE

Silver nitrate	AgCNO	<138	180
Semicarbazide	$NH_2-CO-NH-NH-$ $-CO-NH_2$	< 90	180

CYANIDE

Silver nitrate	AgCN	<237	179

DYSPROSIUM

Oxalic acid	$Dy_2(C_2O_4)_3 \cdot 4H_2O$	45-140	563
idem	$Dy_2(C_2O_4)_3 \cdot 2H_2O$	140-220	563

ERBIUM

Oxalic acid	$Er_2(C_2O_4)_3 \cdot 2H_2O$	175-265	565

EUROPIUM

Aqueous ammonia	Eu_2O_3	>650	556
Oxalic acid	$Eu_2(C_2O_4)_3$	280-315	557

FERROCYANIDE

Benzidine	$3Bzd \cdot H_4[Fe(CN)_6]$	< 52	180

FLUORINE

Bismuth nitrate	BiF_3	50-93	192
Triphenyltin chloride	$(C_6H_5)_3SnF$	<158	193
Lead chloride	PbClF	66-538	193
Uranium (IV) sulphate	U_3O_8	812-945	193

GADOLINIUM

Aqueous ammonia	Gd_2O_3	>650	558
Oxalic acid	$Gd_2(C_2O_4)_3$	300-350	559

GALLIUM

Sodium sulphite	Ga_2O_3	>813	418
Camphoric acid	$Ga_2(CO_2-C_8H_{14}CO_2)_3$	94-125	419

Precipitating reagent	Form weighed	Temperature limits (°C)	See p.
	GERMANIUM		
Molybdate + oxine	$GeO_2 \cdot (MoO_3)_{12} \cdot (C_9H_7ON)_4$	50-115	422
idem	$GeO_2 + 12MoO_3$	500-920	422
	GOLD		
Citarin	Au	45-965	597
Thiophenol	AuC_6H_5S	60-157	598
	HAFNIUM		
Aqueous ammonia	HfO_2	>350	571
Ammonia (gaseous)	HfO_2	>350	571
Selenious acid	$Hf(SeO_3)_2$	<230	572
Mandelic acid	$Hf(C_6H_5-CHOH-CO_2)_4$	<240	572
	HOLMIUM		
Oxalic acid	$Ho_2(C_2O_4)_3 \cdot 2H_2O$	200-240	564
idem	$Ho_2(C_2O_4)_3$	400-500	564
	INDIUM		
Potassium cyanate	In_2O_3	>475	505
Hydrogen sulphide	In_2S_3	94-221	505
Luteocobaltic chloride	$[Co(NH_3)_6][InCl_6]$	< 95	505
Oxine	$In(C_9H_6ON)_3$	100-285	506
Diethyldithiocarbamate	$In[CS_2 \cdot N(C_2H_5)_2]_3$	100-210	506
	IODINE		
Silver nitrate (on I⁻)	AgI	60-900	521
Copper sulphate	Cu_2I_2	60-296	521
Thallium nitrate	TlI	60-473	522
Palladium chloride	PdI_2	85-365	521
Silver nitrate (on IO_3^-)	$AgIO_3$	80-410	522
	IRIDIUM		
Formic acid	Ir	<880	589
	IRON		
Gaseous ammonia	Fe_2O_3	>474	323
Aqueous ammonia + ethanol	Fe_2O_3	>370	324
Ammonium nitrite	Fe_2O_3	>343	324
Aqueous ammonia + hydrazine	Fe_2O_3	>346	324
Ammonium formate	Fe_2O_3	>296	330
Ammonium acetate	Fe_2O_3	>242	330
Cupferron	$Fe[C_6H_5(NO)_2]_3$	<100	330
Neocupferron	$Fe[C_{10}H_7(NO)_2]_3$	<111-113	331
Oxine	$Fe(C_9H_6ON)_3$	<284	331
5,7-Dibromo-oxine	$Fe(C_9H_4Br_2ON)_3$	72-95	332
Dimine	Empirical factor $P \times 0.039$	< 95	333

Precipitating reagent	Form weighed	Temperature limits (°C)	See p.
	LANTHANUM		
Oxalic acid	$La_2O_3 \cdot CO_2$	700-800	539
Oxine	$La(C_9H_6ON)_3$	220-390	540
	LEAD		
Hydrogen peroxide	PbO_2	100-120	619
Iodic acid	$Pb(IO_3)_2$	<400	623
Sodium periodate	$Pb_3(IO_5)_2$	141-280	623
Hydrogen sulphide	PbS	98-108	624
Sulphuric acid	$PbSO_4$	271-959	625
Potassium sulphate	$PbSO_4 \cdot K_2SO_4$	40-906	625
Disodium hydrogen phosphate	$Pb_2P_2O_7$	>355	626
Sodium carbonate	$PbCO_3$	<142	626
Oxalic acid	PbC_2O_4	50-300	628
Sodium phthalate	$Pb(CO_2)_2C_6H_4$	228-320	629
Gallic acid	$Pb_2CO_2 \cdot C_6H_2O_3$	<152	629
Sodium anthranilate	$Pb(C_7H_6O_2N)_2$	<198	630
Salicylaldoxime	$PbC_7H_5O_2N$	45-180	631
Picrolonic acid	$Pb(C_{10}H_7O_5N_4)_2$	58-112	632
Thionalide	$Pb(C_{12}H_{10}NS)_2$	71-134	630
Mercaptobenzothiazole	$Pb(C_7H_4NS_2)_2$	<120	630
Mercaptobenzimidazole	$PbOH \cdot C_7H_5N_2S$	97-172	630
	LITHIUM		
Zinc acetate + uranyl acetate	$Li[Zn(UO_2)_3(CH_3CO_2)_9]$	125-162	168
idem	$ZnU_2O_7 \cdot \frac{1}{2}Li_2U_2O_7$	497-1000	168
	LUTETIUM		
Oxalic acid	$Lu_2(C_2O_4)_3 \cdot 2H_2O$	190-315	570
	MAGNESIUM		
Disodium hydrogen phosphate	$Mg_2P_2O_7$	>477	218
Oxalic acid	$MgC_2O_4 \cdot 2H_2O$	<176	223
idem	MgC_2O_4	233-397	223
idem	MgO	>480	223
Orange IV	$Mg(C_{18}H_{14}O_3N_3S)_2$	103-358	—
Oxine	$Mg(C_9H_6ON)_2$	206-300	223
8-Hydroxyquinaldine	$Mg(C_{10}H_8ON)_2$	88-300	224
Quinaldinic acid	$Mg(C_{10}H_6O_2N)_2 \cdot 2H_2O$	60-145	224
idem	$Mg(C_{10}H_6O_2N)_2$	203-330	224
	MANGANESE		
Sulphuric acid	$MnSO_4 \cdot H_2O$	37-150	316
idem	$MnSO_4$	200-940	316
Disodium hydrogen phosphate	$MnNH_4PO_4 \cdot H_2O$	<120	317
Potassium oxalate	MnC_2O_4	100-214	319
Sodium anthranilate	$Mn(C_9H_6ON)_2$	117-250	319

Precipitating reagent	Form weighed	Temperature limits (°C)	See p.
Pyridine + thiocyanate	$Mn(C_5H_5N)_2(SCN)_2$	84-92	318
Oxine	$Mn(C_9H_6ON)_2 \cdot 2H_2O$	<117	319
idem	$Mn(C_9H_6ON)_2$	188-320	319
	MERCURY		
Hypophosphorous acid	Hg	< 71	601
Nitric acid	HgO	100-200	601
Hydrochloric acid	Hg_2Cl_2	<130	602
Potassium iodide	HgI_2	45-88	602
Potassium iodate	$Hg_2(IO_3)_2$	<175	602
Sodium thiosulphate	HgS	75-220	603
Cobalt thiocyanate	$Co[Hg(SCN)_4]$	50-200	604
Potassium chromate	Hg_2CrO_4	52-256	604
Ammonium bichromate + pyridine	$[Hg(C_5H_5N)_2]Cr_2O_7$	56-66	605
Reinecke's salt	$Hg[Cr(NH_3)_2(SCN)_4]$	77-158	605
Ammoniacal cadmium iodide	$[Cd(NH_3)_4][HgI_3]_2$	< 69	606
Potassium iodide + copper sulphate + propylenediamine	$[Cupn_2][HgI_4]$	<157	606
Oxalic acid	$Hg_2C_2O_4$	<100	606
Sodium anthranilate	$Hg(C_7H_6O_2N)_2$	<113	607
Dithian	$HgCl_2 \cdot C_4H_8S_2$	<100	607
Copper dibiguanide chloride + potassium iodide	$[Cu(C_2H_7N_5)_2][HgI_4]$	60-175	607
Thionalide	$Hg(C_{12}H_{10}ONS)_2$	90-169	608
2-Phenyloxine	$Hg(C_6H_5-C_9H_5ON)_2$	70-275	608
	MOLYBDENUM		
Calcium chloride	$CaMoO_4$	>230	464
Barium chloride	$BaMoO_4$	>320	464
Lead nitrate	$PbMoO_4$	>505	464
Cadmium chloride	$CdMoO_4$	82-250	464
Silver nitrate	Ag_2MoO_4	89-250	464
Oxine	$MoO_2(C_9H_6ON)_2$	160-326	466
5,7-Dibromo-oxine	$MoO_2(C_9H_4Br_2ON)_2$	105-165	467
2-Phenyloxine	$MoO_2(C_6H_5-C_9H_5ON)_2$	100-305	467
	NEODYMIUM		
Aqueous ammonia	Nd_2O_3	>608	549
Oxalic acid	$Nd_2(C_2O_4)_3$	209-350	550
	NICKEL		
Sodium hydroxide	NiO	>250	356
Bromine + potassium hydroxide	NiO	>345	357
Iodide-iodate mixture	$Ni(OH)_2$	100-260	357
Aqueous ammonia (with the sulphate)	$NiSO_4 \cdot 8NiO \cdot 16H_2O$	91-172	358

Precipitating reagent	Form weighed	Temperature limits (°C)	See p.
Aqueous ammonia (with the sulphate)	$NiSO_4 \cdot 8NiO$	200-815	358
Sodium carbonate	$NiCO_3 \cdot NiO$	101-217	361
Thiocyanate + pyridine	$[Ni(C_5H_5N)_4](SCN)_2$	< 63	361
idem	$[Ni(C_5H_5N)_3](SCN)_2$	110-130	361
Anthranilic acid	$Ni(NH_2-C_6H_4-CO_2)_2$	<307	363
5-Bromoanthranilic acid	$Ni(NH_2-C_6H_3Br-CO_2)_2$	60-218	363
3-Aminonaphthoic acid	$Ni(NH_2-C_{10}H_6-CO_2)_2$	<154	363
Oxine	$Ni(C_9H_6ON)_2 \cdot 2H_2O$	<120	364
idem	$Ni(C_9H_6ON)_2$	160-232	364
2-Methyloxine	$Ni(CH_3-C_9H_5ON)_2 \cdot H_2O$	65-215	365
idem	$Ni(CH_3-C_9H_5ON)_2$	245-300	365
2-Phenyloxine	$Ni(C_6H_5-C_9H_5ON)_2$	115-310	366
Pyridine-2-carboxylic acid	$Ni(C_6H_4O_2N)_2 \cdot 4H_2O$	145-230	366
Quinoline-8-carboxylic acid	$Ni(C_{10}H_6O_2N)_2$	200-350	365
Quinoline-2-carboxylic acid	$Ni(C_{10}H_6O_2N)_2$	180-330	366
Dilituric acid	$Ni[C_4H_2O_3N_2(NO_2)_2]$	180-240	366
β-Nitroso-α-naphthylamine	$Ni(C_{10}H_7ON_2)_2$	80-220	367
Dicyanodiamidine	$Ni(C_2H_5ON_4)_2$	123-240	367
Nitroaminoguanidine	$NiO(CH_5O_2N_5)_2$	57-118	367
Dimethylglyoxime	$Ni(C_4H_7O_2N_2)_2$	79-172	368
Diaminoglyoxime	$Ni(C_2H_5O_2N_2)_2 \cdot 2H_2O$	58- 78	368
Nioxime	$Ni(C_6H_9O_2N_2)_2$	60- 85	

NIOBIUM

Cupferron	Nb_2O_5	>650	458

NITRATE

Cinchonamine	$C_{19}H_{24} ON_2 \cdot HNO_3$	39-196	185
Di-(α-methylnaphthyl)-amine	$C_{22}H_{19}N \cdot HNO_3$	60-234	185

OSMIUM

Oxidants	OsO_4	<190	588

PALLADIUM

Carbon monoxide	Pd	<384 (in air)	475
Hydrazinium sulphate	Pd	<384 (in air)	476
Ethylene	Pd	<384	476
Sodium formate	Pd	<242	476
Potassium iodide	PdI_2	84-365	476
Hydrogen sulphide	PdS	150-179	476
Acetylene	PdC_2	71-109	477
Dimethylglyoxime	$Pd(C_4H_7O_2N_2)_2$	45-171	477
Methylbenzoylglyoxime	$Pd(C_{10}H_9O_3N_2)_2$	52-150	478
Salicylaldoxime	$Pd(C_7H_5O_2N)_2$	93-197	478
β-Furfuraldoxime	$Pd(C_4H_3O-CHNOH)_2$	51-156	478
Cyclohexanedionedioxime	$Pd(C_6H_9O_2N_2)_2$	50-169	478
α-Nitroso-β-naphthol	$Pd(NO-C_{10}H_6O)_2$	50-245	479
6-Nitroquinoline	$Pd(C_9H_6O_2N_2)_2$	51-198	479

Precipitating reagent	Form weighed	Temperature limits (°C)	See p.
o-Phenanthroline	$PdCl_2 \cdot C_{12}H_8N_2$	50-390	480
Thionalide	$Pd(C_{10}H_7-NH-CO-CH_2S)_2$	50-195	480
2-Mercaptobenzothiazole	$Pd(C_7H_4NS_2)_2$	50-388	480
Picolinic acid	$Pd(C_6H_4O_2N)_2$	60-380	481
PHOSPHORUS			
Mercuric chloride	Hg_2Cl_2	100-130	242
Zinc sulphate	$Zn_2P_2O_7$	>610	242
Barium acetate	$BaHPO_4$	126-384	243
idem	$Ba_2P_2O_7$	>520	243
Magnesia mixture	$Mg_2P_2O_7$	>477	244
Basic bismuth nitrate	$BiPO_4$	119-961	243
Ferric chloride + acetate	$FePO_4$	>531	243
Zirconyl chloride	$(ZrO)_2P_2O_7$	>600	243
Thallium (I) acetate + silver nitrate	Ag_2TlPO_4	20-720	244
Molybdic reagent	$(NH_4)_3PO_4(MoO_3)_{12}$	180-410	244
Oxine	$3(C_9H_7ON)H_7P(Mo_2O_7)_6 \cdot 2H_2O$	176-225	245
Quinoline	$3(C_9H_7ON) \cdot H_3PO_4 \cdot 12MoO_3$	155-370	246
PLATINUM			
Reducing agents	Pt	100-600	591
Ammonium salt	$(NH_4)_2[PtCl_6]$	<181	592
Potassium salt	$K_2[PtCl_6]$	54-270	592
Rubidium salt	$Rb_2[PtCl_6]$	70-674	592
Cesium salt	$Cs_2[PtCl_6]$	200-409	592
Thallium salt	$Tl_2[PtCl_6]$	65-155	592
Thiophenol	$Pt(C_6H_5S)_2$	230-300	593
POTASSIUM			
Perchloric acid	$KClO_4$	73-653	255
Periodic acid	KIO_4	82-287	257
Molybdic reagent	$K_3(MoO_3)_{12}PO_4$	96-798	—
Cupric nitrate + lead nitrate + nitrous fumes	$K_2Pb[Cu(NO_2)_6]$	87-156	—
Perrhenic acid	$KReO_4$	54-220	265
Hydrochloroplatinic acid	$K_2[PtCl_6]$	54-270	265
Lithium cobaltihexanitrite	$K_3[Co(NO_2)_6] \cdot 2H_2O$	60-220	264
idem	$CoO + 3KNO_3$	300-460	264
Tartaric acid	$HCO_2-(CHOH)_2-CO_2K$	54-200	267
6-Chloro-5-nitrotoluene-m-sulphonic acid	$KSO_3-C_6H_2Cl(NO_2)-CH_3$	54-278	267
Dilituric acid	$KC_4H_2O_5N_3$	66-296	—
Dipicrylamine	$KC_{12}H_4O_{12}N_7$	50-220	268
Sodium tetraphenylboride	$K[B(C_6H_5)_4]$	50-265	268

Precipitating reagent	Form weighed	Temperature limits (°C)	See p.
	PRASEODYMIUM		
Aqueous ammonia	Pr_6O_{11}	>676	546
Oxalic acid	$Pr_2(C_2O_4)_3$	346-500	547
	RHENIUM		
Potassium salt	$KReO_4$	54-220	585
Tetraphenylarsonium chloride	$[As(C_6H_5)_4]ReO_4$	126-183	586
	RHODIUM		
Reducing agent; heat	Rh	<662	472
Thionalide	$Rh(C_{10}H_7\text{-NH-CO-CH}_2S)_3$	80-250	473
2-Mercaptobenzoxazole	$Rh(C_7H_4OSN)_3$	92-155	473
	RUBIDIUM		
Perchloric acid	$RbClO_4$	95-343	433
Hydrochloroplatinic acid	$Rb_2[PtCl_6]$	70-674	435
Dipicrylamine	$[(NO_2)_3C_6H_2]_2NRb$	70-274	436
6-Chloro-5-nitro-*m*-toluenesulphonic acid	$RbSO_3\text{-}C_6H_2Cl(NO_2)\text{-}CH_3$	70-220	436
Sodium tetraphenylboride	$Rb[B(C_6H_5)_4]$	70-240	436
	RUTHENIUM		
Reducing agent; heat	Ru	<850	470
Thionalide	$Ru(C_{10}H_7\text{-NH-CO-}\text{-CH}_2S)_3$	60-175	470
	SAMARIUM		
Aqueous ammonia	Sm_2O_3	>813	553
Oxalic acid	$Sm_2(C_2O_4)_3$	344-360	554
	SCANDIUM		
Aqueous ammonia	Sc_2O_3	>542	286
Oxalic acid	$Sc_2(C_2O_4)_3 \cdot 10H_2O$	67- 89	287
	SELENIUM		
Ammonium sulphite	Se	<370	430
Stannous chloride	Se	<272	430
Hydrogen peroxide + lead salt	$PbSeO_4$	15-330	431
	SILICON		
Potassium fluoride	$K_2[SiF_6]$	60-410	238
Oxine (Ox)	$SiO_2 \cdot 12MoO_3 \cdot 4\ Ox \cdot 2H_2O$	160-200	239
idem	$SiO_2 \cdot 12MoO_3$	593-813	239
Hexamine (Hex)	$SiO_2 \cdot 12MoO_3 \cdot \cdot 4Hex \cdot 4H_2O$	80-150	239
idem	$SiO_2 \cdot 12MoO_3$	473-840	239

Precipitating reagent	Form weighed	Temperature limits (°C)	See p.
Pyramidone (Pyr)	$SiO_2 \cdot 12MoO_3 \cdot$ $\cdot 3Pyr \cdot 4H_2O$	80-100	240
idem	$SiO_2 \cdot 12MoO_3$	400-787	240
Pyridine	$SiO_2 \cdot 12MoO_3$	370-825	239

	SILVER		
Ammoniacal copper chloride	Ag	<950	483
Hypophosphorous acid	Ag	<477	483
Hydroxylamine	Ag	<600	483
Ascorbic acid	Ag	<624	483
Hydrochloric acid	AgCl	70-600	484
Hydrobromic acid	AgBr	70-946	485
Hydriodic acid	AgI	60-900	485
Sodium thiosulphate	Ag_2S	130-650	485
Hydrogen sulphide	Ag_2S	70-615	485
Thallium acetate + arsenate	Ag_2TlAsO_4	20-846	487
Potassium cyanide	AgCN	<237	487
Potassium thiocyanate	AgSCN	<224	488
Potassium chromate	Ag_2CrO_4	92-812	488
Cupric dipropylenediamine sulphate	$[Cupn_2]\,[AgI_2]_2$	<155	488
Cobaltic dithiocyanato diethylenediamine thiocyanate	$[Coen_2\,(SCN)_2]\,[Ag(SCN)_2]$	<144	489
Oxalic acid	$Ag_2C_2O_4$	<100	489
Thionalide	$AgC_{12}H_{10}ONS$	<105	490

	SODIUM		
Perchloric acid	$NaClO_4$	130-471	197
Blanchetiere's reagent	$MgU_2O_7 + \frac{1}{2}Na_2U_2O_7$	360-745	208
Zinc acetate + uranyl acetate	$ZnU_2O_7 + \frac{1}{2}Na_2U_2O_7$	360-674	209
Ball's reagent	$6NaNO_2 \cdot 9CsNO_2 \cdot$ $\cdot 5Bi(NO_2)_3$	160-670	206

	STRONTIUM		
Iodic acid	$Sr(IO_3)_2$	157-680	438
Sulphuric acid	$SrSO_4$	100-300	438
Sodium dihydrogen arsenate	$Sr_2As_2O_7$	410-950	439
Ammonium carbonate	$SrCO_3$	>410	440
Potassium dichromate	$SrCrO_4$	100-400	440
Potassium ferrocyanide	$Sr_2[Fe(CN)_6] \cdot 9H_2O$	20-110	440
idem	$Sr_2[Fe(CN)_6] \cdot 4H_2O$	110-140	440
Potassium oxalate	SrC_2O_4	177-400	440

	SULPHUR		
Silver nitrate (from a sulphide)	Ag_2S	69-615	248
idem (from a thiosulphate)	Ag_2S	129-649	249
Sodium arsenite	As_2S_3	200-265	248

Precipitating reagent	Form weighed	Temperature limits (°C)	See p.
Lead nitrate	PbS	98-107	624
Mercuric nitrate	HgS	<109	603
Calcium chloride	$CaSO_3$	385-410	249
Lead oxide	$PbSO_4$	271-959	625
Benzidine	$C_{12}H_{12}N_2 \cdot H_2SO_4$	72-130	250
Strychnine	$C_{21}H_{22}O_2N_2 \cdot HSO_4$	<122	251
TANTALUM			
Hexamine + pyrogallol	Ta_2O_5	>770	575
TELLURIUM			
Hexamine (on Te(IV))	TeO_2	20-850	519
Lead nitrate (on Te (VI))	$PbTeO_4$	100-400	520
TERBIUM			
Oxalic acid	$Tb_2(C_2O_4)_3 \cdot 5H_2O$	45-140	561
THALLIUM			
Potassium ferricyanide	Tl_2O_3	100-230	610
Hydrochloric acid	TlCl	56-425	612
Potassium iodide	TlI	70-473	612
Iodic acid	$TlIO_3$	83-562	—
Luteocobaltic chloride	$[Co(NH_3)_6][TlCl_6]$	50-210	614
Potassium chromate	Tl_2CrO_4	97-745	614
Hydrochloroplatinic acid	$Tl_2[PtCl_6]$	65-155	615
Oxine	$Tl(C_9H_6ON)_3$	150-165	615
5,7-Dibromo-oxine	$Tl(C_9H_4Br_2ON)_3$	93-156	—
2-Methyloxine	$Tl(CH_3-C_9H_5ON)_3$	80-145	615
2-Phenyloxine	$Tl(C_6H_5-C_9H_5ON)_3$	100-190	615
Thionalide	$TlS-CH_2-CO-NH-C_{10}H_7$	69-156	616
2-Mercaptobenzothiazole	$TlC_7H_4NS_2$	52-217	616
Tetraphenylarsonium chloride	$[As(C_6H_5)_4]TlCl_4$	50-218	—
Sodium tetraphenylboride	$Tl[B(C_6H_5)_4]$	60-180	616
THIOCYANATE			
Silver nitrate	AgSCN	<224	179
Nickel sulphate + pyridine	$[Ni(C_5H_5N)_4](SCN)_2$	110-130	180
THORIUM			
Gaseous ammonia	ThO_2	>472	647
Hydrogen peroxide	$ThO_2 \cdot 2H_2O$	296-450	648
Iodic acid	$Th(IO_3)_4$	200-300	649
Sodium pyrophosphate	ThP_2O_7	540-946	650
Fumaric acid	ThO_2	>405	651
Sebacic acid	$Th[CO_2-(CH_2)_8-CO_2]_2$	70-125	651
m-Nitrobenzoic acid	$Th(NO_2-C_6H_4-CO_2)_4$	70-153	652
Ferron (Loretine)	$Th(C_9H_4O_4NSI)_2$	110-216	656
Picrolonic acid	$Th(C_{10}H_7O_5N_4)_4$	60-200	657

Precipitating reagent	Form weighed	Temperature limits (°C)	See p.
	THULIUM		
Oxalic acid	$Tm_2(C_2O_4)_3 \cdot 2H_2O$	195-335	567
	TIN		
Selenious acid	Se	80-272	508
Cupferron	SnO_2	>750	509
	TITANIUM		
Ammonia (aqueous or gaseous)	TiO_2	>350	289
Guanidinium carbonate	TiO_2	>500	289
Potassium iodate	$Ti(IO_3)_4 \cdot 3KIO_3$	138-295	290
5,7-Dichloro-oxine	$TiO(C_9H_4Cl_2ON)_2$	105-195	292
idem	TiO_2	>480	292
5,7-Dibromo-oxine	$TiO(C_9H_4Br_2ON)_2$	15-160	292
	TUNGSTEN		
Calcium salt	$CaWO_4$	>400	578
Cadmium salt	$CdWO_4$	>290	578
Lead salt	$PbWO_4$	>100	579
Purpurecobaltic chloride	$12WO_3 \cdot 5[CoCl(NH_3)_5]O$	< 70	579
idem	$12WO_3 \cdot ^5/_3Co_3O_4$	500-790	580
β-Naphthoquinoline	WO_3	>475	580
	URANIUM		
Aqueous ammonia	UO_3	480-610	659
Gaseous ammonia	U_3O_8	675-946	660
Diammonium hydrogen phosphate	$U_2(P_2O_7)_3 \cdot 4UO_3$	673-946	666
Oxalic acid (on U^{4+})	$U(C_2O_4)_2$	100-180	667
β-Isatoxime	U_3O_8	408-946	669
Oxine	$UO_2(C_9H_6ON)_2 \cdot C_9H_7ON$	<157	669
idem	$UO_2(C_9H_6ON)_2$	252-346	669
	VANADIUM		
Silver nitrate (pH 4.5)	$AgVO_3$	60-500	296
Luteocobaltic chloride (neutral or alkaline conditions)	$[Co(NH_3)_6](V_6O_{17})_3$	<127	298
Cupferron	V_2O_5	581-946	299
Oxine	$V_2O_3(C_9H_6ON)_4$	<195	299
Strychnine	V_2O_5	390-950	300
	YTTERBIUM		
Oxalic acid	$Yb_2(C_2O_4)_3 \cdot 2H_2O$	175-325	568
	YTTRIUM		
Oxalic acid	$Y_2(C_2O_4)_3 \cdot 2H_2O$	180-300	443
Oxine	$Y(C_9H_6ON)_3$	250-300	446

SUGGESTED METHODS FOR AUTOMATIC INORGANIC ANALYSIS

Precipitating reagent	Form weighed	Temperature limits (°C)	See p.
	ZINC		
Sulphuric acid	$ZnSO_4$	300-788	407
Ammonium phosphate	$ZnNH_4PO_4$	50-167	408
Cyanamide	$ZnCN_2$	105-152	410
Thiocyanate + pyridine	$[Zn(C_5H_5N)_2](SCN)_2$	< 71	410
Mercurithiocyanate	$Zn[Hg(SCN)_4]$	50-270	410
Oxalic acid	$ZnC_2O_4 \cdot 2H_2O$	< 75	411
Anthranilic acid	$Zn(NH_2\text{-}C_6H_4\text{-}CO_2)_2$	105-141	411
5-Bromo anthranilic acid	$Zn(NH_2\text{-}C_6H_3Br\text{-}CO_2)_2$	< 123	411
Oxine	$Zn(C_9H_6ON)_2$	127-284	411
2-Methyloxine	$Zn(C_{10}H_8ON)_2$	100-220	412
Salicylaldoxime	$Zn(C_7H_6O_2N)_2$	25-285	412
Quinaldinic acid	$Zn(C_{10}H_6O_2N)_2 \cdot H_2O$	80-240	413
idem	$Zn(C_{10}H_6O_2N)_2$	300-340	413
	ZIRCONIUM		
Aqueous ammonia	ZrO_2	>400	445
Gaseous ammonia	ZrO_2	>467	446
Aniline	$ZrO_2 \cdot 2H_2O$	105-150	446
Diethylaniline	ZrO_2	>223	446
Mandelic acid	$Zr(C_6H_5\text{–}CHOH\text{–}CO_2)_4$	60-188	453

PART TWO

The Thermolysis Curves

The elements are discussed in the order of increasing atomic numbers. If mentioned in the particular paper, we have given the make and model of the thermobalance, the rate of heating, and the weights of the starting material.

Our own publications have come from studies carried out with the various models of the Chevenard thermobalance. The publications prior to 1949 came from photographically-recording instruments; the rate of heating was that first shown in Fig.14. The precipitates, slightly moist, were heated in aluminite crucibles, the average weights taken were from 250 to 300 mg, thus approximating the amounts encountered ordinarily in gravimetric analysis. They have never been filtered on paper nor were mixtures with filter paper pulp used in any instance. The electrolytic deposits were heated directly in the platinum vessel which served as the electrode.

CHAPTER 6

Lithium

Along with T. Duval[1] we have selected the best methods for the gravimetric determination of lithium. This investigation was completed in the Diplôme d'Études supérieures of Dupuis[2] and in one of my courses[3]. As early as 1934, Kitajima[4] published thermolysis curves of lithium fluoride, sulphate, and trilithium phosphate. Two standards, the carbonate and the citrate, have been investigated by us[5,6], while Terem and Akalan[7] studied the behaviour of monolithium phosphate, and Cappellina and Babani[8] studied the ferrocyanide.

1. Lithium fluoride

The fluoride method of Boy[9] actually comes down to weighing as sulphate. The fluoride is difficult to filter and it is partially soluble in water; it yields a curve with a horizontal between 270° and 440°.

2. Lithium chloride

Lithium chloride may be weighed in the anhydrous condition following evaporation and ignition. The starting material is usually obtained by the action of hydrochloric acid on lithium carbonate and consequently contains notable amounts of water and traces of the acid which are removed at around 80°. An almost horizontal stretch between 80° and 96° corresponds to the composition $LiCl \cdot \frac{1}{2} H_2O$; this hydrate seemingly had not been reported previously. In contrast, we found no evidence on the curve for the hydrates with 3, 2 or 1 H_2O. The half-molecule of water is lost between 96° and 175° and the anhydrous chloride yields a perfectly horizontal stretch between 175° and 606° (melting point). Sublimation becomes appreciable above this latter temperature and amounts to 8 mg per 292 mg at 877°[1].

3. Iron potassium lithium paraperiodate

The thermolysis curve of ferric potassium lithium paraperiodate is steadily downward. The precipitation is completely quantitative with regard to the lithium but the great excess of potassium requisite to the insolubilization of the product is unfavourable to a gravimetric operation as the final step as opposed to a titration or a colorimetric procedure (Fig. 48d).

4. Lithium sulphate

Lithium sulphate is obtained by the action of dilute sulphuric acid on the carbonate. The product subjected to pyrolysis is accordingly a mixture of monohydrate, water, and a slight amount of sulphuric acid. The curve shows initially a rather rapid descent due to the loss of moisture. The monohydrate $Li_2SO_4 \cdot H_2O$ is stable from 50° to 72° (a good horizontal); this compound decomposes between 72° and 160°. Some workers have reported that the dehydration occurs at 100°, others give 180°. The range of existence of lithium sulphate is revealed by a horizontal beginning at 160° and extending to beyond 877°. As a matter of fact, the loss slightly exceeds the value for 1 molecule of water from one horizontal to the other. Kitajima[4] gives 470° as the minimum temperature for obtaining the anhydrous product.

5. Lithium sulphamate

Lithium sulphamate (amidosulphonate) $LiSO_3-NH_2$ was prepared by Capestan[10]. It is anhydrous and stable up to around 250°; it then is transformed into the imidodisulphonate $NH(SO_3Li)_2$ with loss of ammonia, and yields a horizontal between 300° and 480°. Lithium sulphate, Li_2SO_4, then appears around 500°.

6. Lithium nitrate

Lithium nitrate $LiNO_3 \cdot H_2O$ loses water from room temperature on and gives the semi-hydrate with an oblique straight line in the neighbourhood of 91°. The anhydrous salt $LiNO_3$ gives a horizontal between 140° and approximately 500°. It then sharply evolves nitrogen oxides without discontinuity. A horizontal corresponding to the oxide Li_2O is reached between 900° and 910°.

7. Lithium phosphate

Lithium phosphate Li_3PO_4 is obtained by treating lithium chloride with disodium hydrogen phosphate and sodium hydroxide. After evaporation to dryness on the water bath, the residue is taken up in lukewarm water and very dilute ammonium hydroxide. Up to 450° the curve shows a slightly descending portion (loss of water and ammonia). Kitajima[4] gives 400° with a dry product. The phosphate Li_3PO_4, whose presence is revealed by a good horizontal, may be weighed after heating to this latter temperature.

8. Monolithium dihydrogen phosphate

Monolithium dihydrogen phosphate, which is not used in gravimetry, undergoes a dissociation, according to Terem and Akalan[7], in two steps: between 160° and 200° it forms the hydrogen pyrophosphate and between 260° and 400°, the metaphosphate.

9. Lithium arsenite

The precipitation of the arsenite, suggested by Gaspar y Arnal[11], does not appear to be quantitative; the thermolysis curve of this salt shows no horizontal section.

10. Lithium carbonate

The carbonate selected for the preparation of standard solutions was found by us[6] to be stable from room temperatures up to 428°. It should be noted that Souchay and Schaal[12] proposed this salt as a pH buffer.

11. Lithium ferrocyanide

The ferrocyanide $Li_4[Fe(CN)_6]\cdot 8H_2O$ progressively loses water between 60° and 170° where the horizontal of the anhydrous salt starts (Cappellina and Babani[8]).

12. Lithium aluminate

The aluminate precipitate is prepared and washed by strict adherence to the directions given by Grothe and Savelsberg[13]. It loses some water even at room temperatures, and the dehydration then continues, though slowly, up to 471°, where the constant weight corresponding to the theoretical formula $2Li_2O\cdot 5Al_2O_3$ is reached. The total loss of water amounts to 19 molecules of which 17 are probably combined.

13. Zinc uranium lithium acetate

Ever since Miller and Traves[14] proposed that zinc uranium lithium acetate be employed for analytical purposes, the exact formula of the precipitate has been a matter of discussion from time to time. It is not possible to determine the water by direct dehydration. The thermolysis curves obtained by us show a water content between 6 and 8 molecules. Accordingly, the precipitate is a mixture of hydrates (as is also true of the sodium salt). Because of the high molecular weight of this triacetate, the precision of the determination is affected hardly at all if 6, 6.5, 7, or 8 molecules of water are accepted in the calculation. However, when the matter must be handled in a rigorous manner, it is better to weigh as the anhydrous salt $Zn(UO_2)_3Li(CH_3CO_2)_9$ by keeping the precipitate in the portion of the curve between 125° and 162°, and employing the conversion factor 0.00490 rather than 0.00456 for 6 H_2O. The complex decomposes immediately above this latter temperature and very rapidly between 270° and 292°; this decomposition is deep-seated and yields some tetravalent uranium (color of the residue). Reoxidation takes place from 292° to 497°. Above this temperature, there is a mixture of pyro-uranates $ZnU_2O_7 \cdot \frac{1}{2}Li_2U_2O_7$ (confirmed by direct weighing and by infra-red absorption) and this mixture may be weighed as a check. The method has been well studied by others[15] and it is the only procedure which we recommend for the automatic determination of lithium.

14. Lithium citrate

Duval and Wadier[6] have proposed lithium citrate as a standard substance. It contains 4 molecules of water and retains its hydrated condition up to 56°. The salt becomes anhydrous around 251° and then progressively decomposes. Above 950° it produces a residue of more or less carbonated lithium oxide.

REFERENCES

[1] T. DUVAL AND C. DUVAL, *Anal. Chim. Acta*, 2 (1948) 57.
[2] T. DUPUIS, *Diploma of Higher Studies*, Paris, 1948.
[3] C. DUVAL, *Chim. anal.*, 34 (1952) 209.
[4] I. KITAJIMA, *J. Chem. Soc. Japan*, 55 (1934) 199.
[5] C. DUVAL, *Anal. Chim. Acta*, 13 (1955) 36.
[6] C. DUVAL AND C. WADIER, *Anal. Chim. Acta*, 23 (1960) 543.
[7] H. N. TEREM AND S. AKALAN, *Rev. fac. sci. Univ. Istanbul*, 16A (1951) 14.

[8] F. CAPPELLINA AND B. BABANI, *Atti Accad. sci. ist. Bologna*, 11 (1957) 76.
[9] C. BOY, *Z. Erzbergbau u. Metallhüttenw.*, 1 (1948) 112.
[10] M. CAPESTAN, *Ann. Chim.*, 5 (1960) 215.
[11] T. GASPAR Y ARNAL, *Anales soc. españ. fis. y quim.*, 30 (1932) 406.
[12] P. SOUCHAY AND R. SCHAAL, *Mikrochim. Acta*, (1954) 371.
[13] H. GROTHE AND W. SAVELSBERG, *Z. anal. Chem.* 110 (1937) 81.
[14] C. C. MILLER AND F. TRAVES, *J. Chem. Soc.*, (1936) 1395.
[15] Y. A. CHERNIKHOV, T. A. USPENKAYA AND R. S. ANAN'INA, *Zavodskaya Lab.*, 9 (1940) 28.

CHAPTER 7

Beryllium

Up to the present, thermolysis studies have been made of the hydroxide, sulphate, ammonium beryllium phosphate, basic carbonate, metaniobate, and the 1,3-dimethylviolurate. The gravimetric discussion has been due to Duval and Duval[1]; Dupuis[2] has reviewed the various procedures for precipitating the hydrated oxide. No organic reagent has been found as yet which gives a satisfactory complex that is stable at a low temperature. Accordingly, weighing the beryllium pyrophosphate is the best method for the automatic determination.

1. Beryllium oxide hydrate

(i) Precipitation by ammonium chloride   The method most generally employed for obtaining the hydrous oxide $BeO \cdot nH_2O$ is to treat an aqueous solution of beryllium chloride with several grams of ammonium chloride, and then add a slight excess of ammonium hydroxide in the cold. The liquid is decanted and the precipitate is washed on the filter several times with warm water made slightly alkaline with ammonia and containing ammonium nitrate. Care should be exercised to protect the product from the carbon dioxide of the air. The retained water is removed with no discontinuity and this removal is revealed by two more or less sharp changes in curvature at 150° and at 260°. The horizontal of the oxide starts at 950°. There is no need to heat to 1200° in order to obtain a beryllium oxide (glucina) which does not reabsorb water, provided automatic gravimetry is employed. Trials with a sample which in theory should yield 0.1073 g of beryllium oxide gave the following losses in weight at the indicated temperatures:

Temp. (°C)	348	418	545	613	678	748	816	951
Loss (mg)	15.0	9.0	6.0	5.0	4.4	4.0	2.0	0

Kiba and Ikeda[3] specify 880° as the minimum temperature for the formation of the oxide, while Noshida[4] recommends 950°.

(ii) Preparation by electrolysis Our findings are in good agreement with those of Jolibois and Bergès[5] who studied the pyrolysis (with a Chevenard thermobalance) of a hydrate prepared by electrolysis, dried in vacuo, and crystallized. These workers, as did Van Bemmelen[6], noted the transition at 150° which was thought to be the limit between the absorbed water and the water of constitution. In reality, the infra-red spectrum always shows the presence of water molecules and not of OH groups. Terem[7] in turn, using a Guichard balance, also concluded that absorbed water is present. The recording made with our equipment does not indicate that the hydrated oxide maintains a constant weight between 150° and 180°, the rate of dehydration is even quite high between these temperatures in the case of the gelatinous product.

(iii) Other precipitants Dupuis[2] has made a critical study of the various precipitants which have been suggested in the literature. The effect of these different precipitants is to be seen in the rate of precipitation, the ease of filtration, the structure as revealed by the Debye and Scherrer spectra, and the minimum temperature at which the oxide BeO starts to form. None of these precipitates corresponds, after drying, to the formula $Be(OH)_2$. On the contrary, and in opposition to the classical idea, the precipitates which are most amorphous with respect to X-ray examination have been found to filter most easily, all other conditions being the same. Table 3 shows a comparison of the different precipitants. They should be compared with the tabulation of the compounds yielding the various aluminas (See Chapter Aluminium).

TABLE 3

PRECIPITATION OF BERYLLIUM OXIDE HYDRATE

Precipitated by	Min. temp.	Precipitated by	Min. temp.
Aqueous ammonia, cold	900	Iodide–iodate mixture	812
Aqueous ammonia, hot	595	Ammonium carbonate	756
Gaseous ammonia, at 15°	440	α-Picoline	950
Potassium cyanate	809	Monoethylamine	705
Ammonium benzoate	780	Hydrazinium carbonate	950
Hexamine	950	Ammonium sulphide	592
Selenious acid	800	Sodium sulphite	950
Ammonium nitrite	850	Sodium hydroxide*	610
Tannin	571	Sodium hydroxide**	774
Guanidinium carbonate	555	Oxine	789

* Haber and Van Oordt[8]   ** Gmelin[9]; Britton[10]

References p. 173

2. Beryllium sulphate

In the case of the sulphate, we distinguish two cases:

(i) Crystallized beryllium sulphate This sulphate $BeSO_4 \cdot 4H_2O$, studied by Gordon and Denisov[11] and by Terem[12], when subjected to increasing linear heating of 180° per hour produces the dihydrate at 160°, the monohydrate at 220°, and the anhydrous salt at 550°. If the latter is heated at the rate of 240° per hour it loses sulphur trioxide from 700° to 930° without forming the basic sulphate, the size of the particles being without effect. We have found a short horizontal relating to the dihydrate between 88° and 92° but no monohydrate; our rate of heating was greater. The anhydrous product was found to be stable from 346° to 679°, without however presenting a strictly horizontal stretch. The dissociation of the anhydrous sulphate did not occur until 1031° was reached.

(ii) Moist sulphate When a thermolysis study is made of the moist sulphate, *i.e.* the gummy mass resulting from the action of sulphuric acid on a beryllium salt–which is the case encountered in analysis–the curve is of a different kind. The level of the anhydrous sulphate is less distinct and never is horizontal; in other words, the sulphate begins to decompose before all of the water has been lost. Therefore, it is not possible to specify definitely the temperature to which the sulphate must be heated before it is weighed. Our best curves indicate that the theoretical weight of the sulphate is reached somewhere between 544° and 670°. This gravimetric method cannot be recommended as reliable. If the initial product is acid, *i.e.* if it contains an excess of sulphuric acid, the curve changes slope around 1000° and leads to beryllium oxide BeO, which is not obtained in pure condition even at 1050°. In case the starting material is at pH 6, the change in slope is observed in the neighbourhood of 920°. Weighing the oxide derived from the sulphate is therefore not advisable in our opinion.

3. Beryllium nitrate

This salt, which was studied by Wendlandt[13], has the formula $Be(NO_3)_2$ $\cdot 4H_2O$. It begins to lose water between 35° and 70°, and oxides of nitrogen are evolved between 265° and 370°.

4. Ammonium beryllium phosphate

The precipitate of ammonium beryllium phosphate is obtained, according to Moser and Singer[14], by treating a slightly acidic solution of beryllium sulphate with disodium hydrogen phosphate, along with ammonium nitrate and acetate. The precipitate is then dissolved in boiling dilute nitric acid, and the desired precipitate is thrown down by adding 2% ammonium hydroxide. The product is washed with hot 5% ammonium nitrate solution, and then conforms strictly to the formula $NH_4BePO_4 \cdot 6H_2O$. Its thermolysis curve is extremely simple and presents no particular features marking the loss of water and ammonia. The destruction of the crystals begins at 30°, so that weighing of the hydrated or anhydrous phosphate is out of the question. The theoretical weight of the pyrophosphate is obtained starting at 640°. We believe this is the best gravimetric method for determining beryllium.

5. Basic beryllium carbonate

According to Noshida[4], the basic carbonate forms the oxide BeO from 350° on.

6. Beryllium niobate

The niobate $Be_7Nb_{12}O_{37} \cdot 22H_2O$ was studied by Pehelkin, Lapitskii, Spitsyn and Simanov[15]. It is brought to the tetrahydrate state at 105°. The curve indicates that only 0.3 H_2O is still present at 160°. The metaniobate is reached at 610°.

7. Beryllium 1,3-dimethylviolurate

The 1,3-dimethylviolurate has been suggested[16] as a suitable form for the gravimetric determination of beryllium.

REFERENCES

[1] T. DUVAL AND C. DUVAL, Anal, Chim. Acta, 2 (1949) 53.
[2] T. DUPUIS, Thesis, Paris, 1954; Compt. rend., 230 (1950) 957; Mikrochemie, 35 (1950) 477
[3] T. KIBA AND T. IKEDA, J. Chem. Soc. Japan, 60 (1939) 911.
[4] I. NOSHIDA, J. Chem. Soc. Japan, 48 (1927) 520.
[5] P. JOLIBOIS AND M. BERGÈS, Compt. rend., 224 (1947) 79.
[6] J. M. VAN BEMMELEN, J. prakt. Chem., 26 (1882) 232.
[7] H. N. TEREM, Rev. fac. sci. univ. Istanbul, 99 (1943) 111.

[8] F. HABER AND VAN OORDT, *Z. anorg. Chem.*, 38 (1904) 377.

[9] C. G. GMELIN, *Poggendorff's Ann.*, 50 (1840) 176.

[10] H. T. S. BRITTON, *Analyst*, 46 (1921) 359.

[11] B. E. GORDON AND A. M. DENISOV, *Ukrain. Khim. Zhur.*, 19 (1953) 368.

[12] H. N. TEREM, *Compt. rend.*, 222 (1946) 1347.

[13] W. W. WENDLANDT, *Texas J. Sci.*, 10 (1958) 392.

[14] L. MOSER AND J. SINGER, *Monatsh.*, 48 (1927) 679.

[15] V. A. PEHELKIN, A. V. LAPITSKII, V. I. SPITSYN AND Y. P. SIMANOV, *Zhur. Neorg. Khim.*, 1 (1956) 1784.

[16] R. J. ROBINSON, A. BERLIN AND M. E. TAYLOR, *3rd Conf. on Anal. Chem., Prague*, 1959.

CHAPTER 8

Boron

The general thermogravimetric study of boron compounds has been made by Duval[1], whereas Haladjian and Carpeni[2] have given a very extended account of the possibilities afforded by the thermobalance in a field foreign to analysis. The curves obtained by Wendlandt[3] for the tetraphenylborons are discussed in the Chapters Ammonium, Potassium, Rubidium and Cesium.

1. Boric acid

Boric acid, which is used for standardizing titrimetric solutions, has been studied by Duval[4]. With respect to its determination, Partheil and Rose[5] and others have shown that boric acid can be extracted with ether and weighed as such after a stay in a desiccator over sulphuric acid. The acid retains the formula $B(OH)_3$ up to 55°. It is interesting to note that no part of its dehydration curve indicates the formation of pyroboric acid $H_2B_4O_7$; on the other hand, between 133° and 168° there is a horizontal whose ordinate corresponds with the formation region of metaboric acid HBO_2. This therefore provides a simple means of obtaining pure metaboric acid; its position on the chart corresponds to a molecular weight of 43.8 (calculated 43.828). Metaboric acid dehydrates smoothly above 168° and a perfectly horizontal stretch belonging to the anhydride B_2O_3 starts at 443°. Urusov[6] has also followed this dehydration.

The work of Haladjian and Carpeni[2] may be summed up as follows: they dehydrated orthoboric acid at a constant temperature of 74° but used different rates of heating, namely 100°, 50°, and 10° per hour. In the first case, the metaboric acid signaled its formation by an inflection at 150°, in the second the inflection appeared at 122°, and in the third case by a horizontal around 100–120°.

They also conducted an isothermal dehydration at 83° with three grain-sizes; then at 77° with three different weights, namely 3, 6, and 9 millimols of boric acid. The horizontal of metaboric acid can be made to disappear.

The dehydration process depends greatly on the influence of the gaseous atmosphere, wet or dry. It is even possible to prevent all isothermal dehydration at 74°. The water vapour also participates by exercising a certain catalytic action; it markedly lowers the minimum dehydration temperature. The study was completed by investigating reversible re-hydrations, which were even repeated several times (cyclings) and by making density measurements. These workers have been able, in general, to relate the horizontals with the structures and the angular points with the catalytic action of water.

We have directed our attention likewise to the gravimetric determination by distillation. The boric acid arising from the saponification of methyl borate is ordinally heated in the presence of an excess of lime, or sodium tungstate, or ammonium phosphate.

(i) Heating with an excess of lime   In this case (the Gooch procedure[8]), it is found that the horizontal of the thermolysis curve corresponding to metaboric acid has disappeared. The extraneous moisture escapes up to 92°, and the water of constitution is given off from 92° to 685°. Beyond this temperature, there remain calcium metaborate $Ca(BO_2)_2$ and the excess of quick lime. In practice, the heating is ordinarily carried too high to be certain of obtaining a constant weight; examination of our curves reveals also that there is not a trace of carbonation of the lime; otherwise there will be a break in the neighbourhood of 813°, corresponding to the decomposition of calcium carbonate.

(ii) Heating with sodium tungstate   Heated in the presence of sodium tungstate as proposed by Ashman[9], the curve shows that constant weight is obtained at a much lower temperature, approximately between 450° and 455°.

(iii) Heating with ammonium phosphate   The boric acid, which has been set free from its ester, is heated along with diammonium phosphate (Ashman[9]). However, it is much more difficult to obtain a constant weight for the so-called boron phosphate BPO_4. This will require that the crucible be maintained for an hour at least at 1000°. The last traces of water retained by the pyrophosphate formed between 946° and 1000° are expelled but slowly. The infra-red absorption experiments, conducted

in association with Lecomte, in the 6–18μ region show that the resulting salt is not an orthophosphate, and that the formula BPO_4 as well as the analogies drawn from it remain very doubtful.

2. Calcium metaborate

The weighing of boron in the form of calcium metaborate, as suggested by Ditte[10], yields a thermogram showing that a constant weight for this salt is obtained from 475° on.

3. Barium metaborate

The precipitation as barium metaborate is conducted in alcoholic medium with a saturated solution of barium hydroxide, as proposed by Berg[11]. The curve, otherwise identical with the preceding one, indicates that constant weight is not obtained until 745° is reached. The borates of other metals (Ag, Cd, etc.) whose employment has been suggested for the gravimetric determination of boron have not been included here because of their relatively high solubility and the difficulty of handling them.

4. Potassium tetrafluoroboride

The procedure of Stromeyer[12] has been used by us for the well known method of Berzelius in which potassium tetrafluoroboride $K[BF_4]$ is weighed. This complex salt is remarkably stable up to 410°; it then dissociates smoothly and at 980° the crucible contains nothing but potassium fluoride K_2F_2.

5. Nitron tetrafluoroboride

Weighing as nitron tetrafluoroboride is the only method suggested for the automatic determination of boron since the thermolysis curve has an absolutely horizontal stretch extending from 50° to 197° for the complex prepared by Berkovich and Kulyashev[13]. The material then undergoes progressive decomposition with transitions in the vicinity of 300° and 400°. The residue, which has constant weight, and which is formed beyond 980°, consists of boric anhydride; it results from a secondary reaction. Some of the boron is given off as the fluoride.

References p. 178

6. Sodium tetraphenylboride

Sodium tetraphenylboride (kalignost) precipitates not only ammonium, potassium, rubidium, and cesium (see the appropriate chapters) but also amines, quaternary ammonium salts, etc. Thus, Wendlandt, van Tassel and Horton[14] have constructed thermolysis curves with 8-hydroxyquinoline, 2-methyl-8-hydroxyquinoline, 5,7-dichloro-8-hydroxyquinoline. 5,7-dibromo-8-hydroxyquinoline 5,7-diiodo-8-hydroxyquinoline, and likewise the differential thermal analysis curves. The heatings were done on the Campbell and Gordon balance, using 90–100 mg samples and the heating rate of 6° per minute.

The minimal decomposition temperatures are respectively 50°,70°, 150°,127°, and 203° so far as can be judged from the published curves. These workers believe that the initial loss is due rather to sublimation than to decomposition.

REFERENCES

[1] C. DUVAL, *Anal. Chim. Acta*, 4 (1950) 55.
[2] J. HALADJIAN AND J. CARPENI, *Bull. Soc. chim. France*, (1956) 1679.
[3] W. W. WENDLANDT, *Chemist-Analyst*, 26 (1957) 38; *Anal. Chem.*, 28 (1956) 1001.
[4] C. DUVAL, *Anal. Chim. Acta*, 13 (1955) 428.
[5] A. PARTHEIL AND J. ROSE, *Ber.*, 34 (1901) 3611.
[6] V. V. URUSOV, *Doklady Akad. Nauk S.S.S.R.*, 116 (1957) 97.
[7] W. R. OWEN AND M. D. SUTHERLAND, *J. Sci. Food Agr.*, 7 (1956) 88.
[8] F. A. GOOCH, *Am. Chem. J.*, 9 (1887) 23; *Chem. News*, 55 (1887) 7.
[9] C. ASHMAN JR., *J. Soc. Chem. Ind.* (London), (1916) 1263.
[10] A. DITTE, *Compt. rend.*, 80 (1875) 490, 561; *Ann. Chim. phys.*, 4 (1875) 549.
[11] P. BERG. *Z. anal. Chem.*, 16 (1877) 25.
[12] A. STROMEYER, *Liebigs Ann.*, 100 (1856) 82.
[13] V. L. BERKOVICH AND Y. V. KULYASHEV, *J. Appl. Chem. (U.S.S.R.)*, 10 (1937) 192.
[14] W. W. WENDLANDT, J. H. VAN TASSEL AND G. R. HORTON, *Anal. Chim. Acta*, 23 (1960) 332.

CHAPTER 9

Carbon

We have only fragmentary studies of the thermolysis of various specimens of carbon; it is too soon to attempt to get a view of the whole subject from these. We can cite the studies which Waters[1] carried out on coke carbon with the thermobalance which he built, and the investigations by Gill[2] with his balance on samples rich in volatile substances, and the work of Perrot, Bastick and Guérin[3] on sugar carbon, and the determination of the speed of the combustion of carbon by Okada and Ikegawa[4].

1. Carbonate and thiocarbonate ion CO_3^{2-}, CS_3^{2-}

We have no gravimetric method for the CO_3^{2-} ion or the CS_3^{2-} ion but several procedures have been reported for determining the derivatives of cyanogen. Duval[5] has made a thermogravimetric study of these methods.

2. Cyanide ion CN^-

The cyanide ion CN^- is ordinarily determined by the Fresenius[6] procedure of weighing silver cyanide. This salt is stable up to 237°; cyanogen begins to come off beyond this temperature.

3. Thiocyanate ion CNS^-

The thiocyanate ion CNS^- may be determined by weighing as silver thiocyanate, cuprous thiocyanate, barium sulphate or nickel pyridine thiocyanate.

(i) Silver thiocyanate The horizontal section of the curve of silver thiocyanate AgCNS extends up to 224°, the temperature at which loss of sulphur commences. Van Name[7] has reported 115° as the drying temperature and this is correct.

(ii) Weighing as cuprous thiocyanate   Cuprous thiocyanate CuCNS is stable up to 300° and the level which corresponds to it is not strictly horizontal. Instead of weighing it directly, some workers have converted the thiocyanate into cupric sulphate $CuSO_4$ or into cupric oxide CuO.

(iii) Barium sulphate   Barium sulphate, depending on the sample, should be taken to between 780° and 1200° to secure constant weight.

(iv) Nickel pyridine thiocyanate   The material prepared according to Spacu[8] and formulated $Ni(SCN)_2 \cdot (C_5H_5N)_4$ is stable only up to 63°. Accordingly it must be dried in a desiccator as recommended by the present writer. It may also be weighed as $Ni(SCN)_2 \cdot (C_5H_5N)_2$ provided the precipitate is kept between 110° and 130°.

4. Cyanate ion CNO^-

The CNO^- ion of cyanates may be determined gravimetrically in two ways: either as a silver salt[9] or as hydrazodicarbonamide[10]. Silver cyanate maintains constant weight up to 138° but the evolution of cyanogen becomes significant only beyond 240° and metallic silver appears at 465°. The curves of silver cyanide and silver cyanate are similar and both may be employed for automatic determination.

5. Ferrocyanide ion $[Fe(CN)_6]^{4-}$

The determination of the ferrocyanide ion $[Fe(CN)_6]^{4-}$ is based on the weighing of the ferrocyanides of silver, luteocobaltic, benzidine, or as mixed compound with molybdic anhydride.

(i) Silver ferrocyanide   The double decomposition of silver nitrate and potassium ferrocyanide yields a white precipitate which is treated as suggested by Castiglioni[11]. The thermogram of silver ferrocyanide has a good horizontal extending from 60° to 229°, which is satisfactory for an automatic determination on the basis of its formula being $Ag_4[Fe(CN)_6]$. Decomposition occurs beyond 229°, and around 812° there is a horizontal due to a mixture of iron oxides and silver.

(ii) Luteocobaltic ferrocyanide   Hynes, Malko and Yanowski[12] prescribe that the luteocobaltic ferrocyanide $[Co(NH_3)_6]_4[Fe(CN)_6]_3$ be dried at 110° or that the mixture $8Co_3O_4 \cdot 9Fe_2O_3$ be ignited. The first procedure

is not acceptable because the curve obtained from the precipitate shows distinctly that it loses weight up to 356°; the second method is satisfactory for all temperatures between 356° and 1013°.

(iii) Benzidine ferrocyanide When using benzidine we have followed the directions given by Cumming[13] but, in contrast to his report, our white precipitate contained no water of crystallization. Accordingly, its formula is simply $(NH_2-C_6H_4-NH_2)_3H_4[Fe(CN)_6]$. Moreover, we suggest that in the future the precipitate not be destroyed and consequently that the final weighing form not be Fe_2O_3, whose gravimetric factor is too unfavorable. The curve yielded by the precipitate reveals that it is stable up to 52° with a composition in accord with that given above. The practically horizontal stretch observed near 124° doubtless corresponds merely to a mixture with an apparent molecular weight of 650 per gram-atom of iron. The oxide Fe_2O_3 appears at 944°.

(iv) Mixed compound with molybdic anhydride More recently Arribas-Jimeno[14] has suggested that ferrocyanides be weighed in the form of the mixed compound with molybdic anhydride, namely $(NH_4)_4[Fe(CN)_6]\cdot 2MoO_3\cdot 3H_2O$. This compound becomes anhydrous between 100° and 150°; hydrogen cyanide is evolved beyond 150°. A horizontal, which corresponds to the ensemble $CFe_3 + C + 6\ MoO_3$, is observed between 275° and 650°; molybdic anhydride volatilizes beyond 650°.

6. Ferricyanide ion [Fe(CN)$_6$]$^{3-}$

The determination of the ferricyanide ion $[Fe(CN)_6]^{3-}$ likewise is based on the insolubility of its silver, luteocobaltic, and benzidine salts.

(i) Silver ferricyanide Silver ferricyanide is produced by double decomposition between potassium ferricyanide and silver nitrate using the directions of Castiglioni[11]. This salt is not stable and loses some cyanogen at ordinary temperatures; its individuality is in doubt. The thermolysis curve is inclined downward until 500° is reached and then an almost horizontal stretch is given which corresponds to mixtures of metallic silver and iron oxides.

(ii) Precipitation with luteocobaltic chloride In the second method, the precipitation is effected by double decomposition between potassium ferricyanide and luteocobaltic chloride. The golden-yellow precipitate,

produced in accord with the directions of Murgulescu and Dragulescu[15] should be dried according to them in the cold and under reduced pressure. Actually, the product does not decompose under 151°, and above 393° it yields a mixture of iron oxides and cobalt. The straight line obtained steadily rises up to 151°. Oxygen is taken up at the rate of 3 mg per 136 mg. Consequently this compound cannot be used for the automatic determination.

(iii) Precipitation with benzidine We have repeated the technique of Cumming and Good[16] for the third procedure, but in contrast to what occurs in the case of the ferrocyanide, the blue-black precipitate obtained is a mixture which loses weight from ordinary temperatures on. Iron oxide appears even as low as 750°, and this form of determination should be retained.

7. Ferri nitrosopentacyanide (nitroprusside) ion [FeNO(CN)$_5$]$^{2-}$

Up to the present no gravimetric study has been made of the ferri-nitrosopentacyanide (nitroprusside) ion.

REFERENCES

[1] P. L. WATERS, *Nature*, 178 (1956) 324; *Anal. Chem.*, 32 (1960) 853.
[2] A. F. GILL, *Can. J. Research*, 10 (1934) 703.
[3] J. M. PERROT, M. BASTICK AND H. GUÉRIN, *Compt. rend.*, 250 (1960) 351.
[4] J. OKADA AND T. IKEGAWA, *Tôkai Denkyoku Gihô*, 17 (1955) 6; *C.A.*, (1957) 6292.
[5] C. DUVAL, *Anal. Chim. Acta*, 5 (1951) 506.
[6] R. FRESENIUS, *Quantitative Analyse*, 6th Ed. p. 301.
[7] G. VAN NAME, *Am. J. Sci.*, 10 (1901) 451.
[8] G. SPACU, *Bul. soc. stiinte Cluj.*, 1 (1922) 314.
[9] H. E. STAGG, *Analyst*, 71 (1946) 557.
[10] J. LEBOUCQ, *J. pharm. chim.*, 5 (1927) 531.
[11] A. CASTIGLIONI, *Z. anal. Chem.*, 126 (1943) 61.
[12] W. A. HYNES, M. G. MALKO AND L. K. YANOWSKI, *Ind. Eng. Chem.*, *Anal. Ed.*, 8 (1936) 356.
[13] W. M. CUMMING, *J. Chem. Soc.*, (1924) 240.
[14] S. ARRIBAS-JIMENO, *Rev. inform. quím. anal. (Madrid)*, 8 (1959) 3.
[15] I. G. MURGULESCU AND C. DRAGULESCU, *Z. anal. Chem.*, 123 (1942) 272.
[16] W. M. CUMMING AND W. GOOD, *J. Chem. Soc.*, (1926) 1924.

CHAPTER 10

Nitrogen

We will consider in succession the anions derived from nitrogen: hydrazoate (azide) N_3^-, hyponitrite $N_2O_2^{2-}$, nitrite NO_2^-, nitrate NO_3^- and then the various ammonium salts NH_4X.

A. Anions Derived From Nitrogen

1. Azide ion N_3^-

The only gravimetric method known for the azide ion consists in precipitating silver azide AgN_3 as suggested by Dennis and Isham[1] and Copeman[2]. After treating this salt with nitric acid, its silver is precipitated as silver chloride, which has a constant weight from 70° to 600°(Duval and Ng Dat Xuong[3]).

2. Hyponitrite ion $N_2O_2^{2-}$

In analogous fashion, the hyponitrites are converted into the silver salt, which yields metallic silver when heated. The latter is determined as silver chloride (Divers[4]).

3. Nitrite ion NO_2^-

(i) Reduction of silver bromate   The nitrite ion is determined through its reducing action on silver bromate. According to Busvold[5] and Laird and Simpson[6], a given weight of this salt is reduced to silver bromide AgBr which is then weighed. The latter has a constant weight between 70° and 946° and it is not necessary to go above 230° for the bromate.

(ii) Precipitation by 2,4-diamino-6-hydroxypyrimidine   The resulting pink precipitate, according to Hahn[7] and Oelsner[8], does not adhere to the walls of the glass container, it settles quickly, and may be filtered off

at once. Unfortunately, the thermolysis curve includes no horizontal but is always inclined downward. The slope is least pronounced between 127° and 220°. Beyond this latter temperature, the decomposition proceeds vigorously and the crucible is empty at 500°.

4. Nitrate ion NO_3^-

The nitrate ion NO_3^- can be determined gravimetrically by ten procedures.

(i) Drying of ammonium nitrate   Drying of NH_4NO_3 seems to be applicable only to free nitric acid, assumed to be alone in solution (Carnot[9]). When it is pure, ammonium nitrate does not decompose below 151°. Up to this temperature, and like all explosive substances which we have heated, it shows a distinct gain in weight (1/40) because of the fixation of atmospheric oxygen and the probable formation of an unstable pernitrate which initiates the chain decomposition. The latter, which yields water vapour and nitrous oxide, stops suddenly at 276°, leaving the crucible entirely empty.

(ii) Heating with powdered quartz   The nitrate to be determined is mixed with 5 times its weight of powdered quartz and the mixture is heated to determine the loss in weight from which it is then possible to deduce the quantity of actual nitrate in the sample, since of course the nitrogen oxides are the only volatile materials involved. The curve (obtained with potassium nitrate) shows a horizontal after the water has been driven off and extends from 207° to 510°. The dissociation takes place between 510° and 972°. It is easy to measure the distance between the two parallel horizontals and from this to deduce the quantity corresponding to the anhydride N_2O_5.

(iii) Heating with sodium tungstate   An identical technique, devised by Gooch and Kuzirian[10], consists in replacing the powdered quartz with sodium tungstate. The curve (always obtained with potassium nitrate) includes a horizontal extending up to 700°. The destruction of the nitrate occurs chiefly between 700° and 947°.

(iv) Heating with potassium bichromate   If pure potassium bichromate is pulverized and then melted as directed by Persoz[11], decomposition of the potassium nitrate begins at 356° and is complete at 814°. These

ignition methods appear to us to be out-moded; they are not capable of general application and it is essential to start with an absolutely dry product. They have already been criticized by Alberti and Hempel[12] but they may possess some interest from the standpoint of increasing our knowledge of solid state reactions.

(v) Lead nitrate When the material to be determined is free nitric acid, it may be treated with an excess of lead oxide and then the mixture of lead oxide and nitrate is heated to constant weight (Carnot[9]). The curve, analogous to that obtained by Nicol[13], shows that under these conditions lead nitrate does not commence to decompose below 380°. However, automatic determination is not feasible because a very slow rise occurs up to this temperature, as in the case of all of the oxidants which we have heated. Lead oxide is formed from this nitrate above 520° but obviously its weight is of no interest to the analyst.

(vi) Nitron nitrate The traditional procedure is that proposed by Busch[14]. The precipitate of nitron nitrate is stable up to 241°, but above this temperature it explodes and the crucible is entirely empty at 472°. The right portion of the curve extending from 20° to 241° is ascending; the gain in weight is of the order of 8 mg per 421 mg. However, if the heating current is shut off around 228°, a straight cooling line, symmetrical with the heating line, is obtained, and the undestroyed compound regains its initial weight. It thus seems logical to ask why it has been the custom to heat this compound in order to dry it; nitron nitrate can simply be kept in the desiccator prior to the weighing.

(vii) Precipitation with nitron formate Mestrezat and Delaville[15] modified the Winkler[16] procedure for the precipitation with nitron formate (Fornitral). The salt produced in this manner yields a thermogram, which after the removal of the moisture has a horizontal extending from 77° to 228° and is slightly ascending. The crucible is empty at 313°.

(viii) Cinchonamine nitrate The excellent cinchonamine method of Arnaud and Padé[17] yields a precipitate which filters well. Cinchonamine nitrate yields a perfectly horizontal stretch from 39° and 196°. Since the organic matter is completely consumed at 745°, the crucible is empty.

(ix) Precipitation with di-(α-methylnaphthyl)-amine The procedure suggested by Rupe and Becherer[18] in which the nitrate is precipitated

with di-(α-methylnaphthyl)-amine gave a product which washes and filters very well. It is not necessary to wait 12 hours before proceeding. The methods previously published for the preparation of the amine were found to be unsatisfactory; we therefore prepared it by the catalytic reduction (hydrogenation in the presence of Raney nickel) of the corresponding nitrile. The curve lends itself well to the automatic determination of nitrates since it includes a good horizontal extending as far as 234°. Sudden decomposition sets in after that and the resulting tar burns progressively.

(x) Precipitation with p-tolylisothiourea Finally, we have precipitated the nitrate in potassium nitrate by means of *p*-tolylisothiourea as recommended by Arndt[19]. The curve descends continuously from room temperature to 811° where the crucible is empty. Consequently this method cannot be recommended for gravimetric analysis.

B. SALTS DERIVED FROM AMMONIA

1. Ammonium chloride

The salt NH_4Cl has been examined both as precipitate[20] and as standard[21]. This salt loses water up to 40°; constant weight is maintained from this temperature up to 173°. Usually, it is recommended to take 110°; above this, the ammonium chloride sublimes. The following data, in dry air, refer to losses in weight suffered by 522.8 mg of the dry chloride:

Temp. (°C)	173	214	278	340
Loss (mg)	2.35	24.51	116.6	323. 5

2. Ammonium iodide

Ammonium iodide NH_4I used as a standard[22] has been found to be dry and anhydrous, and maintains a constant weight up to 225°. Above this temperature, it sublimes with partial decomposition, and the crucible is completely empty at 377°.

3. Acid ammonium iodate

The acid iodate $NH_4H(IO_3)_2$ which is used in gravimetric analysis is of special interest because it has the very low conversion factor of 0.03311 for ammonia. The procedure given by Riegler[23] produces a precipitate which filters well and dries immediately. Its thermolysis curve has been followed only up to 168° since iodine vapour is given off at this temperature. The acid iodate may be considered to be stable up to 146° so that the suggested temperature 110° is suitable. This method can be proposed for the automatic determination.

4. Ammonium sulphate

The neutral sulphate $(NH_4)_2SO_4$ has been investigated for the determination of sulphur and as a standard material. When it is produced by treating the sulphuric acid under analysis with an excess of ammonia and evaporating to dryness, the thermolysis curve shows a good horizontal up to 266°. The decomposition then begins at once; it is marked by decrepitation. The crucible is empty at 514°.

5. Ammonium persulphate

Ammonium persulphate $(NH_4)_2S_2O_8$ employed as a standard[24] is anhydrous and is stable up to 235°. It then transforms into ammonium sulphate and the remainder of the curve is identical with the one discussed in 4.

6. Ammonium nitrate

Ammonium nitrate NH_4NO_3 has been considered above for the determination of nitrate ion. We have also investigated it as a standard[25]. Water plays a part in determining the velocity and the temperature of decomposition. If the bottle has been opened several times it is advisable to bring the salt to a temperature below 193° since it is hygroscopic. The sublimation and decomposition into water and nitrous oxide begin to make themselves felt at this latter temperature and the decomposition accelerates around 280° and is complete in the vicinity of 327°.

By means of his thermobalance, Guichard[26] has determined traces of water in ammonium nitrate by constructing the isotherm at 119° for example.

7. Ammonium sulphamate

Ammonium amidosulphonate or sulphamate $NH_4SO_3-NH_2$ has been included by Capestan[27] in his general study of salts of this type. At 170° this salt first yields imidodisulphonate:

$$2\ NH_4SO_3-NH_2 \rightarrow NH(SO_3NH_4)_2 + NH_3$$

and then gives the sulphate:

$$3\ NH(SO_3NH_4)_2 \rightarrow 3\ (NH_4)_2SO_4 + 3\ SO_2 + NH_3 + N_2$$

around 300° but without yielding a horizontal because the sulphate decomposes in its turn.

8. Diammonium phosphate

Diammonium phosphate $(NH_4)_2HPO_4$ used as a standard shows a horizontal up to 111°; decomposition and sublimation occur beyond this temperature. The decomposition probably leads to a complex pyrophosphate. When the horizontal is reached around 900°, only half of the expected weight of phosphoric anhydride is found.[21]

9. Ammonium metavanadate

The metavanadate NH_4VO_3 is taken up in Chapter Vanadium. Considered in the dry state for the direct preparation of a standard solution[28], we point out that it is stable up to 173°; the horizontal of the anhydride V_2O_5 starts at 389°.

10. Ammonium bicarbonate

The monoammonium carbonate (bicarbonate) NH_4HCO_3 dissociates, according to Tsuchiya[29], in four stages from 60° to 110°.

11. Ammonium paramolybdate

The paramolybdate $7MoO_3 \cdot 3(NH_4)_2O \cdot 4H_2O$ is discussed in Chapter Molybdenum. We have already pointed out (Chapter 4) the system of isotherms pertaining to this compound. Used as a standard[28], it shows no loss of weight up to 109°. Above this temperature, the salt yields in succession $5MoO_3 \cdot 2(NH_4)_2O$ from 158° to 208°, then $4MoO_3 \cdot (NH_4)_2O$ from 256° to 304°. The horizontal of molybdic anhydride extends from 393° to 751°, and then this oxide sublimes rapidly.

12. The compound $(Hg_5N_2Cl_2)O-CH_3 \cdot 3H_2O$

The compound $(Hg_5N_2Cl_2)O-CH_3 \cdot 3H_2O$ obtained by the procedure of Gerresheim[30] and Buisson[31] attracted our attention because of its low analytical factor for nitrogen, due to the presence of mercury in the molecule. Unfortunately, the curves prepared with this compound show a continual descent. There is no distinction between the loss of water, carbon dioxide, and mercury. Therefore, the method should be discarded.

13. Ammonium chloroplatinate

Ammonium chloroplatinate $(NH_4)_2$ [$PtCl_6$], which serves for determining ammonia, begins to decompose at $181°$ (loss of chlorine); this is slow up to $276°$ and gives rise to platinum dichlorodiammine [$PtCl_2$ $(NH_3)_2$] and hydrogen chloride. The decomposition is complete at $407°$. Up to $538°$, a constant weight of platinum sponge is obtained; the latter then takes up oxygen on its surface (the phenomenon does not occur in nitrogen) and forms PtO. (The latter has been identified through X-ray spectra.) The maximum weight is at $607°$, but the transformation is far from quantitative. At higher temperatures, the oxide decomposes and the initial weight is regained at $811°$. This experiment demonstrates the oxidation of platinum under atmospheric pressure; it does not occur with crucibles, filings, wire, but it does proceed with platinum black. These changes in weight (which may amount to as much as 5 mg per 123 mg) are not always taken into account when using Gooch-Neubauer filtering crucibles and in the determination of the molecular weight of amines, for example, by the classical chloroplatinate method.

14. Ammonium acetate

Ammonium acetate NH_4COOCH_3, which is used as a standard salt for preparing buffer solutions[21], is quite deliquescent and difficult to weigh accurately. The removal of water by heating is not complete below $176°$ when the rate of heating is $300°$ per hour. It is impossible to distinguish between extraneous and chemically bound water, whose loss results successively in the formation of acetamide and acetonitrile.

15. Ammonium oxalate

The neutral oxalate $(NH_4)_2C_2O_4 \cdot H_2O$ which is employed for preparing standard solutions[32], is stable up to around $65°$ when heated linearly at the rate of $300°$ per hour. It loses its molecule of water between $65°$ and $133°$ and is anhydrous and stable up to $182°$. The formation of oxamide begins at this temperature and continues up to $236°$. Cyanogen is evolved very rapidly around $290°$.

Martens[33] found that the salt is dehydrated from $35°$ to $100°$ and that it has a short horizontal up to $130°$. A decomposition sets in then and becomes rapid at $205°$, and the curve shows a change of direction at $234°$, doubtless due to transformation to oxamide. At $275°$ the crucible is filled with a black carbonaceous residue, which is mostly paracyanogen.

16. Ammonium binoxalate

Martens[33] reports that the binoxalate $NH_4HC_2O_4 \cdot H_2O$ loses water from the time the heating begins; at $75°$ a short horizontal indicates the production of the semi-hydrate, and the anhydrous salt appears at $127°$. At $150°$ the latter starts to decompose with change of direction from $239°$ to $264°$. The crucible is empty at $420°$.

17. Ammonium tetroxalate

The tetroxalate $NH_4H_3(C_2O_4)_2 \cdot 2H_2O$ was also studied by Martens[33]. The salt loses water at $148°$. The remainder of the curve, with changes in direction at $203°$, $226°$, and $246°$, has not been interpreted.

18. Diammonium citrate

Diammonium citrate $(NH_4)_2HC_6H_5O_7$ is anhydrous and quite dry. It is stable up to $161°$; at higher temperatures the combustion and decomposition proceed uneventfully. The crucible is empty around $630°$; the final traces of carbon burn with difficulty[25].

19. Ammonium tetraphenylboron compound

The gravimetric determination of the ammonium ion as the tetraphenylboron compound has been the subject of a number of studies, including those by Wendlandt[34] and Howick and Pflaum[35]. The NH_4 $[B(C_6H_5)_4]$ complex starts to sublime around $130°$. When about $2/3$ of the product has disappeared, at a heating rate of $4.5°$ per minute, the remainder leaves a deposit of boric anhydride in the vicinity of $625°$.

REFERENCES

[1] L. M. DENNIS AND H. ISHAM, *J. Am. Chem. Soc.*, 29 (1907) 24, 220.
[2] D. A. COPEMAN, *J. S. African Chem. Inst.*, 10 (1927) 18.
[3] C. DUVAL AND NG DAT XUONG, *Anal. Chim. Acta*, 6 (1952) 245.

[4] E. DIVERS, *J. Chem. Soc.*, 75 (1899) 166.
[5] N. BUSVOLD, *Chemiker-Ztg.* 38 (1914) 28; 39 (1915) 214.
[6] J. S. LAIRD AND T. C. SIMPSON, *J. Am. Chem. Soc.*, 41 (1919) 524.
[7] F. L. HAHN, *Ber.*, 50 (1917) 705.
[8] A. OELSNER, *Z. angew. Chem.*, 31 (1918) 179.
[9] A. CARNOT, *Traité d'analyse des substances minérales*, Dunod, Paris, 1904, Vol 2, p. 79.
[10] F. A. GOOCH AND S. B. KUZIRIAN, *Z. anorg. Chem.*, 71 (1911) 323.
[11] J. PERSOZ, *Dinglers J.*, 161 (1861) 284.
[12] ALBERTI AND HEMPEL, *Z. angew. Chem.*, 5 (1892) 101.
[13] A. NICOL, *Compt. rend.*, 224 (1947) 1824; 226 (1948) 810.
[14] M. BUSCH, *Ber.*, 38 (1905) 861, 4055.
[15] M. MESTREZAT AND M. DELAVILLE, *Bull. Soc. chim. biol.*, 8 (1926) 1217.
[16] L. W. WINKLER, *Z. angew. Chem.*, 34 (1921) 46.
[17] A. ARNAUD AND L. PADÉ, *Compt. rend.*, 98 (1884) 1488; 99 (1884) 190.
[18] H. RUPE AND P. BECHERER, *Helv. Chim. Acta*, 6 (1923) 880.
[19] F. ARNDT, *Liebigs Ann.*, 384 (1911) 322.
[20] T. DUVAL AND C. DUVAL, *Anal. Chim. Acta*, 2 (1948) 103.
[21] C. DUVAL, *Anal. Chim. Acta*, 13 (1955) 427.
[22] C. DUVAL, *Anal. Chim. Acta*, 16 (1957) 546.
[23] E. RIEGLER, *Z. anal. Chem.*, 42 (1903) 677.
[24] C. DUVAL, *Anal. Chim. Acta*, 20 (1959) 264.
[25] C. DUVAL AND C. WADIER, *Anal. Chim. Acta*, 20 (1959) 263.
[26] M. GUICHARD, *Compt. rend.*, 215 (1942) 20.
[27] M. CAPESTAN, *Ann. Chim. (Paris)*, 5 (1960) 207.
[28] C. DUVAL, *Anal. Chim. Acta*, 15 (1956) 224.
[29] R. TSUCHIYA, *J. Chem. Soc. Japan*, 74 (1953) 210.
[30] H. GERRESHEIM, *Liebigs Ann.*, 195 (1879) 373.
[31] A. BUISSON, *Compt. rend.*, 144 (1907) 493.
[32] C. DUVAL, *Anal. Chim. Acta*, 13 (1955) 35.
[33] P. H. MARTENS, *Bull. inst. agron. et stat. recherches Gembloux*, 21 (1953) 134.
[34] W. W. WENDLANDT, *Chemist Analyst*, 26 (1957) 38: *Anal. Chem.*, 28 (1956) 1001.
[35] L. C. HOWICK AND R. T. PFLAUM, *Anal. Chim. Acta*, 19 (1958) 342.

CHAPTER 11

Fluorine

The following forms have been suggested for the gravimetric determination of fluorine: lithium fluoride, calcium fluoride, lanthanum fluoride, bismuth fluoride, thorium fluoride, triphenyltin fluoride, lead chlorofluoride, uranium(IV) oxyfluoride, barium silicofluoride, potassium silicofluoride and thorium silicofluoride. Dupuis and Duval[1] have made an experimental and critical study of these proposals. The general studies by Geyer[2], Beaucourt[3], Mac Vicker[4], Rinck[5] and Frommes[6] have been of great service in this connection.

1. Lithium fluoride

The precipitation as lithium fluoride LiF as suggested by Ehrenfeld[7] and eventual weighing as lithium sulphate will not be discussed here again. See Chapter Lithium.

2. Calcium fluoride

Regarding the weighing as calcium fluoride see Chapter Calcium.

3. Lanthanum fluoride

With regard to the method of Meyer and Schulz[8] in which lanthanum fluoride is brought down, we have repeated this study and found that the white precipitate yields a curve which is inclined downward until 475° is reached, and has an abrupt descent at 322°. A horizontal, which corresponds to the sum of $LaF_3 + La_2O_3$ starts at 475° and extends to at least 946°.

4. Bismuth fluoride

Bismuth fluoride, produced by the method of Domange[9], begins to decompose at 103°; it yields a horizontal which may be used perhaps for the automatic determination of fluorine. A new horizontal, which corresponds to a fluorine-free compound, appears above 163°. The molecular weight 260 of this material agrees fairly well with the hydroxide $Bi(OH)_3$.

5. Thorium fluoride

Thorium fluoride ThF_4, which was produced and washed by the method of Gooch and Kobayashi[10], is shown on the thermolysis curve as a break near 90°. It is not stable and yields no horizontal. It loses fluorine and forms pure hydroxide $ThO_2 \cdot 2H_2O$, which gives the horizontal extending from 242° to 475°. Above 475°, the hydroxide in turn loses water, leaving thorium oxide ThO_2 beyond 946°.

6. Triphenyltin fluoride

Dupuis and Duval tried the new precipitation of the fluoride ion, recommended by Allen and Furman[11], by means of triphenyltin. This reagent was discovered by Krause and Becker[12]. Allen and Furman recommend that the precipitate be dried for 30 minutes at 110°; it has the formula $(C_6H_5)_3SnF$. The curve recorded with this compound shows that it is stable up to 158°. The decomposition then starts abruptly, yielding for each 4 molecules, 1 molecule of stannic fluoride, 2 molecules of stannic oxide SnO_2, which constitute the residue in the crucible, and doubtless also 1 molecule of an organometallic tin compound, which ultimately is converted to SnO_2 and forms a fine network on the walls of the furnace.

7. Lead chlorofluoride

Starck[13] and Adolph[14] have given directions for the precipitation of lead chlorofluoride PbClF and recommend that it be dried at 140–150°. The pyrolysis curve is extremely distinct and includes a good horizontal from 66° to 538° which corresponds to the above formula and which is suitable for the automatic determination of fluorine. Decomposition sets in above 538° and the resulting lead chloride sublimes quickly.

8. Uranium oxyfluoride

Regarding uranium(IV) oxyfluoride see Chapter Uranium.

References p. 194

9. Barium silicofluoride

Barium silicofluoride Ba [SiF$_6$] is prepared by the method of Rose[15]. It yields a constant weight up to 345°, where it begins to decompose into volatile silicon fluoride SiF$_4$ and barium fluoride BaF$_2$, which in turn maintains a constant weight from 542° to 946°. We believe the method has not much to offer.

10. Potassium silicofluoride

Regarding potassium silicofluoride see Chapter Silicium.

11. Thorium silicofluoride

Dupuis and Duval have attempted to repeat the procedure of Deladrier[16] for the preparation of thorium silicofluoride. However, the product resisted all of the filtering processes available to them.

REFERENCES

[1] T. DUPUIS AND C. DUVAL, Anal. Chim. Acta, 4 (1950) 615.
[2] R. GEYER, Z. anorg. Chem., 252 (1943) 42.
[3] J. H. BEAUCOURT, Metallurgia, 38 (1948) 353.
[4] L. D. MAC VICKER, Trans. Illinois State Acad. Sci., 30 (1937) 190.
[5] E. RINCK, Bull. Soc. chim. France, (1948) 305.
[6] M. FROMMES, Z. anal. Chem., 96 (1934) 211.
[7] R. EHRENFELD, Chemiker-Ztg., 29 (1905) 440.
[8] R. J. MEYER AND W. SCHULZ, Z. angew. Chem., 38 (1925) 203.
[9] L. DOMANGE, Bull. Soc. chim. France, 9 (1942) 96; Compt. rend., 213 (1941) 31.
[10] F. A. GOOCH AND M. KOBAYASHI, Am. J. Sci., 45 (1918) 370.
[11] N. ALLEN AND N. H. FURMAN, J. Am. Chem. Soc., 54 (1932) 4625.
[12] E. KRAUSE AND R. BECKER, Ber., 53 (1920) 183.
[13] G. STARCK, Z. anorg. Chem., 70 (1911) 173.
[14] A. ADOLPH, J. Am. Chem. Soc., 37 (1915) 2509.
[15] H. ROSE, Handbuch der analytischen Chemie, 6th Ed., 1871, p. 570.
[16] E. DELADRIER, Chem. Weekblad, 1 (1904) 324.

CHAPTER 12

Sodium

The following compounds have been suggested for the gravimetric determination of sodium: chloride, perchlorate, sulphate, antimonate, sodium cesium bismuth nitrite, sodium magnesium (or zinc) uranyl acetate. The investigation of the precipitates has been conducted by Duval and Duval[1]. Because of their solubility, many sodium salts have been suggested for the preparation of standard solutions by direct weighing, and sodium peroxide has made it possible to conduct a general study of solid state reactions with various oxides.

1. Sodium peroxide

Oxylith or sodium peroxide Na_2O_2 presents the following course in the thermolysis curve studied by Duval[2,3] and in more detail by Viltange[4]. The correct interpretation of the curves could only be carried out after the development of a micromethod for the determination of the quantity of sodium carbonate contained in the peroxide. The curves are not the same for specimens taken from bottles which are opened for the first time as opposed to those which have been unstoppered more than once.

Depending on the sample, there is a slight gain in weight, due to the absorption of moisture and carbon dioxide, from room temperatures up to $63° \pm 5°$ or $140° \pm 10°$. Then, up to $260° \pm 10°$, the loss of weight is due to the release of moisture. Sodium peroxide absorbs only carbon dioxide up to $314° \pm 4°$; the curve includes a horizontal or a line that is slightly inclined from the horizontal. Considerable amounts of oxygen are given off from $314°$ to $660° \pm 40°$ and the peroxide changes into sodium monoxide Na_2O. The oxide is partially carbonated between $660°$ and $850°$; the curve frequently shows a slight increase of weight. Starting at $850°$, all of the sodium carbonate decomposes. The release of the carbon dioxide is not complete at $1050°$ (the temperature limit of the furnaces used) and it is necessary to keep the contents of the crucible at this temperature for at least two hours if this decomposition needs to be

References p. 212

carried to its conclusion. The trials were conducted on samples ranging from 80 to 400 mg.

Heating of sodium peroxide with several oxides   Viltange subsequently carried out the reaction of sodium peroxide in the thermobalance with the following oxides: Al_2O_3, Bi_2O_3, Cr_2O_3, Sb_2O_3, MoO_3, WO_3, UO_2 and UO_3, U_3O_8, V_2O_5, Nb_2O_5, Ta_2O_5 [5-8]. The products formed at various temperatures were identified by infra-red spectrophotometry.

For instance, vanadic anhydride, V_2O_5 yielded a pervanadate $NaVO_4$ between 85° and 307° and an orthovanadate Na_3VO_4. A certain amount of vanadic anhydride and only the orthovanadate remains between 307° and 440°. Ortho- and metavanadate $NaVO_3$ coexist beyond 440° along with the residual anhydride.

This extensive study was undertaken to find out what the analyst is really accomplishing when he makes "a fusion" with a metal oxide and sodium peroxide, and to discover whether the quantity of peroxide used is always suitable to his purpose.

2. Sodium hydroxide

Duval[9] studied the hydroxide NaOH from the standpoint of its use as a standard. The pellets of this material (Merck) absorb moisture up to around 94°, and then return to a material which is essentially anhydrous between 150° and 337° where the product is already molten (m.p. 318°). Water of constitution then comes off and a new horizontal appears above 480°, which corresponds to sodium oxide Na_2O. The latter is stable up to 980°, which marks the upper limit of the trials. Sometimes, the oxide is slightly carbonated.

Mauret[10] has defined the regions on the thermolysis curves over which the various hydrates exist; he used a thermobalance heated by infra-red radiation. He detected and made an X-ray characterization of the two compounds $NaOH \cdot H_2O$, which has been known for a long time, and $2NaOH \cdot H_2O$, which is a new compound, whose existence had hitherto given rise to contradictory investigations.

3. Sodium chloride

Sodium chloride NaCl, prepared by the action of hydrochloric acid on sodium carbonate, smoothly loses the excess of this acid and the water up to 407°. A perfectly horizontal stretch is observed up to 878°.

4. Sodium perchlorate

Sodium perchlorate $NaClO_4 \cdot H_2O$ is prepared by treating a solution of sodium chloride with an excess of 72% perchloric acid, and then removing the excess by heating. The thermolysis curve shows that the dehydration is complete at 130° and a horizontal of constant weight extends from 50° to 81°. (It should be noted that 50° corresponds essentially to the eutectic temperature of the monohydrate and the anhydrous salt.) The latter is stable up to 471°. Ordinarily, the directions call for weighing after drying the salt at 350°, which is correct. However, the existence region of the perchlorate is rather wide so that it is permissible to select some other temperature. Beyond 471°, decomposition sets in and becomes tumultuous at 538°. The anhydrous chloride is obtained[1] above approximately 620°.

5. Sodium iodate

Rocchiccioli[11] constructed the thermolysis curve of sodium iodate $NaIO_3 \cdot H_2O$ as part of a general investigation of the structure of the IO_3^- ion. The dehydration begins at 110° and is complete around 152° where the horizontal of the anhydrous iodate starts. The study was not continued beyond 250°.

6. Sodium hydrogen paraperiodate

Sodium acid paraperiodate $Na_2H_3IO_6$, which is used as a standard substance, was found by Duval[12], to be anhydrous and stable up to 207° where decomposition sets in and gives rise to water, oxygen, and sodium iodate, iodide, and oxide. The iodate and iodide sublime above 820°.

7. Sodium sulphide

Colourless sodium sulphide, used as a standard substance, contains almost 9 molecules of water when taken from a bottle that is being opened for the first time[13] but it is very hygroscopic. The direct preparation of a standard solution of this salt is most difficult because it begins to lose water around 30°. The recorded curve indicates that the anhydrous material gives an almost horizontal stretch between 240° and 307°; from then on the curve ascends in an essentially linear fashion and sodium sulphite is obtained around 950° but it is not entirely pure (traces of sulphate).

8. Sodium sulphate

Sodium sulphate $Na_2SO_4 \cdot 10H_2O$ is prepared by the action of sulphuric acid on sodium carbonate. The crystals begin to lose weight at room temperature. The eutectic point ($32.45°$) is not marked by any feature in the curve. At $72°$, the curve has a point of inflection corresponding to the approximate composition $Na_2SO_4 \cdot 4.75H_2O$. The dehydration is complete at $90°$ and the horizontal corresponding to the anhydrous salt remains level up to above $880°$[1].

9. Sodium pyrosulphate

Duval[12] investigated the behaviour of sodium pyrosulphate, which occurs in somewhat hygroscopic plates; it is not advisable to weigh them as they come from the bottle. Stability is maintained up to around $197°$, in contrast to sodium acid sulphate $NaHSO_4$, which is not a standard but which Duval examined for comparison purposes. As to the pyrosulphate $Na_2S_2O_7$, it slowly loses SO_3 up to around $800°$, and the horizontal of the neutral sulphate Na_2SO_4 appears from $840°$ on.

The equation:

$$Na_2S_2O_7 \rightarrow SO_3 + Na_2SO_4$$

has not been rigorously verified and the melting point of a sample taken from a fresh bottle is $147°$. This is not in agreement with the data published by Cambi and Bozza[14] who reported $400°$ and who in addition state that the material is not decomposed at this temperature. Our figures are likewise in disagreement with those of a recent study by Tomkova, Jirů and Rosický[15]. They are of the opinion that the bisulphate $NaHSO_4$ is converted into the pyrosulphate at $205°$ and that the latter is wholly changed into sulphate at $370°$. This finding may be a result of the low sensitivity of their thermobalance because the curve of the loss of SO_3 is almost a horizontal.

10. Sodium bisulphate

Sodium bisulphate or acid sulphate $NaHSO_4 \cdot 1H_2O$ was heated at the rate of $300°$ per hour and then $150°$ per hour. It has also been studied at constant temperature, etc. It starts by losing its water of crystallization up to around $100°$, where we have found a break in the curve. Then water of constitution is given off from $100°$ to $340°$:

$$2 NaHSO_4 \rightarrow Na_2S_2O_7 + H_2O$$

with abatement and an oblique level around 260°. The resulting pyro-sulphate is stable up to 430° at least. The release of sulphuric anhydride then proceeds up to 620°. The neutral sulphate is obtained:

$$Na_2S_2O_7 \rightarrow Na_2SO_4 + SO_3$$

It is difficult to weigh the bisulphate to prepare a standard solution because it loses water even at room temperature (starting at 30° in the case of a fresh bottle).

11. Sodium thiosulphate

Duval[9] and also a Russian worker[16] have studied sodium thiosulphate $Na_2S_2O_3 \cdot 5H_2O$. The salt does not retain its 5 molecules of water beyond 37°. Already, at 47°, the loss is very obvious. A short horizontal near 114° corresponds closely to a loss of 2 molecules of water. A slightly inclined horizontal, but much longer, extends from 215° to 387°, and agrees essentially with the anhydrous salt. The decomposition then occurs vigorously at 475°. The residue contains sodium sulphate especially, and variable amounts of sulphite and sulphide. It is not possible to give a single equation representing this disintegration.

12. Sodium sulphamate

Sodium amidosulphonate (sulphamate) $NH_2\text{--}SO_3Na$ has been heated by Capestan[17] and like the potassium and lithium salts yields the imidodi-sulphonate:

$$2\ NaSO_3\text{--}NH_2 \rightarrow NH(SO_3Na)_2 + NH_3$$

13. Sodium selenite

Sodium selenite Na_2SeO_3 (commercial grade) contains variable quanti-ties of hydroselenite. Therefore, in her general study Rocchiccioli[11] neutralized selenious acid with sodium carbonate or *vice versa*. The neutral salt is produced at pH 10.5 and the hydrogen selenite (biselenite) at pH 5.3. If the neutral salt is heated at the rate of 300° per hour it remains stable up to 690° and then begins to oxidize, though incompletely, to selenate.

14. Sodium hydrogen selenite

Sodium hydrogen selenite $NaHSeO_3$ loses one-half molecule of water per molecule between 100° and 230°, and is converted into pyroselenite as shown by the equation:

$$2\,NaHSeO_3 \rightarrow Na_2Se_2O_5 + H_2O$$

This new salt, which is stable up to 400°, then decomposes to yield the neutral selenite. The thermolysis curve of crystals of the latter, contaminated with biselenite, is intermediate between the two preceding ones[11].

15. Sodium nitrite

Sodium nitrite $NaNO_2$, dry, white, non hygroscopous, taken from a fresh bottle, is perfectly stable up to 538°. A slow decomposition then sets in and continues up to around 750°, becomes more rapid between 792° and 956°, and around 1000° the residue consists of sodium oxide Na_2O. These findings[12] are therefore not in agreement with those of Oswald[18] who reported 320° for the start of the decomposition, nor with the results of Rây[19] who observed a transformation into nitrate around 500°. The corresponding gain in weight is not shown on our curve.

16. Sodium nitrate

(i) Normal thermolysis Sodium nitrate $NaNO_3$, which we have heated either at the rate of 150° or 300° per hour, proved to be stable up to 670°. Sometimes, as do all oxidants, its curve showed a very distinct rise up to this temperature; the gain amounted to 2 mg per 265 mg. The oxygen is given off smoothly and the horizontal of sodium nitrate is reached between 980° and 1000°, temperatures which are lower when lead is present.

(ii) Thermolysis under reduced pressure or in gas atmosphere Furthermore, Cases-Casanova[20] working either under reduced pressure or in an atmosphere of argon, oxygen, or nitrogen oxides, between 500° and 900°, made a simultaneous study of sodium nitrite and nitrate, whose pyrolyses are very sensitive to the rate of flow of the scouring gas, the pressure, the shape of the container, and to the weight of the salt employed. In an initial phase, the nitrate undergoes dissociation into oxygen and nitrite; in a second phase, the nitrite disproportionates with reformation of nitrate. The writer has also investigated the influence of traces of oxides, such as Fe_2O_3, Al_2O_3, SiO_2, SnO, on the thermolysis.

17. Sodium hypophosphite

Sodium hypophosphite NaH_2PO_2 has been carefully studied by Merlin[21] in the Eyraud balance. The sample weighed 150 mg; the heating in nitrogen was at the rate of 1.5° per minute, and the pressure was equivalent to 5 mm of mercury. Up to about 180°, we note the stability level of the pure salt, and a slow loss between 200° and 270°, which becomes very rapid up to 300°, corresponding to the disproportionation:

$$14\,NaH_2PO_2 \rightarrow 6\,PH_3 + 4\,Na_2HPO_3 + 2\,NaPO_3 + Na_4P_2O_7 + 3\,H_2O$$

A new and slightly inclined level between 300° and 370° is related to the reaction:

$$8\,Na_2HPO_3 \rightarrow 2\,PH_3 + 4\,Na_3PO_4 + Na_4P_2O_7 + H_2O$$

18. Trisodium phosphate

Duval[22] and Zettlemoyer et al.[23] have provided information regarding the thermolysis of trisodium phosphate $Na_3PO_4.6H_2O$ when heated at the rate of 300° per hour. A specimen taken from a bottle which had been opened several times showed 2.3 molecules of water per molecule. By exactly neutralizing phosphoric acid with sodium carbonate and recrystallizing the product, the resulting salt showed 10.4 molecules of water. Therefore, it probably was a mixture of 10 and 12 or 12 and 7 H_2O. In addition, it is difficult to prevent these products from taking up carbon dioxide. The thermolysis curve has a horizontal for the trisodium phosphate starting at 230°. Of course, if this curve is examined in the vicinity of 840° it will be discovered whether it has a slight jog corresponding to the dissociation of any sodium carbonate which may be present. Here again, it will be much better to employ the anhydrous salt when preparing a standard solution.

19. Disodium phosphate

The disodium phosphate Na_2HPO_4 has been the subject of many studies. The recording made point by point by Marcel Guichard in 1923 has been discussed in Chapter 1. This study has been repeated by Duval and Dupuis[9], Murthy, Bharadwaj and Mallya[24], Wendlandt[25], Waters[26], and others.

The commercial product taken from the bottles never contains 12 molecules of water of crystallization but rather a number near 7. The material which Duval[22] used for this investigation carried 6.7 H_2O. The dehydration starts at room temperature; it appears to be complete at 180° when the heating rate is 300° per hour, and 180° marks the beginning of the good horizontal of anhydrous disodium phosphate. This salt is stable over a range of 100° at least. It is known that at this point two molecules lose another molecule of water, and above 355° this loss results in the formation of sodium pyrophosphate. If therefore, when preparing a buffer solution, the operator prefers to start with the anhydrous salt rather than commercial materials, it will be necessary beforehand to keep the salt between 180° and 280° for a time. However, the salt will become difficultly soluble in water.

20. Sodium pyrophosphate

Sodium pyrophosphate (commercial grade) contains 10 molecules of water (calculated 9.96) and its thermolysis curve shows that this decahydrate is stable only up to 46°. The dehydration proceeds up to 156°; here the horizontal of the anhydrous product begins and extends, with no change in weight, as far as 826°, the upper limit of our experiment. We therefore believe it preferable to weigh out the anhydrous material if a standard solution is to be prepared.

This salt has been heated in mixtures by Griffith[28].

21. Sodium hydrogen triphosphate

Morin[27] studied the behaviour of the sodium triphosphate $Na_3H_2P_3O_{10} \cdot 1.4H_2O$ when heated at the rate of 100° per hour. It showed a rapid loss of water between 140° and 165° and then a very slow loss between 165° and 255°. The product obtained by quenching at 250° corresponds to the removal of 0.9 molecule of water. It is pure β-disodium pyrophosphate $Na_2H_2P_2O_7$. The removal of water again becomes considerable from 255° on; it results from the decomposition of this latter species.

22. Sodium triphosphate

Griffith[28] found that the triphosphate $Na_5P_3O_{10} \cdot 6H_2O$ is completely dehydrated in the vicinity of 100°. However, it does not yield the anhydrous salt but instead a complex mixture of ortho- and pyrophosphates since only 5 molecules of the 6 come off easily. The printed account contains the isotherm at 100°, and also the curve resulting from the heating of the two salts $Na_4P_2O_7 \cdot 10H_2O + Na_5P_3O_{10} \cdot 6H_2O$, and some mixtures with sodium carbonate.

23. Sodium trimetaphosphate

We wish also to point out that Salih Hisar[29] has made a study with the thermobalance of solid state reactions between trimetaphosphate $(NaPO_3)_3$ and various salts, such as NaI, NaBr, NaCl, NaF.

24. Sodium arsenates

Guerin *et al.*[30] have studied the sodium arsenates. The acid salt $Na_2O \cdot 2As_2O_5 \cdot 7H_2O$ is completely dehydrated at 200° in vacuo. The slight excess of As_2O_5, which it invariably contains, comes off at 300°, and then, always in vacuo, it changes at 550° into the metarsenate:

$$Na_2O \cdot 2As_2O_5 \rightarrow Na_2O \cdot As_2O_5 + As_2O_3 + O_2$$

The metarsenate melts at 625° and its decomposition, detectable at this temperature, leads rapidly, beginning at 800°, to the sodium pyroarsenate $Na_4As_2O_7$. The latter may also be prepared by the dehydration at 220° of the tribasic orthoarsenate.

If the dodecahydrate $Na_2HAsO_4 \cdot 12H_2O$ is placed *in vacuo* it yields, at ordinary temperature, the trisodium orthoarsenate Na_3AsO_4. The latter on being heated to 950° and higher decomposes at the same time that it distinctly sublimes.

25. Sodium antimonate

Sodium antimonate is used for the determination of sodium. According to Pauling[31] the salt should be written Na [Sb(OH)$_6$], and this formula has been confirmed by Dupuis[32, 33] on the basis of the Raman and infra-red spectra. The salt retains variable amounts of water depending on the conditions under which the precipitation is conducted. The specimen examined in our general study[1] was prepared by the reaction of a hot solution of potassium "pyroantimonate" with sodium chloride; it has the composition $Na_2O \cdot Sb_2O_5 \cdot 6.07H_2O$. It starts to lose water at 128°. The curve shows that two-thirds of the water are eliminated around 225° where a distinct break in the curve appears. The final third is lost more slowly and the dehydration is complete at 600°. The residue consists of $NaSbO_3$ and maintains a constant weight from this temperature up to 950° and shows itself to be different from the potassium salt.

26. Sodium borohydride

The borohydride $NaBH_4$ was investigated by Ostroff and Sanderson[34].

They heated it in air, nitrogen, and hydrogen in the Mauer thermo-balance. The final product is the metaborate $NaBO_2$. The behaviour on heating has also been followed by differential thermal analysis. The starting decomposition temperatures are 290° (in air), 503° (in nitrogen), and 512° (in hydrogen). The metaborate is completely formed at 420° (in air) and at 620° (in hydrogen).

An identical study has been based on the potassium compound KBH_4.

27. Sodium tetraborate

Sodium tetraborate (borax) $Na_2B_4O_7 \cdot 10H_2O$ has engaged the attention of a number of workers, for example, Duval[22], Dupuis and Duval[35], Griffith[28]. The decahydrate was heated at the rate of 150° per hour and proved to be stable up to 35–38°. It then slowly loses half of its water up to around 108° where an oblique level or simply a point of inflection appears. After this, the loss of water proceeds almost regularly up to 172°. The last molecule of water is very firmly attached and it is necessary to carry the temperature to at least 500° or even 525° to obtain a constant weight of the anhydrous salt. The starting material does not contain precisely 10 H_2O but 9.45. Another specimen, which was obtained by crystallization of a solution prepared from the anhydrous salt had only 9.1 H_2O but the shape of the dehydration curve was as before. Conse-quently, it is not advisable to weigh this compound for preparing buffer solutions on the assumption that it is the decahydrate.

28. Sodium carbonate

Sodium carbonate has been thoroughly studied from the titrimetric standpoint by Duval[22], Dupuis and Duval[35], Newkirk and Aliferis[36], and Easterbrook[37]. The latter believes that the sesquicarbonate yields a sufficiently pure carbonate Na_2CO_3 between 210° and 270°.

(i) Heating of reagent grade sodium carbonate Sodium carbonate has both supporters and detractors with regard to its use as a standard for acidic titrimetric solutions. We have made a critical study with samples weighing from 200 to 500 mg in aluminite crucibles and salts of reagent grade (R.P.). The first set of experiments show that the carbonate is really dry when it is taken from a fresh bottle. The recorded curve is horizontal up to 840° ± 5° (m.p. 851°). Decomposition then sets in ab-ruptly and continues up to around 1000°. The loss at this temperature rose to 23.0 mg when the initial weight of the dry salt was 559.4 mg.

(ii) Heating of reagent grade salt with water   The second set of experiments dealt with the same material to which a little water was added intentionally to learn at what precise temperature the latter has disappeared. This water is given off between 76° and 100°, and therefore, in practice, when the bottle has been opened a number of times, the operator can be sure that the moisture will be removed and that the salt will serve for direct weighing if it is kept between 100° and 840° for 15 minutes. We wish to note that between these temperatures a loss of 0.1 mg is not observable by recording on the thermobalance for a weight of material between 200 and 500 mg.

(iii) Heating in gas atmosphere   In a very important study, Newkirk and Aliferis[36] carried out thermolyses of sodium carbonate in crucibles made of porcelain, alumina, platinum, or gold, in dry or moist air, in nitrogen, and carbon dioxide. They observed a most important fact, namely that mixtures of sodium carbonate and silica lose weight from 500° on. This solid state reaction has often produced errors when sodium carbonate was "desiccated" in quartz vessels.

The decomposition of sodium carbonate at temperatures going up to 1000° is completely suppressed by using an atmosphere of carbon dioxide.

These workers draw the conclusion that sodium carbonate may be heated without damage to any temperature between 250° and 700°, at least in a stream of dry air or carbon dioxide.

(iv) Sodium carbonate decahydrate   Sodium carbonate decahydrate $Na_2CO_3 \cdot 10H_2O$ has been studied by Reisman[38] under a heating rate of 0.5° per minute. He conducted thermal differential analyses and thermogravimetric measurements simultaneously to construct the isobaric dissociation curve on the modified Ainsworth balance.

29. Sodium bicarbonate

Sodium hydrogen carbonate (bicarbonate) was studied by Duval[22], Dupuis and Duval[35], and Waters[39]. If heated at the rate of 300° per hour, this salt maintains its weight up to 79°, the temperature at which slow decomposition sets in. Around 113° the material has lost about 4 mg out of 236 mg. The curve then changes direction and the conversion into carbonate is complete, namely at 186° in an open crucible and with ready escape of the carbon dioxide. The horizontal of sodium carbonate discussed above then begins, and the decomposition of the latter commences at 840°. Therefore, sodium bicarbonate may be employed for preparing a standard sodium solution.

The mixed sodium and potassium salt was discussed at the 1st Congress on thermogravimetry held at Kazan[16].

30. Sodium silicate

Crystalline sodium silicate $Na_2SiO_3 \cdot 9H_2O$ was investigated by Manvelyán et al.[40] who found that 5 H_2O were lost between 80° and 90°. The dehydration is not completed below 500°, but there is a preliminary horizontal corresponding approximately to a content of 1 molecule of water.

31. Sodium chromate

Sodium chromate was studied by Borchardt[41] and later by Dupuis[42]. If the thermolysis curve is recorded in the air, and in an uncovered crucible, and at a heating rate of 200° per hour, the salt $Na_2CrO_4 \cdot 4H_2O$ yields an intermediate horizontal around 100–110°, which corresponds to the composition $Na_2CrO_4 \cdot 3H_2O$, a hitherto unreported hydrate. It was studied by infra-red spectrography and by deuteration.

32. Sodium tungstate

Sodium tungstate $Na_2WO_4 \cdot 2H_2O$, used as a standard, was studied by Duval[43] and by Kolli, Pirogava and Spitsyn[44]. It is quite dry and may be weighed directly. If, however, a trace of extraneous moisture is suspected, the drying temperature should not go beyond 80°. The removal of water is complete at 150°, the temperature which marks the beginning of the horizontal of the anhydrous material.

33. Sodium ferrocyanide

Sodium ferrocyanide $Na_4[Fe(CN)_6] \cdot 10H_2O$ was found by Cappellina and Babani[45] to be dehydrated between 30° and 125°.

34. Sodium cesium bismuth nitrite

The triple nitrite $6NaNO_2 \cdot 9CsNO_2 \cdot 5Bi(NO_2)_3$, which is known as

Ball's salt[46], has been suggested for the determination of sodium in biological media. This writer suggests that the yellow needles be dried for 15 minutes at 100°. As a matter of fact, the recorded curve is still descending at this temperature; the horizontal of constant weight does not begin below 160°. It is remarkable that this triple nitrite, which undoubtedly is a complex salt, is found to be stable up to 670°. Here it turns golden yellow. We had expected that this material would conform to the formula $Cs_9Na_6[Bi(NO_2)_6]_5$ but in a recent study by Puget[47] it has been proved through infra-red spectrography that the bismuth atom is not encircled by six nitrite groups. The initial loss of weight of the crystals is of the order of 1/50th; some of those who have worked with this material have not discovered this loss. Accordingly, this decrease in weight can only involve water of crystallization corresponding to 3 or 4 molecules, a fact based on the infra-red spectrum, and on the temperature at which this water begins to be lost, and on the sharpness of the crystals. Because of the high molecular weight, namely 3759, it is not possible for us to be more precise; no matter what the number of molecules at the start, the weighing of the residue constitutes an excellent though little known method for determining sodium.

35. Sodium uranyl selenite

Claude[48] has studied the behaviour of the sodium uranyl selenites on heating. The monouranyl diselenite $Na_2UO_2(SeO_3)_2$ is stable up to around 450° and loses one molecule of selenious anhydride as shown by the equation:

$$2\ Na_2UO_2(SeO_3)_2 \rightarrow Na_2SeO_3 + Na_2U_2O_7 + 3\ SeO_2$$

This loss is ended around 650°.
The diuranyl triselenite likewise suffers a loss of one molecule of SeO_2 between 480° and 530°:

$$Na_2(UO_2)_2(SeO_3)_3 \rightarrow Na_2U_2O_5(SeO_3)_2 + SeO_2$$

and then loses a second molecule between 530° and 750°, indicating a double loss:

$$Na_2U_2O_5(SeO_3)_2 \rightarrow Na_2U_2O_7 + 2\ SeO_2$$

The acid monosodium diuranyl diselenite $NaHUO_2(SeO_3)_2$ if dried in vacuo is stable up to 200°; it contains no water of crystallization. Between 200° and 400° it loses one molecule of water and one of selenious anhydride and yields one of the preceding compounds:

$$2\ NaHUO_2(SeO_3)_2 \rightarrow SeO_2 + H_2O + Na_2\ (UO_2)_2(SeO_3)_3$$

36. Streng's salt (Sodium magnesium uranyl acetate)

We have studied the precipitate of Streng's salt[49] produced by the action of the Blanchetière reagent[50] or Kahane's reagent[51].

(i) Precipitation by Blanchetière's reagent The molecular weight of 1500 has always made it difficult to learn the precise water content, and furthermore, as our recordings show, it is impossible to prepare the anhydrous salt without losing acetic acid. Electrophoresis experiments carried out by Charonnat[52], show that this salt corresponds to the complex formula $Na[Mg\{UO_2(CH_3COO)_3\}_3]$. The precipitated crystals do not have a constant composition; they contain between 6 and 9 molecules of water and, in accord with the usual rule, the water content rises with lowering of the temperature at which they are precipitated. We worked at $20°$ with samples containing essentially $6.5\ H_2O$. Up to $91°$, the weight remains constant, but at $110°$, the temperature which is usually stated to be proper for drying, the loss, though slight, is still measurable on the photographic paper. We have not taken into account the observation reported by Wyrouboff[53] that the salt is not yet completely dehydrated at $200°$; actually at this temperature a goodly portion of the acetic acid has already disappeared subsequent to the break in the curve at $170°$. The removal of 9 molecules of acetic acid along with the simultaneous gain of 10.5 atoms of oxygen proceed up to $360°$, and in a distinctly explosive fashion from $275°$ on. The composition of the residue conforms exactly to that given by Kahane, namely $\frac{1}{2}Na_2U_2O_7 \cdot MgU_2O_7$; the residue can be weighed in this form up to $745°$. At that temperature, we encountered a fact which we believe to be new: it regains some weight (and the gain can be due only to oxygen) up to $882°$ through reoxidation of the uranium, which has been partially reduced in the furnace by carbon monoxide and ketene. From this point on, the curve descends slowly; the pyrouranates decompose, and the uranium is partly converted to U_3O_8.

These findings lead to the conclusion that Streng's salt is a variable mixture of hydrates. The analytical factors obtained from it are not precise. If the precipitation is conducted at $20° \pm 1°$, the composition conforms rather well with $6.5\ H_2O$, and the global mixture of hydrates may be weighed after drying at up to $91°$. It is more accurate to heat the precipitate between $360°$ and $745°$ and to weigh the mixture of the two pyrouranates, whose composition is not dependent on the initial precipitation temperature. The analytical factor for sodium, namely

0.0247, is less favourable but, as compensation, the composition of the material being weighed is known.

(ii) Precipitation by Kahane's reagent   When employing the Kahane reagent, we have carefully followed the directions laid down by this chemist[51]. The precipitation was made from a sodium chloride solution, at $20° \pm 1°$, by adding the water–alcohol solution of the reagent, which was not more than 48 hours old and recently filtered. The curves obtained with different sodium contents are comparable with each other and with the preceding one. The precipitates do not contain precisely 8 molecules of water. These disappear from 103° upward; no horizontal corresponding to the anhydrous salt is observed, but merely a break at 156°. The acetic acid is lost, slowly at first, and up to 250°; then the decomposition becomes almost instantaneous. A horizontal relating to the double pyrouranate starts at 305°. On one of our recordings, the loss of acetic acid was so rapid that the gain of oxygen required for the formation of the U_2O_7 groups did not occur simultaneously and the curve had to ascend to reach its horizontal level. As in the previous case, we have not observed up to 880° an ascent indicating a gain of oxygen, the heating rates being the same. Once again, we believe that the more precise value of this classic method of determination is obtained by weighing the double pyrouranate $\frac{1}{2}Na_2U_2O_7 \cdot MgU_2O_7$.

37. Sodium zinc uranyl acetate

The precipitate, whose formula is $Na[Zn\{UO_2(CH_3COO)_3\}_3] \cdot nH_2O$, is prepared by the procedure given in ref.[54] and taken from the paper of Barber and Kolthoff[55]. The salt was precipitated from a neutral solution of sodium chloride. The pyrolysis curve shows a stretch that is strictly rectilinear up to 75°; this is the region of the mixed hydrates. The complex salt loses its water from 75° to 118°. The horizontal of the anhydrous salt is so short that weighing in this form becomes a delicate matter. It therefore is better to work from 118° to 125°; the acetic acid then disappears slowly up to 219°; and then suddenly, and the removal is complete at 360°. The residue $\frac{1}{2}Na_2U_2O_7 \cdot ZnU_2O_7$ maintains its weight constant up to 674°. The curve then rises a little up to 948° doubtless for the same reasons that prevailed in the case of Streng's salt. The curve which we have recorded[1] is that of a complex prepared at 20° and which contained not 6 but 7.2 molecules of water. The water content at the beginning varies between 6 and 9 molecules. Consequently we are dealing with a mixture; accordingly, the analytical factors ordinarily given are no more than approximate. The only reliable portion of the graph is the region from 360° and 674° which corresponds to the double pyrouranate. We suggest that it alone be used for automatic determination of sodium.

38. Sodium acetate

Sodium acetate, used as a standard, has been studied by Duval[22]. This hydrate, $NaC_2H_3O_2 \cdot 3H_2O$, loses some water even at room temperature. In none of our trials was it found to contain precisely 3 molecules of water, but only 2.77–2.80 H_2O. The dehydration is complete between 155° and 170°, according to the rate of heating. There are two distinct inflection points, one around 55°, the other near 95°. If therefore this compound is to be used for preparing a standard solution for buffer purposes, it is necessary to take the anhydrous salt, i.e. to keep the material between 170° and 440° for some time before weighing it. The acetate begins to decompose, above 440°, an action which seems to end at 540°. The horizontal belonging to sodium carbonate begins at this point, though this product sometimes may be contaminated with carbon. Burriel-Marti et al.[56] found that the anhydrous acetate is stable up to at least 350°.

39. Sodium oxalate

Neutral sodium oxalate, used as a reagent[2], is anhydrous and not hygroscopic. It undergoes no change in weight up to 450°; the decomposition begins very slowly and is not really important up to 500°. The velocity increases between 570° and 590° and ends abruptly and gives rise to sodium carbonate, which often is contaminated with carbon, coming from the decomposition of carbon monoxide. The reaction:
$$Na_2C_2O_4 \rightarrow Na_2CO_3 + CO$$
is confirmed on the recording with a precision of 1/75. Martens[57], using a sample weighing 400 mg, found that this loss of carbon monoxide occurs between 450° and 542°.

40. Sodium binoxalate

Martens[55] also found that sodium binoxalate $NaHC_2O_4 \cdot H_2O$ begins to lose water at 100°. The anhydrous salt shows itself by a horizontal extending from 155° to 175°; the decomposition of the anhydrous salt becomes rapid at 200° and is ended at 270°, producing the neutral oxalate which then behaves as described above.

41. Sodium tartrate

Neutral sodium tartrate, used as a standard[58], contains two molecules of water at the beginning. It retains these up to 57°, where the salt then melts in its own water of crystallization. Thereupon, water is lost progressively up to 154° with change of direction when one molecule of water has disappeared. The anhydrous salt is stable up to 255°; here is a massive loss of carbon monoxide and water of constitution; a mixture of carbon and sodium carbonate is obtained in the neighbourhood of 309°. This carbon burns progressively up to around 700°; the residual sodium carbonate in turn starts to decompose around 875° (instead of 840° when this compound is heated by itself at the same rate).

42. Seignette (Rochelle) salt

Seignette (or Rochelle) salt is hydrated potassium sodium tartrate. When heated under these same conditions it yields, around 40°, a curve that always slopes downward, and though uneven there is no horizontal stretch. Since there is no reference point on the curve, it would appear difficult to recommend this salt for a precise weighing. However, it is conceded that it contains precisely 4 molecules of water. It must not be heated above 40°.

43. Sodium citrate

Sodium citrate, taken as a standard[2], has the formula $Na_3C_6H_5O_7 \cdot 5.5H_2O$, which it retains up to 145°. Beyond this, the salt loses some water and the horizontal which appears between 218° and 309° corresponds approximately to a dihydrate. It is not possible to obtain the anhydrous salt without decomposition, and beyond 517° the residue consists of sodium carbonate mixed with carbon.

44. Sodium terephthalate

Sodium terephthalate[13] is stable at least up to 300° and proves to be anhydrous. The loss of weight which it suffers then up to 617° is truly minimal. However, decomposition sets in suddenly between 617° and

670°. The residual carbon burns extremely slowly, and around 840° the crucible contains nothing but sodium carbonate.

45. Sodium disulphoresorcinolate

A critical study of sodium disulphoresorcinolate by Peltier[59] showed that the precipitate is obtained only with sodium chloride, and that the precipitation is not quantitative, and hence useless for quantitative purposes.

The recorded curve shows a horizontal stretch up 405°. Around 407° there is abrupt decomposition and a residue of neutral sulphate Na_2SO_4 is obtained from 744° on.

REFERENCES

[1] T. DUVAL AND C. DUVAL, Anal. Chim. Acta, 2 (1948) 97.
[2] C. DUVAL, Anal. Chim. Acta, 15 (1956) 224.
[3] C. DUVAL, Proc. Intern. Symposium on Reactivity of Solids, Gothenburg, 1952, p. 511.
[4] M. VILTANGE, Thesis, Paris, 1960; Ann. Chim. (Paris), 5 (1960) 1037.
[5] M. VILTANGE, Compt. rend., 242 (1956) 781.
[6] M. JACQUINOT, Compt. rend., 238 (1954) 105.
[7] M. VILTANGE, Compt. rend., 239 (1954) 61.
[8] M. VILTANGE, Compt. rend., 244 (1957) 1215.
[9] C. DUVAL, Anal. Chim. Acta, 16 (1957) 221.
[10] P. MAURET, Compt. rend., 240 (1955) 2151.
[11] C. ROCCHICCIOLI, Thesis, Paris 1960; Ann. Chim. (Paris), 5 (1960) 1013.
[12] C. DUVAL, Anal. Chim. Acta, 20 (1959) 21.
[13] C. DUVAL AND C. WADIER, Anal. Chim. Acta, 23 (1960) 257.
[14] L. CAMBI AND G. BOZZA, Ann. chim. appl., 13 (1923) 224.
[15] D. TOMKOVA, P. JIRŮ AND J. ROSICKÝ, Collection Czechoslov. Chem. Communs., 25 (1960) 957.
[16] 1st. Congr. Thermography. Kazan, 1953, p. 91, 96.
[17] M. CAPESTAN, Ann. Chim. (Paris), 5 (1960) 215.
[18] M. OSWALD, Ann. Chim. (Paris), [9] 1 (1914) 100.
[19] P. RÂY, J. Chem. Soc., 87 (1905) 183.
[20] J. CASES-CASANOVA, Thesis, Paris, 1958.
[21] J. C. MERLIN, Bull. Soc. chim. France, (1955) 927.
[22] C. DUVAL, Anal. Chim. Acta, 13 (1955) 32.
[23] A. C. ZETTLEMOYER, C. H. SCHNEIDER, H. V. ANDERSON AND R. J. FUCHS, J. Phys. Chem., 61 (1957) 991.
[24] A. R. V. MURTHY, D. S. BHARADWAJ AND R. M. MALLYA, Chem. & Ind. (London), (1956) 300.
[25] W. W. WENDLANDT, Anal. Chem., 30 (1958) 56.
[26] P. L. WATERS, Nature, 178 (1956) 324.
[27] C. MORIN, Thesis, Paris, 1960.
[28] E. J. GRIFFITH, Anal. Chem., 29 (1957) 198.
[29] R. SALIH HISAR, Bull. Soc. chim. France, (1955) 916.
[30] H. GUÉRIN, J. MASSON, P. MATTRAT, R. BOULITROP, C. DUC-MAUGÉ, R. MARTIN AND R. MASS, Bull. Soc. chim. France, (1953) 440.
[31] L. PAULING, J. Am. Chem. Soc., 55 (1933) 1895, 3052.

32 T. DUPUIS, *Thesis*, Paris, 1954.
33 T. DUPUIS, *Rec. trav. chim.*, 71 (1952) 111.
34 A. G. OSTROFF AND R. T. SANDERSON, *J. Inorg. & Nuclear Chem.*, 4 (1957) 230.
35 T. DUPUIS AND C. DUVAL, *Chim. anal.*, 33 (1951) 189.
36 A. E. NEWKIRK AND I. ALIFERIS, *General Electric Co. Phys. Chem. Research*, No. 405; *Anal. Chem.*, 30 (1958) 982.
37 W. C. EASTERBROOK, *Analyst*, 82 (1957) 383.
38 A. REISMAN, *Anal. Chem.*, 32 (1960) 1566.
39 P. L. WATERS, *Anal. Chem.*, 32 (1960) 852.
40 M. G. MANVELYÁN, G. G. BABAYAN AND A. A. ABRAMYAN, *Izvest. Akad. Nauk Armyan S.S.R.*, 11 (1958) 159; *C.A.*, (1959) 957.
41 H. J. BORCHARDT, *J. Phys. Chem.*, 62 (1958) 166.
42 T. DUPUIS, *Compt. rend.*, 256 (1960) 1237.
43 C. DUVAL, *Anal. Chim. Acta*, 13 (1955) 427.
44 I. D. KOLLI, G. N. PIROGAVA AND V. I. SPITSYN, *Zhur. Neorg. Khim.*, 1 (1956) 2470; *C.A.*, (1957) 925.
45 F. CAPPELLINA AND B. BABANI, *Atti accad. sci. ist. Bologna*, 11 (1957) 76.
46 W. C. BALL, *J. Chem. Soc.*, 95 (1909) 2126; 97 (1910) 1408.
47 Y. PUGET, *Diploma of Higher Studies*, Paris, 1960.
48 R. CLAUDE, *La chimie des hautes températures, 2e Colloq. Natl. C.N.R.S.* 1957, p. 147; *Ann. chim.* (Paris), 5 (1960) 192.
49 A. STRENG, *Z. anal. Chem.*, 23 (1884) 186.
50 A. BLANCHETIÈRE, *Bull. Soc. chim. France*, 33 (1923) 807.
51 E. KAHANE, *Bull. Soc. chim. France*, 47 (1930) 382.
52 R. CHARONNAT, *Bull. Soc. chim. France*, 5 (1938) 205.
53 G. WYROUBOFF, *Bull. Soc. franc. minéral.*, 24 (1901) 93.
54 HOPKINS AND WILLIAMS LTD., *Organic Reagents for Metals*, 4th Ed., London, 1946, p. 158.
55 H. H. BARBER AND I. M. KOLTHOFF, *J. Am. Chem. Soc.*, 50 (1928) 1625; 51 (1929) 3233.
56 F. BURRIEL-MARTI, S. JIMENEZ-GOMEZ AND C. ALVAREZ-HERRERO, *Anales edafol. y fisiol. vegetal (Madrid)*, 14 (1955) 222.
57 P. H. MARTENS, *Bull. inst. agron. et stas. recherches Gembloux*, 21 (1953) 134.
58 C. DUVAL, *Anal. Chim. Acta*, 20 (1959) 264.
59 S. PELTIER, *Anal. Chim. Acta*, 2 (1948) 328.

CHAPTER 13

Magnesium

In 1947 Duval and Duval[1] made a thermogravimetric study of brucite and magnesium hydroxide, and also of magnesium fluoride, sulphate, phosphate, arsenate, magnesium ammonium carbonate, oxalate, oxinate. The information obtained has now been supplemented by investigations of the chloride, perchlorate, bromide, iodide, amidosulphonate, selenite, nitrate, carbonate, the basic carbonates and dolomite, various silicates, ferrocyanide, niobate, molybdate, formate, acetate, methyloxinate, picolinate, quinaldate, etc.

1. Magnesium hydroxide

(i) Precipitation with sodium hydroxide Magnesium hydroxide $Mg(OH)_2$ was prepared by treating an aqueous solution of magnesium chloride with strong sodium hydroxide solution; the suspension was diluted with water and allowed to stand for 2 hours. The thermolysis curve descends continuously until the temperature reaches 820°. The absorbed water begins to come off at room temperature, the loss is rapid near 62°, begins to slow down at 134°, and is complete near 290°. An abrupt change in the loss of weight is then observed, and the greater part of the combined water is eliminated at 420°; the rest of the water then departs very slowly. The losses, as measured on the graph, suffered by a calculated weight of magnesium oxide amounting to 277.3 mg, at a number of temperatures, were:

Temp. (°C)	416	481	548	616	681	751	819
Loss (mg)	9.0	7.2	4.0	3.0	2.4	1.9	0

(ii) Preparation by electrolysis These results are in good agreement with the findings by Jolibois and Bergès[2] who studied a magnesium hydroxide prepared by electrolysis and dried in vacuo. Accordingly, when magnesium is to be determined by ignition of its hydroxide, it is essential to heat the residue to above 800°.

(iii) Precipitation with potassium hydroxide The oxide free of Cl^- ion was produced above 800° from a specimen prepared by precipitation with potassium hydroxide[3]. This method or the hydration of the oxide MgO yields a true base $Mg(OH)_2$ as proved by the infra-red spectrum. The same is true of brucite. We will see that this is not the case with iron hydroxide. Finally, Gibaud and Geloso[4], using the Guichard balance, observed that magnesium hydroxide begins to lose water only at 380°.

(iv) Dehydration of brucite We have compared this finding with the dehydration of a specimen of brucite $MgO \cdot H_2O$ (kindly offered by Mlle. Caillère) coming from the Lows Mine in Fulton Lancaster, Texas, and corresponding to No. 63-115 of the Museum de Paris. This sample was manganese-free, and contained 0.50% iron (calculated as FeO) and 1.27% carbon dioxide. The weight remained constant up to 283°; beyond this temperature the carbon dioxide and the water came off but it was impossible to distinguish sharply between the loss of these two compounds. A practically constant weight was obtained at 815°. The curve rises slightly beyond 950° (0.15 mg); this increase is doubtless the result of the capture of oxygen by the trace of residual iron oxide.

A personal communication No. 1064, DMQ.PR 305 from the General Electric Company (Schenectady, N.Y.) contains three curves which demonstrate the influence of the rate of heating, namely from 2.5° to 5° per minute.

Brucite has been heated likewise by Gill[5], and by Claisse, East and Abesque[6] in magnesia-bearing rocks.

2. Magnesium fluoride

Magnesium fluoride, prepared according to the scanty directions given by Nicolardot and Dandurand[7], appears to be a mixture of the acid fluoride and the neutral fluoride. It is obtained by treating a hot solution of magnesium chloride (containing ammonium chloride and ammonium hydroxide to bring the pH to 7) with a solution of ammonium fluoride. The gelatinous precipitate is allowed to stand 12 hours before filtering, and is then washed with water, ethanol, and ether. The weight curve slopes downward until the temperature reaches 420–425°. The neutral MgF_2 can be weighed above this temperature. The two changes in direction observed around 75° and 285° do not relate to any simple quantitative result.

References p. 224

3. Magnesium chloride

Magnesium chloride, after recrystallization, contains 5.5 molecules of water, which it starts to lose at ordinary temperatures[8]. Accordingly, it is impossible to prepare a standard solution by direct weighing of this salt and furthermore, the dehydration does not result in the horizontal of the anhydrous salt because under the conditions at which we worked the water vapour released produced decomposition. As a result, we obtained the horizontal of magnesium oxide from 615° on.

4. Magnesium perchlorate

According to Zinov'ev and Chudinova[9], the perchlorate $Mg(ClO_4)_2 \cdot 6H_2O$ loses its water in two stages, at 141° and 185°. Around 410° a third horizontal, which was also followed by differential thermal analysis, agrees with the decomposition:

$$2\ Mg(ClO_4)_2 \rightarrow Cl_2O_7 + MgO \cdot Mg(ClO_4)_2$$

Finally, near 547°, a loss of oxygen sets in with formation of an oxychloride:

$$MgO \cdot Mg(ClO_4)_2 \rightarrow MgO \cdot MgCl_2 + 4\ O_2$$

5. Magnesium bromate

Magnesium bromate $Mg(BrO_3)_2 \cdot 6H_2O$ begins to lose water at room temperature. Around 165° a short horizontal appears; it belongs to the dihydrate. This salt is not very stable; it readily loses oxygen. The residue puffs up considerably; it consists of almost pure anhydrous magnesium bromide (traces of MgO) beginning at 400° (Rocchiccioli[10]).

6. Magnesium iodide hexamine

Yosida[11] has shown that the complex magnesium iodide hexamethylenetetramine $MgI_2 \cdot 2(CH_2)_6N_4 \cdot 10H_2O$ is stable only below 40°.

7. Magnesium iodate

Rocchiccioli[10] heated magnesium iodate $Mg(IO_3)_2 \cdot 4H_2O$ at different rates and found that it is dehydrated from 136° on. The complete removal of the water is slowly accomplished at 270°.

8. Magnesium sulphate

(i) Heating at the rate of 300° per hour   Magnesium sulphate, just taken from a fresh bottle, contains 7 molecules of water but does not retain all of them above 36° if heated at the rate of 300° per hour. The distinct horizontals for the various hydrates are not observed during the dehydration, except perhaps for the monohydrate. The removal of the last traces of water[1, 12] is difficult and is accomplished between 300° and 400° as shown first by Hackspill and Kieffer[13]. The horizontal of the anhydrous salt persists up to at least 880°; we have not observed any decomposition starting at 750° as has been reported. See also Gill[5], Gordon and Denisov[14], and Kamecki and Palej[15].

(ii) Heating at the rate of 0.6° per minute   Fruchart and Michel[16] have again taken up this matter and employed thermal analysis and thermogravimetry. They had a different objective: to heat the salt slowly at the rate of 0.6 per minute so as to be able to observe its horizontals. The heptahydrate yielded the hexahydrate around 60° and it in turn underwent an allotropic transformation near 90° and an aqueous fusion. There were then formed in succession: the pentahydrate at 103°, the trihydrate at 108°, the dihydrate at 127°, and the monohydrate at 149°. The latter then gives rise to the anhydrous sulphate between 178° and 285°. The oxide does not appear until a temperature above 1100° is reached.

9. Magnesium sulphamate

The amidosulphonate (sulphamate) $Mg(NH_2-SO_3)_2 \cdot 4H_2O$, prepared by Capestan[17], rapidly loses 2.5 molecules of water from 80° to 135°; the rest of the water comes off more slowly between 135° and 215°. The anhydrous $Mg(NH_2-SO_3)_2$ changes into the imidodisulphonate $MgNH(SO_3)_2$ at 350°. Finally, the anhydrous sulphate is obtained around 500°.

10. Magnesium selenite

If magnesium selenite $MgSeO_3 \cdot 6H_2O$ is heated at the rate of 300° per hour, it begins to lose water at 100–105°. The dehydration accelerates smoothly and is complete near 260°. The decomposition of the anhydrous salt likewise occurs without special incident between 660° and 850°. At

References p. 224

higher temperatures, the horizontal of magnesium oxide begins. According to Rocchiccioli[10], there is no oxidation to selenate.

11. Magnesium nitrate

We have heated magnesium nitrate $Mg(NO_3)_2 \cdot 6H_2O$ at various rates without observing the intermediary hydrates except through changes in direction. The dehydration begins around 55° and the formation of the oxide MgO is complete around 475°. The production of a basic salt $2Mg(NO_3)_2 \cdot MgO$ may be deduced to some extent from the inflection observed near 345°. See also Wendlandt[18].

12. Magnesium ammonium phosphate

Magnesium ammonium phosphate $MgNH_4PO_4 \cdot 6H_2O$ was precipitated in accord with the directions given in the well-known Treadwell text. The curve is inclined downward until a temperature is reached which depends on the weight of substance involved, but varying from 380° to 480° when the weight goes from 100 mg to 400 mg. The magnesium pyrophosphate $Mg_2P_2O_7$ results. Consequently it is needless to ignite at 900° as is sometimes prescribed. The theoretical weight of 150 mg (which we used in one instance) was attained within 1 mg even at 400°. At no time did we observe the separate departures of the water and the ammonia except when the cover was kept on the crucible (inflection around 170° when heating at 300° per hour). If the dry hexahydrate is weighed in air, it is advisable not to go beyond 64°. The curve undergoes a distinct change of direction near 205°. Contrary to all expectation, this does not correspond to the appearance of the monohydrogen orthophosphate $MgHPO_4$.

Our findings parallel and agree for certain weights with those found by Sagortschew[19] who used the emanation method of Hahn.

13. Magnesium ammonium arsenate

Magnesium ammonium arsenate $MgNH_4AsO_4 \cdot 6H_2O$ was produced by treating a solution of magnesium chloride with several grams of ammonium chloride and then adding ammonium arsenate. Hydrochloric acid is added drop by drop until the precipitate redissolves; phenolphthalein is introduced and then 2% ammonia until the color change occurs, and then concentrated ammonium hydroxide is added.

The curve greatly resembles the preceding one; it is difficult to start with a specimen having precisely 6 molecules of water. This is why we do not recommend weighing the hexahydrate after drying in air. Even at 50° the loss amounts to 2 mg in 223 mg. The pyroarsenate is obtained from 415° on. Around 750° an arsenical odor becomes noticeable but there is no significant loss of weight up to 885°. In agreement with Ducru[20], we find that the loss of water and ammonia is continuous from 40° to 415°. Yosida[11] gives the region from 850° to 950° as that of the pyroarsenate. Although various writers have suggested weighing $MgNH_4AsO_4 \cdot 1.5H_2O$ from 82° to 93°, or the monohydrate at 100°, or the anhydrous compound at 150°, none of these weighing forms deserves serious consideration. However, as in the case of magnesium ammonium phosphate, there is a not very distinct change in the slope of the curve beginning at 98°. The level is almost horizontal around 288° and corresponds well to the compound $MgHAsO_4$. A sudden inflection is seen near 350°, and one of the two molecules of water of this material is lost. In our opinion, the determination of magnesium by means of magnesium ammonium arsenate is not as good as the procedure involving magnesium ammonium phosphate.

14. Magnesium ammonium carbonate

Magnesium ammonium carbonate $MgCO_3 \cdot (NH_4)_2CO_3 \cdot 4H_2O$ results from the action of Schaffgotsch reagent[21], namely a water–alcohol solution of ammonium carbonate, on a neutral solution of magnesium chloride. The mixed solutions are stirred for 5 minutes and the suspension allowed to settle for 15 minutes, after which the precipitate is filtered off and washed with Schaffgotsch reagent. The curve was recorded by Duval and Duval[1] and again by Dupuis and Dupuis[22]. With a heating rate of 300° per hour, the slope is constantly downward until about 500° where the horizontal of the oxide MgO appears. An inflection or a sloped level appears between 200° and 300° but it does not correspond to magnesium carbonate. No distinction can be detected between the escape of the water or the ammonia. Automatic determination of magnesium in dolomitic rocks has been accomplished on the basis of this curve.

15. Magnesium carbonate

Both natural and artificial specimens of magnesium carbonate have been examined.

References p. 224

(i) Giobertite and magnesite Caillère and Pobeguin[23] found that when giobertite or magnesite are pyrolyzed, decomposition sets in at 425° in air and at 480° in nitrogen. Therefore, among the natural carbonates, it decomposes at the lowest temperature. Gill[5], Tsuzuki and Nagasawa[24], Collari and Galimberti[25], Sensale[26], and Saito[27] have also heated this material. Saito worked in air with grains which passed a 200-mesh sieve, and at a flow rate of 100 ml per hour; he found that the decomposition starts at 320°, becomes appreciable at 350°, and slowly ends at 590°. In carbon dioxide, flowing at 200 ml per minute, and with a heating rate of 2° per minute, the decomposition sets in at 510°, becomes vigorous at 560°, and is complete at 650°.

(ii) Artificial magnesium carbonate A number of workers have studied artificial magnesium carbonate. Visinti and Cignetti[28] investigated the trihydrate $MgCO_3 \cdot 3H_2O$ particularly, heating it in nitrogen flowing at 0.5 liter per minute and with a heating rate of 1.7° per minute. Under these conditions the decomposition is complete at 510°. They also used the following:

$$0.35 \text{ l/min of nitrogen} + 0.15 \text{ l/min of } CO_2$$
$$0.30 \text{ l/min of nitrogen} + 0.20 \text{ l/min of } CO_2$$
$$0.50 \text{ l/min of carbon dioxide}$$

and found the decomposition to be finished between 550° and 600°. Deller and Weller[29] report 560°.

The dihydrate $MgCO_3 \cdot 2H_2O$ was prepared by Gibaud and Geloso[4] by adding alcohol to an aqueous solution of commercial $4MgCO_3 \cdot Mg(OH)_2 \cdot 6H_2O$ charged with carbon dioxide. This dihydrate loses its water between 160° and 250°, leaving MgO (in an atmosphere of carbon dioxide).

The basic carbonate $5MgCO_3 \cdot 4CO_2 \cdot 5H_2O$ was heated by Schwob[30] who found to undergo the following transformations:

$$5MgO \cdot 4CO_2 \cdot 5H_2O \rightarrow 5MgO \cdot 3CO_2 \cdot 3H_2O + CO_2 + 2\,H_2O \ (250\text{--}350°)$$
$$5MgO \cdot 3CO_2 \cdot 3H_2O \rightarrow 5MgO \cdot 1.5\,CO_2 + 1.5\,CO_2 + 3\,H_2O \ (400\text{--}450°)$$
$$5MgO \cdot 1.5CO_2 \rightarrow 5MgO + 1.5\,CO_2 \ (450\text{--}500°).$$

Visinti and Cignetti[28] found that the same basic carbonate had completed this reaction at 400° if heated in nitrogen (0.5 l/min), or around 550° in nitrogen (0.3 l/min) plus carbon dioxide (0.2 l/min) or in nitrogen (0.2 l/min) plus carbon dioxide (0.3 l/min), and finally always around 550° in pure carbon dioxide (0.5 l/min).

Another basic carbonate $3MgCO_3 \cdot MgO \cdot 4H_2O$ was found by Orosco[31] to yield three dehydration levels at 240°, 340°, and 395°, respectively.

(iii) Dolomite The important matter of dolomite and the analysis of dolomitic rocks will now be taken up.

Saito[27] found that the decomposition became energetic around 600° and appeared to be completed near 780°. He used 500 mg of a natural product $CaMg(CO_3)_2$ which had been ground to pass a 200-mesh sieve; he heated this fine material in the air at the rate of 2° per minute. In carbon dioxide, flowing at the rate of 200 ml per minute, the following reaction still began at 600° but almost stopped at 700°:

$$MgCa(CO_3)_2 \rightarrow CO_2 + MgO + CaCO_3$$

Then the calcium carbonate in turn underwent decomposition:

$$MgO + CaCO_3 \rightarrow CO_2 + MgO + CaO$$

Several workers repeated this work but without making reference to it: Papailhau[32], Gill[5], Schwob[30], Gritton, Gregg and Winsor[33], Gibaud and Geloso[4]. These latter workers deduced a simple method for the automatic analysis of a mixture of dolomite, carbonate, lime, and magnesia from their curves.

Furthermore, Dupuis and Dupuis[22] examined a typical dolomite from Bathonien de Lodève (Hérault). In the air, a horizontal is observed from room temperature up to around 450°. The decomposition is slow at first from 450° to 620°; then the escape of the carbon dioxide becomes more rapid and is complete in the vicinity of 900°. A horizontal corresponding to a mixture of the two oxides is observed beyond 900°, but the curve does not directly reveal the calcium and the magnesium content of the rock. It is necessary to heat the sample in an atmosphere of carbon dioxide, as directed by Gibaud and Geloso[4], if two stages are to be observed in the decomposition of dolomite.

This is the reason why Dupuis and Dupuis, faced with this procedural complication, have preferred to develop a method for analyzing the rock in which the mixed products are weighed. The steps in their procedure are: solution of the sample, removal of iron, precipitation of the two carbonates by means of the Schaffgotsch reagent (see above), filtration, heating at 520–550° at the rate of 300° per hour, weighing to obtain the sum of $MgO + CaCO_3$, and then keeping the mixture at 950° which yields the sum of the $MgO + CaO$. Their paper contains several supporting analyses of four dolomites from the Jurassic region of Causses and from Poitou.

16. Magnesium niobate

According to Pehelkin, Lapitskii, Spitsyn and Simanov[34], the complex niobate $Mg_7Nb_{12}O_{37} \cdot 44H_2O$ loses water and leaves the compound carrying 6 H_2O at 140°, and 5 H_2O at 160°, the monohydrate at 340°, while the metaniobate is formed from 555° on.

17. Magnesium silicates

Among the numerous studies of various silicates, it is necessary to cite especially the investigation of asbestos by Duval[35] and by Gill[5]. It should be pointed out that this material suffers an abrupt loss of water at 283°. Caillère and Rouaix[36] heated a palygorskite from Senegal employing the techniques of thermogravimetric analysis and differential thermal analysis. This mineral, which has the formula:

$$(Si_{7 \cdot 35}Al_{0 \cdot 65}) (Al_{2 \cdot 26}Fe^{3+}_{0 \cdot 15}Mg_{1 \cdot 48}) O_{21}Ca_{0 \cdot 21}$$

loses a total of 20% of water in three stages: 6% up to 130°, 5% between 150° and 250°, and 9% beyond 250° up to 1000°. Gill[5] and Tsuzuki[24] made studies of the serpentines, and also of talc, montmorillonite, the chlorites, kaolinites, holloysites (on the Kinij and Swata thermobalance).

18. Magnesium ferrocyanide

Cappellina and Babani[37] report that the ferrocyanide $Mg_2[Fe(CN)_6] \cdot 9H_2O$ loses two molecules of water from 25° to 85°, two more from 85° to 110°, and the remaining five from 110° to 175°.

19. Calcium magnesium ferrocyanide hexamine

A more complex ferrocyanide, namely 3 CaMg $[Fe(CN)_6] \cdot 4(CH_2)_6 N_4 \cdot 40H_2O$, was prepared by Yosida[38] with hexamethylenetetramine. It decomposes from 25° on.

20. Magnesium molybdate

According to Teller[39], the hydrated molybdate $MgMo_3O_{10}$, when heated in the Eyraud balance at the rate of 25° per hour, yields hydrates containing 12 or 13, 10, 7, 5 H_2O and finally the anhydrous compound.

21. Magnesium formate

Zapletal, Jedlička and Růžička[40] investigated the formate. It decomposes after being dehydrated. Our curve indicates that the dihydrate

$Mg(COOH)_2\cdot2H_2O$ begins to lose water at 158° and yields the anhydrous salt from 220°. We have been able to preserve the latter up to 310–320°.

22. Magnesium acetate

Magnesium acetate $Mg(CH_3CO_2)_2\cdot4H_2O$ has been heated by Burriel-Marti, Jimenez-Gomez and Alvarez-Herrero[41] in order to determine the composition of mineral waters. The salt loses water in such fashion as to yield the anhydrous compound between 100° and 220° (no indication) and then the decomposition leads to magnesium oxide MgO around 330°.

23. Magnesium oxalate

Magnesium oxalate $MgC_2O_4\cdot2H_2O$ was prepared by the Classen procedure[42]. The pyrolysis curve[1] made it possible for Peltier and Duval[43] to determine calcium and magnesium without preliminary separation. The dihydrate is stable up to 180° but the 2 molecules of water disappear between 180° and 250°. The level due to the anhydrous salt is not strictly horizontal; it extends from 250° to 400°. Carbon monoxide and carbon dioxide are then released simultaneously (in contrast with the breakdown of calcium oxalate). In other words, the anhydrous carbonate does not appear nor do the various basic carbonates reported in the chemical literature. The anhydrous magnesium oxide appears from 500° on (see Kawagaki[44], Erdey and Paulik[45], Wiedemann and Nehring[46].) From 500° and up to 1015° (the upper limit of the experiment) the loss is insignificant and does not amount to 1/200.

Wiedemann and Nehring[46] also studied the mixed crystals containing magnesium and nickel oxalate $NiC_2O_4\cdot2H_2O$; they worked in vacuo, hydrogen, nitrogen, and in the air.

24. Magnesium oxinate

Magnesium oxinate $Mg(C_9H_6ON)_2$ was first described as carrying 4 molecules of water but the investigations by Dupuis[47] and by Borrel and Pâris[48] have shown that the dihydrate appears on drying at 105–110°. The anhydrous salt proved to be stable from 200° to 375° and then yielded magnesium oxide MgO progressively. The results obtained by Miller and McLennan[49] with regard to a monohydrate have not been confirmed.

According to Dupuis, there is danger that the dihydrate may be contaminated with an excess of the reagent; the temperature at which the organic matter undergoes combustion depends on the weight of the substance. It is better to weigh the anhydrous salt rather than the di-hydrate, either in the usual procedure or when automatic determination is used. Dupuis described a procedure for each of these cases, and these methods have been satisfactorily tested on tens of determinations.

25. Magnesium methyloxinate

The bright yellow methyloxinate $Mg(CH_3-C_9H_5NO)_2 \cdot H_2O$, which was isolated by Borrel and Pâris[50], decomposes from 134° to 182° and yields the horizontal of the anhydrous chelate. The latter decomposes from 310° on and finally yields magnesium oxide MgO.

26. Magnesium picolinate

Thomas[51] found that magnesium picolinate, which likewise was originally reported by Borrel and Pâris, contains two molecules of water, which it retains up to 145°.

27. Magnesium quinaldate

Magnesium quinaldate $Mg(C_{10}H_6O_2N)_2 \cdot 2H_2O$ is colorless. It was heated to 700° at the rate of 180° per hour and kept at that temperature for two hours by Thomas[51]. She found that it is stable up to 145°. The dehydration is complete at 230° and the anhydrous salt is stable over a range of 100 degrees. Magnesium oxide is produced above 576°.

28. Magnesium diliturate and 1,3-dimethylviolurate

The diliturate and the 1,3-dimethylviolurate have thermolysis curves which made it possible for Robinson, Berlin and Taylor[52] to suggest that these salts be used for the gravimetric determination of magnesium.

REFERENCES

[1] T. Duval and C. Duval, *Anal. Chim. Acta*, 2 (1948) 45.
[2] P. Jolibois and M. Bergès, *Compt. rend.*, 224 (1947) 79.
[3] C. Cabannes-Ott, *Thesis*, Paris, 1958.
[4] M. Gibaud and M. Geloso, *Chim. anal.*, 36 (1954) 153.
[5] A. F. Gill, *Canad. J. Research*, 10 (1934) 703.
[6] F. Claisse, F. East and F. Abesque, *Quebec Dept. Mines, Prelim Repts.*, No. 305, 1954.

7 P. NICOLARDOT AND F. DANDURAND, *Rev. Mét.*, 16 (1919) 193.
8 C. DUVAL AND C. WADIER, *Anal. Chim. Acta*, 23 (1960) 257.
9 A. A. ZINOV'EV AND L. I. CHUDINOVA, *Zhur. Neorg. Khim.*, 1 (1956) 1722; *C.A.*, (1957) 2439.
10 C. ROCCHICCIOLI, *Thesis*, Paris, 1960.
11 Y. YOSIDA, *J. Chem. Soc. Japan*, 60 (1940) 915.
12 C. DUVAL AND C. WADIER, *Anal. Chim. Acta*, 23 (1960) 541.
13 L. HACKSPILL AND A. P. KIEFFER, *Ann. Chim. (Paris)*, 14 (1930) 267.
14 E. GORDON AND A. M. DENISOV, *Ukrain. Khim. Zhur.*, 19 (1953) 368.
15 J. KAMECKI AND S. PALEJ, *Rocznicki Chem.*, 29 (1955) 691.
16 R. FRUCHART AND A. MICHEL, *Compt. rend.*, 246 (1958) 1222.
17 M. CAPESTAN, *Ann. Chim. (Paris)*, 5 (1960) 218.
18 W. W. WENDLANDT, *Texas J. Sci.*, 10 (1958) 392.
19 B. SAGORTSCHEW, *Z. physik. Chem. (Leipzig)*, B, 182 (1938) 31.
20 O. DUCRU, *Ann. Chim. Phys.*, 22 (1901) 179.
21 SCHAFFGOTSCH, *Ann. phys.*, 104 (1858) 482.
22 J. DUPUIS AND T. DUPUIS, *Mikrochim. Acta*, (1958) 186.
23 S. CAILLÈRE AND T. POBEGUIN, *Bull. soc. franç. minéral*, 83 (1960) 36.
24 Y. TSUZUKI AND K. NAGASAWA, *J. Earth Sci. Nagoya Univ.*, 5 (1957) 153; *C.A.*, (1958) 13552.
25 N. COLLARI AND L. GALIMBERTI, *Boll. sci. fac. chim. ind. Bologna*, 18 (1940) 253; *C.A.*, (1943) 3304.
26 R. SENSALE, *Periodico mineral. (Rome)*, 20 (1951) 165.
27 H. SAITO, *Sci. Repts. Tôhoku Imp. Univ.*, 16 (1927) 54.
28 B. VISINTI AND M. CIGNETTI, *Gazz. chim. ital.*, 90 (1960) 863.
29 R. M. DELLER AND S. W. WELLER, *Trans. Faraday Soc.*, 55 (1959) 2203.
30 Y. SCHWOB, *Rev. matériaux construct. et trav. publ.*, No. 413 (1950) 33, 85.
31 E. OROSCO, *Ministerio trabalho ind. ecom., (Inst. nac. tecnocol.) Rio de Janeiro*, (1940); *C.A.*,(1941) 3485.
32 J. PAPAILHAU, *Bull. soc. franç. minéral.*, 82 (1959) 367.
33 H. T. S. GRITTON, S. J. GREGG AND G. W. WINSOR, *Trans. Faraday Soc.*, 48 (1952) 63, 73.
34 V. A. PEHELKIN, A. V. LAPITSKII, V. I. SPITSYN AND Y. P. SIMANOV, *Zhur. Neorg. Khim.*, 1 (1956) 1784; *C.A.*, (1957) 2445.
35 C. DUVAL, *Anal. Chim. Acta*, 3 (1949) 163.
36 S. CAILLÈRE AND S. ROUAIX, *Compt. rend.*, 246 (1958) 1442.
37 F. CAPPELLINA AND B. BABANI, *Atti. accad. sci. ist. Bologna*, 11 (1957) 76.
38 Y. YOSIDA, *J. Chem. Soc. Japan*, 61 (1940) 130.
39 G. TELLER, *Thesis*, Strasbourg, 1959.
40 V. ZAPLETAL, J. JEDLIČKA AND V. RŮŽIČKA, *Chem. Listy*, 50 (1956) 1406; *C.A.*, (1957) 2438.
41 F. BURRIEL-MARTI, S. JIMENEZ-GOMEZ, C. ALVAREZ-HERRERO, *Anales edafol. y fisiol. vegetal (Madrid)*, 14 (1955) 221.
42 A. CLASSEN, *Z. anal. Chem.*, 18 (1879) 373.
43 S. PELTIER AND C. DUVAL, *Anal. Chim. Acta*, 1 (1947) 408.
44 K. KAWAGAKI, *J. Chem. Soc. Japan*, 72 (1951) 1079.
45 L. ERDEY AND F. PAULIK, *C. A.*, (1956) 3952.
46 H. G. WIEDEMANN AND D. NEHRING, *Z. anorg. Chem.*, 304 (1960) 137.
47 T. DUPUIS, *Thesis*, Paris, 1954.
48 M. BORREL AND R. PÂRIS, *Anal. Chim. Acta*, 4 (1950) 267.
49 C. C. MILLER AND I. C. MCLENNAN, *J. Chem. Soc.*, (1940) 656.
50 M. BORREL AND R. PÂRIS, *Anal. Chim. Acta*, 5 (1951) 573.
51 G. THOMAS, *Thesis*, Lyon, 1960.
52 R. J. ROBINSON, A. BERLIN AND M. E. TAYLOR, *3rd Conf. on Anal. Chem.*, Prague, 1959.

CHAPTER 14

Aluminium

The interest in the use of the thermobalance appears especially in the study of important minerals and in the determination of aluminium as its oxide. This matter has engaged the attention of chemists for a long time. It is known in fact that if the sulphate is the starting material and if it is necessary to remove the crucible from the furnace in order to weigh to make the weighing after cooling, it is essential to carry the temperature to at least 1200°. Consequently it is only natural to wish to replace the weighing method by a written recording. See Dupuis and Duval[1].

1. Aluminium hydroxide

The precipitation as hydrous alumina $Al_2O_3 \cdot nH_2O$ has given rise to many and varied procedures, both in the choice of the salt employed (chloride, nitrate, sulphate, etc.) and in the choice of the alkaline precipitant. We have examined a large number of these variants with the threefold objective of finding a definite product, one that filters rapidly, and one that will yield the oxide Al_2O_3 at the lowest temperature possible and free of water and impurities due to the anions of the starting salt. The infra-red spectrum has provided answers to various questions. (See Dupuis and Duval[1] and Cabannes-Ott[2].)

None of the precipitates actually has the formula $Al(OH)_3$. Only the precipitate produced by the action of hydrazine on aluminium chloride has a structure with OH groups corresponding to böhmite.

The precipitates obtained from aluminium sulphate (or the alums) always contain sulphate ions SO_4^{2-} which are not removed by washing, and not always revealed by adding a barium salt, but they do appear in the infra-red spectrum of the powdered specimen. See also Glemser and Rieck[3], Groot and Troutner[4], Erdey and Paulik[5]. Of all the precipitants, ammonia water has the least to recommend it. It is better to use potassium cyanate, gaseous ammonia, or hexamethylenetetramine. Table 4 shows

TABLE 4

PRECIPITATION OF ALUMINIUM HYDROXIDE

Precipitated by	Quality	Minimum temp. (C°)	Reference
Aqueous ammonia		1031	6-8
Gaseous ammonia	+ +	475	9
Urea	+	672	10
Urea + succinate	+	611	10
Mercury chloramide	+	676	11
Hexamethylenetetramine	+ +	475	12
Pyridine	+ +	478	13
Ammonium acetate	+ +	475	14, 15
Ammonium formate	+ +	539	15
Ammonium succinate	+ +	509	6
Ammonium benzoate	+ +	607	16
Sodium salicylate	+	650	17
Ammonium sulphide	+ +	414	18
Ammonium carbonate	+ +	409	19
Ammonium bicarbonate	+ +	514	20
Hydrazinium carbonate	+ +	524	21
Potassium cyanate	+ +	510	22
Ammonium nitrite	+ +	480	23
Sodium thiosulphate	+	675	24
Potassium iodide–iodate	+	880	25
Sodium bisulphite	+ +	412	26
Carbon dioxide	+	945	27
Bromine	+ +	280	28
Tannin	+	898	29
Hydrazine	+ +	350	8

the minimum temperature required to produce the oxide (whose infra-red spectrum no longer includes bands due to water). The slightly moist samples, whose initial weights ranged from 250 to 300 mg, were heated at the rate of 300° per hour. In one instance we found methods for producing a material which filtered rapidly but which acquired a constant weight only at a high temperature (thus excluding the use of glass crucibles), and in two other cases the resulting hydrous aluminas filtered well and could be satisfactorily ignited at low temperatures. It should be pointed out once more, that if the chemist persists in using the earlier methods, it is essential that he heat the alumina to a temperature such that it will not take up water while it is cooling and while it is being weighed on an ordinary balance. Such dangers are eliminated in our automatic procedures since the crucible is not taken from the furnace. We pointed out these facts in July 1948[1] and it is regrettable that numerous criticisms have been directed against us by persons who seemingly had not read what we stated at that time.

References p. 235

In particular, we are not certain how we should judge the work of Milner and Gordon[30] who doubted our conclusions. They precipitate a hydrous alumina, wash it on paper, burn the latter at 525°, heat the residue on the thermobalance for 45 minutes at 600°, and then conclude that in the conventional methods it is necessary to carry the residue to 1200°, a fact which I have never contested, and which has been known for more than a century.

2. Natural aluminium hydroxides and oxides

A few words regarding the natural hydroxides and oxides of aluminium are in order here.

(i) Böhmite   Ott[31] reports that böhmite decomposes from 350° on after a horizontal extending from 250° to 350°. She obtained the same result with deuterated böhmite prepared in a box provided with sleeves. Imelik, Petitjean and Prettre[32] working under reduced pressure (0.001 mm mercury) found that this böhmite is much more stable than the corresponding gel and that it starts to lose water only at 300°, although the gel has already reached the composition $2Al_2O_3 \cdot H_2O$ at 260°. An abrupt departure of water leading to a practically anhydrous alumina starts only at 360° in vacuo and at 400° in the air. The infra-red spectrum completely substantiates this finding and shows no water bands in the residue. In another study, Eyraud, Trambouze, Tran Huu The and Prettre[33] combined the thermogravimetric analysis with enthalpic analysis. Working always under a pressure of 10^{-3} mm of mercury, Eyraud and Goton[34] constructed the 442°, 470°, 488°, 498°, 512°, and 542° isotherms for böhmite $2Al_2O_3 \cdot H_2O$. Since the infra-red spectrum of böhmite does not contain the bands due to water, but only those due to OH, it is perhaps better to write its formula as $Al_2O_3 \cdot 2AlO(OH)_2$.

(ii) Hydrargillite or gibbsite   A number of workers have heated hydrargillite or gibbsite. A product satisfying the formula $Al_2O_3 \cdot 3H_2O$ or better $Al(OH)_3$ from Muzo (Colombia) and corresponding to No. 138-14 of the collection of the Museum de Paris, has shown us[1] that the dehydration is smooth, commences at the ordinary temperature, undergoes an abrupt fall around 72° when half of the water has escaped. The dehydration is total at 320°; then the horizontal of the oxide begins. Paulik and Erdey[35] in their turn, studied a hydrargillite from Istria. See also Sato[36] and the comparison with ordinary alumina and bayerite. Eyraud, Goton and Prettre[37] found that the decomposition proceeds in

two stages. During the first, the mineral loses essentially 5/6th of its water within a very short temperature interval; the final product is böhmite. Under a high vacuum (10^{-3} mm mercury) the speed is measurable in the vicinity of 175–180° and the reaction is practically completed in 2 or 3 days. According to Eyraud and Goton[34] it is a zero order reaction. They also traced the isotherms at 206°, 215°, 226°, 235°, and 243°, whereas Eyraud, Goton, Trambouze, Tran Huu The and Prettre[33] used a combination of thermogravimetry and differential enthalpic analysis, along with differential thermal analysis and calorimetry. Regarding gibbsite see also Tertian and Papée[38] and Brown, Clark and Elliott[39].

(iii) Diaspore   The diaspore which we heated came from Chester (Massachusetts) and is housed under No. 67-131 of the collection at the Museum de Paris. It first undergoes a smooth increase in weight up to 282°. This involves a phenomenon whose effects are slight since the gain is only 5 mg in 366 mg. We believe that we are seeing here the consequence of an oxidation or carbonation of ferruginous impurities. The dehydration sets in suddenly around 520°; it is complete at 685°; the level of the oxide is perfectly horizontal[1]. Ott[2] found that the thermolysis curve of diaspore (and the infra-red spectrum) are identical with those given by the material precipitated in the reaction between aluminium chloride and hydrazine. See also Paulik and Erdey[35].

(iv) Bauxite   Because of the impurities it is difficult to obtain comparable results in the case of bauxite. See, for instance, Orosco[40] and Paulik and Erdey[35] (derived thermogram).

(v) Bayerite   Bayerite was studied by Eyraud *et al.*[33] by differential thermal analysis and calorimetry.

3. Aluminium chloride

Aluminium chloride has the formula $AlCl_3 \cdot 6H_2O$ when taken from a fresh bottle. However, the handling necessary to place the sample into the crucible allows the salt to pick up a little moisture, whose departure up to 38° is signaled on the thermolysis curve. The hexahydrate appears stable between 38° and 118°; then hydrolysis and dehydration lead to the oxide Al_2O_3, rather rapidly up to 500°, and very slowly thereafter. The conversion seems to be complete around 800°[41]. See also Gill[42].

From among the many variants of the precipitation of the chloride in organic media, we have selected that of Minning[43], who uses a mixture of acetone and acetyl chloride (4 : 1), the latter being freshly distilled and free from phosphorus. The requisite hydrogen chloride is expelled from its aqueous solution (commercial) by adding concentrated sulphuric acid. The pyrolysis curve shows the loss of washing liquids up to 39°. We do not recommend weighing the hexahydrate between 39° and 118°, since the corresponding level slopes slightly downward.

Aluminium chloride–ether Hagenmuller and Rouxel[44] studied the system: aluminium chloride–ether, reported by Menzel and Fröhlich. This molecular association is stable up to 150°. Under reduced pressure, the solid residue consists of aluminium chloride, one or more high polymers with ethylenic linkages, and a hydroxyethoxy aluminium chloride $HO-AlCl-OC_2H_5$, which decomposes above 200° in two ways:

$$HO-AlCl-OC_2H_5 \rightarrow AlO(OH) + C_2H_5Cl$$
$$HO-AlCl-OC_2H_5 \rightarrow AlOCl + C_2H_5OH$$

which occur simultaneously in the vicinity of 300°. The amorphous AlOCl decomposes near 600° into alumina and aluminium chloride.

4. Aluminium perchlorate

Zinov'ev and Chudinova[45] made a study of aluminium perchlorate. The salt $Al(ClO_4)_3 \cdot 6H_2O$ melts in its own water of crystallization around 82°. This is followed by an hydrolysis which yields a basic product $HO-Al(ClO_4)_2$, which in turn begins to decompose at 178°. The residue consists of alumina formed as follows:

$$4 \, Al(ClO_4)_3 \rightarrow 2 \, Al_2O_3 + 6 \, Cl_2 + 21 \, O_2$$

5. Aluminium sulphate

Extremely diverse temperatures have been reported for the onset of the decomposition of aluminium sulphate: 200°, 250°, 316°, 450°, etc. The material which we[67] employed contained close to 16 molecules of water. When a heating rate of 300° per hour is used, the commercial product must be kept between 450° and 610° to obtain an accurately standard solution; beyond this temperature the salt loses sulphuric anhydride, and the decomposition is almost complete at 1000°. See also Yamamotò and Bito[46]. Tanabe[47] made a study of the basic aluminium sulphates, employing chemical, thermogravimetric, spectrographic (infra-red and X-ray) techniques. Three groups of curves are found, depending on the Al : SO_4 ratio, and two types of water: one of these disappeared at 500°. The OH groups remained above 700°.

6. Aluminium ammonium alum

The ammonium alum may have the formula $NH_4[Al(H_2O)_2(SO_6H_4)_2]$·$6H_2O$ in analogy with the chromium alum, whose formula has been established experimentally, but in the case of the aluminium alum it is assuredly a more imperfect complex than in the case of the chromium compound. When ammonium alum is heated to 1030° at the rate of 130° per hour, it is stable up to 54°. It loses all of its water around 310° in three stages which are not very clearly separated. First there is a loss of 6 molecules of water of crystallization, up to about 126°, then 4 molecules of water combined with sulphuric acid pass off up to around 170°, and finally the 2 molecules of water, which are located at the extremities of the quaternary axis of the Werner octahedron. The product which makes its appearance after the almost horizontal level between 308° and 524° is a complex of the $NH_4[Al(SO_4)_2]$ type. In fact, according to preliminary studies, the ammonium sulphate may already have disappeared by sublimation and dissociation. This ammonium sulphate begins to be lost between 524° and around 620°. There then remains aluminium sulphate $Al_2(SO_4)_3$, which starts to decompose progressively around 620°. The formation of Al_2O_3 is not complete[41] even at 1030°.

See also Erdey[48] regarding the thermolysis of the potassium salt.

7. Aluminium nitrate

The aluminium nitrate $Al(NO_3)_3$·$9H_2O$, which we used, was slightly deliquescent. It loses its water and decomposes without transition. A change in direction at 154° is in conformity with the weight of the anhydrous salt. The horizontal of the oxide Al_2O_3 is reached sluggishly between 750° and 1000° at a heating rate of 300° per hour. See also Wendlandt[49] who reports 50° for the onset of the dehydration; the horizontal of the oxide begins at 460°, a figure that appears improbable in our opinion.

8. Aluminium phosphate

D'Yvoire[50] studied two phosphates, for which he has given the preparation procedures, namely $2P_2O_5 \cdot Al_2O_3 \cdot 9H_2O$ and $2P_2O_5 \cdot Al_2O_3 \cdot 5H_2O$. He heated them at the rate of 150° per hour. The first lost 6 molecules of water from 95° to 150° to give $2P_2O_5 \cdot Al_2O_3 \cdot 3H_2O$, which is extremely hygroscopic and decomposes around 200° into $AlPO_4$ (with cristobalite structure) and an amorphous phase which loses water and progressively crystallizes to yield $Al(PO_3)_3$.

The second compound, a pentahydrate, loses 2 molecules of water between 140° and 195°. The product, if quenched at 195°, has the global composition $2P_2O_5 \cdot Al_2O_3 \cdot 3H_2O$ but the X-ray spectra show that we are not dealing here with the trihydrate noted above but with a mixture of $AlPO_4$ (berlinite) and $Al(H_2PO_4)_3$. Consequently, only this latter compound is formed from 200° on. A mixture of $AlPO_4$ and $Al(PO_3)_3$ remains at the close of the heating.

From the standpoint of chemical analysis, we[1] prepared a precipitate in a buffered medium (pH 5.0–5.4) from disodium phosphate, using the procedure suggested by Lundell and Knowles[51]. The slope of the curve changes very distinctly at 228°. Even though the residue shows a constant weight from 743° on, which corresponds to the formula $AlPO_4$, it is probable that the initial precipitate is a mixture containing, without doubt, a little of the acid phosphate $Al_2(HPO_4)_3$, which accounts for the singularity of the curve between 126° and 228°.

9. Aluminium arsenates

With respect to the aluminium arsenates, we have found a study by Martin, Masson, Duc-Maugé and Guérin[52] dealing with the mono-aluminium derivative $Al(H_2AsO_4)_3$. This compound decomposes from 550° on and yields directly the aluminium orthoarsenate $AlAsO_4$, which these workers have likewise distinguished on the equilibrium diagram. It is stable up to 950°; beyond this temperature it yields aluminium oxide with no discernible intermediate compound. It is essential to heat above 1050° to remove the last traces of arsenic anhydride. The thermolysis curve does not confirm the existence of the pyroarsenate.

10. Cryolite

Cryolite from Greenland and a commercial product (a mixture of cryolite and aluminium oxide) have engaged the attention of Paulik and Erdey[35]. Tananaev and Lelchuk[53] showed that cryolite does not have the commonly accepted formula $Na_3[AlF_6]$, but rather

11NaF·4AlF$_3$. This compound has been suggested for the gravimetric determination of aluminium and Paulik and Erdey direct that it be dried at 125–130°. The recorded curve shows that the alcohol retained from the washing comes off up to 66°; the horizontal corresponding to the double salt noted above extends only to 82°; then aluminium fluoride starts to come off. If the material is weighed after it has been dried at 125°, there is danger that the results will be low. The filtering crucible should be kept between 66° and 82°.

11. Lithium aluminate

See Chapter Lithium regarding lithium aluminate.

12. Sodium aluminate

With respect to sodium aluminate, Viltange[54] studied the reaction of various mixtures of aluminium oxide and sodium peroxide. Below 309°, the only losses of weight recorded were those corresponding to a loss of water. At 309° ± 18°, there began a release of oxygen and this continued up to 738° ± 48°. The heating curve contains an irregularity at 328° ± 16°; the reaction is very rapid at this temperature. The weight remains constant from 738° to 1000°. Only two reactions have been recorded: the one at the moment of spontaneous decomposition of the sodium peroxide, where there is formation of $Na_n(AlO_2)_n$ in which n denotes 2 or 3 (analysis by infra-red absorption); the other, occurring at a higher temperature and producing sodium carbonate and the residual oxide.

13. Aluminium silicates

Numerous studies have been made of the aluminium silicates. Hénin[55] dealt with synthetic clays made from silica gel as the starting material. Paulik and Erdey[35] heated a natural clay, Zettlitz kaolin, kaolinite from Arkansas, while Papailhau[56] studied kaolin, Saito[57] heated clay at two rates, Galimberti[58] dealt with nontronite, allophane, clay, beidellite, montmorillonite, halloysite, nacrite, kaolinite, anauxite, and dickite. Mielenz, Schieltz and King[59] were the first to combine thermogravimetry and differential thermal analysis in the study of dickite, antigorite, chrysotile, kaolinite, halloysite, and several micas: biotite, lepidolite, talc, muscovite, pyrophyllite, prochlorite. This investigation was completed by studies of hectorite, montmorillonite, nontronite, celadonite, vermiculite, illite, jefferisite, sepiolite, and palygorskite.

Cordierite from Madagascar $(Fe^{2+},Mg)_2Al_3Si_5O_{18}$ was heated at the rate of 300° per hour by Iiyama[60] with grains several tens of microns across (20 to 80), and at 50° per hour with grains from 300 to 800 microns across. A continuous loss of water is observed; it accelerates sharply around 500°. The author suggests the existence of two kinds of water here. See Chapter Magnesium regarding the work on a palygorskite from Senegal by Caillère and Rouaix[61].

14. Aluminium cupferronate

The precipitation of aluminium by cupferron in a weakly basic medium is due to Pinkus and Martin[62]. The recorded curve shows that it is impossible to obtain a horizontal for the anhydrous complex. The decomposition begins around 72°, and is very rapid between 150° and 230°. The combustion of the carbon then ensues and is finished between 250° and 750° depending on the quantities involved. Beyond this temperature the horizontal corresponding to aluminium oxide appears, and the latter is the weighing form usually employed in this case.

15. Aluminium oxinate

Aluminium oxinate $Al(C_9H_6ON)_3$ is prepared by the directions given by Berg[63]. It is anhydrous but it usually retains traces of moisture up to 110–115°, or sometimes even 135° according to Borrel and Pâris[64]. The horizontal of the anhydrous complex may be used for the automatic determination of aluminium; it is stable up to 375°; beyond this temperature, the oxine decomposes and, depending on the amounts involved, the tar burns off between 700° and 1000°, prior to giving the horizontal of the oxide Al_2O_3.

16. Aluminium 5,7-dibromooxinate

If instead of oxine, its 5,7-dibromo derivative is used as precipitant, then according to Sanko and Burssuk[65] and Berg and Küstenmacher[66], the precipitate $Al(C_9H_4Br_2ON)_3$ obtained in acetone solution, yields a shorter horizontal, extending from 130° to 190°. Even at 210° there is a loss of weight amounting to 1/70th. In our opinion, there is no need to complicate the determination of aluminium by injecting this procedure;

it would be far better to retain the well developed oxine method, which lends itself so well to the appended separations and determinations, and which yields a horizontal of constant weight that is twice as long. We have not carried the pyrolysis of the dibromo-oxinate to the point where the aluminium oxide is obtained.

REFERENCES

[1] T. DUPUIS AND C. DUVAL, *Anal. Chim. Acta*, 3 (1949) 191.
[2] C. CABANNES-OTT, *Thesis*, Paris, 1958.
[3] O. GLEMSER AND G. RIECK, *Proc. 3rd Intern. Symposium on Reactivity of Solids, Madrid*, Vol. 1, 1956, p. 361.
[4] C. GROOT AND V. H. TROUTNER, *Anal. Chem.*, 29 (1957) 835.
[5] L. ERDEY AND F. PAULIK, *C.A.*, 50 (1956) 3952.
[6] C. R. FRESENIUS, *Introduction à l'analyse*, Vol. 1, 1903, p. 212.
[7] W. BLUM, *J. Am. Chem. Soc.*, 38 (1916) 1295.
[8] M. BERGÈS, *Compt. rend.*, 225 (1947) 241.
[9] F. TROMBE, *Compt. rend.*, 215 (1942) 539; 216 (1943) 888; 225 (1947) 1156.
[10] H. H. WILLARD AND N. K. TANG, *J. Am. Chem. Soc.*, 59 (1937) 1190; *Ind. Eng. Chem., Anal. Ed.*, 9 (1937) 357.
[11] B. SOLAJA, *Chemiker-Ztg.*, 49 (1925) 337.
[12] P. RÂY AND A. K. CHATTOPADHYA, *Z. anorg. Chem.*, 169 (1928) 99.
[13] E. A. OSTROUMOFF, *Z. anal. Chem.*, 106 (1936) 170.
[14] O. BRUNCK, *Chemiker-Ztg.*, 28 (1904) 514.
[15] W. FUNK, *Z. anal. Chem.*, 45 (1906) 181, 489.
[16] I. M. KOLTHOFF, V. A. STENGER AND B. MOSKOVITZ, *J. Am. Chem. Soc.*, 56 (1934) 812.
[17] K. YOUNG AND HSIAO-CHING LAY, *C.A.*, 30 (1936) 695.
[18] W. F. HILLEBRAND AND G. E. F. LUNDELL, *Applied Inorganic Analysis*, 1927, p. 59.
[19] C. MEINEKE, *Angew. Chem.*, (1888) 224.
[20] T. KOZU, *Bull. Chem. Soc. Japan*, 70 (1935) 356.
[21] A. JILEK AND J. LUKAS, *Collection Czechoslov. Chem. Communs.*, 2 (1930) 63, 113.
[22] R. RIPAN, *Bull. soc. stiinte Cluj*, 3 (1927) 311.
[23] E. SCHIRM, *Chemiker-Ztg.*, 33 (1909) 877; 35 (1911) 980.
[24] J. E. CLENNELL, *Metal Ind. (N.Y.)*, 21 (1922) 273.
[25] A. STOCK, *Ber.*, 33 (1900) 548.
[26] M. BARBIER, *Bull. Soc. chim. France*, 4 (1910) 1027.
[27] R. FRICKE AND K. MEYRING, *Z. anorg. Chem.*, 188 (1930) 127.
[28] W. JAKOB, *Anz. Krakau Akad., Ser. A.*, (1913) 56.
[29] W. R. SCHOELLER AND H. W. WEBB, *Analyst*, 54 (1929) 709.
[30] O. I. MILNER AND L. GORDON, *Talanta*, 4 (1960) 115.
[31] C. OTT, *Compt. rend.*, 240 (1955) 68.
[32] B. IMELIK, M. PETITJEAN AND M. PRETTRE, *Compt. rend.*, 236 (1953) 1278.
[33] C. EYRAUD, R. GOTON, Y. TRAMBOUZE, TRAN HUU THE AND M. PRETTRE, *Compt. rend.*, 240 (1955) 862, 1082.
[34] C. EYRAUD AND R. GOTON, *J. Chim. phys.*, 51 (1954) 430.
[35] F. PAULIK AND L. ERDEY, *Acta Chim. Acad. Sci. Hung.*, 13 (1957) 117.
[36] T. SATO, *J. Appl. Chem. (London)*, 9 (1959) 331.
[37] C. EYRAUD, R. GOTON AND M. PRETTRE, *Compt. rend.*, 238 (1954) 1028.
[38] R. TERTIAN AND D. PAPÉE, *Compt. rend.*, 241 (1955) 1575.
[39] J. F. BROWN, D. CLARK AND W. ELLIOTT, *J. Chem. Soc.*, (1953) 84.
[40] E. OROSCO, *Ministerio trabalho ind. ecom., Inst. nac. tecnol. (Rio de Janeiro)*, 1940. *C.A.*, 35 (1941) 3485.

[41] C. Duval, *Anal. Chim. Acta*, 20 (1959) 21.
[42] A. F. Gill, *Can. J. Research*, 10 (1934) 703.
[43] H. D. Minning, *Am. J. Sci.*, 40 (1915) 482.
[44] P. Hagenmuller and J. Rouxel, *Compt. rend.*, 247 (1958) 1623.
[45] A. Zinov'ev and L. I. Chudinova, *Zhur. Neorg. Khim.*, 1 (1956) 1722.
[46] K. Yamamotò and K. Bito, *Bull. Dept. Appl. Chem. Waseda Univ.*, 9 (1929) 12.
[47] H. Tanabe, *J. Pharm. Soc. Japan*, 77 (1957) 50.
[48] L. Erdey, *Periodica Polytech.*, 1 (1957) 35.
[49] W. W. Wendlandt, *Texas J. Sci.*, 10 (1958) 392; *C.A.*, (1959) 12082.
[50] F. D'Yvoire, *Compt. rend.*, 247 (1958) 297.
[51] G. E. F. Lundell and H. B. Knowles, *Ind. Eng. Chem.*, 14 (1922) 1136.
[52] R. Martin, J. Masson, C. Duc-Maugé and H. Guérin, *Bull. Soc. chim. France*, (1959) 412.
[53] I. V. Tananaev and J. L. Lelchuk, *Zhur. Anal. Khim.*, 2 (1947) 93.
[54] M. Viltange, *Thesis*, Paris, 1960; *Ann. Chim. (Paris)*, 5 (1960) 1037.
[55] S. Hénin, *Compt. rend.*, 244 (1957) 225.
[56] J. Papailhau, *Bull. soc. franç. minéral.*, 82 (1959) 370.
[57] H. Saito, *Sci. Repts. Tôhoku Imp. Univ.*, 16 (1927) 57.
[58] L. Galimberti, *Boll. sci. fac. chim. ind. Bologna*, 18 (1940) 263.
[59] R. C. Mielenz, N. C. Schieltz and M. E. King, *Proc. 2nd Nat. Conf. Clays Clay Minerals, N.R.C. Publ.*, No. 327, Washington D.C., 1954, p. 285.
[60] J. T. Iiyama, *Bull. soc. franç. minéral*, 83 (1960) 155.
[61] S. Caillère and S. Rouaix, *Compt. rend.*, 246 (1958) 1442.
[62] A. Pinkus and F. Martin, *Chim. & Ind. (Paris)*, 17 (1927) 182.
[63] R. Berg, *Das Oxin*, Stuttgart, 1937, p. 93.
[64] M. Borrel and R. Pâris, *Anal. Chim. Acta*, 4 (1950) 267.
[65] A. M. Sanko and A. J. Burssuk, *Chem. Zentr.*, 107 (1936-II) 3928.
[66] R. Berg and H. Küstenmacher, *Z. anorg. Chem.*, 204 (1932) 215.
[67] C. Duval, *Anal. Chim. Acta*, 15 (1956) 224.

CHAPTER 15

Silicon

There has been but little work on silicon apart from the methods of determination in the form of silica, potassium silicofluoride, and various silicomolybdates (with oxine, hexamethylenetetramine, pyramidone, pyridine)[1].

1. Silicon dioxide

(i) Gelatinous silica Gelatinous silica may be weighed after triple "insolubilization". In the general case, it is found that the greater part of the retained water is removed at 160° where the curve shows a sudden break. The last traces of water and hydrochloric acid then come off very slowly from 160° to 358°. A good horizontal is then obtained which extends beyond 1000°. Kato *et al.*[2] report that constant weight is not obtained until 900° is reached. Yamamoto and Bito[3] studied the silica gel after modifying their thermobalance.

(ii) Preparation with addition of gelatin It is also beneficial, according to Stross[4], to add at 60° some 0.25% gelatin sol to the solution to be analyzed. The recorded curve is identical with the preceding one but the constant weight for silica is not attained below 505°.

2. Mixture of silica and alumina

Shinkai[5] directs that the silica to be determined be mixed with aluminium chloride and that a mixture of alumina and silica flocculated by each other be weighed. The recorded curve is still as before with constant weight reached at 475°. If in the course of these experiments, the slightest trace of iron is introduced into the material being determined or if the hydrochloric acid contains iron, the corresponding curves ascend again at 675° and the oxidation of the iron is not complete below 950°.

3. Mixture of silica and sodium peroxide

Duval and Lecomte[6] studied the solid state reaction between silica and sodium peroxide (oxylith):
$$2 Na_2O_2 + SiO_2 \rightarrow Na_4SiO_4 + O_2$$
It is practically finished at 315° but the temperature must be taken to 500° to bring the reaction to actual completion. There is much analogy between the pyrolysis curve of oxylith alone and that of a mixture of sodium peroxide and silica.

4. Silicon diimide

The diimide $Si(NH)_2$ of Glemser and Neumann[7] prepared from $SiCl_4$ and liquid ammonia, was decomposed under an ammonia pressure equivalent to 50 mm of mercury. At 400° the following reaction occurs:
$$6 [Si(NH)_2] \rightarrow 2 [Si_3(NH)_3N_2] + 2 NH_3$$
and then, at 600°:
$$2 [Si_3(NH)_3N_2] \rightarrow 3 [Si_2(NH)N_2] + NH_3$$
and finally, at 1250°:
$$3 [Si_2(NH)N_2] \rightarrow 2 Si_3N_4\alpha + NH_3$$
The intermediate products are highly hygroscopic; they exist in the polymerized state.

5. Potassium silicofluoride

The weighing of silicon as potassium silicofluoride $K_2[SiF_6]$ indicates that this compound is anhydrous between 60° and 410°, if the washing was finished with ethanol and the heating conducted in a platinum crucible. Above 410° there is gradual decomposition; the following data were obtained from one of the recorded curves when 239.40 mg of the dry silicofluoride was heated rapidly (1000° in 2 hours). The course of the decomposition is quite comparable to that of potassium fluoborate:

Temp. (°C)	410	542	719	813	906
Loss (mg)	0	2.0	5.2	13.2	35.9

Matsuura[8] reported that the horizontal of the anhydrous compound extends from 150° to 500°.

6. Mixture of silica and molybdic anhydride

Duval[9], Duval and Lecomte[6], Dupuis[10], and Kiba and Ikeda[11] have made studies of the mixture $SiO_2 \cdot 12MoO_3$. See Chapter Molybdenum.

7. Oxine silicomolybdate

Dupuis[10] restudied oxine silicomolybdate prepared by the method of Volinetz[12] and Brabson et al.[13]. The starting material, which could not be used in gravimetric analysis, had the composition $(C_9H_7NO)_4H_8$ $[Si(Mo_2O_7)_6]\cdot4H_2O$. The precipitate loses some water from ordinary temperature on even after it has been dried several times in the air. The first part of the thermolysis curve varies from product to product; there is a horizontal stretch from 155–160° to 225–250°, which corresponds to the composition of a dihydrate $4C_9H_6NO\cdot SiO_2\cdot12MoO_3\cdot2H_2O$. Then the oxine decomposes along a curve similar to that encountered for the phosphomolybdate. The final level extends from 440° to 770° and agrees with the presence of the system $SiO_2\cdot12MoO_3$. The molybdic anhydride MoO_3 sublimes rapidly above 770°.

8. Hexamine silicomolybdate

The complex with hexamethylenetetramine was prepared by the method given by Duval[14]; its formula is $(C_6H_{13}N_4)_4[Si(Mo_2O_7)_6]\cdot4H_2O$. The thermolysis curve shows that the heating rate plays a significant role. At 300° per hour, the complex first loses 4 molecules of water up to 200° which marks the beginning of the horizontal corresponding to the compound $(C_6H_{13}N_4)_4H_4[Si(Mo_2O_7)_6]$ or else $4C_6H_{13}N_4\cdot SiO_2\cdot12MoO_3 = 2420$; this is suitable for the automatic determination of silicon, the factor for Si being 0.0115. The complex then decomposes suddenly, giving off combined water and about half of the amine up to 250°, where occasionally a short horizontal begins and continues as far as 275°. The rest of the amine then burns quietly up to 450°, save for a distinct bend in the curve near 363°. The curve is almost horizontal from 451° to 520°, but the mixture $SiO_2\cdot12MoO_3$ reaches a constant weight only at 520° and maintains it up to 680°. The molybdic anhydride MoO_3 sublimes from 680° to 1180° and the crucible finally contains only silica.

9. Pyridine silicomolybdate

Pyridine silicomolybdate, prepared as described by Babko[15] has the formula $(C_5H_6N)_4H_4[Si(Mo_2O_7)_6]\cdot xH_2O$ or else $4C_5H_5N\cdot SiO_2\cdot12MoO_3\cdot (4 + x)H_2O$. This precipitate cannot serve for the gravimetric determination of silicon since the content of water of crystallization is not constant from one preparation to another. In several recordings, the value of x has in fact been found to be 6 whereas in others it was not far from 2. In general, the composition is constant up to $158°$; the combined water and $1/4$ of the pyridine escape slowly up to $354°$, and the rest of the pyridine is rapidly released between $354°$ and $415°$. The level corresponding to the mixture $SiO_2 + 12MoO_3$ begins at $415°$. Molybdic anhydride MoO_3 starts to sublime at $742°$.

10. Pyramidone silicomolybdate

The King and Watson[16] method was used with pyramidone. They, like Hecht and Donau[17], were not sure of the formula of the precipitate, and hesitated between $(C_{13}H_{17}N_3O)_3H_8[Si(Mo_2O_7)_6] \cdot xH_2O$ (with x probably is equal to 6) or else $3\,(C_{13}H_{17}N_3O) \cdot SiO_2 \cdot 12MoO_3 \cdot (4+x)H_2O$, when the drying temperature was $60–70°$. The thermolysis curve of the precipitate includes a level from $137°$ to $180°$ but it agrees rather with the composition $3(C_{13}H_{17}N_3O) \cdot SiO_2 \cdot 12MoO_3 \cdot 4H_2O$. A second level goes from $200°$ to $250°$ but no formula can be satisfactorily fitted to it. Finally, the level of the mixture $SiO_2 \cdot 12MoO_3$ begins at $500°$ and stops at $753°$ where the molybdic anhydride starts to sublime. This method should be discarded.

11. Antipyrine silicomolybdate

Dupuis[10] attempted to substitute antipyrine for pyramidone, a method of determination which had not been previously suggested. The precipitation is quantitative; the precipitate is stable up to $237°$. Some supplementary trials have shown that at $600°$ the $SiO_2 : MoO_3$ ratio is not equal to $1/12$ but rather is closer to $1/6$. Furthermore, some trials on precipitates obtained from measured volumes of standard solutions of alkali silicates that had been stored in paraffined vessels and at a suitable pH, have not yielded reproducible results.

REFERENCES

[1] T. DUPUIS AND C. DUVAL, Compt. rend., 229 (1949) 51; Anal. Chim. Acta, 4 (1950) 50.
[2] H. KATO, H. HOSIMIJA AND S. NAKAZIMA, J. Chem. Soc. Japan, 60 (1939) 1115.
[3] K. YAMAMOTO AND K. BITO, Bull. Dept. Appl. Chem. Waseda Univ., 9 (1929) 12; C.A., (1930) 1276.
[4] W. STROSS, Metallurgia, 38 (1948) 63.
[5] S. SHINKAI, J. Soc. Chem. Ind. Japan, 46 (1943) 234.
[6] C. DUVAL AND J. LECOMTE, Compt. rend., 234 (1952) 2445.
[7] O. GLEMSER AND P. NEUMANN, Z. anorg. Chem., 298 (1958) 134.
[8] K. MATSUURA, J. Chem. Soc. Japan, 52 (1931) 730.

9   C. Duval, *Proc. Intern. Symposium Reactivity of Solids Gothenburg*, 1952, p. 505, 511.
10 T. Dupuis, *Compt. rend.*, 228 (1949) 841; *Thesis*, Paris, 1954.
11 T. Kiba and T. Ikeda, *J. Chem. Soc. Japan*, 60 (1939) 911.
12 M. J. Volinetz, *C.A.*, 30 (1936) 7497.
13 J. A. Brabson, H. C. Mattraw. G. E. Maxwell, A. Darrow and M. F. Needham, *Anal. Chem.*, 20 (1948) 504.
14 C. Duval, *Anal. Chim. Acta*, 1 (1947) 33.
15 A. K. Babko, *J. Appl. Chem. U.S.S.R.*, 10 (1937) 374.
16 E. J. King and J. L. Watson, *Mikrochemie*, 20 (1936) 49.
17 F. Hecht and J. Donau, *Anorganische Mikrogewichtsanalyse*, Springer, Vienna, 1940, p. 244.

CHAPTER 16

Phosphorus

This chapter will deal with several anions containing phosphorus. See Dupuis and Duval[1].

1. Hypophosphite ion $PO_2H_2^-$

We have not found any gravimetric study in the literature.

2. Phosphite ion $PO_3H_2^-$

The only method suggested depends on heating mercurous chloride. The latter sublimes from 130° on. See Chapter Mercury.

3. Pyrophosphite ion $P_2O_5^{4-}$

No gravimetric procedure has been discovered in the literature.

4. Hypophosphate ion $P_2O_6^{4-}$

We have not found any gravimetric method in the literature.

5. Metaphosphate ion PO_3^-

All of the methods reported are based on the determination of orthophosphates.

6. Pyrophosphate ion $P_2O_7^{4-}$

Most of the published methods involve determining orthophosphates. See, for instance, the ferric ammonium alum procedure of Raewsky[2], or that of Berthelot and André[3], which employs magnesium salts. The pyrophosphate ion may likewise be precipitated directly by a zinc salt, as proposed by Von Knorre[4] and used by Travers and Chu[5], and Bell[6]. The latter suggests igniting the precipitate at 500–600°, and then at 900°. The recorded curve shows that the loss of water continues up to 610° after starting slowly at 100°. The horizontal due to zinc pyrophosphate $Zn_2P_2O_7$ begins at 610°.

7. Orthophosphate ion PO_4^{3-}

(i) Barium phosphate Apparently the precipitation of barium phosphate has not been restudied since it was first published by Wackenroder and Ludwig[7]. The compound was thrown down by adding barium acetate to disodium phosphate. The monoacid salt $BaHPO_4$ precipitates; it filters nicely. The thermolysis curve has a horizontal from 126° to 384° which may be used for the automatic determination of phosphorus. One molecule of water is lost by two molecules of the phosphate and the corresponding pyrophosphate $Ba_2P_2O_7$ in turn gives a horizontal starting at 520°.

(ii) Magnesium ammonium phosphate Regarding the precipitation as magnesium ammonium phosphate see Chapter Magnesium.

(iii) Tin phosphate The precipitation as tin phosphate devised by Reynoso[8] serves particularly for separation purposes. The product can be converted to magnesium pyrophosphate as the weighing form.

(iv) Bismuth phosphate Bismuth phosphate, precipitated according to the procedure given by Keschan[9], is dry from 119° on, and its weight is constant up to 961°.

(v) Ferric phosphate We adopted the method of Raewsky[10] as modified by Mohr[11] for preparing ferric phosphate, *i.e.* precipitation by triammonium phosphate in the presence of ammonium acetate. The thermolysis curve shows that after the water and the retained ammonium salts have been driven off the phosphate is constant in weight from 531° on.

(vi) Zirconyl phosphate Close adherence to the directions given by Stumper and Mettelock[12] produced for us a precipitate of zirconyl phosphate $ZrOHPO_4$. It loses moisture, bound water, and retained hydrochloric acid up to 596°, which marks the start of a horizontal conforming

to the formula $(ZrO)_2P_2O_7$, namely zirconyl pyrophosphate. Nothing further happens if the material is then heated to 1050° as suggested by these workers.

(vii) Mercurous phosphate   The method given by Munroe[13] for mercurous phosphate is for separation purposes; the final weighing form is magnesium pyrophosphate.

(viii) Silver thallium phosphate   The method of Spacu and Dima[14] was employed for precipitating silver thallium phosphate Ag_2TlPO_4. It dries at once and may be weighed after being brought to any temperature between 20° and 720°, the curve showing a horizontal extending over this entire temperature interval. At 946°, the loss of weight amounts to no more than 2 mg in 350 mg. The Spacu and Dima method is excellent for the automatic determination of phosphorus.

(ix) Pentamminecobaltinitrate phosphomolybdate   We have found that this compound $[CoNO_3(NH_3)_5]$ $H_3PMo_{12}O_{41}$, produced and washed in accord with the directions of Furman and Sate[15], is not stable. Actually, it is necessary to dry this material under reduced pressure in a desiccator. From 72° on it decomposes and leads, between 484° and 780°, to a mixture of cobalt phosphate and molybdic anhydride. The latter begins to sublime at 780°. See Dupuis[16].

(x) Uranium phosphate   See Chapter Uranium regarding uranium phosphate.

(xi) Ammonium phosphomolybdate   Ammonium phosphomolybdate is the principal method of determining phosphorus along these lines. Special attention is directed to the work of Dupuis and Duval[1], Dupuis[16], and Duval[17]. The numerous curves which we have recorded were obtained with dry or moist products prepared by the methods laid down by Treadwell [18]. He directs that the weighing in the form of $(NH_4)_3PO_4 \cdot 12MoO_3$ should follow prolonged heating at 170°. According to Woy[19], gentle ignition yields the mixture $P_2O_5 \cdot 24MoO_3$. The precipitate, which usually is represented by the formula $(NH_4)_3PO_4(MoO_3)_{12} \cdot 2HNO_3 \cdot H_2O$, loses water and nitric acid up to 180°. If the recording of the same curve is begun again, but on a precipitate dried in the air, the loss of weight up to 180° corresponds to $2\ HNO_3$ and not to $2HNO_3 + H_2O$. The

difference is not significant and does not affect the determination of the phosphorus in any way. A perfectly horizontal section extends from 180° to 410°; it agrees very well with the formula $(NH_4)_3PO_4 \cdot (MoO_3)_{12}$.

Starting at 410°, 2 molecules of this material lose 6 molecules of ammonia and 3 molecules of water up to 540°. However, all of our recordings show that the decomposition is even more deep-seated since the curves have a distinct bend and then rise again. Accordingly, a transient reduction must be assumed followed by a reoxidation of the molybdenum by the air to yield the mixture $P_2O_5 \cdot 24MoO_3$. At 540°, the material in the crucible is green; it consists of a mixture of "molybdenum blue" and yellow phosphomolybdate. The renewed rise of the curve is very slow. Theoretically, the horizontal is only reached between 812° and 850°, but the ignition zone may extend over the interval 600–850° without significant damage to the determination of phosphorus, but the molybdic anhydride sublimes rapidly beyond 850°. The slow ascent of the curve between 540° and 812° now provides the reason why so many workers have obtained erratic results when determining phosphorus. Evidently, in proportion to this ascent, the composition changes from one temperature to another, and thus gives rise to the bizarre correction formulas that appear in certain books dealing with quantitative analysis. Finally, it should be noted that the determination of phosphorus after drying the precipitate between 180° and 410° is more attractive as regards the analytical factor for the anhydride P_2O_5 than the method suggested by Woy.

Employing the Stockdale method[20], Wendlandt[21] heated 70–90 mg of the precipitate on the thermobalance at the rate of 300° per hour. The precipitate had been washed with ammonium nitrate or with nitric acid. The compound after being air-dried lost extraneous water up to 60°; a horizontal corresponding to $(NH_4)_2H[P(Mo_3O_{10})_4] \cdot H_2O$ extends from 160° to 415°. Above 415°, a renewed loss of weight leads to the horizontal of $P_2O_5 \cdot 24MoO_3$, starting at 500°. We have not been able to reproduce this statement and we stand by our earlier findings. See also, in this same paper, our remarks on Wendlandt's communication[17].

(xii) Oxine phosphomolybdate The phosphomolybdate of 8-hydroxy-quinoline (oxine) comes down in a variety of compositions depending on the conditions of precipitation. The curves obtained with precipitates produced in accord with the directions of Hecht and Donau[22] are typical of the destruction of all oxine compounds. A horizontal from 236° to 268° corresponds to the composition reported by these workers, namely $3(C_9H_7ON)H_7[P(Mo_2O_7)_6] \cdot 2H_2O$, which is contrary to the finding by

Kiba and Ikeda[23] who give the region as 115° to 135°. Around 341°, the oxine disappears with no evidence of the production of the anhydrous compound. Then, at 600°, there is an irregularity in the curve analogous to the bend in the ammonium phosphomolybdate curve. A slow rise up to 765° follows. The stretch corresponding to the ensemble $P_2O_5 \cdot 24MoO_3$ is perfectly horizontal from 765° to 820°; beyond this temperature, the molybdic anhydride begins to sublime.

In some cases, Dupuis[16] found that the horizontal, which is located between 235° and 300° as an average, corresponds to the composition of the anhydrous compound; in other instances, it corresponds to a product containing an excess of reagent.

(xiii) Quinoline phosphomolybdate Wendlandt and Hoffman[24] studied the thermolysis curve of quinoline phosphomolybdate, suggested by Perrin[25] and Fennell and Webb[26]. After the precipitate has dried in the air for 24 hours, it starts to lose weight at 107° (heating rate 4.5° per minute with a specimen weighing 95–100 mg). The loss amounts to very near 1 molecule of water per molecule of $(C_9H_7N)_3 \cdot H_3PO_4 \cdot 12MoO_3 \cdot 1H_2O$. A horizontal is observed from 155° to 370° and corresponds to the anhydrous product. (Perrin suggested drying at 160°, and Fennell and Webb proposed 250°.) Decomposition sets in beyond 370° and $P_2O_5 \cdot 24MoO_3$ results at 500°. These workers have also traced the isotherms at 100°, 150°, and 200°, as well as the thermal analysis curves. The precipitate can be dried in less than 15 minutes at 200°.

(xiv) Strychnine phosphomolybdate We have applied the Embden method[27], which was repeated by Antoniani and Jona[28], for precipitating the phosphomolybdate by means of strychnine sulphate. The recorded curve includes a horizontal stretch from 54° to 225° and still another from 525° to 827°, after which the molybdic anhydride sublimes. Embden recommends drying at 105–110° and assumes that the weight of the precipitate is 39 times that of the phosphoric anhydride P_2O_5. We do not agree with this; the residue conforming to the second horizontal accords with the composition $P_2O_5 \cdot 33MoO_3$, in other words the residue contains an excess of molybdic anhydride. Even though Embden claims that his method is very precise, we suggest that it be discarded. A study should be made to determine the exact formula of the precipitate.

REFERENCES

[1] T. DUPUIS AND C. DUVAL, *Anal. Chim. Acta*, 4 (1950) 256; *Compt. rend.*, 229 (1949) 51.
[2] RAEWSKY, *Compt. rend.*, 26 (1848) 205.

[3] M. Berthelot and G. André, *Compt. rend.*, 123 (1896) 773; 124 (1897) 261.

[4] G. Von Knorre, *Z. angew. Chem.*, 5 (1892) 639.

[5] A. Travers and Y. K. Chu, *Helv. Chim. Acta*, 16 (1933) 913.

[6] R. N. Bell, *Anal. Chem.* 19 (1947) 97.

[7] H. Wackenroder and H. Ludwig, *Arch. Pharm.*, (2) 56 (1848) 265, 283.

[8] A. Reynoso, *Compt. rend.*, 33 (1851) 385.

[9] A. Keschan, *Z. anal. Chem.*, 128 (1948) 215.

[10] Raewsky, *Compt. rend.*, 24 (1847) 681.

[11] F. Mohr, *Z. anal. Chem.*, 2 (1863) 250.

[12] R. Stumper and P. Mettelock, *Compt. rend.*, 224 (1947) 122.

[13] C. E. Munroe, *Bull. Soc. chim. France*, (2) 16 (1871) 90.

[14] G. Spacu and L. Dima, *Z. anal. Chem.*, 120 (1940) 317.

[15] N. H. Furman and H. M. State, *Ind. Eng. Chem. Anal. Ed.*, 8 (1936) 420.

[16] T. Dupuis, *Thesis*, Paris, 1954.

[17] C. Duval, *Anal. Chim. Acta*, 20 (1959) 270.

[18] F. P. Treadwell, *Manuel de chimie analytique*, 2nd French Ed., Dunod, Paris, Vol. 2, p. 401.

[19] R. Woy, *Chemiker-Ztg.*, 21 (1897) 441.

[20] D. Stockdale, *Analyst*, 83 (1958) 24.

[21] W. W. Wendlandt, *Anal. Chim. Acta*, 20 (1959) 267.

[22] F. Hecht and J. Donau, *Anorganische Mikrogewichtsanalyse*, Springer, Vienna, 1940 p. 420.

[23] T. Kiba and T. Ikeda, *J. Chem. Soc. Japan*, 60 (1939) 911.

[24] W. W. Wendlandt and W. M. Hoffman, *Anal. Chem.*, 32 (1960) 1011.

[25] C. H. Perrin, *J. Assoc. Offic. Agr. Chemists*, 41 (1958) 758.

[26] T. R. F. Fennell and J. R. Webb, *Talanta*, 2 (1959) 105.

[27] G. Embden, *Z. physiol. Chem.*, 113 (1921) 138.

[28] C. Antoniani and R. B. Jona, *Giorn. chim. ind. appl.*, 10 (1928) 203

CHAPTER 17

Sulphur

1. Free sulphur

Occasions may arise in practice when it is necessary to weigh the sulphur extracted with carbon disulphide for instance (see Frühling[1]) or the sulphur collected after treating a thiosulphate with hydrochloric acid (see Gmelin[2]). The two types of curves are identical. When heated in air under atmospheric pressure, sulphur loses no weight below 135°. The combustion then proceeds very smoothly; the crucible is perfectly empty at 256°.

2. Sulphide ion S^{2-}

Since sulphides are readily determined by titrimetric procedures, there is not much interest in their gravimetric determination. Some direct methods have been proposed, however. With respect to silver sulphide Ag_2S, there is no use in treating the precipitate, obtained with silver nitrate, with hydrogen in a Rose crucible. This sulphide is stable in the air from 69° to 615°.

Arsenic sulphide As_2S_3 should be kept between 200° and 265°. The horizontal in the lead sulphide curve is very short; it extends only from 97.5° to 107.2°. The weighing as cupric sulphide in reality is based on that of cupric oxide. As to mercuric sulphide HgS, the procedure is the one devised by Volhard[3] who directed that the product be dried at 110°, whereas Krustinsons[4] notes that the temperature should not exceed 175°. The curve indicates that this latter temperature is manifestly too high. The precipitated sulphide begins to decompose at 109°. The crucible is empty at 478°. Dupuis and Duval[5] advise that this sulphide be dried at a temperature below 100°.

3. Sulphoxylate ion $S_2O_2{}^{2-}$

4. Dithionite, hydrosulphite, hyposulphite ion $S_2O_4^{2-}$

5. Pyrosulphite, metabisulphite ion $S_2O_5^{2-}$

6. Thionate ions $S_{2+n}O_6^{2-}$

No special gravimetric procedures for these ions have been found in the literature. In the case of the *dithionates* $M_2S_2O_6$ or *hyposulphates*, none of the known purpureocobaltic salts $[CoX(NH_3)_5]X_2$ has yielded a precipitate which is sufficiently insoluble for gravimetric purposes. There is only a single weighing form for all of these ions, namely barium sulphate, following oxidation of the sulphur to the hexavalent state.

7. Sulphite ion SO_3^{2-}

The usual weighing form is again barium sulphate. However, according to Romero and Gijon[6], and Gaspar y Arnal and Poggio Mesorana[7], calcium sulphite $CaSO_3 \cdot \frac{1}{2}H_2O$ can be quantitatively precipitated in an aqueous alcoholic medium. It gives no horizontal at 50°; it loses weight up to 385° to give the anhydrous salt between 385° and 410°. The retained water is doubtless combined in the form $Ca_2H_2S_2O_7$. The curve rises from 410° to 946°, the sulphite slowly takes up oxygen, and calcium sulphate $CaSO_4$ is obtained quantitatively at this latter temperature.

Regarding lead sulphite, see Chapter Lead.

8. Thiosulphate ion $S_2O_3^{2-}$

(i) Weighing as sulphur First, it is possible to weigh the sulphur liberated when a thiosulphate is decomposed by an acid, but it is difficult to collect this colloidal product.

(ii) Mercuric sulphide It is also possible to heat the thiosulphate with mercuric cyanide and to weigh the mercuric sulphide resulting from the reaction:

$$H_2S_2O_3 + H_2O + Hg(CN)_2 \rightarrow H_2SO_4 + 2\ HCN + HgS$$

However, the method has little practical value because it can only be applied to the solution of the free acid and furthermore it has been found to be dangerous. See Kessler[8].

(iii) Silver sulphide A third method has been proposed by Gmelin[2]

and Faktor[9]. It should be remembered that the silver salt obtained in this procedure should be kept below 615°.

These methods are seldom employed and have been displaced almost entirely by iodometric procedures.

9. Sulphamate ion NH_2–SO_3^-

Recrystallized sulphamic acid NH_2–SO_3H has been found to be stable up to 186°. Beyond this temperature, it sublimes and decomposes, the crucible being empty at 447°. See Duval[10].

Capestan[11] made a study of the fate of numerous metal sulphamates (amidosulphonates). The findings will be found in the corresponding chapters.

10. Sulphate ion SO_4^{2-}

(i) Barium sulphate (see Chapter Barium)

(ii) Lead sulphate (see Chapter Lead)

(iii) Sodium sulphate (see Chapter Sodium)

(iv) Ammonium sulphate   If the weighing form is to be ammonium sulphate, the sulphuric acid to be determined is treated with ammonium hydroxide and the solution taken to dryness. See Marchand[12] and Weinig[13]. As pointed out in Chapter Ammonium, ammonium sulphate yields an excellent horizontal up to 200°. Decomposition then sets in immediately with decrepitation and bumping. The crucible is entirely empty at 500°.

(v) Luteocobaltic bromosulphate   In the method devised by Mahr and Krauss[14], the sulphate ion is precipitated as hexamminecobalt(III) (luteocobaltic) bromosulphate. These workers recommend drying the precipitate at 80°. The recorded curve shows a horizontal stretch up to 56°. (We advise that this temperature not be exceeded.) A slight increase in weight then follows up to 191°. Decomposition sets in with loss of ammonia and bromine. From 423° to 881°, the residue consists of almost pure cobalt(II) sulphate, which gradually loses sulphur if heated above the latter temperature.

(vi) Benzidine sulphate   The precipitation of sulphate as the benzidine salt is now rather well known. According to Müller[15], the precipitate has the formula $C_{12}H_{12}N_2 \cdot H_2SO_4$. Its quantity is determined by titration. Although it is partially soluble, the precipitate may nevertheless be employed in gravimetry since the thermolysis curve includes a horizontal extending from 72° to 130°. This satisfies the formula just given and is suitable for the automatic determination of sulphate ion. At higher temperatures, the compound disintegrates completely into water, sulphuric anhydride, carbon dioxide, nitrogen, and carbon.

(vii) Tolidine sulphate   The method of Bieringer and Borsum[16] in which tolidine is used in place of benzidine, appears to be much less favourable. In order to obtain a precipitate which filters well, it is necessary to use the reagent in alcoholic solution, and the excess of alcohol may cause the precipitation of the metal sulphate plus adsorbed tolidine. The curves obtained with different lots of the precipitate are not reproducible; we have not thought it worthwhile to show them here. We recommend that this method be discarded. *o,o'-Dianisidine* is not more suitable for determining sulphate ion.

Hegedus and Fukker[17] made a study on the Chevenard thermobalance of the reduction of various sulphates.

11. Persulphate ion $S_2O_8{}^{2-}$

The usual practice is to weigh strychnine persulphate, using the procedure developed by Vitali[18]. Sometimes we have added strychnine chloride rather than nitrate to the standard solution of potassium persulphate. The suspension is allowed to stand overnight, the precipitate is filtered, and then washed with a very little cold water. The curve of the colorless precipitate includes a horizontal up to 122°; beyond this the two acid and basic constituents of the molecule break down and the crucible is empty at 810°.

REFERENCES

[1] S. FRÜHLING, *Z. angew. Chem.*, 2 (1889) 242.
[2] L. GMELIN, *Handbuch der anorganischen Chemie*, 7th Ed. Vol. Schwefel, 586. p.
[3] J. VOLHARD, after F. TREADWELL, *Traité de chimie analytique*, French Ed., Dunod, Paris, 1920, Vol. 2 p. 160.
[4] J. KRUSTINSONS, *Z. anal. Chem.*, 125 (1943) 98.
[5] T. DUPUIS AND C. DUVAL, *Anal. Chim. Acta*, 4 (1950) 623.

⁶ M. S. ROMERO AND M. S. GIJON, *Inform. quím. anal. (Madrid)*, 1 (1947) 69; *Analyst*, 73 (1948) 579.
⁷ T. GASPAR Y ARNAL AND J. M. POGGIO MESORANA, *Anales soc. españ. fís. y quím.*, 43 (1947) 439.
⁸ F. KESSLER, *Ann. Phys.*, 74 (1949) 269.
⁹ F. FAKTOR, *Pharm. Post*, 33 (1900) 169.
¹⁰ C. DUVAL, *Anal. Chim. Acta*, 16 (1957) 224.
¹¹ M. CAPESTAN, *Ann. Chim. (Paris)*, 5 (1960) 207; *Thesis*, Paris, 1960.
¹² J. MARCHAND, *Ann. Physik*, 42 (1837) 556.
¹³ M. WEINIG, *Z. angew. Chem.*, 5 (1892) 204.
¹⁴ C. MAHR AND K. KRAUSS, *Z. anal. Chem.*, 128 (1948) 477.
¹⁵ W. MÜLLER, *Ber.*, 35 (1902) 1587.
¹⁶ J. BIERINGER AND W. BORSUM, *Chemiker-Ztg.*, 30 (1906) 721.
¹⁷ A. J. HEGEDUS AND K. FUKKER, *Z. anorg. Chem.*, 284 (1956) 20.
¹⁸ D. VITALI, *Boll. chim. farm.*, 42 (1903) 273, 321

CHAPTER 18

Chlorine

1. Chloride ion Cl⁻

(i) Silver chloride Precipitated silver chloride, formed and collected in the absence of any reducing material, has led to a very simple recording. A perfectly horizontal stretch extends from 70° to 600° and is suitable for automatic determination. There is a gradual slight increase in weight (amounting to 2 mg in 318 mg) from 600° to 946°; we have not been able to explain this observation since the bromide remains intact under these same conditions whereas the iodide loses some weight. See Dupuis and Duval[1]. The silver chloride turns violet on exposure to light but Duval[2] could detect no change in weight accompanying this photochemical reaction. See Chapter Bromine. When heated in a stream of hydrogen, silver chloride is reduced from ordinary temperatures on, and notably between 300° and 480°; it yields metallic silver. The break in the curve at 450° is in conformity with the composition Ag_2Cl, but it is visible even to the naked eye that this is a mixture of silver (carried along by nitric acid) and unaltered silver chloride.

(ii) Mercurous chloride Congdon *et al.*[3] suggest that mercurous chloride be dried in a desiccator, either in the cold or at 105°. These results are acceptable, but most of the temperatures given in the literature regarding the volatility of mercurous chloride are much too high. The weight remains unchanged up to 130°. Gradual heating of this salt provided us with the following data regarding the temperatures and corresponding losses of weight suffered by 417.6 mg of dry precipitated calomel:

Temp. (°C)	130	167	283	410
Loss (mg)	0	5.8	184.0	366.6

Other thermolysis curves of chlorides can be found in the Chapters Lithium, Sodium, Ammonium, Potassium, etc.

2. Hypochlorite ion ClO⁻

We have found no special gravimetric method for this ion.

3. Chlorite ion ClO_2^-

Lasègue[4] advises that the yellow crystalline precipitate of lead chlorite $Pb(ClO_2)_2$ be kept in a desiccator. When heated it gives a horizontal up to an explosion temperature whose height depends on the rate of heating. We have been able to locate this temperature between 77° and 118°. The salt disproportionates even in the cold (with no change of weight) and yields lead chloride and lead chlorate (as shown by its powder spectrum in the infra-red).

4. Chlorate ion ClO_3^-

We have not found any special gravimetric method for this ion. However, some thermolysis curves of chlorates have been recorded. See Chapter Potassium.

5. Perchlorate ion ClO_4^-

The precipitation as nitron perchlorate by the Fichter[5] method has been found to be excellent for the automatic determination of perchlorates. The recorded curve has a good horizontal (as opposed to nitron nitrate) between 40° and 232°. Beyond this latter temperature, the salt decomposes with swelling and explosion.

REFERENCES

[1] T. DUPUIS AND C. DUVAL, *Anal. Chim. Acta*, 4 (1950) 615.
[2] C. DUVAL, *Mikrochemie*, (1956) 1430.
[3] L. A. CONGDON et al., *Chem. News*, 129 (1924) 302, 317, 334.
[4] G. LASÈGUE, *Bull. Soc. chim. France*, 11 (1912) 884.
[5] F. FICHTER, *Z. anal. Chem.*, 68 (1926) 298.

CHAPTER 19

Potassium

The classic methods of determining potassium include weighing it as chloride, perchlorate, sulphate, cobaltinitrite, perrhenate, chloroplatinate, 6-chloro-5-toluene-*m*-sulphonate, bitartrate, picrate, dipicrylaminate, tetraphenylboron compound. (Duval and Duval[1].) To these have now added certain potassium compounds, some soluble, some not soluble, using as standards such compounds as the bromate, permanganate, acid phthalate, bisulphate, acid phosphate, acetate, etc.

1. Potassium chloride

The potassium chloride KCl studied by Duval and Duval came from the action of hydrochloric acid on potassium carbonate; the solution was taken to dryness. The salt loses weight up to 219°, and then remains stable. Usually, 500° is taken as the drying temperature but the salt may be heated to 815° with no observable loss of weight.

2. Potassium chlorate

The potassium chlorate $KClO_3$ used by Duval[2] was heated at the rate of 300° per hour; it was stable up to 495°. Then oxygen was gradually given off up to 552° where the horizontal corresponding to the chloride began.

Duval also made a thermolytic study of the familiar solid state reaction between $KClO_3 + MnO_2$, which is so often employed for preparing oxygen in the laboratory. See also Osada[3].

3. Potassium perchlorate

The evaporation residue from the action of 20% perchloric acid on potassium carbonate yields a curve which slopes downward until 73° is reached (loss of water and acid). A horizontal of constant weight corresponding to the anhydrous perchlorate continues as far as 655°, where the evolution of oxygen begins and continues rapidly[1,4].

Hogan, Gordon and Campbell[5] studied the solid state reaction between potassium perchlorate and barium nitrate. The latter catalyzes the decomposition of the perchlorate, which when heated at the rate of 15° per minute decomposes at 520° (in the case of a 1 : 1 molecular mixture) in other words, at a lower temperature than the perchlorate alone.

These workers likewise investigated such pyrotechnical mixtures as: perchlorate 30%, aluminium 40%, barium nitrate 30%.

4. Potassium bromide

Potassium bromide KBr is invariably anhydrous and dry. Its weight does not change up to 800°. See Duval and Wadier[6].

5. Potassium bromate

Potassium bromate $KBrO_3$ does not change weight up to 366°, where distinct decomposition sets in. Duval[7] found that this salt may be weighed directly just as it comes from the bottle.

6. Potassium iodide

Potassium iodide KI is anhydrous and maintains a constant weight up to 715°. Since its melting point is 773°, it has already decomposed slightly when it melts. See Duval[7].

7. Potassium iodate

Potassium iodate KIO_3 undergoes no loss of weight if heated to 500° at the rate of 300° or 500° per hour. The recordings are straight horizontals. Duval[7] points out that the salt may therefore be weighed directly from the bottle.

8. Potassium biniodate

The use of potassium biniodate $KH(IO_3)_2$ is being advocated more and more as a standard acidic substance and as an oxidant. It is anhydrous and its weight remains constant up to 148°. Two molecules of the salt then lose 1 molecule of water but there is no evidence of the production of corresponding anhydroiodic acid. Beginning at 180° and continuing up to 370°, there is a horizontal corresponding to the mixture 2 $KIO_3 + I_2O_5$. The iodic anhydride then disappears by decomposition and sublimation and a short inclined stretch begins around 500° indicating that nothing is left except potassium iodate. The latter however soon decomposes as noted above. See Duval[8].

9. Potassium metaperiodate

Potassium metaperiodate KIO_4 is anhydrous, quite dry as it comes from the bottle, and has been found to be stable up to 264°. It then decomposes very slowly, but the rate of decomposition increases abruptly around 350°. The potassium iodate, formed quantitatively according to Rammelsberg, remains as such between 350° and 524°. Consequently, the melting point of 582°, reported by Carnelley, cannot apply to the periodate or even to the iodate. Beyond 524°, there is renewed release of oxygen and potassium iodide is eventually obtained. See Duval[8].

10. Potassium sulphate

The sulphate K_2SO_4 used in analysis is prepared by the action of sulphuric acid on potassium carbonate; the moist and acidic product does not acquire a rigorously constant weight until the temperature is brought to 408°. From then until 880°, this weight remains unchanged. Duval and Duval[1] observed no loss by volatilization from 800°. If specimens of potassium sulphate taken from different bottles are heated, the maximum loss of weight does not exceed 1/200th on heating to around 800°. Any such loss doubtless can be ascribed to a trace of occluded sulphuric acid.

11. Potassium bisulphate

With regard to potassium acid sulphate (bisulphate) $KHSO_4$, Tomkova, Jirů and Rosický[9] estimate that it is converted to the pyrosulphate $K_2S_2O_7$ in the vicinity of 210°, and the latter in turn yields the neutral sulphate around 395°.

12. Potassium persulphate

Anhydrous potassium persulphate $K_2S_2O_8$ does not change weight up to 230°. See Duval[10]. It then decomposes abruptly, losing oxygen, and forming potassium pyrosulphate, which is stable up to 488°. Another decomposition then occurs yielding sulphuric anhydride and potassium sulphate. This decomposition is completed around 900°.

13. Potassium sulphamate

Potassium amidosulphonate (sulphamate) KSO_3-NH_2, first prepared by Capestan[11], is stable up to 250°. It then loses ammonia up to around 300°:

$$2 KSO_3-NH_2 \rightarrow NH(SO_3K)_2 + NH_3$$

forming potassium imidodisulphonate, which is stable up to about 450°. An abrupt decomposition into sulphate then begins:

$$3 NH(SO_3K)_2 \rightarrow 3 K_2SO_4 + 3 SO_2 + NH_3 + N_2$$

the level of the sulphate being perfectly horizontal.

14. Potassium nitrate

Potassium nitrate KNO_3 is stable from room temperature up to 737–745°. However, a close examination of the curve appears to indicate a slight decomposition, amounting to about 1/200th of the weight of the specimen, from 140° on. The decomposition becomes appreciable only around 843°. Various authors have stated that potassium nitrite is formed along with nitrogen oxide vapours but Duval and Wadier[12] found that the passage through the melting point near 339° is not signalled by any distinct decomposition.

15. Potassium dihydrogen phosphate

Potassium diacid phosphate KH_2PO_4 was discussed at the 1st Congress on Thermogravimetry held at Kazan[4] and has been studied by Boullé and Dominé-Bergès[13]. If heated in air, at 100° per hour, the salt begins to decompose around 220° and continues to do so until an almost horizontal level is established, extending from 255° to 275°. If the heating rate is increased to 150° per hour, this horizontal lies between 265° and 285°. It corresponds to a fusion. The decomposition begins at 220°:

$$2 KH_2PO_4 \rightarrow H_2O + K_2H_2P_2O_7$$

The formation of the triphosphate begins around 240°:

$$KH_2PO_4 + K_2H_2P_2O_7 \rightarrow H_2O + K_3H_2P_3O_{10}$$

Around 315°, this triphosphate loses water to form the metaphosphate:

$$K_3H_2P_3O_{10} \rightarrow 3 KPO_3 + H_2O$$

The study of the above $K_3H_2P_3O_{10} \cdot H_2O$ was made by Morin[14]; she used 800 mg and raised the temperature at the rate of 100° per hour. There was a very slow loss of weight from 120° to 137° which became more marked between 215° and 280°, depending on the specimens involved. This decomposition leads directly to the crystalline Kurrol polyphosphate $(KPO_3)_x$.

16. Potassium arsenates

Duc-Maugé[15] made a minute study of the potassium arsenates.

(i) Monopotassium triarsenate This compound $K_2O \cdot 3As_2O_5 \cdot 4H_2O$ is stable at the ordinary temperature. When heated at the rate of 150° per hour in the Chevenard thermobalance it yields a curve with four distinct levels: the tetrahydrate up to 180–200°, the dihydrate from 250° to 340°, the monohydrate from 370° to 400°, and the anhydrous salt beyond 420°. Duc-Maugé obtained the dihydrate at 170° under a pressure of 5 mm of mercury, but the dehydration is very slow after the first molecule of water has been lost, which leads to the conclusion that a trihydrate exists, a finding which however has not been confirmed by any other method. Under reduced pressure, the tetrahydrate begins to lose water from 80° to 100°, and forms the dihydrate rather quickly at 130–140°. The latter is stable up to 200°, where in turn it produces the monohydrate. The latter gives up its water at 300° and is converted into the anhydrous salt.

(ii) Monopotassium binarsenate The monopotassium binarsenate $K_2O \cdot 2As_2O_5 \cdot 5H_2O$ gives three perfectly horizontal levels when heated in the thermobalance. It is stable up to 100°; the monohydrate exists between 250° and 350°, and the anhydrous salt is present beyond 400°. In addition, a slightly inclined level between 150° and 200° reveals the presence of the trihydrate, which has but little stability. Under a pressure of 8 mm of mercury, the pentahydrate changes into the trihydrate at 100°. Duc-Maugé obtained the monohydrate at 140° and under a pressure of 10 mm. He was not able to reach the dihydrate stage at 120° and 3 mm. This pentahydrate decomposes in vacuo from 75° on, and apparently the trihydrate is formed near 90–110°, and the dihydrate around 120–130°. But it is difficult to demonstrate the presence of these two compounds in an impressive fashion. The monohydrate is stable between 150° and 300°; it loses its water at 320–350°.

(iii) Monopotassium orthoarsenate The monopotassium orthoarsenate $K_2O \cdot As_2O_5 \cdot 2H_2O$ or KH_2AsO_4 loses (by one or another of the three modes of investigation) its two molecules of water simultaneously with formation of the metarsenate but with no evidence of the production of the acid pyroarsenate, either in vacuo, where the dehydration begins at 150°, or on the thermobalance (190–200°).

(iv) Dipotassium orthoarsenate The dipotassium orthoarsenate $2K_2O \cdot As_2O_5 \cdot 3H_2O$ or $K_2HAsO_4 \cdot H_2O$ decomposes slowly in vacuo at 20° and is converted directly into the anhydrous salt without going through the monohydrate, even though the latter is stable and only starts to lose water near 50° to give the anhydrous salt. The latter loses a half-molecule of water and produces the hemihydrate of the pyroarsenate at 120–150°. Finally, the latter yields the anhydrous pyroarsenate at 150°. When the trihydrate is heated on the thermobalance at the rate of 25° or 50° per hour, it melts and under these conditions the results are no longer comparable. However, a very short level extending from above 150° to 160° is characteristic of the anhydrous salt, and the monohydrate of the pyroarsenate is found between 180° and 250°. When the monohydrate was heated at 75° per hour it likewise melted. However, the levels corresponding to the anhydrous salt and to the hemihydrate of the pyroarsenate are very distinct. Duc Maugé has not been able to find the monohydrate by keeping the trihydrate at 40° under a pressure of 3 mm of mercury. But when nuclei of the monohydrate were introduced at the start of the operation, he was able to produce this compound under a pressure of approximately 4 mm. The monohydrate likewise decomposed at 70° under 10 mm pressure; the anhydrous salt resulted and the latter gave the hemihydrate at 135° under 10 mm pressure.

(v) Tripotassium orthoarsenate The tripotassium orthoarsenate $K_3AsO_4 \cdot 7H_2O$ is converted at 20° in vacuo into the trihydrate, which in turn yields the monohydrate at 50°. It is difficult to secure the monohydrate because it starts to lose its water at 50° and the conversion into the anhydrous salt is very rapid at 70°. Even though the heating rate is lowered to 25° per hour, the salt melts on the thermobalance. However, the end of the curve shows, though not in very distinct fashion, both the monohydrate and obviously also the anhydrous salt.

The heptahydrate changes into the trihydrate at 45° and under 10 mm of mercury, and the trihydrate, under these same conditions, yields the monohydrate distinctly. Finally, the latter gives the anhydrous salt at 130°.

(vi) Pentapotassium arsenate No definite compound was obtained by dehydrating the pentapotassium arsenate $5K_2O \cdot As_2O_5 \cdot 18H_2O$. The loss of weight is variable unless the compound has been heated to a temperature above 150° where the dihydrate is invariably obtained.

17. Potassium "pyroantimonate"

Dupuis[16] made a lengthy study of the compound known as potassium "pyroantimonate." The curve starts horizontally but the salt begins to lose water at 30°; the loss continues rather regularly up to 750°. Then a horizontal begins and continues up to at least 950°; it corresponds to the compound $K_2H_2Sb_2O_7$, *i.e.* to the true pyroantimonate. Consequently, the starting material cannot be a pyro-compound. Calculation reveals that the loss of weight between the two levels corresponds exactly to $5H_2O$ (on the basis of 2 gram atoms of antimony). Accordingly, the initial compound can be written as $K[Sb(OH)_6]$ in conformity with the equation:
$$2 \ K[Sb(OH)_6] \rightarrow K_2H_2Sb_2O_7 + 5 \ H_2O$$
If now the same starting material is heated up to 375° in 3 hours, the resulting curve has the same shape; at 207°, an inflection point indicates the loss of three-fifths of the water. Another and very distinct inflection point is observed at 500°, marking the beginning of the departure of the rest of the water which is slowly released up to 750°.

18. Potassium carbonate

Potassium carbonate crystals usually carry 1.5 H_2O. The specimen we used probably was a mixture of the latter and the hexahydrate. Calculations from the curve agree with the formula $K_2CO_3 \cdot 3H_2O$. This water is gradually lost and no horizontal appears up to 158°. Then the level stretch of the anhydrous carbonate begins and extends to about 875°. Reisman[17] believes that the initial salt has the formula $5K_2CO_3 \cdot 8H_2O$.

19. Potassium bicarbonate

Tsuchija[18] studied potassium bicarbonate between 140° and 200°; the horizontals due to the following compounds are observed:
$$KHCO_3 \rightarrow 5 \ KHCO_3 \cdot K_2CO_3 \rightarrow 2 \ KHCO_3 \cdot K_2CO_3 \rightarrow K_2CO_3$$

20. Potassium borates

There has also been an important study of the potassium borates.

(i) Potassium borax Haladjian and Carpéni[19] investigated the tetraborate (potassium borax) $K_2B_4O_7 \cdot 4H_2O$. They heated the samples at the rate of $10°$ per hour in air containing water vapour (saturated either at $20°$ or at $95°$). No matter what the partial pressure of the water vapour, the dehydration began at around $112°$. The more humid the reactional atmosphere, the lower the temperature at which the anhydrous compound is obtained. For example, in the second instance:

$$K_2B_4O_7 \cdot 4H_2O \rightarrow K_2B_4O_7 \cdot 2H_2O + 2\ H_2O \text{ (at } 140°)$$
$$K_2B_4O_7 \cdot 2H_2O \rightarrow K_2B_4O_7 \cdot 1H_2O + 1\ H_2O \text{ (at } 200°)$$
$$K_2B_4O_7 \cdot 1H_2O \rightarrow K_2B_4O_7 + 1\ H_2O \qquad \text{(at } 250°)$$

These workers have likewise obtained a dehydration isotherm at $111° \pm 1°$.

(ii) Potassium pentaborate They have also studied the monopotassium pentaborate $KB_5O_8 \cdot 4H_2O$. They found that 3 molecules of water are lost simultaneously beginning either at $120°$ (atmosphere saturated with water at $20°$) or starting at $170°$ (atmosphere saturated with water at $90°$). A slight entrainment of boric acid by the water vapour is inevitable. It is impossible to obtain a precisely stoichiometric horizontal for the monohydrate. The anhydrous salt is eventually obtained near $450°$.

(iii) Polyborate Carpéni, Haladjian and Pilard[20] studied the polyborate $K_2HB_5O_9 \cdot 2H_2O$. They traced the dehydration and rehydration curves after four cycles, and were then able to write the following equations, the first two representing reversible reactions:

$$K_2HB_5O_9 \cdot 2H_2O \rightleftharpoons K_2HB_5O_9 \cdot H_2O + H_2O \text{ (at } 140°)$$
$$K_2HB_5O_9 \cdot H_2O \rightleftharpoons K_2HB_5O_9 + H_2O \qquad \text{(at } 360°)$$
$$2\ K_2HB_5O_9 \rightarrow K_4B_{10}O_{17} + H_2O \qquad \text{(at } 450°)$$

21. Potassium hexachlorotitanate

Morozov and Toptygin[21] studied the thermal stability of potassium hexachlorotitanate $K_2[TiCl_6]$.

22. Potassium thiocyanate

Potassium thiocyanate (sulphocyanate) KSCN is a very stable salt and remains unaltered up to around 480°. It then suffers a slow decomposition, losing sulphur and yielding potassium cyanide. See Duval[8].

23. Potassium permanganate

Potassium permanganate $KMnO_4$ crystallizes with no water nor does it hold moisture. Duval[2] traced its thermolysis curve and found it to be stable up to 245°; the curve always begins horizontally. A slight gain in weight is observed only after 110° is reached. There is a considerable loss of oxygen between 243° and 295°. A level that slowly slopes downward follows and continues to about 630°. A loss of approximately one atom of oxygen per molecule occurs between 245° and 300° and the total loss up to 700° rises to 3 atoms of oxygen per 2 molecules of permanganate. The black material obtained beyond 700° is undoubtedly a mixture of manganese oxides and potassium oxide. See Duval[2] and ref. 4.

24. Potassium alum

Potassium alum, taken from a bottle which had been opened several times, was found to be not deliquescent and agreed with the gross formula $K[Al(SO_4)_2] \cdot 12H_2O$ (with no suppositions about the role of the molecules of water). It is not stable for very long when heated; loss of water becomes evident at 48°. There is no intermediate horizontal during the dehydration; however, a point of return is observed on the curve at 131° when half of the water of crystallization has been driven off. The departure of the six other molecules (water of constitution) is extremely slow and the level of two anhydrous sulphates is truly horizontal only between 336° and 716°. Decomposition sets in beyond this temperature, SO_3 is lost, and the residue conforms very closely to $Al_2O_3 \cdot K_2SO_4$, though the quantity of SO_3 may vary from one trial to another. If it is desired to prepare a standard solution from this alum, it is advisable that the salt, if heated, be kept below 40° prior to the weighing, and that it be assumed to contain 12 H_2O. Another procedure is to bring the alum to constant weight at 336° and to weigh the salt as the double sulphate $K_2SO_4 \cdot Al_2(SO_4)_3$. (Duval[33])

25. Potassium chromate

Duval[8] found that potassium chromate K_2CrO_4 shows no appreciable change in weight up to 975°.

References p. 268

26. Potassium bichromate

Potassium bichromate $K_2Cr_2O_7$ is always anhydrous and quite dry no matter what sample is taken. Duval[22] has found that the thermolysis of this salt is extremely simple and the curve consists of a horizontal straight line. The precise temperature at which decomposition commences is difficult to determine because the disintegration begins slowly between 625° and 650°. It is not appreciable below 700° (loss approximately 1/100).

27. Potassium ferricyanide

Duval[8] found that potassium ferricyanide $K_3[Fe(CN)_6]$ is anhydrous, stable, and maintains a constant weight up to 312°. An extremely complex decomposition starts at this temperature; there is a reoxidation and also a loss of cyanogen.

28. Potassium ferrocyanide

Duval[8] found potassium ferrocyanide $K_4[Fe(CN)_6] \cdot 3H_2O$ to be stable up to 45°. The water escapes up to 115°. Then the horizontal of the anhydrous salt is observed, extending up to around 370°. An oxidation of the iron ensues, followed by a loss of cyanogen. A complex carbide of iron is the final product.

29. Potassium cobaltinitrites

(i) Precipitation with sodium cobaltinitrite Duval and Duval[1] treated a saturated solution of potassium chloride in the cold with sodium cobaltinitrite; the product had the approximate formula $Na_3K_3[Co(NO_2)_6] \cdot nH_2O$. The recorded curve confirmed this formula, which had been in doubt for many years. The curve slopes downward without interruption; in particular it is not possible to observe a horizontal marking the production of a hydrate at 100° (where the descent is very accentuated) or a horizontal marking the production of the anhydrous salt. The latter begins to decompose before all of the water has been driven off.

The thermolysis of the true potassium cobaltinitrite is discussed in Chapter Cobalt.

(ii) Precipitation with lithium cobaltinitrite Dupuis[23] has modified the

determination of potassium by using lithium cobaltinitrite as reagent; it yields a yellow lithium-free precipitate $K_3[Co(NO_2)_6]\cdot 2H_2O$. If its temperature is raised linearly at the rate of 300° per hour, and 20 mg of the product is taken for the trial, it is found that the latter maintains a constant weight up to around 200°. Beyond this temperature, all of the water escapes along with oxides of nitrogen up to 300°, which marks the beginning of a second horizontal. The latter corresponds to the ensemble $CoO + 3\ KNO_3$, which is stable up to 460° at least. The two horizontals are suitable for the automatic determination of potassium.

30. Potassium perrhenate

The (meta)perrhenate was studied by Tribalat. (See Duval[1].) Their reagent was perrhenic acid prepared by dissolving metallic rhenium in nitric acid (sp. gr. 1.2), evaporating on the water bath, and taking up in water. The precipitate $KReO_4$ loses water and the alcohol used for washing up to 54°; beyond this there is a perfectly horizontal stretch as far as 220°, the temperature at which the first sputtering is noted. Therefore, the temperature 110° prescribed for drying is not exclusive.

31. Potassium chloroplatinate

Potassium chloroplatinate is prepared from a piece of platinum wire dissolved in *aqua regia*; the platinic chloride is taken up in water, and the solution is treated with potassium chloride. The golden yellow precipitate should not be weighed after it has been heated above 270°. Between this temperature and 672° there is an almost imperceptible decomposition, but the loss of chlorine then increases and at 880° it amounts to 1/3rd of the theoretical.

32. Potassium acetate

Burriel-Marti, Jimenez-Gomez and Alvarez-Herrero[24] found that anhydrous potassium acetate $KC_2H_3O_2$ is stable up to 410–420°. It is the most stable of all of the acetates studied by them.

33. Potassium oxalate

According to Martens[25], potassium oxalate $K_2C_2O_4\cdot H_2O$ is stable up to 50° when heated at the rate of 150° per hour. The dehydration begins at 50° and the anhydrous salt is obtained at 115°. The latter is stable up to 425°; its conversion into carbonate is completed when the temperature has reached 550°. The loss of weight corresponds well to the elimination of carbon monoxide.

Duval[26] has found that if the anhydrous oxalate is heated at the rate of 300° per hour, the horizontal extends from 168° to 500°. The level corresponding to the carbonate is maintained to at least 914°.

34. Potassium binoxalate

Potassium binoxalate KCO_2-CO_2H (KHC_2O_4) was found by Martens[25] to be stable up to 200° when it was heated at the rate of 150° per hour. The subsequent decomposition, which is rapid at 255°, is complete at 300°. The residue consists of neutral potassium oxalate, which decomposes as described above.

35. Potassium tetroxalate

According to Martens[25], potassium tetroxalate $KH_3(C_2O_4)_2 \cdot 2H_2O$ begins to lose water at 60°. Another decrease in weight occurs at 122° and intensifies around 150°, with a slight inflection near 165°. This change ends at 215°. The residue consists of the binoxalate, which decomposes as just described.

In summary, the following equations show the behaviour of the three preceding compounds:

$$KH_3(C_2O_4)_2 \cdot 2H_2O \rightarrow KH_3(C_2O_4)_2 + 2\ H_2O \ (60\text{--}122°)$$
$$KH_3(C_2O_4)_2 \rightarrow KH(C_2O_4) + H_2C_2O_4 \ (125\text{--}215°)$$
$$2\ KHC_2O_4 \rightarrow K_2C_2O_4 + H_2C_2O_4 \ (215\text{--}280°)$$
$$K_2C_2O_4 \rightarrow K_2CO_3 + CO \ (425\text{--}550°)$$

36. Potassium palmitate

Duval and Wadier[12] investigated potassium palmitate. The slightly moist material became dry and showed a constant weight from 40° to 275°. Decomposition set in at 275° and proceeded gradually up to 485°. Then the carbon burned slowly and the crucible contained only potassium carbonate at 820°. Accordingly, this compound, which has a soapy appearance, should be dried before it is weighed.

37. Potassium bitartrate

The directions given by Casamajor[27] were followed for precipitating potassium acid tartrate (bitartrate). The anhydrous product CO_2K–$CHOH$–$CHOH$–CO_2H maintains a constant weight up to 200°. Decomposition begins near 215° and soon becomes almost explosive. It is difficult to follow the curve because of the intumescence of the carbonaceous matter. This compound is of no interest from the standpoint of gravimetric analysis.

38. Potassium 6-chloro-5-nitrotoluene-*m*-sulphonate

Duval and Duval [1] followed the procedure given in "*Organic Reagents for Metals*"[28] when precipitating potassium 6-chloro-5-nitrotoluene-*m*-sulphonate by means of the corresponding sodium salt. A 10% aqueous solution of the reagent was allowed to react with a solution of potassium chloride. The recorded curve stays level to beyond 278°. The temperature of 120° ordinarily suggested for drying the precipitate is therefore not limiting. The compound decomposes explosively at 345° and yields a black swollen product.

39. Potassium biphthalate

Potassium hydrogen phthalate (biphthalate) is anhydrous and quite dry. When it is continuously heated in linear ascending manner, Duval[22] found that the decomposition starts at 240° if the heating rate is 150° per hour and at 236° for 300° per hour. He also traced the isotherms at 150°, 160°, and 170°. In no case did he observe the loss of weight mistakenly reported by Caley and Brundin[29]. Duval's experiments were made with R.A.L.* products taken from fresh bottles.

40. Potassium picrate

Duval[1] heated potassium picrate very slowly (3rd speed of the Chevenard thermobalance). No change in weight was observed up to 217°. Consequently, the temperature of 105° prescribed by the manuals is too restrictive. The explosion (due to heating) does not occur until 310° when the Duval method of working is employed.

---

* Abbreviation for Roux, Aiguillon, Legroux (Messrs. Kuhlmann, Paris).

41. Potassium dipicrylaminate

The procedure of Van Nieuwenburg and Van der Hoek[30] was used for precipitating potassium dipicrylaminate. The recorded line shows that the precipitate may be dried up to 220° with no change in weight or risk of explosive decomposition. The temperature of 110° is correct but too restrictive[1].

42. Potassium tetraphenylboron

Potassium tetraphenylboron $K[B(C_6H_5)_4]$, prepared by the method of Raff and Brotz[31], was heated by Wendlandt[32] and others. The complex is stable up to 265° and then gradually decomposes. The organic matter burns and the horizontal of potassium metaborate KBO_2 begins at 715° if the heating rate is 4.5° per minute. Potassium may be determined automatically by means of the first horizontal, but the simultaneous determination of potassium, rubidium, and cesium is not possible or at least risky since the decomposition temperatures (265°, 240°, and 210°, respectively) are too close to each other.

REFERENCES

[1] T. DUVAL AND C. DUVAL, *Anal. Chim. Acta*, 2 (1948) 105.
[2] C. DUVAL, *Anal. Chim. Acta*, 16 (1957) 224.
[3] H. OSADA, *J. Ind. Explosives Soc. Japan*, 15 (1954) 313; *C.A.*, (1955) 10712.
[4] *1st Congr. Thermography.*, Kazan, 1953, p. 112.
[5] D. HOGAN AND S. GORDON, *J. Phys. Chem.*, 62 (1958) 1433; V. D. HOGAN, S. GORDON AND C. CAMPBELL, *Anal. Chem.*, 29 (1957) 306.
[6] C. DUVAL AND C. WADIER, *Anal. Chim. Acta*, 23 (1960) 541.
[7] C. DUVAL, *Anal. Chim. Acta*, 15 (1956) 223.
[8] C. DUVAL, *Anal. Chim. Acta*, 16 (1957) 546.
[9] D. TOMKOVA, P. JIRŮ AND J. ROSICKÝ, *Collect. Czechoslov. Chem. Communs.*, 25 (1960) 957.
[10] C. DUVAL, *Anal. Chim. Acta*, 20 (1959) 21.
[11] M. CAPESTAN, *Ann. Chim. (Paris)*, 5 (1960) 207.
[12] C. DUVAL AND C. WADIER, *Anal. Chim. Acta*, 23 (1960) 257.
[13] A. BOULLÉ AND M. DOMINÉ-BERGÈS, *Colloq. I.U.P.A.C.*, Münster, 1954, p. 258.
[14] C. MORIN, *Thesis*, Paris, 1960.
[15] C. DUC-MAUGÉ, *Ann. Chim. (Paris)*, 3 (1958) 815.
[16] T. DUPUIS, *Thesis*, Paris, 1954; *Rec. trav. chim.*, 71 (1952) 111.
[17] A. REISMAN, *Anal. Chem.*, 32 (1960) 1566.
[18] R. TSUCHIJA, *J. Chem. Soc. Japan*, 74 (1953) 97.
[19] J. HALADJIAN AND G. CARPÉNI, *Bull. Soc. chim. France*, (1960) 1629.
[20] G. CARPÉNI, J. HALADJIAN AND M. PILARD, *Bull. Soc. chim. France*, (1960) 1634.
[21] I. S. MOROZOV AND D.Y. TOPTYGIN, *Zhur. Neorg. Khim.*, 5 (1960) 88.
[22] C. DUVAL, *Anal. Chim. Acta*, 13 (1955) 35.
[23] T. DUPÚIS, *Anal. Chim. Acta*, 9 (1953) 493.

24 F. BURRIEL-MARTI, S. JIMENEZ-GOMEZ AND C. ALVAREZ-HERRERO, *Anales edafol. y fisiol. vegetal (Madrid)*, 14 (1955) 222.
25 P. H. MARTENS, *Bull. inst. agron. et stas. recherches de Gembloux*, 21 (1953) 133.
26 C. DUVAL, *Anal. Chim. Acta*, 20 (1959) 263.
27 P. CASAMAJOR, *Chem. News*, 34 (1876) 231, 342.
28 HOPKIN AND WILLIAMS LTD., *Organic Reagents for metals*, 4th Ed., London, p. 138.
29 E. R. CALEY AND R. H. BRUNDIN, *Anal. Chem.*, 25 (1953) 142.
30 C. J. VAN NIEUWENBURG AND T. VAN DER HOEK, *Mikrochemie*, 18 (1935) 175.
31 P. RAFF AND W. BROTZ, *Z. anal. Chem.*, 133 (1951) 241.
32 W. W. WENDLANDT, *Chemist Analyst*, 26 (1957) 38; *Anal. Chem.*, 28 (1956) 1001.
33 C. DUVAL, *Anal. Chim. Acta*, 13 (1955) 429.

CHAPTER 20

Calcium

In addition to the precipitates serving for the determination of calcium: fluoride, iodate, sulphate, arsenate, carbonate, mixed ferrocyanides, molybdate, tungstate, nickelohexanitrite, oxalate, racemic tartrate, oxinate, loretinate and picrolonate studied by Peltier and Duval[1], we have also found in the literature various investigations dealing with the thermal behaviour of the oxide, oxychloride, perchlorate, sulphamate, selenite, pyroselenite, selenate, nitrate, phosphates, apatites, dichromate, formate, acetate, etc.

1. Calcium oxide

(i) *Laboratory studies*   The hydroxide studied by Cabannes-Ott[2] was prepared either by treating a calcium salt with potassium hydroxide, or by hydrating the oxide (from ignition of the oxalate) by means of water vapour or liquid water. The curves produced by these preparations are similar; there is a horizontal up to 400° corresponding to the hydroxide $Ca(OH)_2$, sometimes accompanied by a little carbonate $CaCO_3$ and also accompanied by the loss of a little moisture at the start. If the heating rate is 300° per hour, a second horizontal begins at around 500–505°, it may be observed subsequent to the loss of the water of constitution. It is attributable to the oxide CaO and possibly to still undecomposed carbonate. The latter then decomposes at a temperature which varies with the amount present but always near 810°. Hence the final horizontal corresponds to the oxide CaO.

(ii) *Industrial studies*   There have been other studies of calcium oxide especially from the industrial standpoint. These include the work of Biffen[3] on the silicates of hydrated limes, the work of Rémy-Genneté, Mauret and Audouze[4] on fatty limes, of Gill[5], and also the investigation of gypsum and dolomite by Roy-Buckle[6], etc.

Biffen used the Chevenard balance with a heating rate of 8° per minute and worked in an atmosphere devoid of moisture and carbon dioxide. A sample weighing 0.104 g lost its water from 425° to 534°; another specimen weighing 1.223 g became anhydrous from 395° to 649°.

2. Calcium fluoride

Calcium fluoride CaF_2 is prepared by treating a solution of calcium chloride with an excess of potassium fluoride. After standing over night, the precipitate is washed several times by decantation. The pyrolysis was conducted in an aluminite crucible which had suffered no loss in weight when heated to 950°. The recorded curve shows a slow loss of moisture which is complete at 400°. A constant weight is maintained from this point on. In view of the difficulty ordinarily encountered in this method of determining calcium (difficult filtration, taking up of moisture during weighing) we recommend the use of automatic determination rather than the ordinary volumetric procedure employing ferric chloride and zinc iodide. Matsuura[7] suggests heating the calcium fluoride to 800° at least.

3. Calcium oxychloride

Binka and Satava[8] studied the oxychloride $Ca(OH)_2 \cdot CaCl_2 \cdot 10.5H_2O$ in conjunction with differential thermal analysis. They report that at 55° 3 molecules of the hydroxide carry $10H_2O$; at 125°, the approximate composition is $3Ca(OH)_2 \cdot CaCl_2$; from 600° to 1050°, there is pyrolysis to yield $3 CaO + H_2O + 2 HCl$.

4. Calcium perchlorate

The perchlorate $Ca(ClO_4)_2 \cdot 4H_2O$ melts at 57° and loses its water at 256°, and undergoes a polymorphic transformation at 340°. It decomposes between 470° and 478°, and finally yields calcium chloride and oxygen.

5. Calcium iodate

Precipitated calcium iodate $Ca(IO_3)_2$ has been found to vary in composition from one trial to another; actually it is contaminated with the monohydrate. Rocchiccioli[9] heated the hexahydrate prepared specially. The dehydration occurs between 92° and 170-175° when the heating rate is 100° per hour. The horizontal related to the anhydrous iodate then extends up to $540°$[1, 10]; it is suitable for automatic determination of calcium.

If the heating is continued, the reaction discovered by Rammelsberg[11] sets in and gives rise to the paraperiodate $Ca_5(IO_6)_2$ which is red-brown. The reaction:

$$5 Ca(IO_3)_2 \rightarrow Ca_5(IO_6)_2 + 4 I_2 + 9 O_2$$

has been quantitatively verified by us. We have found that it begins at $550°$, that it proceeds very rapidly between $680°$ and $700°$, and that it is completed when the temperature reaches $770°$. See also Bestougeff[12]. The horizontal of the iodide then begins at $850-855°$.

6. Calcium sulphate

(i) Preparation from calcium chloride Calcium sulphate precipitated in the cold has the formula $CaSO_4 \cdot 2H_2O$, which also represents gypsum. It is prepared by adding sulphuric acid and ethanol to a solution of calcium chloride below $60°$ and filtering after two hours. If heated at the rate of $300°$ per hour, the precipitate is stable up to $107-110°$, and loses its two molecules of water up to $186°$ where the horizontal of the anhydrous salt begins. This level is maintained up to $990°$ at least.

(ii) Gypsum The purest gypsum we were able to secure gave the following analysis:

Water (moisture)	0.3%	SO_4	53.9%
Water (combined)	20.2%	CO_3	1.5%
Insoluble (HCl)	0.1%	$Al_2O_3 + Fe_2O_3$ traces	
Calcium	23.4%		

The curve exhibits no discontinuity in the course of the dehydration when plaster of Paris is formed, in accord with the well-known study by Jolibois and Lefèvre[13]. Gypsum has been heated by a number of investigators, notably Gill[5].

The curve of the natural product has an irregularity between $740°$ and $820°$ which is directly explained in the light of what has been learned about the decomposition of calcium oxalate. We are dealing here with the slight quantity of calcium carbonate present in the gypsum and consequently with the release of 4 mg of carbon dioxide which corresponds to the 1.69% of CO_3. Thus the thermobalance has given us the means of making a gasometric study without being obliged to set up special equipment and of putting the sample into solution.

In a study involving both gravimetry and spectrography, we, along

Lecomte and Pain[14], have used some gypsum in leaflets, and some in powder form, also the artificial product $CaSO_4 \cdot 2H_2O$, plaster of Paris $CaSO_4 \cdot \frac{1}{2}H_2O$, deutero gypsum $CaSO_4 \cdot 2D_2O$, deutero plaster of Paris $CaSO_4 \cdot \frac{1}{2}D_2O$, the mixed product $CaSO_4 \cdot \frac{1}{2}D_2O \cdot \frac{3}{2}H_2O$, the anhydrous sulphate which had been heated to 300°, 800° and 1005°, as well as the natural anhydrite.

(iii) Plaster of Paris To obtain the light plaster of Paris, we heated gypsum in the furnace of the thermobalance with the air saturated with light water vapour and stopping at 150°.

The deutero plaster of Paris (heavy plaster) was prepared by keeping the anhydrous sulphate, initially heated to 300°, in a water-tight vessel along with 99% heavy water. The deutero plaster of Paris was prepared in the same manner as the light plaster but keeping it in heavy water vapour in the furnace of the thermobalance. When slaked with distilled light water, this deutero plaster eventually yielded the mixed compound $CaSO_4 \cdot \frac{3}{2}H_2O \cdot \frac{1}{2}D_2O$. The objective of this investigation was especially to demonstrate that the water is not water of crystallization but that it is bound to sulphur in all of these compounds. Consequently, gypsum thus becomes a salt of orthosulphuric acid, namely CaH_4SO_6 and the deutero or heavy gypsum is CaD_4SO_6.

7. Calcium sulphamate

Calcium amidosulphonate (sulphamate) $Ca(NH_2-SO_3)_2 \cdot 4H_2O$ was prepared by Capestan[15] who precipitated the product from its concentrated aqueous solution by adding alcohol, and dried it under reduced pressure over phosphoric anhydride. It loses its water in a single step from 40° to 220°; the thermolytic conversion into the sulphate proceeds via the intermediate imidodisulphonate.

8. Calcium selenite

Calcium selenite $CaSeO_3 \cdot 1H_2O$ is prepared by double decomposition $(Na_2SeO_3 + CaCl_2)$. Rocchiccioli[9] found it to be stable as far as 185°. Water is lost up to 300°; hence the water must be bound to sulphur. With a heating rate of 300° per hour, the anhydrous salt is stable up to 1000° at least, the limit of the experiment.

9. Calcium pyroselenite

Calcium pyroselenite $CaSe_2O_5 \cdot 2H_2O$ is prepared by the action of selenious acid on calcium carbonate. It was heated by Rocchiccioli[9] at the rate of 300° per hour. She found it to be stable up to around 120°. Complete dehydration is achieved at 320° but the level of the anhydrous salt is not horizontal. At this temperature, the loss of SeO_2 is already considerable; it accelerates at 350°, slows down between 400° and 500°, again accelerates, and 545° marks the beginning of the level of the neutral selenite $CaSeO_3$, which is perfectly horizontal as we have pointed out above. There is no oxidation to selenate.

10. Calcium selenate

Calcium selenate $CaSeO_4 \cdot 2H_2O$ is less stable than gypsum $CaSO_4 \cdot 2H_2O$. It decomposes to give the selenite from 650° on; the reduction is complete around 800°. See Selivanova, Schneider and Strel'tsov[16].

11. Calcium nitrate

If calcium nitrate $Ca(NO_3)_2 \cdot 4H_2O$ is heated slowly (3rd speed) it is stable up to 90°. It dehydrates smoothly up to 160°. The horizontal of the anhydrous salt extends as far as 475°; beyond this temperature, reddish vapours are given off. See also Wendlandt[17].

12. Calcium monohydrogen phosphate

The monoacid phosphate $CaHPO_4 \cdot 2H_2O$, also known as dicalcium phosphate, has been studied by Boullé[18]. If heated in the air it does not yield the anhydrous salt but the pyrophosphate is formed directly:
$$2\ CaHPO_4 \rightarrow Ca_2P_2O_7 + H_2O$$
The dehydration slows down noticeably around 180° and then becomes rapid again up to 220°. At this temperature, 19.5% of water has been removed (formation of $CaHPO_4$ requires the removal of 20.9%). The remaining hydrated phosphate disappears only slowly beyond 220° at the same time that the formation of the pyrophosphate begins and then becomes more rapid starting at 400°. On the basis of three specimens, Boullé confirmed the finding that the form of the dehydration curve changes with the mode of preparation.

13. Calcium dihydrogen phosphate

Morin[19] studied 1 g samples of the diacid phosphate $Ca(H_2PO_4)_2$. Water begins to be lost around 250° and this loss continues rapidly up to about 320°. The pyrophosphate $Ca_3(HP_2O_7)_2$ can be detected at 295° accompanied by a little of the triphosphate $Ca_2HP_3O_{10}$. At 345°, the latter becomes much more conspicuous than the pyrophosphate. Starting at 420°, the removal of the water is hastened by decomposition of the two salts in the preceding mixture. The crystalline phases at 445° are principally the β-polyphosphate accompanied by a little of the γ-poly-phosphate. At 500°, the γ-polyphosphate is almost completely converted into the β-polyphosphate. In this interesting carefully executed research, Morin has also worked out the curves pertaining to the hydrated amor-phous calcium polyphosphate where $n = 15$ and $n = 2$. See also the studies by Vyzgo and Saibova[20].

14. Tricalcium phosphate

Tricalcium phosphate $Ca_3(PO_4)_2$ or better, according to Brasseur[21], $[Ca_3P_2O_8]_3 \cdot 2H_2O$, shows a continual and total loss of water of about 6% with an inflection near 750° (loss of $\frac{1}{2}$ molecule of water per 3 molecules of $Ca_3P_2O_8$). It is thought that $^3/_2$ molecules of water are present in the zeolithic form.

The nature of this phosphate is still a matter of lively discussion. It has also been heated by Chaudron[22] and Wallaeys[23].

15. Calcium pyrophosphate

Boullé and Dubost[24] isolated a new pyrophosphate which they de-signate as α; its formula is $2CaH_2P_2O_7 \cdot 3H_2O$. If 500 mg of this compound are heated at the rate of 150° per hour, in a dry atmosphere, it loses water rapidly between 200° and 250°. Above 370°, it forms condensed polyphosphates but the loss of water stops. The anhydrous product obtained at 600° undergoes a transformation at 800° as shown by the X-ray diagram. These workers have found that the pyrophosphate is the principal constituent of the mixtures resulting from the dehydration of the monocalcium orthophosphate at $\leqslant 310°$.

Guérin and Artur[25] heated the pyrophosphate above 1200°. After 7 hours, at 1215° for instance, the transformation into orthophosphate amounts to 93.1%.

16. Calcium monofluorophosphate

Calcium monofluorophosphate $CaPO_3F \cdot 2H_2O$ was prepared by Vu Quang Kinh and Montel[26]. It loses weight from 80° on, releasing essentially water vapour up to 180°, and the solid becomes amorphous as shown by the X-ray plate. Starting at 180°, there is evolution of hydrogen fluoride and a crystalline phase is not formed until the temperature reaches 400°. The α-form of calcium pyrophosphate is metastable and is irreversibly transformed into the stable β-calcium pyrophosphate near 550°.

17. Apatites

Various interesting studies of the apatites have been published. Chaudron[22] reports that it is possible to follow a solid state reaction occurring around 600° between calcium fluoride and tricalcium phosphate:
$$3\ Ca_3(PO_4)_2 + CaF_2 \rightarrow [Ca_3(PO_4)_2]_3 \cdot CaF_2$$
or with lime, at 850°, to yield a hydroxylapatite:
$$3\ Ca_3(PO_4)_2 + Ca(OH)_2 \rightarrow [Ca_3(PO_4)_2]_3 \cdot Ca(OH)_2$$
If the latter compound is heated with calcium fluoride, it loses water at 600°, which makes its determination possible.

(i) Mixed apatites   Chaudron[22] likewise prepared mixed apatites on the thermobalance. The reaction between calcium fluoride and the pyrophosphate sets in at 800° and ends at 920°. In this instance, phosphorus oxyfluoride is given off:
$$9\ Ca_2P_2O_7 + 7\ CaF_2 \rightarrow \tfrac{5}{2}[Ca_3(PO_4)_2]_3 \cdot CaF_2 + 3\ POF_3$$
This reaction has been followed also by Montel[27] isothermally at 900° and 700°.

(ii) Hydroxylapatite   Wallaeys[28] studied the reaction of carbon dioxide at 900° on a hydroxylapatite. The resulting carbonated apatite was then subjected to continuous heating in the presence of calcium fluoride up to 1000°.

(iii) Wagnerite   Vincent[29] made a study of a wagnerite $Ca_3(PO_4)_2 \cdot CaCl_2$; it began to decompose at 900° and particularly beyond 1000°. The isotherm at 700° revealed no change.

(iv) Other apatites   The apatite $3\ Ca_3(PO_4)_2 \cdot CaCl_2$ begins to lose

chlorine at 650°. The following findings relate to the chlorophosphate $Ca(H_2PO_4)_2 \cdot CaCl_2 \cdot 2H_2O$: the anhydrous salt is obtained at 150°; a mixture of the anhydrous salt and a new phase appears at 250°; the new phase is present alone at 350°; an amorphous product is found at 450°; calcium pyrophosphate is present at 500°. The pyrolyses of the corresponding bromine and iodine derivatives are much less clear-cut.

18. Calcium hydrogen arsenate

The monoacid calcium arsenate $CaHAsO_4$ is precipitated by treating a slightly ammoniacal solution of calcium chloride with a slight excess of monoacid sodium arsenate. The resulting amorphous product retains moisture up to 157° and then the curve shows a change of slope. The water of constitution begins to come off at this temperature:

$$2\ CaHAsO_4 \rightarrow H_2O + Ca_2As_2O_7$$

The conversion into pyroarsenate is complete at 350°. From this temperature on and up to 946° there is a perfectly horizontal level. Consequently, the weighing should be made between these temperatures if acceptable results are to be obtained.

19. Calcium niobate

Calcium niobate $Ca_7Nb_{12}O_{37} \cdot 36H_2O$, like all the salts of this type, has been studied by Pehelkin et al.[30]. It forms the hexahydrate at 140–150°, the dihydrate at 340°, the monohydrate at 460°, and the metaniobate $Ca(NbO_3)_2$ at 640°.

20. Calcite

Calcite has been studied by various workers, especially by Saito[31] who worked in air and in carbon dioxide. He used 500-mg samples, the material passed a 200-mesh sieve, and the heating rate was 120° per hour. The air flowed at 100 ml per minute, the carbon dioxide at 200 ml per minute. The decomposition occurred in the first case from 600° to 800–810°, and from 900° to 920° in the second case. The artificial product behaved in almost the same fashion.

Papailhau[32], who heated 200 mg in a silica crucible in carbon dioxide flowing at 60 ml per minute, found the decomposition to begin at 870° and to be very active between 890° and 980°.

Caillère and Pobeguin[33] found that calcite starts to break down at

675° in air and at 700° in nitrogen. They report that the corresponding temperatures for aragonite are 645° and 660° at this same rate of heating.

Calcium carbonate and calcite have been also heated in air, nitrogen, and carbon dioxide by Richer and Vallet[34] who overlooked the fact that (a) Saito had made the same experiments before them, and (b) that it was Honda and not Guichard who invented the thermobalance. In this same paper, Richer and Vallet criticized the present writer harshly, selecting from among his publications a general lecture delivered under the chairmanship of M. Chevenard. In this research, these workers, using a nitrogen atmosphere (30 liters per hour), compared the temperature marking the end of the decomposition of powdered artificial calcium carbonate (783°), with that of pulverized calcite (802°), of calcite in lumps (891°), using 352-mg samples and a heating rate of 150° per hour. They studied the influence of the mass of the material on the final temperature, employing the artificial material and the same rate of heating as before. Their findings were:

Mass (mg)	25	50	75	150	175	200	250	300	350
Temp. (°C)	692	714	731	746	750	765	770	776	780

They then studied the effect of the heating rate:

Rate (°C/h)	19	38	100	158	304
Temp. start t_1	500	517	537	545	557
Temp. finish t_2	683	710	752	771	810
$t_2 - t_1$	183	193	215	226	253

An identical trial was made using carbon dioxide in place of nitrogen. Although the threshold temperature of the decomposition of calcium carbonate is well defined in carbon dioxide (900°) and rather well defined in nitrogen (around 500°), the same cannot be said of the temperature at which the decomposition ends, namely 914° to 1034° in carbon dioxide, and between 683° and 891° in nitrogen.

(i) Carbonate and lime content of hydrated calcium silicate The carbonate and lime content of samples of hydrated calcium silicate have been determined by thermogravimetry. The author[3] developed the calculations required to make this automatic determination.

(ii) Calcium carbonate in building materials and in marble Other studies of calcium carbonate have been made by Wanmaker and Verheyke[35], by Schwob[36] in building materials with the aid of the Longchambon thermobalance, and in marble by Orosco[37], Roy-Buckle[6]. See also[38]. The discussion of calcium oxalate (see below) should also be

consulted in this connection since it yields calcium carbonate transitorily. Consult also Peltier and Duval[1].

(iii) Mixtures of calcium and magnesium carbonate The mixtures of $CaCO_3 + MgCO_3$ have been discussed in Chapter Magnesium. Mere reference will be given here to the work of Somiva and Hirano[39] on these mixtures, and to the work of Gill[5] and Dupuis and Dupuis[40] on dolomite. See also Wright, Hoffman and Schnitzer[41].

(iv) Reaction between calcium carbonate and ammonium sulphate Burriel-Marti and Garcia-Clavel[42] studied the solid state reaction between calcium carbonate and ammonium sulphate. They also investigated the analogous reactions of strontium carbonate and barium carbonate. As expected, the corresponding alkaline earth sulphate resulted. The presence of ammonium sulphate lowers the decomposition threshold of the carbonates. These investigators also heated these mixtures isothermally at 150°, 200°, 225°, 275°, and 300° in the molecular ratios of 1 : 1.

21. Calcium silicate

Van Bemst[43] made an investigation involving thermogravimetry, X-ray analysis, infra-red absorption, and differential thermal analysis of the β-silicate Ca_3SiO_5 and the γ-silicate Ca_2SiO_4 between 4° and 60° to gain insights into the $CaO–SiO_2.H_2O$ system. Burriel-Marti and Gaspar-Tébar[44] studied thermogravimetrically a number of reactions, including those between calcium oxide and silica (+ oxides of iron and aluminium).

22. Calcium bichromate

Durand and Michel[45] used differential thermal analysis conjointly with thermogravimetry in their study of calcium bichromate $CaCr_2O_7 \cdot 5H_2O$. When heated at the rate of 0.6° per minute, this salt yields a tetrahydrate which is stable up to 81°. Aqueous fusion occurs at this temperature, and then the tetrahydrate yields the trihydrate $CaCr_2O_7 \cdot 3 H_2O$, which exists up to 130°, where the monohydrate appears. The latter, in turn, starts to break down at 147° to yield the anhydrous salt. At this same heating rate, the production of the chromate occurs at around 440–460°; finally the chromite appears between 1067° and 1500°:

$$CaCr_2O_7 \rightarrow CaCrO_4 + CrO_3$$
$$4 CaCrO_4 \rightarrow 2 Cr_2O_3.CaO + 3O_2 + 2CaO.$$

References p. 284

23. Calcium molybdate

Calcium molybdate $CaMoO_4$ is anhydrous; it may be prepared by treating a neutral boiling solution of calcium chloride with 2 drops of concentrated ammonium hydroxide and adding a slight excess of a solution of ammonium molybdate. The boiling is continued for 10 minutes and the precipitate is washed with hot water. The pyrolysis curve is very simple. The moisture is removed smoothly and constant weight is observed from 230° on. Peltier and Duval[1] found a break in some of the recordings around 750° and spontaneous reduction.

24. Calcium tungstate

Saint-Sernin[46] believes that weighing as the tungstate $CaWO_4$ is the most accurate gravimetric method for determining calcium. Calcium tungstate is prepared by adding sodium tungstate to a hot, slightly ammoniacal (pH 8) solution of calcium chloride. The precipitate is crystalline and settles rapidly. It is rather difficult to dry (between 400° and 500° depending on the quantity, but not 800° as suggested by some workers). It may be weighed even at 130–150°; the error due to excess of moisture has reached 1.21%. Peltier[10] used this salt for determining calcium in the presence of a large excess of magnesium.

25. Calcium ferrocyanide

Cappellina and Babani[47] report that calcium ferrocyanide $Ca_2[Fe(CN)_6] \cdot 10H_2O$ is dehydrated between 30° and 140°.

26. Calcium potassium (or ammonium) ferrocyanide

The double calcium potassium (or ammonium) ferrocyanide $CaK_2[Fe(CN)_6]$ or $Ca(NH_4)_2[Fe(CN)_6]$, prepared by Flaschka and Spitzy[48], has been suggested for the gravimetric determination of calcium. The ammonium salt is stable up to 280–320°; the decomposition becomes explosive beyond the latter temperature. The potassium salt remains unchanged up to 325–330° where the odour of cyanogen becomes apparent.

27. Calcium potassium nickelohexanitrite

The nickelohexanitrite $CaK_2[Ni(NO_2)_6]$ is prepared by treating a calcium salt with a reagent composed of suitable proportions of acetic acid, nickel nitrate, and potassium nitrite. This compound has been known for a long time and has been studied by Astruc and Mousseron[49]. They used it for an indirect determination of calcium through measurement of the nitrite: (a) colorimetrically, (b) by reduction to ammonia, (c) by oxidation with permanganate.

The pyrolysis curve shows the elimination of three molecules of water up to 80°; they are probably bound to the calcium through secondary valences. The horizontal of the compound with the above formula extends from 80° to 250°, after which the interpretation of the curve does not lead to any simple result. If it is assumed that the final product is a mixture of oxides, there must be a transitory oxidation of NO_2 and NiO. Accordingly, the formula of the complex would be $[Ca(H_2O)_3]K_2[Ni(NO_2)_6]$. We have thought then that along with the methods of Mousseron and Astruc, there is a place for a gravimetric method since the foregoing complex contains precisely one atom of nickel per atom of calcium, and nickel can be determined with high precision by electrolysis. See Peltier[10].

28. Calcium formate

Zapletal, Jedlička and Rüžička[50] have heated calcium formate.

29. Calcium acetate

Calcium acetate $Ca(CH_3CO_2)_2 \cdot 4H_2O$ was studied by Burriel-Marti, Jimenez-Gomez and Alvarez-Herrero[51] during an investigation of ion exchangers. It loses its water near 80°. The anhydrous product yields a level that is essentially horizontal from 200° to 380°, where its starts to give off acetone and water of constitution. The experiment was not carried beyond 450°. This study was recently repeated by Walter-Lévy and Laniepce[52]; they characterized three distinct varieties of anhydrous calcium acetate by X-ray analysis.

30. Calcium oxalate

Calcium oxalate $CaC_2O_4 \cdot H_2O$ was prepared by the classic method of treating a slightly ammoniacal solution of calcium chloride with ammonium chloride and then adding ammonium oxalate. The thermolysis curve is shown in fig. 55a; it includes four parallel levels. These may be utilized in four methods of determining calcium by automatic gravimetry:

$$CaC_2O_4 \cdot H_2O \text{ from } 20° \text{ to } 100–105°$$
$$CaC_2O_4 \quad \text{from } 240° \text{ to } 400–410°$$
$$CaCO_3 \quad \text{from } 500° \text{ to } 660°$$
$$CaO \quad \text{starting at } 830°$$

These figures were obtained from 174.2 mg of the precipitate heated in an aluminite crucible at 300° per hour.

We again call attention to the fact that this curve can be employed in a student exercise for checking the atomic weight of carbon (see p.102) and that if used in connection with the thermolysis curve of magnesium oxalate, it provides a convenient means of determining calcium in the presence of magnesium without preliminary separation. See Peltier and Duval[10].

Of course, if the crucible is taken out of the furnace, no note is any longer taken of the hygroscopicity of the resulting materials. This was not understood by Miller[53]. Nevertheless, Villamil[54] and Moles and Villamil[55] believed that the anhydrous oxalate (stable up to 380°) may be weighed after being dried in a current of dry air. We are in agreement with them.

Calcium oxalate has been pyrolyzed also by Auméras[56] in the Guichard balance, by Waters[57], Erdey and Paulik[58], Markova and Filippova[59], Terem and Tuccarbasi[60], etc.

Furthermore, Lecomte, Pobeguin and Wyart[61] have studied the trihydrate of calcium oxalate by X-rays and infra-red. The forms obtained are subjected to increasing temperatures without stating whether thermogravimetry was involved.

31. Calcium tartrate

The calcium tartrate usually encountered in analytical practice has the formula $Ca(CO_2–CHOH–CHOH–CO_2) \cdot 4H_2O$. It is formed by treating a solution of calcium chloride made slightly acid with acetic acid with a 10% solution of sodium acetate and a 10% solution of racemic tartaric acid as suggested by Brönsted[62]. The heating curve is rather complicated, at least in its central portion. Two molecules of water are given off up to 100°; then the two others are evolved up to 226° with a change of slope at 136° corresponding to the formation of the compound $CaC_4H_4O_6 \cdot \frac{1}{4}H_2O$. The latter had already been identified by Chatterjee and Dhar[63] at 170°. A horizontal corresponding to calcium carbonate was observed by Duval between 430° and 750°; it may be used for an accurate automatic determination of calcium.

32. Calcium oxinate

Calcium oxinate $Ca(C_9H_6ON)_2$ is prepared from an ammoniacal solution of calcium chloride heated to 60° and then treated with an alcoholic solution of oxine. It was heated at 50°, 100°, 150°, and 300° per hour. Although the curves obtained from these various heating rates have the same general form, they lead to quite different temperature values. At 150° per hour, constant weight is not attained below 171° with 116 mg of sample; the slightly oblique level of the chelate extends up to 380°; the decomposition of the oxine then is completed up to 750–760°. The resulting lime may contain slight particles of tar which are held tenaciously.

33. Calcium ferronate

We used the procedure suggested by Schoorl[64] for preparing the calcium complex of loretine (yatren, ferron). It had the composition $3Ca$ $[C_9H_4NI(OH)SO_3]_2 \cdot 10H_2O$ when dried below 80° (see the 1st edition, p. 159, issued 1953). More recently, Gillis, Van der Stock and Hoste[65], employing a seemingly purer reagent, and heating at the rate of 350° per hour, found:

    from   60° to 130° $Ca[C_9H_4NI(OH)SO_3]_2 \cdot 3H_2O$
    from 165° to 235° $Ca[C_9H_4NI(OH)SO_3]_2$(anhydrous complex)
    from 540° to 960° $CaSO_4$

We no longer have any of the reagent which we used in 1946, and consequently we accept the findings of Gillis, Van der Stock and Hoste.

34. Calcium quinaldate

The quinaldate or calcium quinoline-2-carboxylate $Ca(C_{10}H_6O_2N)_2 \cdot H_2O$ was prepared by Thomas[66]. The colourless material begins to lose water at 190°. The anhydrous salt starts to decompose at 345°. The temperature at which the oxide CaO is obtained is not given.

35. Calcium picrolonate

The yellow picrolonate has the formula $Ca(C_{10}H_7O_5N_4)_2 \cdot 7H_2O$. It is formed overnight when a hot aqueous solution of picrolonic acid is treated with a warm slightly ammoniacal solution of calcium chloride. Ordinarily, the formula is given as though the material carries 8 H_2O, but the curve shows only 7 H_2O, and in this we are in agreement with Robinson and Scott[67]. The compound is usually weighed at ordinary temperature after passing a stream of filtered dry air over the precipitate on the filter. Kiba and Ikeda[68] report that this procedure is strictly valid only up to 40°. A hemihydrate is stable beyond 150–175° and the horizontal of the anhydrous material (already contaminated with carbon) appears from 200° to 230°. It is not advisable to weigh the compound after it has been heated between these two temperatures.

REFERENCES

[1] S. PELTIER AND C. DUVAL, *Anal. Chim. Acta*, 1 (1947) 345.
[2] C. CABANNES-OTT, *Thesis*, Paris, 1958.
[3] F. M. BIFFEN, *Anal. Chem.*, 28 (1956) 1133.
[4] P. RÉMY-GENNETÉ, P. MAURET AND B. AUDOUZE, *Bull. Soc. chim. France*, (1956) 955.
[5] A. F. GILL, *Can. J. Research*, 10 (1934) 703.
[6] E. ROY-BUCKLE, *J. Phys. Chem.*, 63 (1959) 1231.
[7] K. MATSUURA, *J. Chem. Soc. Japan*, 52 (1931) 730.
[8] J. BINKA AND V. SATAVA, *Silikáty*, 1 (1957) 174; *C.A.*, (1958) 16967.
[9] C. ROCCHICCIOLI, *Thesis*, Paris, 1960.
[10] S. PELTIER, *Diploma of Higher Studies*, Paris, 1947; S. PELTIER AND C. DUVAL, *Anal. Chim. Acta*, 1 (1947) 408.
[11] C. RAMMELSBERG, *Poggendorfs Ann.*, 44 (1838) 577.
[12] N. BESTOUGEFF, *Diploma of Higher Studies*, Paris, 1960.
[13] P. JOLIBOIS AND H. LEFÈVRE, *Compt. rend.*, 176 (1923) 1317.
[14] C. DUVAL, J. LECOMTE AND C. PAIN, *Compt. rend.*, 237 (1953) 238.
[15] M. CAPESTAN, *Ann. Chim. (Paris)*, 5 (1960) 215.
[16] N. M. SELIVANOVA, V. A. SCHNEIDER AND I. S. STREL'TSOV, *Zhur. Neorg. Khim.*, 4 (1959) 1481; *C.A.*, (1960) 9461.
[17] W. W. WENDLANDT, *Texas J. Sci.*, 10 (1958) 392; *C.A.*, (1959) 12082.
[18] A. BOULLÉ, *Colloq I. U. P. A. C.*, Münster, 1954 p. 217.
[19] C. MORIN, *Thesis*, Paris, 1960.
[20] V. S. VYZGO AND M. T. SAIBOVA, *Doklady Akad. Nauk. Uzbek. S.S.R.*, (1959) 28; *C.A.*, (1960) 10484.
[21] H. A. L. BRASSEUR, *Colloq. I.U.P.A.C.*, Münster, 1954, p. 199.
[22] G. CHAUDRON, *ibid.*, p. 172.
[23] R. WALLAEYS, *Thesis*, Paris, 1951.
[24] A. BOULLÉ AND M. P. DUBOST, *Compt. rend.*, 247 (1958) 1864.
[25] H. GUÉRIN AND A.ARTUR, *Colloq. Mineral Chem. I.U.P.A.C.*, Münster, 1954, p. 190.
[26] VU QUANG KINH AND G. MONTEL, *Compt. rend.*, 249 (1959) 117.
[27] G. MONTEL, *Colloq. I. U. P. A. C.*, Münster, 1954, p. 178.
[28] R. WALLAEYS, *ibid.*, p. 183.
[29] J. P. VINCENT, *Ann. Chim. (Paris)*, 5 (1960) 579.
[30] V. A. PEHELKIN, A. V. LAPITSKII, V. I. SPITSYN AND Y. P. SIMANOV, *Zhur Neorg. Khim.*, 1 (1956) 1784; *C.A.*, (1957) 2445.
[31] H. SAITO, *Sci. Rep. Tôhoku Imp. Univ.*, 16 (1927) 54.
[32] J. PAPAILHAU, *Bull. Soc. franç. minéral.*, 82 (1959) 367.

33 S. Caillère and T. Pobeguin, *Bull. soc. franç. minéral.*, 83 (1960) 36.
34 A. Richer and P. Vallet, *Bull. Soc. chim. France*, (1953) 148; *Compt. rend.*, 249 (1959) 680; 246 (1958) 2133.
35 W. L. Wanmaker and M. L. Verheyke, *Philips Research Repts.*, 11 (1956) 1.
36 Y. Schwob, *Rev. matériaux Construct. et trav. publ.*, No. 411 (1949) 49; No. 413 (1950) 33, 85.
37 E. Orosco, *Ministério trabalho ind. ecom., (Inst. nac. tecnocol. Rio de Janeiro)*, 1940; *C.A.*, 35 (1941) 3485.
38 *Ist Congr. Thermography, Kazan*, 1953, p. 7.
39 T. Somiva and S. Hirano, *J. Soc. Chem. Ind. Japan.*, 34 (1931) 381.
40 T. Dupuis and J. Dupuis, *Mikrochim. Acta*, (1958) 186.
41 J. R. Wright, I. Hoffman and J. Schnitzer, *J. Sci. Food Agric.*, 11 (1960) 163.
42 F. Burriel-Marti and M. E. Garcia Clavel, *Proc. Intern. Symposium on Reactivity of Solids, Madrid*, 1956, p. 211.
43 A. Van Bemst, *Bull. soc. chim. Belg.*, 64 (1955) 333.
44 F. Burriel-Marti and D. Gaspar-Tébar, *ibid.*, p. 177. *Proc. intern. Symposium on Reactivity of Solids, Madrid*,1956,
45 R. Durand and A. Michel, *Compt. rend.*, 246 (1958) 1864.
46 A. Saint-Sernin, *Compt. rend.*, 156 (1913) 1019.
47 F. Cappellina and B. Babani, *Atti accad. sci. ist. Bologna*, 11 (1957) 76.
48 H. Flaschka and H. Spitzy, *Mikrochemie*, 35 (1949) 269.
49 A. Astruc and M. Mousseron, *Compt. rend.*, 190 (1930) 1556.
50 V. Zapletal, J. Jedlička and V. Růžička, *Chem. Listy*, 50 (1956) 1406.
51 F. Burriel-Marti, S. Jimenez-Gomez and C. Alvarez-Herrero, *Anales edafol. y fisiol. vegetal (Madrid)*, 14 (1955) 213.
52 L. Walter-Lévy and J. Laniepce, *Compt. rend.*, 250 (1960) 332.
53 C. C. Miller, *Analyst.*, 78 (1953) 186.
54 C. D. Villamil, *Anales soc. españ. fís. y qguím.*, 22 (1924) 74.
55 E. Moles and C. D. Villamil, *Anales soc. españ. fís. y quím.*, 22 (1924) 174.
56 M. Auméras, *Compt. rend.*, 181 (1925) 214.
57 P. Waters, *Nature*, 178 (1956) 324.
58 L. Erdey and F. Paulik, *C.A.*, 50 (1956) 3952.
59 G. A. Markova and A. G. Filippova, *Zavodskaya Lab.*, 10 (1941) 481; *C.A.*, (1941) 7870.
60 H. N. Terem and S. Tuccarbasi, *Rev. fac. sci. univ. Istanbul, Sér. C*, 23 (1958) 14; *C.A.*, (1959) 6835.
61 J. Lecomte, T. Pobeguin and J. Wyart, *Compt. rend.*, 216 (1943) 808.
62 J. N. Brönsted, *Z. anal. Chem.*, 42 (1903) 15.
63 K. P. Chatterjee and N. R. Dhar, *J. Phys. Chem.*, 28 (1924) 1023.
64 N. Schoorl, *Pharm. Weekblad*, 76 (1939) 620.
65 J. Gillis, J. Van der Stock and J. Hoste, *Mikrochim. Acta*, (1956) 760.
66 G. Thomas, *Thesis*, Lyon, 1960.
67 P. L. Robinson and W. E. Scott, *Z. anal. Chem.*, 88 (1932) 417.
68 T. Kiba and T. Ikeda, *J. Chem. Soc. Japan*, 60 (1939) 911.

CHAPTER 21

Scandium

Dupuis and Duval[1] had previously studied the thermolysis of scandium hydroxide, oxalate, and basic tartrate. Some additional salts and chelates can now be added to this list: chloride, nitrate, cupferronate, mandelate, oxinate and its substitution products.

The starting material for our investigations (65 g Sc_2O_3) had been extracted from 100 kg of thortveitite by the method developed by Boulanger and Urbain[2]. In this procedure, the scandium is finally obtained as the acetylacetonate, which can be adequately purified by distillation under reduced pressure. This product is converted into an inorganic form by refluxing it with concentrated sulphuric acid for several hours on a water bath. The purity of the oxide obtained from the resulting sulphate was checked spectrographically by the late J. Bardet.

1. Scandium hydroxide

Scandium hydroxide is prepared by treating the sulphate with warm ammonium hydroxide. The adsorbed water and part of the water of constitution are rapidly given off up to 155°. A new loss of water is observed from 153° to 542° but this removal is at a much lower rate. Scandium oxide Sc_2O_3 is stable from 542° on. The recorded curve is very similar to those given by beryllium and aluminium oxides, but the horizontal of the oxide starts at a lower temperature in the present case.

2. Scandium chloride

Scandium chloride $ScCl_3 \cdot 6H_2O$ was thermolyzed by Wendlandt[3]. It does not yield an oxychloride horizontal. The decomposition sets in at 60°, and the oxide Sc_2O_3 appears at 710°.

3. Scandium nitrate

Scandium nitrate $Sc(NO_3)_3 \cdot 6H_2O$ does not give a horizontal for either the anhydrous salt or the oxynitrate. Wendlandt[4] reports that the horizontal of the oxide Sc_2O_3 starts at 510°.

4. Scandium oxalate

A precipitate of scandium oxalate was obtained by treating an aqueous solution of scandium chloride on the water bath (60°) with a saturated solution of oxalic acid. This product lost retained water up to 67°. The rather short horizontal beginning at this temperature corresponds to the decahydrate $Sc_2(C_2O_4)_3 \cdot 10H_2O$, which we have isolated and analyzed. It begins to decompose at 89° and an abrupt change in slope occurs at 277° when only approximately five molecules of water remain, but we have not been able to isolate the pentahydrate. Likewise, with our rate of heating, we have found it impossible to obtain a horizontal for the anhydrous oxalate. The loss of water, carbon dioxide, and carbon monoxide sets in imperceptibly at 450° and continues up to 608°. The horizontal corresponding to the oxide Sc_2O_3 is observed from 608° on.

Wendlandt[5], contrary to our findings, obtained a hexahydrate which began to lose water at 50°, and a horizontal corresponding to a dihydrate was observed from 185° to 220°. However, we have not been able to reproduce these findings with different heating rates but which were linearly increasing.

5. Scandium mandelate

Alimarin and Shen Han-Si[6] prepared scandium mandelate by the action of mandelic acid. The precipitate $H_3[Sc(C_8H_6O_3)_3] \cdot nH_2O$ is stable up to 280°.

6. Scandium tartrate

The basic tartrate, which is without precise formula, is formed according to Meyer and Winter[7] by the action of ammonium tartrate on scandium sulphate. It loses water and decomposes with continuous loss of weight up to 542°. The horizontal of the scandium oxide then begins. The analytical data are not in accord with a simple formula for the dry initial material. It should be pointed out that this precipitation provides an excellent means of separating scandium from thorium.

References p. 288

7. Scandium oxinate

Wendlandt[8] made a study of scandium oxinate containing an excess of oxine. It starts to decompose at 125°; the horizontal of the scandium oxide begins at 600°.

8. Scandium 5,7-dihalogenooxinates

The 5,7-dichlorooxinate and the 5,7-dibromooxinate were likewise prepared by Wendlandt[8]. The curves are identical with those given by the oxinate, and the horizontal of the scandium oxide starts at 600°. The 5,7-diodooxinate $Sc(C_9H_4I_2ON)_3$ is stable up to 75°, and the resulting Sc_2O_3 is formed, according to Wendlandt[9], above 525°.

9. Scandium methyloxinate

The methyloxinate $Sc(C_{10}H_8ON)_3 \cdot C_{10}H_8NOH$, which was prepared by Wendlandt[10], begins to lose weight at 78°. There is a break in the curve at 120–160°, and another at 225°. The latter corresponds closely to the compound $Sc(C_{10}H_8ON)_3$, which decomposes at once, and the horizontal of the scandium oxide Sc_2O_3 begins above 660°.

10. Scandium ammonium carbonate

Spieyn, Komissarova, Schacky and Pushkina[11] have published a continuous curve for ammonium scandium carbonate $NH_4Sc(CO_3)_2 \cdot 2H_2O$.

REFERENCES

[1] T. DUPUIS AND C. DUVAL, *Anal. Chim. Acta*, 3 (1949) 183.
[2] C. BOULANGER AND G. URBAIN, *Compt. rend.*, 174 (1922) 1442.
[3] W. W. WENDLANDT, *J. Inorg. & Nuclear Chem.*, 5 (1957) 118.
[4] W. W. WENDLANDT, *Anal. Chim. Acta*, 15 (1956) 435.
[5] W. W. WENDLANDT, *Anal. Chem.*, 30 (1958) 56.
[6] I. P. ALIMARIN AND SHEN HAN-SI, *Zhur. Anal. Khim.*, 15 (1960) 34.
[7] R. J. MEYER AND H. WINTER, *Z. anorg. Chem.*, 67 (1910) 399.
[8] W. W. WENDLANDT, *Anal. Chem.*, 28 (1956) 499.
[9] W. W. WENDLANDT, *Anal. Chim. Acta*, 15 (1956) 533.
[10] W. W. WENDLANDT, *Anal. Chim. Acta*, 17 (1957) 274.
[11] V. I. SPIEYN, L. N. KOMISSAROVA, V. M. SCHACKY AND G. J. A. PUSHKINA, *Zhur. Neorg. Khim.*, 5 (1960) 2223.

CHAPTER 22

Titanium

Titanium is usually precipitated as the hydroxide by ammonia, guanidinium carbonate, or tannin, or as the iodate, selenite, phosphate, cupferronate, oxinate, 5,7-dichlorooxinate, 5,7-dibromooxinate, for gravimetric determination. Dupuis and Duval[1] have discussed the resulting precipitates and their behaviour when subjected to the action of heat.

1. Titanium hydroxide

(i) Precipitation by aqueous ammonia   The gelatinous precipitate of tetravalent titanium hydroxide is prepared by the procedure of Bornemann and Schirmeister[2]. Its curve reveals the loss of water at two stages: up to 112°, where the moisture or imbibed water is released rapidly, and from 112° to 350°, where there is a slow loss of what undoubtedly is water of constitution. The good horizontal observed above 350° represents the oxide TiO_2. The weight theoretically demanded by the formula $Ti(OH)_4$ is found at 112°; however, this temperature is not marked by any horizontal.

(ii) Precipitation by gaseous ammonia   The method of Trombe[3] employs gaseous ammonia; above pH 0.8 we obtained from 0.01 M solution a precipitate of the hydroxide which can be filtered off immediately. The curve has exactly the same form as the preceding one, but the constant weight for TiO_2 begins at 330°.

(iii) Precipitation by guanidinium carbonate   If the method of precipitation is that of Jilek and Kot'a[4] in which the reagent is guanidinium carbonate, the curve again has the same form, but the constant weight for the titanium dioxide does not appear until the temperature has reached 500°.

(iv) Precipitation with tannin   We found that the Shemyakin[5] pro-

cedure, which uses tannin as the precipitant, produces a mixture which must be taken to 600° before the horizontal due to the oxide TiO_2 is attained. The curve has an oblique tangential inflection at 125°; here the dry precipitate weighs approximately 12 times as much as the residual oxide.

2. Titanium potassium iodate

A precipitate $Ti(IO_3)_4 \cdot 3KIO_3$ results when the Beans and Mossman[6] method is followed in which potassium iodate is added to a salt of tetravalent titanium. The product is subsequently reduced with sulphur dioxide. It would be interesting to learn whether this precipitate because of its high molecular weight and marked insolubility might be employed advantageously for direct gravimetry. So far as we can discover, there has been no report along this line. After the moisture has been driven off, there is a horizontal extending from 138° to 295° and this agrees with the formula just given. This horizontal can be employed for the automatic determination of titanium. The decomposition of the double iodate follows a rather complicated course, and above 860° it leads to a mixture of the potassium and titanium oxides whose "proximate molecular weight" is 296 on the basis of one molecular weight of TiO_2.

3. Titanium selenite

Berg and Teitelbaum[7] prepared a material with the approximate formula $SeO_2 \cdot TiO_2 \cdot H_2O$ by treating a tetravalent titanium salt with selenious acid. The product is destroyed by the action of heat and leaves a residue of titanium dioxide above 880°. The method possesses apparently little interest from a number of viewpoints and the evolution of selenium compounds is not entirely without danger to the analyst.

4. Titanium phosphate

Titanium phosphate was prepared by the procedure of Jamieson and Wrenshall[8] who advised that the precipitate be "ignited at the highest available temperature". The curve reveals a rapid release of water up to 89°, and then more slowly from 89° to 345°, and extremely slow from 345° to 400–410°. (The curve is practically horizontal in this interval and the precise point cannot be determined.) The phosphate $TiPO_4$ is obtained from 410° on and yields an excellent horizontal. Murayama[9] reports 650°. The gravimetric factor is 0.335. The formula $Ti_2P_2O_7$ previously given by Etterle[10] is better from the point of view of valence than $TiPO_4$ but the analytical factor 0.317 is less exact. The phosphate method is not worth much consideration as a gravimetric procedure but it may have some interest with regard to separations.

5. Titanium peroxy oxalate

The peroxyoxalate of Kharkar and Patel[11] $TiO_2 \cdot C_2O_4 \cdot 3.5H_2O$ has three changes in direction in its thermolysis curve. They are: from 80–100°, at 250°, and at 320°, where the horizontal of the TiO_2 begins.

6. Titanium cupferronate

The cupferronate precipitated as directed by Bellucci and Grassi[12] is not stable and loses weight from the ordinary temperature on. (See Fig. 63.) It is completely destroyed at 643°, the temperature at which the titanium dioxide horizontal starts.

7. Titanium oxinate

The procedure given by Ishibashi and Shinagawa[13] was employed for precipitating the oxinate in a slightly ammoniacal medium. Its thermolysis curve slopes slightly downward as far as 718°; after that, the horizontal of TiO_2 makes up the remainder of the curve. The proposed complex formula $TiO(C_9H_6ON)_2$ might be observable at 115° judging from the weight of the residue but there is no horizontal in this locality, and consequently the gravimetric determination could only be approximate. Borrel and Pâris[14] have substantiated these findings. From 40° to 50° there is a loss of water, which sometimes is less than 2 molecules, sometimes more. The salt does not correspond to a definite hydrate. No horizontal is found in the vicinity of 100°. An almost horizontal portion of the curve begins at 140–150° and it might give the impression of a definite compound but all the evidence indicates that the latter is not stable since it slowly but smoothly loses weight up to 320°. The decomposition is more rapid beyond this latter temperature and titanium dioxide is finally obtained around 420°.

It is better to finish the determination by a bromide-bromate titration, or to make use of one of the following gravimetric methods.

References p. 292

8. Titanium 5,7-dichlorooxinate

The precipitate produced by means of 5,7-dichlorooxine as described by Fresenius[15], who recommended the interval 120–140° as the proper drying range, lends itself well to the automatic determination. On the other hand, Ishimaru[16] gives 142–169° as the limits of the horizontal. Actually, both of these workers are correct; a careful examination of the thermolysis curve reveals that the horizontal corresponding to the formula $TiO(C_9H_4Cl_2ON)_2$ extends from 105° to 195°. After the complex has been decomposed, the titanium dioxide TiO_2 is produced from 480° on. Ishimaru found 443° with his thermobalance.

9. Titanium 5,7-dibromooxinate

The 5,7-dibromooxinate, which we prepared by the method of Berg and Küstenmacher[17], conforms to the formula $TiO(C_9H_4Br_2ON)_2$. Its thermolysis curve includes a horizontal extending distinctly up to 160°. In fact, the material may be dried up to 192° with a loss below 0.5 mg in 65 mg. These workers suggested 140–150°, whereas Ishimaru[16] gives 116–170°. The product may even be placed in a desiccator in the cold; it will dry very quickly. We suggest the horizontal of this compound for the automatic determination of titanium. The titanium dioxide appeared on our curve from 609° on, and from 400° upward on Ishimaru's curve.

10. Titanium p-hydroxyphenylarsonate

Titanium p-hydroxyphenylarsonate when prepared in accord with the directions of Simpson and Chandlee[18] does not appear to have a definite formula; moreover, its thermolysis curve reveals no horizontal for the anhydrous compound. An explosion at 138° eliminates most of the arsenic; the carbon burns up to 555° and then the horizontal of titanium dioxide starts. This is a separation method; it has little interest as a gravimetric procedure.

REFERENCES

1 T. DUPUIS AND C. DUVAL, *Anal. Chim. Acta*, 4 (1950) 180.
2 K. BORNEMANN AND H. SCHIRMEISTER, *Z. anal. Chem.*, 51 (1912) 499.
3 F. TROMBE, *Compt. rend.*, 215 (1942) 539; 216 (1943) 888; 225 (1947) 1156.
4 A. JILEK AND J. KOT'A, *Collection Czechoslov. Chem. Communs.*, 4 (1932) 72.
5 F. M. SHEMYAKIN, *Zavodskaya Lab.*, 3 (1934) 986.
6 H. T. BEANS AND D. R. MOSSMAN, *J. Am. Chem. Soc.*, 54 (1932) 1905.

[7] R. BERG AND M. TEITELBAUM, *Z. anorg. Chem.*, 189 (1930) 101.
[8] G. S. JAMIESON AND R. WRENSHALL, *Ind. Eng. Chem.*, 6 (1914) 203.
[9] K. MURAYAMA, *J. Chem. Soc. Japan*, 51 (1930) 786.
[10] G. H. ETTERLE JR., *Chemist Analyst*, 27 (1918) 10.
[11] D. P. KHARKAR AND C. C. PATEL, *Proc. Indian Acad. Sci. A*, 44 (1956) 287; *C.A.*, 51 (1957) 7211.
[12] I. BELLUCCI AND L. GRASSI, *Atti accad. nazl. Lincei*, 22 (1913) 30.
[13] M. ISHIBASHI AND M. SHINAGAWA, *J. Chem. Soc. Japan*, 59 (1938) 1027.
[14] M. BORREL AND R. PÂRIS, *Anal. Chim. Acta*, 4 (1950) 267.
[15] W. FRESENIUS, *Z. anal. Chem.*, 96 (1934) 433.
[16] S. ISHIMARU, *J. Chem. Soc. Japan*, 55 (1934) 201.
[17] R. BERG AND H. KÜSTENMACHER, *Mikrochemie, Emich Festschr*, 1930, p 26.
[18] C. T. SIMPSON AND G. C. CHANDLEE, *Ind. Eng. Chem., Anal. Ed.*, 10 (1938) 642.

CHAPTER 23

Vanadium

With respect to the gravimetric determination of vanadium, it has been proposed that this element be precipitated in one of the following forms: vanadates of ammonium, mercury(I), barium (acid and basic medium), lead, silver (acid and basic medium), uranium(VI), manganese(II), cobaltihexammine (acid, neutral, basic medium), internal complexes with α-nitroso-β-naphthol, cupferron, oxine, 2-methyloxine, dicyanodiamidine, and strychnine vanadate. The thermolysis curves of most of these materials have been traced by Duval and Morette[1,2], and the findings have enabled Morette[3] subsequently to make a critical investigation of the automatic determination of vanadium. It should be noted that we have omitted the precipitation with ammonium sulphide as proposed by Norblad[4] which yields inconstant results, and also the procedure suggested by Shemyakin, Adamovich and Pavlova[5] employing ammonium benzoate and cinnamate, which involves the technical defect that the high acidity of the prescribed milieu brings about the precipitation of the organic acids used. We have also found in the literature some studies in which infra-red spectroscopy was combined with thermogravimetry. These investigations included studies of vanadyl hydroxide by Cabannes-Ott[6], the solid reaction products formed by sodium peroxide and vanadic anhydride and studied by Viltange[7], the study of ammonium metavanadate as a standard by Duval[8], vanadyl sulphate by Roch[9], Weychert and Leyko[10], etc.

1. Vanadyl hydroxide

Vanadyl hydroxide, studied by Cabannes-Ott[6], agrees with the formula $VO(OH)_2$ with respect to the pink precipitate produced by the decomposition of vanadyl sulphite. It was heated in air at 300° per hour, then in nitrogen at 150° per hour. The formula just given is retained up to 190°. This has been checked by infra-red absorption. The product loses water from 190° to 230° and yields the oxide V_2O_4 (in nitrogen) and the anhydride V_2O_5 (in air).

2. Vanadyl sulphate

Vanadyl sulphate, used by Roch[9], actually was a mixture of the hydrated sulphates, namely $VOSO_4 \cdot 5H_2O$ and $VOSO_4 \cdot 4H_2O$ in practically equal proportions. The thermogram indicates the existence of hydrates with 4, 3, 1 molecules of water and also the anhydrous salt. Vanadic anhydride is produced at higher temperatures, with colour changes from blue to green, to yellow-brown, to colourless. The temperatures were not determined precisely.

3. Ammonium vanadate

Ammonium metavanadate, used as a standard[8], is discussed in Chapter Ammonium. When heated in the slightly moist condition, as in an actual determination[1,2], it gives a horizontal from 45° to 134°, which agrees well with the formula NH_4VO_3. The fall of weight from 134° to 198° corresponds to the loss of all of the ammonia, leaving metavanadic acid HVO_3, and this finding suggests a possible method of obtaining the latter. However, it is unstable and begins to lose water at 206°; the anhydride V_2O_5 is obtained quantitatively at 448°. This product is stable up to 951°. As is well known, this method may not be employed for the accurate determination of vanadium.

The metavanadate was also studied by Satava[11] who used a new recording balance, and by Sesbes[12] from 16° to 500°. He found an intermediate compound $NH_3(V_2O_5)_2 \cdot H_2O$ from 280° to 320°, but this is not in agreement with our findings. Tashiro[13] reported 700° as the lowest temperature for obtaining the anhydride V_2O_5, whereas Kato, Hosimija and Nakazima[14] give 650°.

4. Mercurous vanadate

Mercurous vanadate has been precipitated in an almost neutral medium. The curve given by the resulting mixture shows a loss of weight starting at 60° (water, mercury, nitric acid) and continuing up to 675°. The horizontal of vanadic anhydride then extends from 675° to 946°. This method undoubtedly is the most accurate for determining vanadium; however, it cannot be employed for the automatic determination of this element because the precipitate has no definite formula; it retains excess mercury salts. Yosida[15] gives 700° as the lowest temperature for obtaining vanadic anhydride from the mercurous precipitate.

5. Barium vanadate

(i) Precipitation at pH 4.2   Carrière and Guiter[16] found that if ammonium metavanadate is treated at pH 4.2 with barium chloride, the
monohydrate $Ba(VO_3)_2 \cdot H_2O$ is precipitated. The latter is stable up to
128°. Water is given off from 128° to 371°; the horizontal of the anhydrous salt extends from 371° to 950°. This method is not so accurate as
the preceding one for determining vanadium; it gave us a relative error
of — 0.99%.

(ii) Precipitation at pH > 10   We followed the directions of Carrière
and Guiter[16] and precipitated the barium salt at a pH above 10. The
product was washed with dilute ammonium hydroxide. The curve corresponding to this precipitate is not reproducible; it shows a horizontal of
constant weight beyond 951° but the corresponding material is ill-defined
(in particular it does not accord with the formula of an orthovanadate).
We suggest that this method, which likewise is not quantitative, be
discarded. Tashiro[13] gave 620° as the minimum temperature for obtaining
a barium vanadate $Ba_2V_2O_7$.

6. Lead vanadate

According to Carrière and Guiter[16], precipitation of lead vanadate
at pH 4.5 should lead to the orthovanadate $Pb_3(VO_4)_2$, while Moser and
Brandl[17] and also Tashiro[13] state that the product is the pyrovanadate
$Pb_2V_2O_7$. The infra-red absorption spectrum is in accord with the latter
supposition; however, the thermolysis curve is difficult to interpret. This
pyrovanadate apparently is responsible for an oblique level from 350°
to 400° (in accord with Tashiro); the curve ascends slowly as far as 954°.
We suggest that this method be rejected, since in addition it is liable to
yield an error of as much as + 1.5%.

7. Silver vanadate

(i) Precipitation at pH 4.5   Silver metavanadate was precipitated, as
directed by Carrière and Guiter[16], by means of silver nitrate at pH 4.5.
The resulting $AgVO_3$ loses its moisture, and then gives an essentially
horizontal stretch from 60° to around 500°. From then on, it shows a
slight gain in weight with a maximum near 650°, and then a rather
distinct loss beginning at 819°. The residue contains some silver.

(ii) Precipitation in alkaline medium When precipitating with silver nitrate in an alkaline medium, we used the procedure of Moser and Brandl[17] which was generalized for microanalytical purposes by Kroupa[18]. The curve rises from 85° to 953° although Tashiro[13] suggests weighing after drying above 350°. This method has given us a minus error of 5.4% relative to the formula Ag_3VO_4; the filtrate contains traces of vanadium. It seems to us that it will be difficult to prevent the precipitate from being contaminated with hydroxide in view of the presence of ammonia.

8. Ammonium uranyl vanadate

Carnot[19], Blair[20], and Lewis[21] have precipitated ammonium uranyl vanadate $NH_4UO_2VO_4$ with a view of determining uranium. They report different amounts of water. We have found the precipitate to be the hemihydrate (0.5 H_2O), and it is recommended that the precipitate be dried at 105°. The curve shows a slow descent at this temperature and our control analysis then showed a plus error of 2.1%. The pyrovanadate $2UO_3 \cdot V_2O_5$ yields an almost horizontal level from 560° to 950°. Tashiro[13] gives 450–700°. The plus error for this new weighing form is then 2.3%. The method has little to recommend it.

9. Manganese vanadate

Manganese vanadate has been precipitated by the method of Carnot[19] in the presence of ammonium chloride and from boiling solution. The precipitate should be filtered off at once to avoid oxidation. There is a tendency for the precipitate to pass through the filters. The pyrolysis curve shows no evidence of the formation of a stable ammonium manganous vanadate. Water and ammonia are lost up to 346°; beyond this temperature it is difficult to obtain a horizontal for the pyrovanadate $Mn_2V_2O_7$ because the manganese oxidizes. However, Tashiro[13] gives the limits 400–900°. We suggest that this method be discarded since the plus error with this weighing form is 2.9%.

10. Luteocobaltic vanadate

(i) Precipitation at pH 5.1 An orange, somewhat gelatinous product, which coagulates poorly and is hard to filter, was obtained by the method of Parks and Prebluda[22] which employs luteocobaltic chloride in a

medium containing acetic acid (pH 5.1). It seems difficult to conceive the possibility of determining vanadium under such conditions. No horizontal is found in the curve in the region of 100°. At this temperature, the given formula $[Co(NH_3)_6]_4(V_6O_{17})_3$ is exact only within about —1.7%. More accurate figures are obtained by drying at 127° but decomposition then sets in at once. A horizontal is reached at 382° from which the above workers deduced the system $4CoO·9V_2O_5$ (rather than $2CoO·9V_2O_5$ as was incorrectly given in the original text). We, in turn, have arrived at the mixture $4CoO.9V_2O_5$, within close to —2.77%.

(ii) Precipitation at pH 7 The yellow-pink precipitate produced by luteocobaltic chloride in a neutral medium, as directed by Parks and Prebluda[22], conforms very well to the reported formula $[Co(NH_3)_6]$ $(VO_3)_3$. It loses wash water rapidly up to 58°, and then the curve shows a horizontal going from 58° to 143° which agrees with the formula just given within about —0.74%. We suggest that this horizontal be employed for the automatic determination of vanadium. The residue, whose formula is assumed to be $2CoO·3V_2O_5$, forms another horizontal extending from 600° to 950° after a rise due to an oxidation. Actually, our determinations deviate from this latter formula by —2.1%.

(iii) Precipitation at pH 9 The original paper[22] gave no details about the precipitation in an alkaline medium with this same luteocobaltic chloride. We used the following procedure: the metavanadate to be determined was treated with ammonia water in the presence of ammonium acetate so that the pH always remained near 9. The curve is identical with the one obtained from the product formed in neutral surroundings. The analytical results bear this out.

11. Vanadyl α-nitroso-β-naphthol

Quantitative results are not obtained when the vanadium is brought down by α-nitroso-β-naphthol as suggested by Terrisse and Lorréol[23]. If the residue is taken to be V_2O_5, the results are off by —3.5%. Accordingly, the curve has no real significance. The destruction of the organic matter occurs at temperatures up to 419° with an inflection between 100° and 127°. The stretch between 100° and 127° is not horizontal. There is no doubt that this method should be rejected.

12. Vanadyl cupferronate

Turner[24] proposed the use of cupferron to separate vanadium from phosphorus, arsenic, etc. We have followed his procedure, and after the gradual destruction of the inner complex, the curve shows a good horizontal, extending from 581° to 946°. We consider this method to be satisfactory. The minus error is around —0.4% at most. We also suggest that the horizontal be used for the automatic determination of vanadium when using a porcelain filtering crucible.

13. Vanadyl oxinate

The precipitation of $V_2O_3(C_9H_6ON)_4$ with oxine in a solution made acidic with acetic acid is not quantitative. A number of workers have dealt with this method: Berg[25], Jilek and Vicovsky[26], Back and Trelles[27], Ishimaru[28], and Borrel and Pâris[29]. The pyrolysis curve shows a slow descent as far as 195°, whereas Ishimaru reported a horizontal between 120° and 154°. The destruction of the organic matter is accompanied by definite losses of vanadium; above 814° the residue consists of a mixture of oxides. Ishimaru reported 468° as the minimum temperature for obtaining the anhydride, but he was dealing here really with a mixture of vanadium oxides. The final horizontal must not be employed in this method of determining vanadium.

14. Vanadyl 2-methyloxinate

Borrel and Pâris[30] obtained a gelatinous, yellowish, slightly soluble precipitate with 2-methyloxine. This product, $VO(OH)[C_9H_5(CH_3)NO]_2 \cdot 4H_2O$, is stable up to around 100°. It loses its water between 100° and 160°. The anhydrous compound decomposes from 280° upward and finally yields the anhydride V_2O_5 after giving an intermediate product $V_2O_3[C_9H_5(CH_3)NO]_4$. The tetrahydrate can be employed for the automatic determination of vanadium.

15. Vanadyl dicyanodiamidine

Fidler[31] claimed to have obtained practically theoretical figures when he precipitated vanadium with dicyanodiamidine. We used his directions but our results were much too low because the precipitation is not complete. The pyrolysis curve has a horizontal starting at 434°; it corresponds to a mixture of oxides.

References p. 300

16. Strychnine vanadate

The precipitate with strychnine, produced by the procedure of Jilek and Vicovsky[26] filters well but it is difficult to determine when the material has been sufficiently washed. The curve corresponding to this compound has a horizontal stretch between $100°$ and $140°$, which accords with an acid vanadate of strychnine. A second and parallel level begins at $390°$ and accords with the oxide V_2O_5. Weighing leads to an error of around $\pm 0.6\%$; consequently the method is worth considering.

REFERENCES

[1] C. DUVAL AND A. MORETTE, *Compt. rend.*, 230 (1950) 545.
[2] C. DUVAL AND A. MORETTE, *Anal. Chim. Acta*, 4 (1950) 490.
[3] A. MORETTE, *Bull. Soc. chim. France*, (1950) 526.
[4] J. A. NORBLAD, *Bull. Soc. chim. France*, 23 (1875) 64.
[5] F. M. SHEMYAKIN, V. V. ADAMOVICH AND N. P. PAVLOVA, *Zavodskaya Lab.*, 5 (1936) 1129.
[6] C. CABANNES-OTT, *Thesis*, Paris, 1958.
[7] M. VILTANGE, *Thesis*, Paris, 1960.
[8] C. DUVAL, *Anal. Chim. Acta*, 15 (1956) 224.
[9] J. ROCH, *Compt. rend.*, 248 (1959) 3549.
[10] S. WEYCHERT AND J. LEYKO, *Bull. acad. polon. sci. Classe (III)*, 3 (1955) 329.
[11] V. SATAVA, *Silikáty*, 1 (1957) 188; *C.A.*, (1958) 16806.
[12] T. A. SESBES, *Rev. fac. sci. univ. Istanbul.*, C. 20 (1955) 272; *C.A.*, (1956) 16497.
[13] T. TASHIRO, *J. Chem. Soc. Japan*, 52 (1931) 727.
[14] H. KATO, H. HOSIMIJA AND S. NAKAZIMA, *J. Chem. Soc. Japan*, 60 (1939) 1115.
[15] Y. YOSIDA, *J. Chem. Soc. Japan*, 61 (1940) 130.
[16] E. CARRIÈRE AND H. GUITER, *Compt. rend.*, 204 (1937) 1339.
[17] L. MOSER AND O. BRANDL, *Monatsh.*, 51 (1929) 169.
[18] E. KROUPA, *Mikrochemie*, 32 (1944) 245.
[19] A. CARNOT, *Compt. rend.*, 104 (1887) 1803; 105 (1887) 1850.
[20] A. A. BLAIR, *Proc. Am. Phil. Soc.*, 52 (1913) 201.
[21] D. T. LEWIS, *Analyst*, 65 (1940) 560.
[22] W. G. PARKS AND H. N. PREBLUDA, *J. Am. Chem. Soc.*, 57 (1935) 1676.
[23] TERRISSE AND LORRÉOL, after M. BÉARD, *Ann. chim. anal.*, 10 (1905) 41.
[24] W. A. TURNER, *Am. J. Sci.*, 41 (1916) 339.
[25] R. BERG, *Z. anal. Chem.*, 71 (1927) 369.
[26] A. JILEK AND V. VICOVSKY, *Collection Czechoslov. Chem. Communs.*, 4 (1932) 1.
[27] J. M. BACH AND R. A. TRELLES, *Anales asoc. quím. arg.*, 28 (1940) 111.
[28] S. ISHIMARU, *J. Chem. Soc. Japan*, 55 (1934) 201.
[29] M. BORREL AND R. PÂRIS, *Anal. Chim. Acta*, 4 (1950) 267.
[30] M. BORREL AND R. PÂRIS, *Anal. Chim. Acta*, 5 (1951) 573.
[31] J. FIDLER, *Collection Czechoslov. Chem. Communs.*, 14 (1949) 28.

CHAPTER 24

Chromium

Dupuis and Duval[1] made a critical study of the precipitates used for determining chromium and also announced a new finding regarding mercurous chromate. A solid state reaction for the production of chromites is now available, also a study of the arsenates, and a fact of interest with regard to the structure of chrome alum has been uncovered, namely the detection of three kinds of water.

They were not successful in their efforts to precipitate chromium by the Jilek and Kot'a[2] procedure employing selenious acid. They followed the directions as outlined in the abstract of the paper and also modified certain details, especially with respect to the temperature, but acceptable results were never obtained. The outline of the Jilek and Kot'a paper in *Chemical Abstracts*[3] contains an error. Trials in which ammonium sulphite, in large excess, was employed in place of the selenite to destroy the chromite produced, were without success; it is not possible to obtain quantitative results. Finally, from among the reagents recommended by Hanus and Lukas[4], Dupuis and Duval have given serious attention only to thiosemicarbazide, which Hanus and Lukas themselves favoured because it provides the most rapid procedure.

1. Chromium oxide hydrate

(i) Precipitation by aqueous ammonia   When the method of Treadwell[5] was used for precipitating hydrous chromium oxide, care was taken to avoid an excess of ammonium hydroxide. The water and ammonium salts retained by the precipitate continue to be lost up to 812° at least; the theoretical weight of Cr_2O_3 is reached within 1% (on the upper side) at 744°. Occasionally, a point of inflection is observed near 130°, followed by a less pronounced slope, but no definite compound is indicated. If the specimen is heated thermostatically near 100°, there is a horizontal stretch, but a new loss of weight sets in as soon as the temperature rises. An essentially horizontal stretch is found between 350° and 410°. The

References p. 310

loss of weight is almost constant between this latter level and the level due to the final oxide; it is of the order of 18 g per mole of Cr_2O_3. The corresponding dark brown product is therefore a definite product, and the only one which can be stressed in the thermogravimetric study. It had previously been reported by Manchot and Kraus[6] who assigned to it the empirical formula CrO_2, but it may be looked on as being a chromium chromate $Cr_2O_3 \cdot CrO_3$ in view of its chemical properties. More recently, Dominé-Bergès[7], in the course of a study of hydroxides precipitated by electrolysis, showed that we are dealing here with an oxidation product CrO_2, produced during the heating and while the dehydration is proceeding, rather than with a monohydrate $Cr_2O_3 \cdot H_2O$. Cabannes-Ott[8] made an infra-red absorption study of this product. The other precipitates observed up to 380° proved to be $Cr_2O_3 \cdot nH_2O$.

(ii) Precipitation by gaseous ammonia In the Trombe[9] method, an aqueous solution of chrome alum or green chromium chloride is treated with ammonium chloride and then a slow stream of air charged with ammonia gas is passed in. The hydrous chromic oxide begins to appear when the pH reaches 5.5. The current of air plus ammonia is cut off when the pH has risen to 11. The precipitate is collected and washed immediately. The curve obtained with this product includes a rather long horizontal extending from 440° to 475°; it conforms quite well to the formula $Cr_2O_3 \cdot 3H_2O$ rather than $Cr(OH)_3$. This horizontal may be used for the automatic determination of chromium. The horizontal due to the oxide Cr_2O_3 begins above 845°.

(iii) Precipitation by aniline The precipitate obtained by means of aniline, as directed by Schöller and Schraut[10], was found to give a curve which is much like the preceding but the inflection corresponding to the hydrate $Cr_2O_3 \cdot 3H_2O$ has no horizontal tangent. Consequently it is rather difficult to ascertain the exact region between 473° and 540° where this hydrate exists. The horizontal of the green oxide Cr_2O_3 starts near 820°.

(iv) Precipitation by hydroxylamine The method of Jannasch and Mai[11] employs hydroxylamine. The violet precipitate can be washed immediately after it has come down. The thermogram reveals complete disintegration of the complex, particularly an extensive loss of nitrogen at 100°. The hydrate is not shown on the curve as in the two previous cases; instead there is a rather indistinct inflection or retardation at 410° and this corresponds to the composition $Cr_2O_3 \cdot 2H_2O$. The temperature

has to reach 850° before the oxide Cr_2O_3 appears. The results are satisfactory and they are in agreement with those obtained by Friedheim and Hasenclever[12].

(v) Reduction of chromate by thiosemicarbazide Hanus and Lukas[4] report that thiosemicarbazide is able to reduce chromates quantitatively and produce the hydrous oxide. The resulting precipitate coagulates nicely and is washed readily. Its thermolysis curve descends until the temperature reaches 380°. The horizontal, extending from 380° to 410°, agrees well with the formula $Cr_2O_3 \cdot H_2O$, a compound which previously had been reported as being formed under high pressures. The horizontal of the oxide Cr_2O_3 starts at 475° and extends as far as 950° at least.

(vi) Precipitation by cyanate The excellent cyanate method is due to Ripan[13]; it yields a granular precipitate which filters easily. Its thermolysis curve likewise includes a horizontal, between 320° and 370°, which agrees rather well with the composition $Cr_2O_3 \cdot H_2O$. The latter rapidly releases its water and the green oxide Cr_2O_3 is obtained from 475° upward. The method is as accurate as the preceding procedure, but the reagent is more available.

(vii) Precipitation by ammonium nitrite The hydrous oxide is precipitated well by ammonium nitrite, or more conveniently by a mixture of sodium nitrite and ammonium chloride, in the method as reported by Schirm[14] and developed by Treadwell[5]. However, the curve slants downward until the temperature reaches 880° where the oxide Cr_2O_3 appears. The curve has an inflection point near 490° and a point of retrogression at 646°, which corresponds to complicated compositions, such as $Cr_2O_3 \cdot 1.7H_2O$ in the first instance.

(viii) Precipitation by potassium iodide–iodate The hydrous precipitate obtained by means of a potassium iodide–iodate mixture, as directed by Stock and Massaciu[15], filters nicely. However, the thermogram shows no region of constant weight until the temperature has reached 850° where the oxide Cr_2O_3 is present. A level which is short and also not very distinct appears between 343° and 408° but it does not conform to any simple composition, though it is not far from $Cr_2O_3 \cdot 2H_2O$.

2. Ammonium chromate

Ammonium chromate is prepared by neutralizing free chromic acid with ammonium hydroxide. When it is heated, water is given off rapidly up to 50°. The resulting anhydrous salt immediately loses ammonia and this loss continues up to 152°. The residue at this temperature consists of $Cr_2O_3 \cdot H_2O$, which then loses water up to 188°. Then a sudden loss of oxygen leads to Cr_2O_3. It is difficult to obtain precise data because particles of the solid are often swept out of the crucible. This method, of limited use, can scarcely be recommended. It should be noted that the decomposition temperature of the oxide CrO_3 has been given as 330°, but this appears to us to be an unlikely figure.

3. Inorganic chromites

Duval and Wadier[16] developed a solid state reaction for the production of inorganic chromites. In a first series of trials, they intimately mixed the anhydrous chloride, usually prepared just before using, and potassium bichromate, in equal molecular proportions, the total weight being 200 mg. The mixture was prepared in a porcelain crucible, which fitted the ring of the Chevenard thermobalance. When the reaction took place, it could be expressed:

$$2 \, K_2Cr_2O_7 + 2 \, MCl_2 \rightarrow 4 \, KCl + 2 \, MCr_2O_4 + 3 \, O_2$$

Because of the loss of oxygen and the volatilization of the potassium chloride, it is thus possible to record the change in weight of the ensemble as a function of the temperature. Moreover, since the resulting chromite is insoluble in water or even boiling hydrochloric acid, methodical washing will leave the expected product in a pure state.

The heating was linear at the rate of 300° per hour up to 1000–1050°; the furnace was then kept at this temperature for about 30 minutes. As an example, when anhydrous cobalt chloride and potassium bichromate are heated in equimolecular proportions, there is (but not always) a slight loss of moisture up to 168°, and then a horizontal stretch of varying length. In the case in point, this extends up to 372°. The reaction then starts slowly and becomes more rapid around 580°. The yields have been calculated from the weight of the residue after it has been dried in an oven, and on the assumption that the formula of the product is MCr_2O_4 and with respect to the weight of the metal M used (Table 5).

They have also varied the experimental conditions in two ways:
(a) By changing the nature of the salt reacting with the bichromate.
(b) By keeping the chloride the same but varying the proportions. For instance, Table 6 gives the results obtained with cadmium salts reacting with potassium bichromate in equimolecular proportions.

TABLE 5

SOLID STATE REACTION OF METAL CHLORIDES AND POTASSIUM BICHROMATE

Chloride	Color of the product	Yield (%)
$ZnCl_2$	Chocolate brown	82.3
$MgCl_2$	Ochre	73.8
$MnCl_2$	Brownish black	78.2
$FeCl_2$	Brownish black	64.7
$CdCl_2$	Green	60.7
$NiCl_2$	Very deep green	78.2
$CoCl_2$	Deep green	82.2
$CaCl_2$	Light green	34.7
$SrCl_2$	Light green	15.7
$BaCl_2$	Light green	9.3
$BeCl_2$	Light green	35.5
$PbCl_2$	Light green	31.5
$CrCl_3$	Light green	86.8

TABLE 6

REACTION OF CADMIUM SALTS WITH POTASSIUM BICHROMATE
(mixed in equimolecular proportions)

Cadmium salt	Color of the product	Yield (%)
$CdCl_2$	Green	60.7
$CdBr_2$	Green	38.6
CdI_2	Green	47.5
$CdSO_4$	Green	91.3

The Table 7 shows some examples of the yields when the compounds were not mixed in equimolecular proportions.

TABLE 7

REACTION OF METAL CHLORIDES WITH POTASSIUM BICHROMATE
(not mixed in equimolecular proportions)

Metal chloride	Bichromate	Color of the product	Yield (%)
1 AgCl	2 $K_2Cr_2O_7$	Green	99.6
2 $CrCl_3$	3 $K_2Cr_2O_7$	Light green	64.0
2 $CdCl_2$	1 $K_2Cr_2O_7$	Green	77.2
1 $NiCl_2$	2 $K_2Cr_2O_7$	Dark green	85.1
1 $CoCl_2$	2 $K_2Cr_2O_7$	Dark green	91.0

In the special case of chromium(III) chloride, it is interesting to point out that chemical analysis gives Cr_2O_3 as the composition of the resulting compound. In fact, if the formula of the hypothetical chromous acid is

$H_2Cr_2O_4$, its chromium(III) salt should have the formula $Cr_2(Cr_2O_4)_3$, in other words 4 Cr_2O_3.

This research was completed by making an infra-red spectrographic study of the resulting chromites.

4. Chromic chloride hydrates

The three hydration isomers of chromic chloride, namely

$[Cr(H_2O)_6]Cl_3$ Recoura's violet-grey chloride
$[Cr(H_2O)_5Cl]Cl_2 \cdot H_2O$ Bjerrum's chloride
$[Cr(H_2O)_4Cl_2]Cl \cdot 2H_2O$ Recoura's green chloride

have been studied under atmospheric pressure and under reduced pressure by Király, Zalatnai and Beck[17] with the torsion thermobalance of Szabo and Király up to 400–450° at the heating rate of 10° per minute.

(i) Recoura's violet-grey chloride   The findings with the violet chloride were:

$$[Cr(H_2O)_6]Cl_3 \rightarrow Cr(H_2O)Cl_3 + 5\ H_2O \quad (100–250°)$$
$$Cr(H_2O)Cl_3 \rightarrow Cr(OH)Cl_2 + HCl \quad\quad (250–330°)$$
$$Cr(OH)Cl_2 \rightarrow CrOCl + HCl \quad\quad\quad\ (340–400°)$$

(ii) Bjerrum's chloride   For Bjerrum's chloride:

$$[Cr(H_2O)_5Cl]Cl_2 \cdot H_2O \rightarrow Cr(H_2O)_5Cl(OH)Cl + HCl\ (\ 80–130°)$$
$$Cr(H_2O)_5Cl(OH)Cl \rightarrow Cr(OH)Cl_2 + 5\ H_2O \quad\quad (140–350°)$$
$$Cr(OH)Cl_2 \rightarrow CrOCl + HCl \quad\quad\quad\quad\quad\quad (350–440°)$$

(iii) Recoura's green chloride   For the green Recoura's chloride:

$$[Cr(H_2O)_4Cl_2]Cl \cdot 2H_2O \rightarrow Cr(H_2O)_4Cl_2Cl \cdot H_2O + H_2O\ (\ 60–110°)$$
$$Cr(H_2O)_4Cl_2Cl \cdot H_2O \rightarrow Cr(H_2O)_4Cl_2(OH) + HCl \quad (120–190°)$$
$$Cr(H_2O)_4Cl_2(OH) \rightarrow Cr(OH)Cl_2 + 4\ H_2O \quad\quad\quad (190–320°)$$
$$Cr(OH)Cl_2 \rightarrow CrOCl + HCl \quad\quad\quad\quad\quad\quad\quad\ (320–410°)$$

It is probable that the ultimate term in the final horizontal, near 800°, is the oxide Cr_2O_3. The oxychloride CrOCl is scarcely known at all.

5. Chromic sulphate

The violet chromic sulphate $Cr_2(SO_4)_3 \cdot 16.5–17H_2O$ was studied by Mlle Harmelin (unpublished researches). When heated at the rate of 100° per hour, it retains its water completely only up to 35–40°; it then loses 1.5–2 H_2O and, starting at 80°, it yields the hydrate carrying 14, or 14.5, or 15 H_2O. About 9 H_2O is then lost up to 180°, and finally 6 molecules of water are removed up to 500°. The latter doubtless come from OH groups bound to sulphur. The anhydrous sulphate is stable from 500° to around 570°; beyond this temperature it decomposes with loss of SO_3, and the residual Cr_2O_3 gives a horizontal beginning near 720°. See also[18].

6. Chrome alum

After Harmelin[19] established that chrome alum is a binary electrolyte, with chromium in the anion, Harmelin and Duval[20] showed by thermogravimetry and infra-red absorption that this alum should be represented by the formula

$$K[Cr(H_2O)_2(H_4SO_6)_2] \cdot 6H_2O,$$

the 6 molecules of so-called water of crystallization being probably coordinated to the potassium.

When the material is heated at the rate of 150° per hour, the thermolysis curve shows that six molecules of water are given off between 20° and 90°. Then four molecules of water are removed between 100° and 200° because two H_4SO_6 groups are converted into SO_4 groups. Two final molecules of water are removed very slowly between 200° and 450°; they were located at the 1–6 *trans*-positions of the Werner octahedron. The residue, "anhydrous alum" is likewise a complex $K[Cr(SO_4)_2]$. If this complex is heated more strongly, it loses sulphuric anhydride between 550° and 740°, leaving potassium sulphate and chromic oxide Cr_2O_3.

Chrome ammonium alum If chrome ammonium alum is heated under the same conditions, it yields an identical curve, except for the decomposition between 450° and 560° of the ammonium sulphate (and also its sublimation). Since the complex structure $K[Cr(SO_4)_2]$ no longer corresponds to the case of the chrome ammonium alum, the loss of the sulphuric anhydride occurs within temperature limits analogous to those applying to pure chromium sulphate. Harmelin[19] found that the residue from the heating of chrome ammonium alum consists entirely of chromic oxide Cr_2O_3.

Mauret and Vicaire[21] and Mauret and Vorsanger[22] have constructed the dehydration isotherms of these two alums under reduced pressure (10.75 mm and 14–17 mm of mercury, with the heating rates ranging from 80° to 300° per hour).

References p. 310

7. Chromic nitrate

When chromium(III) nitrate is heated at the rate of 300° per hour, it decomposes smoothly from ordinary temperature up to 680°. See also Wendlandt[23]. The formula of the original red salt closely approximates $Cr(NO_3)_3 \cdot 9H_2O$. The curve reveals no difference between the loss of water and the reddish vapours, but merely a slowing down, between 280° and 400°. This does not correspond to the anhydrous nitrate but rather to a green basic material whose formula approximates $Cr(OH)_2NO_3$.

8. Chromic phosphate

Carnot[24] reports that chromic phosphate has a hexahydrate at 100°. The heating curve is in accord with this statement but the alleged hydrate does not show any particular feature on the curve which slopes downward as far as 950°. Beyond this temperature, the anhydrous phosphate produces a horizontal. Therefore it is necessary to heat the specimen high enough to obtain a constant weight. Furthermore, the precipitate coagulates slowly and adheres to the container.

9. Chromium arsenates

The essential properties and existence regions of the chromium arsenates have been determined by Bastick and Guérin[25] and Guérin and Masson[26]. The findings have been obtained with 200-mg samples; the nitrogen used as the sweeping gas flowed at the rate of 500 ml per minute. The more acid of the chromium arsenates is the metaarsenate $Cr_2O_3 \cdot 3As_2O_5$, which loses As_2O_5 continuously. The pyroarsenate $2 Cr_2O_3 \cdot 3As_2O_5$ appears near 850° and from 950° on it decomposes rapidly yielding Cr_2O_3 directly without giving any evidence of the formation of chromium orthoarsenate $Cr_2O_3 \cdot As_2O_5$. These findings agree with those obtained by pyrolysis under reduced pressure[26].

10. Silver chromate

Silver chromate Ag_2CrO_4, prepared by the procedure of Gooch and Weed[27], is stable from 92° to 812°, between which temperatures there is a level that is strictly horizontal. Consequently, it is difficult to understand why Gooch and Weed specified that the precipitate be dried at precisely 135°. This silver chromate loses exactly one molecule of oxygen per mole from 812° to 945°, leaving a mixture of silver chromite and elementary silver. The latter can be removed by means of nitric acid. The decomposition of silver chromate in this manner represents a new reaction in our opinion; it may be formulated:

$$2 \, Ag_2CrO_4 \rightarrow 2 \, Ag + 2 \, O_2 + Ag_2Cr_2O_4$$

11. Mercurous chromate

The mercurous chromate method, quite properly recommended by Treadwell[5], is due to Fichter and Österheld[28]. Its use, like that of the preceding method, is limited to solutions which do not contain any anions except chromates and nitrates. These workers direct that the mercurous chromate be destroyed and the residual chromic oxide Cr_2O_3 be used as the final weighing form. We thought that there was little point in using this method unless advantage was taken of the high molecular weight of the compound, *i.e.* of the high atomic weight of mercury. Actually, the curve shows that mercurous chromate is stable up to 256° after it has been dried at up to 52°. Accordingly, we suggest weighing the anhydrous mercurous chromate produced between these two temperatures. In this case:

Weight of chromium = weight of precipitate × 0.100

The usefulness of this procedure in microgravimetry is therefore quite apparent.

Yosida[29] suggested 180° as the maximum temperature. The oxide Cr_2O_3 may be employed as the weighing form beyond 670°. The conversion of the mercurous chromate into the chromic oxide shows two singular points, one at 404°, the other at 604°. Apparently, they do not lend themselves to a simple interpretation.

12. Barium chromate

Regarding barium chromate see Chapter Barium.

13. Lead chromate

The precipitate of lead chromate $PbCrO_4$ was prepared in accord with the directions of Gibbs[30] who recommends drying at 100°, whereas Sarudi[31] prescribed 120° as the drying temperature. The curve we recorded indicates that anhydrous lead chromate is stable from 90° to 900°. Beyond this latter temperature it loses oxygen, but previously, and from

675° on, it gained weight slightly (1 mg per 400 mg). This is a fact which we have discovered with other compounds prior to their losing oxygen; they begin by taking it from the air. This increase in weight does not occur in a nitrogen atmosphere.

14. Chromic oxinate

The precipitate obtained with oxine and prepared by the method of Austin[32] has the formula $Cr(C_9H_6ON)_3$. After it has been dried, it may be weighed, between 70° and 150°. Therefore, the temperature 105–110° suggested by Austin is suitable. The oxine slowly decomposes between 156° and 360°; the disintegration then accelerates abruptly and the destruction is complete at 500°. The horizontal of Cr_2O_3 is obtained beyond this temperature. Because of the existence of these two parallel and very distinct horizontals, chromium oxinate is suitable for the automatic determination of this metal.

REFERENCES

[1] T. DUPUIS AND C. DUVAL, Compt. rend., 227 (1948) 772; Anal. Chim. Acta, 3 (1949) 345.
[2] A. JILEK AND J. KOT'A, Chem. Listy, 32 (1938) 30.
[3] C.A., 32 (1938) 5327.
[4] T. HANUS AND T. LUKAS, C.A., 6 (1912) 3249.
[5] F. P. TREADWELL, Manuel de chimie analytique Dunod, Paris, 1922, Vol. 2, p. 100.
[6] W. MANCHOT AND R. KRAUS, Ber., 37 (1906) 3512.
[7] M. DOMINÉ-BERGÈS, Compt. rend., 228 (1949) 1435.
[8] C.CABANNES-OTT, Thesis, Paris, 1958.
[9] F. TROMBE, Compt. rend., 215 (1942) 539; 216 (1943) 888; 225 (1947) 1156.
[10] W. SCHÖLLER AND W. SCHRAUTH, Chemiker-Ztg., 33 (1909) 1237.
[11] P. JANNASCH AND J. MAI, Ber., 26 (1893) 1786.
[12] C. FRIEDHEIM AND P. HASENCLEVER, Z. anal. Chem., 44 (1905) 614.
[13] R. RIPAN, Bul. soc. stiinte Cluj, 4 (1928) 28.
[14] E. SCHIRM, Chemiker-Ztg., 33 (1909) 1237.
[15] A. STOCK AND C. MASSACIU, Ber., 34 (1901) 467.
[16] C. DUVAL AND C. WADIER, Proc. 3rd Intern. Symposium on Reactivity of Solids, Madrid, 1957, p. 285; 20th Congr. G.A.M.S., 1957.
[17] D. KIRÁLY, K. ZALATNAI AND M. T. BECK, J. Inorg. & Nuclear Chem., 11 (1959) 170.
[18] I. MUZIKA AND O. SHOR, Zhur. Obshclei Khim., 26 (1956) 1863.
[19] M. HARMELIN, Diploma of Higher Studies, Paris, 1958.
[20] M. HARMELIN AND C. DUVAL, Compt. rend., 246 (1958) 1123.
[21] P. MAURET AND P. VICAIRE, Bull. Soc. chim. France, (1958) 1083.
[22] P. MAURET AND J. J. VORSANGER, Compt. rend., 246 (1958) 3450.
[23] W. W. WENDLANDT, Texas J. Sci., 10 (1958) 392.
[24] A. CARNOT, Bull. Soc. chim. France, 37 (1882) 482.
[25] M. BASTICK AND H. GUÉRIN, La chimie des hautes températures, 2e Colloq. natl. C.N.R.S., 1957, p. 139.
[26] H. GUÉRIN AND J. MASSON, Compt. rend., 245 (1957) 429.
[27] F. A. GOOCH AND L. H. WEED, Z. anorg. Chem., 59 (1908) 94.

28 F. FICHTER AND G. ÖSTERHELD, *Z. anorg. Chem.*, 76 (1912) 347.
29 Y. YOSIDA, *J. Chem. Soc. Japan*, 61 (1940) 130.
30 W. GIBBS, *Z. anal. Chem.*, 12 (1873) 309.
31 I. SARUDI, *Österr. Chemiker-Ztg.*, 43 (1940) 200.
32 E. T. AUSTIN, *Analyst*, 63 (1938) 710.

CHAPTER 25

Manganese

Dupuis, Besson and Duval[1] reported, in 1949, their findings regarding the pyrolysis of the precipitates which have been suggested for the gravimetric determination of manganese. Since then, an important investigation has been conducted by Brenet, Gabano and Seigneurin[2] to establish the individuality of manganese dioxides; to this we will add reports of some researches on the chloride, manganite, dithionate, sulphamate, rhodochrosite, formate, acetate, methyl- and phenyloxinates, picolinate, and the quinaldate. A reading of this chapter will clearly reveal that thermogravimetry by itself is not capable of resolving the problems of pyrolysis; it is well to use it along with other techniques such as infra-red absorption, thermomagnetism, and differential thermal analysis.

1. Manganese(II) oxide hydrate

Manganese hydroxide or hydrous oxide, used in analysis, is gelatinous and readily oxidizable. Its formula is $MnO \cdot xH_2O$. We prepared it by the method given by Fresenius[3] from manganous sulphate and potassium hydroxide in recently boiled water. The form of the curves change of course according to the quantity of water retained and the state of oxidation, but it should be pointed out that the oxide MnO can be obtained between 436° and 610° in a rather high state of purity even in contact with the air. Beyond this temperature, the curve slowly ascends but there is no level for the oxide Mn_2O_3. The salt-like oxide Mn_3O_4 ($2MnO \cdot MnO_2$) is gradually reached between 946° and 1000°. This latter temperature is suitable for bringing the system to constant weight. The Trombe method in which air or an inert gas charged with ammonia is passed into the manganese(II) solution to precipitate the hydrous oxide cannot be applied here since the manganese is not brought down completely in the presence of ammonium ions.

2. Manganese(III) oxide

When pure manganese(III) oxide Mn_2O_3 is heated in lots of 400 mg and at a rate of 300° per hour, it is stable up to 845–850°. It begins to decompose then to form Mn_3O_4, a reaction which is complete at around 950°.

3. Manganite

The manganite, which should preferably be written $MnO(OH)$ rather than $Mn_2O_3 \cdot H_2O$ as indubitably shown by the infra-red spectrum, has been found by Cabannes-Ott[4] to remain unchanged up to 550°. The horizontal starting at 615° corresponds to the oxide Mn_2O_3. The loss of weight in the neighbourhood of 950° reveals the slow change into Mn_3O_4.

4. Manganese(IV) dioxide

As early as 1908, Meyer and Rötgers[5] reported that manganese dioxide undergoes the double transformation:

$$\overset{530°}{MnO_2 \to} \overset{940°}{Mn_2O_3 \to} Mn_3O_4$$

Later, Honda and Stone[6] deduced the triple transformation from magnetic measurements:

$$\overset{535°}{MnO_2 \to} \overset{933°}{Mn_2O_3 \to} \overset{1160°}{Mn_3O_4 \rightleftharpoons} MnO$$

Saito[7] repeated this work by continuous heating on the Honda balance with an artificial product derived from carbonate that had been passed through a 200-mesh sieve; 1-g samples were heated at the rate of 1° per minute in a stream of air flowing at 100 ml per minute. The loss of weight begins at 350°, and becomes appreciable near 450°. The first level observed starts at 535° and doubtless is due to the oxide Mn_2O_3; it slopes slightly. The formation of the oxide Mn_3O_4 begins around 850° and is completed at 940°. When Saito carried the temperature to 1200° he found neither increase or decrease in weight, and consequently did not detect the oxide MnO. Later, Ukai[8] gave 950° for the complete conversion of MnO_2 into Mn_3O_4. Using the Chevenard thermobalance at the heating rate of 300° per hour, we[1] found 946°. However, we have not found the oxide Mn_2O_3 on any of our curves.

(i) Preparation methods   We have prepared hydrous manganese dioxide by all of the methods given up till now in the literature[1]. The curves obtained from the various pyrolyses are all similar. It has not been

possible to detect the hydrate $MnO_2 \cdot H_2O$ on any of the curves, nor the dry peroxide MnO_2, nor the oxide Mn_2O_3. All that can be stated is that although the oxide MnO_2 does not yield a horizontal, a change of direction is nevertheless observed around 100° when the retained or combined water has disappeared. The temperatures above which it is necessary to bring the precipitates to obtain Mn_3O_4 as the weighing form are much lower if MnO_2 is the starting material. These minimum temperatures are given in Table 8.

TABLE 8

PRECIPITATION OF HYDROUS MANGANESE DIOXIDE

Precipitated by	Minimum temp. (°C)	Reference
Potassium permanganate	600	9
Aqueous ammonia + hydrogen peroxide	622	10
Potassium chlorate	475	11
Bromine + yellow mercuric oxide	610	12
Potassium persulphate	608	13
Sodium hypochlorite	611	14
Bromine + potassium carbonate	606	15
Bromine + aqueous ammonia	538	16
Bromine + sodium acetate	538	17
Electrolysis of manganous sulphate:		
in presence of chrome alum	456	18
in presence of sulphuric acid	456	19

Accordingly, the oxide Mn_3O_4 appears at a low temperature when it originates from an anodic deposit or from a chloric oxidation. However, in general, it is necessary to count on a temperature of 550–600°.

From among the auxiliary curves which had to be traced for this study of manganese, we call special attention to the material prepared by Rollet[20] by means of an alternating current acting in contact with a nickel salt. It is entirely analogous to the preceding products; we have not distinctly observed any horizontal level applying to the presumed composition of a manganous acid. Although Rollet gave 400° for the transformation into Mn_3O_4, our curve clearly indicates 476° as the start of the horizontal related to the oxide Mn_3O_4.

(ii) Thermal transformations   Dubois[21] has made a close thermogravimetric study of the problem of the thermal transformations of the oxides of manganese, and has discovered and characterized the α-variety of MnO_2 as compared with pyrolusite or β-MnO_2 and the γ-non-stoichiometric variety. Brenet *et al.*[2,22] made an investigation in which differential thermal analysis, thermomagnetic analysis, and in some cases also X-ray analysis were used conjointly.

The study which interests us was made, in air, with the Chevenard thermobalance, and a heating rate of $180°$ per hour. In the cases of α- and γ-MnO_2, the levels corresponding to the $MnO_{1.55}$ and $MnO_{1.50}$ degrees of oxidation are especially distinct. The curve related to the β-MnO_2 showed only $MnO_{1.50}$ at $680°$. All three yielded $MnO_{1.33}$ at $1000°$. These workers have, in addition, traced the isothermal systems for α- and γ-MnO_2 for the following transformations:

$$\alpha\text{- or } \gamma\text{-}MnO_2 \text{ into } \beta\text{-}MnO_2 \text{ (between } 290° \text{ and } 450°)$$
$$\beta\text{-}MnO_2 \text{ (from } \gamma\text{-}MnO_2) \text{ into } MnO_{1.55} \text{ (between } 480° \text{ and } 680°)$$
$$MnO_{1.55} \rightarrow MnO_{1.50} \text{ (between } 750° \text{ and } 850°)$$
$$MnO_{1.50} \rightarrow MnO_{1.33} \text{ (or } Mn_3O_4) \text{ (between } 850° \text{ and } 1000°)$$

The β-variety of MnO_2 *(pyrolusite)* is the only one which accords with a stoichiometric formula and which is nearly inactive as a catalyst. The activation energy values of decomposition are perfectly reproducible. The manganese is entirely in the tetravalent state. Since the passage of the γ-form into the β-form occurs with no magnetic anomaly, it is normal to assume that in this form the manganese is already tetravalent. Since there are OH groups with hydrogen bridges, an electrostatic neutrality of the lattice will require the assumption of lacunas in Mn^{4+}. This is the reason why the Curie constant should be quite different at low and high temperature as has been observed.

What then can be said about the δ- and the ε-varieties of MnO_2?

5. Manganous chloride

Dubois[21] has studied the chloride $MnCl_2 \cdot 2H_2O$. This salt yields its monohydrate at $150°$ and the anhydrous chloride is obtained at $300°$. The oxides Mn_2O_3 and Mn_3O_4 appear beyond $650°$. The anhydrous chloride melts at this temperature and is strongly retained by these oxides even after washing. The composition of the resulting oxychlorides is too variable to warrant any belief in the existence of definite basic salts.

6. Manganous iodate

The iodate $Mn(IO_3)_2$ was prepared as directed by Minovici and Kollo[23] and washed with 95% alcohol. Its curve is much like those of the alkaline earth iodates and various bromates. Manganese(II) iodate slowly gains weight from room temperature up to 340°. This change, which is quite general among oxidants, is due to an uptake of oxygen and increases with increase in the surface in contact with the air. The gain amounts to 4 mg per 216 mg of the iodate at 342°. The above workers suggested that the salt be dried at 100° and at this temperature the gain in weight is scarcely 1 mg. The slow ascent of the curve precludes the use of the iodate for automatic determinations of manganese. The destruction of the iodate and of the transient periodate is complete at 880°; the oxide Mn_3O_4 is present beyond this temperature.

7. Manganous sulphide

The pink manganous sulphide is precipitated by ammonium sulphide in an ammoniacal medium. When heated at 300° per hour, a specimen of the initial product weighing 273 mg yielded a curve which reveals a rapid loss of moisture up to 40°. The weight of the sulphide remains almost constant up to 100°; then the sulphur gradually disappears yielding MnO but not quantitatively. The minimum point at 140° relates to a mixture of x MnS + y MnO, which is subsequently oxidized to $MnSO_4$ + Mn_3O_4. This oxidation proceeds rapidly from 154° to 360° and then continues very slowly up to around 500°. The anhydrous sulphate is stable over an interval of 100°. Its decomposition is well advanced at 750°; finally the oxide Mn_3O_4 is obtained near 950°. Ukai[8] also reported 950°.

8. Manganous sulphate

Manganous sulphate monohydrate $MnSO_4 \cdot H_2O$ or preferably $MnSO_5$ H_2 is stable up to 150°; this finding is in agreement with Guareschi[24]. The dehydration is entirely finished at 250° and the horizontal of the anhydrous sulphate $MnSO_4$ extends as far as 700°. The decomposition really begins in an observable manner at 940°, and at 1010°, the upper limit of our trials, we obtained a mixture of undecomposed sulphate and the oxide Mn_3O_4. See also Ukai[8], Ostroff and Sanderson[25], and Yamamoto and Bito[26].

9. Manganous dithionate

The dithionate $MnS_2O_6 \cdot 6H_2O$ has but little thermal stability; it loses water from ordinary temperatures upward to yield the tetrahydrate[27] and then a dihydrate which is stable up to 100°; the trihydrate does not appear. Then sulphur dioxide begins to come off and the very short horizontal of the monohydrate is found near 130°. The anhydrous sulphate is obtained near 200° and conforms to the curve of the sulphate described above.

11. Manganous sulphamate

Capestan[28] obtained the sulphamate (amidosulphonate) $Mn(NH_2-SO_3)_2 \cdot 4H_2O$ in crystalline form at 60°. It loses 2.5 molecules of water between 100° and 150°. The hydrate $Mn(NH_2-SO_3)_2 \cdot 1.5H_2O$ is not very stable and is marked only by an inflection. A product corresponding to about $^2/_3H_2O$ per molecule is obtained at 210°. The sulphate $MnSO_4$ is obtained around 480°, but first, as in the case of nickel sulphamate, there is chemical change of the hydrated salt into ammonium sulphate and manganous sulphate.

12. Manganous nitrate

Dubois[29] heated manganous nitrate $Mn(NO_3)_2 \cdot 6H_2O$; toward 300° it changes into impure dioxide with the approximate formula $MnO_{1.95}$. We have not been able to detect either the intermediate hydrates, or the anhydrous salt, nor of course the oxide MnO when we used 300° per hour or 150° per hour. The final product of the reaction at 700° is Mn_3O_4.

13. Manganous ammonium phosphate

The method of Strebinger and Pollak[30] was used to prepare manganous ammonium phosphate $MnNH_4PO_4 \cdot H_2O$. The pyrolysis curve is very similar to that of magnesium ammonium phosphate. The monohydrate is stable only up to 120–125°. Consequently, drying for 2 hours at 100° as recommended by Winkler[31] is appropriate. Ammonia and water vapour are lost, and the change in direction of the curve near 300° conforms rather well with the formation of the slightly stable $MnHPO_4$, which is quantitatively converted into the pyrophosphate $Mn_2P_2O_7$ between 610° and 700°. This finding is in agreement with Ukai[8]; see also Papailhau[32].

14. Manganous carbonate

The carbonate $MnCO_3$ is precipitated by treating a solution of a manganous salt with lithium, sodium, rubidium, potassium, or guanidinium carbonate. According to the critical study by Jilek and Lukas[33], the best results are given by potassium carbonate. The decomposition of the moist precipitate begins near 100° where there is a break in the curve. The oxide Mn_3O_4 starts to appear at 580° after a sudden fall in the curve from 540° on. There is no indication of the formation of the oxide Mn_2O_3. Other studies have been made by Ukai[8], Razouk, Mikhail and Habashy[34], Kissinger, McMurdie and Simpson[35], and Wanmaker and Verheyke[36].

Dialogite   Dialogite, natural manganous carbonate, also known as rhodochrosite, when heated in the thermobalance decomposes at 355° in nitrogen, at 520° in the air, and at 400° in argon, according to Caillère and Pobeguin[37] who made a differential thermal analysis of this mineral.

15. Manganese pyridine thiocyanate

The manganese tetrapyridine thiocyanate $Mn(C_5H_5N)_4(SCN)_2$ is due to Spacu and Dick[38] who suggested that it be dried in the desiccator. Actually, it begins to lose weight above 40° due to loss of pyridine. A short level, which is quite horizontal, extends from 84° to 92°, and signals the existence of the pyridine complex $(C_5H_5N)_2Mn(SCN)_2$, which is much less stable than the corresponding cadmium salt. There is simultaneous loss of cyanogen and pyridine. The experiment was not carried beyond 150° after all the pyridine has been eliminated.

16. Manganous formate

The formate $Mn(COOH)_2 \cdot 2H_2O$ was found by Guareschi[24] to lose its water of crystallization around 70° to yield the anhydrous salt. This compound has also been studied by Kornienko[39], Zapletal *et al.*[40], and by Doremieux and Boullé[41].

17. Manganous acetate

According to Burriel-Marti, Jimenez-Gomez and Alvarez-Herrero[42], the acetate $Mn(CH_3COO)_2 \cdot 1.5H_2O$ seems to lose its water near 100°, and yields an inclined level of the anhydrous compound between 100° and 300°. The oxide Mn_3O_4 appears around 320°. The curve has not been described accurately enough.

18. Manganous oxalate

Gibbs[43], Classen[44] and Fresenius[45] developed the method for precipitating the oxalate $Mn(C_2O_4)_2 \cdot 2H_2O$ by treating a concentrated solution of manganous sulphate with potassium oxalate in the presence of 96% alcohol and acetic acid. The dihydrate is stable only up to 54°; it then loses its two molecules of water up to 100°. The essentially horizontal level of the anhydrous salt extends from 100° to 215°. Carbon monoxide and carbon dioxide are given off suddenly with no formation of the carbonate. The oxide MnO is obtained at about 280° and is stable because of the reducing nature of the surrounding atmosphere. It gradually gains weight and is converted into the oxide Mn_3O_4. We recommend that the weighing form be prepared in the temperature interval 670–945°. See also Kawagaki[46], Erdey and Paulik[47], Doremieux and Boullé[41].

19. Manganous anthranilate

The procedure of Funk and Demmel[48] was followed for the preparation of the anthranilate $Mn(C_7H_6O_2N)_2$. When this precipitate is heated, it loses its imbibed water up to 70° and then is anhydrous and stable up to 200–220°. The corresponding level may be used for the automatic determination of manganese. The anthranilate decomposes abruptly at 260° and a constant weight of the oxide Mn_3O_4 is present from 750° on.

20. Manganous oxinate

Neelakantam[49] prepared the oxinate $Mn(C_9H_6ON)_2 \cdot 2H_2O$ between pH 5.7 and 10.0. It is orange-yellow and does not darken in the air. It is stable up to 117° according to Borrel and Pâris[50] and yields the anhydrous chelate characterized by the good horizontal between 188° and 320°. We recommend this level for the automatic determination of manganese. If the heating is continued, the organic matter is destroyed with an almost linear decrease in weight; then a level appears which does not correspond precisely to Mn_3O_4 because, as has been established previously by other workers, the tar partially reduces this oxide. A more correct formula results if the residue is heated isothermally for 2 hours in the vicinity of 900°.

21. Manganous methyloxinate

Bocquet and Pâris[51] found that the methyloxinate $Mn(CH_3-C_9H_5ON)_2$ $\cdot H_2O$ starts to lose its water at 125°.

References p. 320

22. Manganous 2-phenyloxinate

Bocquet and Pâris[51] prepared the light brown 2-phenyloxinate $Mn(C_6H_5-C_9H_5ON)_2 \cdot H_2O$. It is stable up to 60°. The horizontal of the anhydrous complex extends from 90° to 315°.

23. Manganous picolinate

The picolinate or pyridine-2-carboxylate $Mn(C_6H_4O_2N)_2 \cdot 2H_2O$ was prepared by Thomas[52]. It starts to lose water at 100°.

24. Manganous quinaldate

The quinaldate or quinoline-2-carboxylate $Mn(C_{10}H_6O_2N)_2 \cdot 2H_2O$ was prepared by Thomas[52]. This yellowish-white compound is stable up to 135°. There is a break in the pyrolysis curve near 432°. When 300 mg are heated at 180° per hour, the horizontal of the oxide Mn_3O_4 starts at 604°. The residue should be kept at 700° for 2 hours.

REFERENCES

[1] T. DUPUIS, J. BESSON AND C. DUVAL, Anal. Chim. Acta, 3 (1949) 599.
[2] J. P. BRENET, J. P. GABANO AND M. SEIGNEURIN, 16th Intern. Congr. Pure and Appl. Chem., Paris, 1957, Mineral Chem., p. 69.
[3] R. FRESENIUS, Z. anal. Chem., 11 (1872) 295.
[4] C. CABANNES-OTT, Thesis, Paris, 1958.
[5] R. T. MEYER AND K. RÖTGERS, Z. anorg. Chem., 57 (1908) 104.
[6] K. HONDA AND T. STONE, Sci. Repts. Tôhoku Imp. Univ., 3 (1914) 139.
[7] H. SAITO, Sci. Repts. Tôhoku Imp. Univ., 16 (1927) 47.
[8] T. UKAI, J. Chem. Soc. Japan, 52 (1931) 461.
[9] E. DONATH, Ber., 14 (1881) 982; Z. anal. Chem., 22 (1883) 245.
[10] G. ROSENTHAL, Z. anal. Chem., 17 (1878) 364.
[11] F. BEILSTEIN AND L. JAWEIN, Ber., 12 (1879) 1528, 2096.
[12] J. VOLHARD, Ann., 198 (1879) 360.
[13] L. BARZAGHI, Anais assoc. quím. Brasil, 2 (1943) 187.
[14] E. REICHARDT, Z. anal. Chem., 5 (1866) 61.
[15] V. I. PETRASHEN, C.A., 35 (1941) 1346.
[16] C. MEINEKE, Lehrbuch der chemischen Analyse, 1896, Vol. 1, p. 492.
[17] V. EGGERTZ, Z. anal. Chem., 7 (1868) 495.
[18] E. WOHLMANN, Z. anal. Chem., 89 (1932) 321.
[19] J. BESSON, Ann. Chim. (Paris), 2 (1947) 578.
[20] A. P. ROLLET, Compt. rend., 189 (1929) 34.
[21] P. DUBOIS, Compt. rend., 198 (1934) 1502; Thesis, Paris, 1935.
[22] J. BRENET AND A. GRUND, Compt. rend., 240 (1955) 1210.
[23] S. MINOVICI AND C. KOLLO, Chim. & Ind., (Paris), 8 (1922) 499.
[24] I. GUARESCHI, Gazz. chim. ital., 49 (1919) 134.
[25] A. G. OSTROFF AND R. I. SANDERSON, J. Inorg. & Nuclear Chem., 9 (1958) 45.
[26] K. YAMAMOTO AND K. BITO, Bull. Dept. Appl. Chem. Waseda Univ., 9 (1929) 12.

27 J. SCHREIBER, *Ann. Chim. (Paris)*, 1 (1934) 88.
28 M. CAPESTAN, *Ann. Chim. (Paris)*, 5 (1960) 225.
29 P. DUBOIS, *Ann. Chim. (Paris)*, 5 (1936) 432.
30 R. STREBINGER AND J. POLLAK, *Mikrochemie*, 4 (1926) 16.
31 R. WINKLER, *Z. angew. Chem*, 35 (1922) 234.
32 J. PAPAILHAU, *Bull. soc. franç. minéral.*, 82 (1959) 371.
33 A. JILEK AND J. LUKAS, *Collection Czechoslov. Chem. Communs.*, 3 (1931) 187.
34 I. RAZOUK, R. S. MIKHAIL AND G. M. HABASHY, *Egypt. J. Chem.*, 1 (1958) 223; *C.A.*, (1959) 4874.
35 H. KISSINGER, H. MCMURDIE AND B. S. SIMPSON, *J. Am. Ceram. Soc.*, 39 (1956) 168.
36 W. L. WANMAKER AND M. L. VERHEYKE, *Philips, Research Repts.*, 11(1956).1
37 S. CAILLÈRE AND T. POBEGUIN, *Bull. soc. franç. minéral.*, 83 (1960) 36.
38 G. SPACU AND J. DICK, *Bull. soc. stiinte Cluj.*, 4 (1929) 431; *Z. anal. Chem.*, 74 (1928) 188.
39 P. KORNIENKO, *Referat. Zhur. Khim.*, (1956), *Abstr. No.* 38984.
40 V. ZAPLETAL, J. JEDLIČKA AND V. RÜŽIČKA, *Chem. Listy*, 50 (1956) 1406; *C.A.*, (1957) 2438.
41 J. L. DOREMIEUX AND A. BOULLÉ, *Compt. rend.*, 250 (1960) 3184.
42 F. BURRIEL-MARTI, S. JIMENEZ-GOMEZ AND C. ALVAREZ-HERRERO, *Anales edafol. y fisiol. vegetal. (Madrid)* 14 (1955) 220.
43 W. GIBBS, *Am. J. Sci.*, 22 (1856) 214.
44 A. CLASSEN, *Z. anal. Chem.*, 16 (1877) 315, 470.
45 R. FRESENIUS, *Z. anal. Chem.*, 11 (1872) 415.
46 K. KAWAGAKI, *J. Chem. Soc. Japan*, 72 (1951) 1079.
47 L. ERDEY AND F. PAULIK, *C.A.*, 50 (1956) 3952.
48 H. FUNK AND M. DEMMEL, *Z. anal. Chem.*, 96 (1934) 385.
49 K. NEELAKANTAM, *Current Sci. (India)*, 10 (1941) 21.
50 M. BORREL AND R. PÂRIS, *Anal. Chim. Acta*, 4 (1950) 267.
51 G. BOCQUET AND R. PÂRIS, *Anal. Chim. Acta*, 14 (1956) 201.
52 G. THOMAS, *Thesis*, Lyon, 1960.

CHAPTER 26

Iron

We are completing the critical study made by Duval and Nguyen Dat Xuong[1] of the precipitates serving for the gravimetric determination of iron by investigations of the oxidation of the metal, of some minerals, and certain soluble salts employed as titrimetric standards.

We have omitted the very early method of Fuchs[2] in which copper is precipitated from a solution of its sulphate by the iron being determined. The method is not accurate. Moreover, we have not tried the precipitations with α- and β-naphthalenesulphinic acids as proposed by Krishna and Singh[3] since they employed these reagents only for a titrimetric determination and also because Dubsky, Gravec and Langer[4] have been critical of the method. We, ourselves, have obtained not more than slightly favourable results with benzenesulphinic acid, not only with iron but also with thorium, because of the ease with which these precipitates oxidize. On the other hand, we have added to the procedures already employed, the precipitation of iron hydroxide by means of a stream of air charged with ammonia vapours as suggested by the Trombe[5] technique, and this ordinarily has given us excellent results. We likewise were the first to report a finding which was important as well as practical, namely the precipitation of the cupferronate and the neocupferronate.

1. Metallic iron

(i) Oxidation of metallic iron   Bénard and Talbot[6] employed the thermobalance to study the kinetics of the oxidation reaction of metallic iron. The recording cylinder had a high speed of rotation (0.4 mm per second at the periphery). They used strips with comparatively large surface (14 cm^2) and slight thickness (0.3 mm). The curves obtained under these conditions between 850° and 1050° may be broken down into three successive periods: (1) an initial period of around 20 seconds with these specimens corresponding to the establishment of the thermal equilibrium; (2) a linear period likewise lasting around 20 seconds; (3) a period,

parabolic in shape. Hence there is a transient period in which the speed of taking up oxygen is constant and independent of the law governing the diffusion of the elements of the reaction. The slope of the linear portion which represents the actual speed of oxidation of the iron, not inhibited by the diffusion, undergoes an important increase when the temperature rises. The variation curve shows a pronounced discontinuity near 900°, the transformation temperature of $\alpha \rightleftharpoons \gamma$ iron, and the rate of oxidation increases proportionally more in the α-zone than in the γ-zone.

(ii) Oxidation of Armco iron   De Carli and Collari[7] followed the oxidation in a simplified thermobalance; the specimen of Armco iron was taken to 445°, 480°, 590°, 600°, 700°, and 800°. Every thirty minutes, they determined the weight of oxygen taken up in order to learn the variations of K as a function of $1/T$ and from these variations to deduce the chemical nature of the superficial oxide coating.

(iii) Other oxidations   Hubner-Kossan[8] made a study with the thermobalance of the oxidation of reduced iron, piano wire, cast iron, and also of ferrous sulphate and hydrous ferrous oxide.

2. Ferric oxide hydrate

Hydrous ferric oxide has been precipitated in a number of ways which lead to such formulas as $Fe_2O_3 \cdot nH_2O$ with the water bound more or less firmly. To find OH groups, it is necessary to investigate minerals or certain kinds of rust.

(i) Precipitation by aqueous ammonia   The precipitation by means of aqueous ammonia was carried out as directed by Treadwell[9] and the curve shows a rapid loss of water of imbibition up to around 105°. The loss of water is slower after the break and the sesquioxide Fe_2O_3 is present in pure form starting at 470°. The infra-red spectrum of this oxide no longer shows bands due to water, however bands due to carbonates are present. See also Paulik and Erdey[10], Kato *et al.*[11].

(ii) Precipitation by gaseous ammonia   When the method of Trombe[5] was used, the iron in the solution being analyzed was brought to the trivalent state, the pH was adjusted to 2, and the solution then treated in the cold with a slow current of air charged with ammonia gas. The precipitation sets in at about pH 2.2 if around 50 mg of oxide are present

in 150 ml of water; it is complete after 30 minutes and the product may be filtered at once on paper or on a No. 3 fritted glass crucible. The curve is very similar to the preceding one. There is a break near 111° and the bound water is given off smoothly from 111° to 474° without showing any level for the hydroxide. The horizontal of the sesquioxide starts at 474°. Accordingly, the difference in the two variants (i) and (ii) is much less pronounced than in the case of the precipitation of alumina.

(iii) Precipitation by ammonia and ethanol We followed the method of Davidson[12] exactly when using aqueous ammonia and ethanol, a procedure which is recommended fully for obtaining a precipitate of hydrous ferric oxide easy to filter. The curve closely resembles the preceding ones, but the horizontal of the Fe_2O_3 starts as low as 370°.

(iv) Precipitation by ammonium nitrite The procedure employed with ammonium nitrite was that of Schirm[13] except that ready-made ammonium nitrite was replaced by an equimolecular mixture of sodium nitrite and ammonium chloride. Again the form of the curve is identical with that of the two preceding curves; the sesquioxide appears in a pure state from 345° on.

(v) Precipitation by ammonia and hydrazine If the ferric solution under analysis is treated with 10% ammonium hydroxide and several drops of hydrazine, as proposed by Schirm[14], the precipitate settles immediately. It may be filtered off at once; it dries very fast and the horizontal of the Fe_2O_3 begins at 346°. We have quite properly chosen this procedure for the automatic determination of iron.

(vi) Precipitation by iodide–iodate A precipitate of hydrous ferric oxide which is readily filtered is likewise obtained by the classic method of Moody[15] and Glassmann[16] in which an iodide–iodate mixture is the precipitating agent. This is one of the most ingenious means of gradually lowering the H^+ ion content of a solution. The curve shows a distinct break at 103° signaling the end of the loss of the loosely bound water. There is no horizontal corresponding to the hydroxide, but instead a slowly descending branch between 103° and 400°. The dehydration then slowly accelerates, and beginning at 600° the residue consists of pure Fe_2O_3.

(vii) Precipitation by mercury chloroamide The procedure due to

Solaja[17] employs mercury chloroamide NH_2HgCl. The ferric salt being determined is treated with an excess of the reagent. To us, this application seems more interesting than in the case of $Al_2O_3 \cdot nH_2O$. The precipitate forms at once, settles immediately, and may be filtered off after several minutes. Its curve shows a rather rapid loss of water up to 143°, and then a still more rapid departure of mercury up to 384°, where an excellent horizontal due to the Fe_2O_3 begins. This procedure may be recommended for ordinary gravimetric use but not for automatic determinations.

(viii) Precipitation by pyridine If, in line with the directions given by Mendez and Pinto[18], pyridine is used as the precipitant in place of aqueous ammonia, the precipitate yields a curve which slopes downward until 620° is reached, the temperature which marks the beginning of the horizontal due to the Fe_2O_3. This procedure appears to have little interest in comparison with the preceding methods.

3. Göthite

Göthite, which was heated at the rate of 300° per hour by Cabannes-Ott[19], and by Loisel[20], and Claisse *et al.*[21], is a definite hydroxide FeO(OH), which is stable up to 340° $\pm$ 5°. The dehydration is complete at 500° when this mineral is heated at the rate of 300° per hour.

4. Lepidocrocite

Loisel[20] obtained an identical curve with lepidocrocite, with rusts prepared in distilled water, and in Vittel water*, and with deuterogöthite and deuterolepidocrocite FeO(OD).

5. Ferrous chloride

Ferrous chloride $FeCl_2 \cdot 4H_2O$, investigated by Shen-Chia-Shu and Chang-Miu-Shiao[22], is stable up to 103°. It then yields the dihydrate up to 110°, which in turn gives the monohydrate around 158°. The final molecule of water is removed near 212°. This study was supplemented by a differential thermal analysis.

6. Ferrous sulphide

The ferrous sulphide method, included in various of the older books,

---

* A famous mineral water from Vittel (Département Vosges, France).

References p. 334

requires that the black sulphide obtained initially be heated in a stream of hydrogen, or that it be heated in contact with the air followed by weighing of the resulting ferric oxide. The curve given by the sulphide shows a loss of sulphur and water up to 180°; the resulting mixture of sulphide and oxide then takes up oxygen. The transformation into sulphate appears to be complete at 610°, and the latter begins to decompose around 690°, losing sulphur dioxide and oxygen. Only the sesquioxide Fe_2O_3 remains from 800° on. There is no simple weight relationship between the sulphate formed and the sulphur of the initial precipitate. We recommend that this method be discarded.

7. Pyrites

Iron pyrites has been heated by Saito[23], Gill[24], Truchot[25], etc. From the 28 curves he traced, Saito concluded that the primary reaction may be expressed:

$$2\ FeS_2 + 5\ O_2 \rightarrow 2\ FeO + 4\ SO_2$$

and the temperature at which it begins depends on the size of the grains, namely 410° below 0.074 mm, 440° for 0.295–0.417 mm, 460° for 1.168–2.362 mm. Immediately following this reaction, there is production of Fe_2O_3 and $FeSO_4$; the percentages of these products depends on the size of the grains and the heating rate. The slower the heating, the greater the amount of sulphate.

(i) Pyrrhotine Saito[23] has obtained equally interesting results from pyrrhotine. The primary reaction is the formation of sulphate near 360°, which continues though very slowly in the neighbourhood of 400°. The oxidation becomes very intense at 400–410°:

$$a\ FeS + b\ O_2 \rightarrow c\ FeSO_4 + d\ Fe_2O_3 + e\ SO_2$$

The formation of the sulphate is otherwise more complex than in the case of pyrites. The fritting begins at 430°. Rapid heating must be employed if the sulphur is to be completely removed or else a considerable quantity of sulphate will be formed.

(ii) Marcasite Marcasite was likewise heated by Saito[23]. It yields thermolysis curves of the same form but decomposes at lower temperatures than pyrites. For example, the apparent decomposition point is 370° with 0.074 mm grains heated at 1°/min, or also at 2°/min, whereas grains from 0.295–0.417 mm give 445° at 2°/min, etc.

8. Ferrous sulphate

The perfectly green specimens of ferrous sulphate taken from a fresh bottle have been found by Duval[26], Saito[23], Gill[24], and Ostroff and Sanderson[27] to carry not 7 but close to 6 molecules of water. The thermolysis curve shows that the loss of weight occurs from room temperature on. The molecules of water are not removed all at one time and the shape of the curve depends greatly on the rate of heating. In general, it may be said that 3 molecules leave between 26° and 96°, already with partial oxidation, and that 3 more molecules disappear between 100° and 140°. (Several workers who preceded us reported 110–120° or 160°.) The final molecule of water is lost at the same time that a deep-seated change occurs. The latter does not result in anhydrous ferrous sulphate nor in anhydrous ferric sulphate, but nevertheless a very distinct horizontal is observed. By heating linearly at 1° per minute, Saito found 540–550° for the existence range of this basic sulphate. The latter, which had been described previously, may be written $Fe_2O_3 \cdot 2SO_3$. It then loses sulphur trioxide and the level of the oxide Fe_3O_4 begins above 720°.

9. Ferric sulphate

Ferric sulphate, whose curve has been traced in order to interpret the curves of Mohr's salt and ferric ammonium alum, begins to decompose around 600°:

$$Fe_2(SO_4)_3 \rightarrow Fe_2O_3 + 3 SO_3$$

However, since the reaction is not conducted in vacuo or in a highly reducing milieu, but in the air, the final product, above 750°, is the oxide Fe_3O_4. See also Saito[23].

10. Mohr's salt (ferrous ammonium sulphate)

Mohr's salt $FeSO_4 \cdot (NH_4)_2SO_4 \cdot 6H_2O$ as taken from the bottle shows no loss of moisture on our recordings. It is not necessary to go beyond 80° to dry it. The loss of 6 molecules of water is complicated by the departure of a little ammonia and an oxidation of iron, to such an extent that between 192° and 310° there results $NH_4[Fe(SO_4)_2]$, corresponding in a way to the anhydrous alum, a triple operation which proceeds up to about 450°. It is not possible to obtain levels that are strictly horizontal. Between 450° and 510°, two molecules of the above complex share the loss of one molecule of ammonium sulphate. There remains then a new level extending from 510° to around 600° and corresponding to ferric sulphate $Fe_2(SO_4)_3$ (see above) which, in addition, is not strictly pure.

11. Ferric ammonium alum

Ferric ammonium alum may not be weighed directly since it loses water from room temperature on. Consequently, assuming that the formula is $NH_4[Fe(H_2O)_2(SO_6H_4)_2]\cdot 6H_2O$, we do not find, as in the case of chrome alum, three successive levels for the loss of the 6 molecules of water of crystallization, and 4 molecules of water linked to the sulphur, and 2 molecules of water of the complex. When the heating rate is 300° per hour, this water is lost in a continuous manner up to around 210°. The beginning of a rectilinear but not horizontal level agrees rather well with the formula of the anhydrous alum $NH_4[Fe(SO_4)_2]$; then beyond 435°, this new complex in its turn breaks down with loss of ammonium sulphate, and as in the preceding instances we arrive at ferric sulphate. The latter decomposes between 750° and 770° to yield the oxide Fe_3O_4.

12. Ferrrous ethylenediamine sulphate

Ferrous ethylenediamine sulphate $FeSO_4.(CH_2-NH_3)_2SO_4.4H_2O$ has been suggested as a primary standard for redox titrations. Its thermolysis has been investigated by Wendlandt and Ewing[28]. This tetrahydrate may be heated in the air between 105° and 110° without loss. The anhydrous salt is stable from 238° to 295°. At higher temperatures the iron(II) sulphate yields the oxide.

13. Ferrous sulphamate

Ferrous sulphamate (amidosulphonate), prepared with exclusion of air and dried at 60° in a stream of nitrogen, contains between 4 and 5 molecules of water according to Capestan[29]. It loses about one-half of its water below 110° and forms the dihydrate, which in turn gives the anhydrous salt from 160° to 195°. The thermolysis with production of sulphate then proceeds from 275° to 445°.

14. Ferric nitrate

Duval[30] found that ferric nitrate loses its water of crystallization from 38° on and consequently it is nearly impossible to prepare a standard solution of this salt by direct weighing. It begins to decompose at 120°, giving off reddish vapours, and the reaction even becomes explosive around 144°. It then abates near 160° and the final traces of nitrogen oxides are gradually evolved. The resulting Fe_3O_4 is obtained around 310° and the weight stays constant up to 1000°. Although it appears to have a hygroscopic look, the initial salt, according to the calculations made from the graph, contains approximately 9 molecules of water (molecular weight 392 as opposed to theoretical 398). See also Wendlandt[31].

15. Ferric phosphate

We have adopted Mohr's method[32] for the preparation of ferric phosphate $FePO_4$, *i.e.* precipitation by triammonium phosphate in the presence of ammonium acetate. The thermolysis curve indicates that after the water and the ammonium salts have been driven off, the expected phosphate has constant weight from 530° on.

16. Mispickel

Saito[23] has traced five curves for samples of mispickel (arsenopyrite) from Kap-san, Korea. The mineral, FeAsS, oxidizes rapidly, beginning at 500°, with sublimation of arsenious anhydride. This latter compound is obtained best by rapid oxidation at 600°. If the specimen is heated slowly and to a low temperature, iron arsenate appears and lessens the sublimation of the anhydride. No decrepitation has been observed.

17. Siderite

Siderite $FeCO_3$ has been investigated by Schwob[33] and by Caillère and Pobeguin[34]. This mineral begins to decompose at 425° in air, and at 475° in nitrogen. The study was supplemented by differential thermal analysis.

18. Ferrous formate

Ferrous formate was investigated by Kornienko[35] and Doremieux and Boullé[36]. When 250 mg were heated at 60° per hour in oxygen, the promoter effect of this gas was exhibited. The method of preparation and the age of the product apparently have an influence. Thus a freshly prepared specimen or one kept for 15 days in vacuo (the salt oxidizes in the air) shows a fall from 280° to 260° and from 320° to 300° in an atmosphere of nitrogen, or an increase from 150° to 185° and from 180° to 230° in an oxygen atmosphere. The sample behaved like an aged specimen when prepared by the action of iron(II) sulphate on barium formate.

19. Basic ferric formate

Basic iron formate $Fe(OH)_2(HCO_2)$, which is employed in gravimetric analysis, is conveniently prepared by the method outlined by Funk[37]. The precipitate filters very quickly and the curve shows a rapid decrease in weight up to 182°, corresponding to loss of water and carbon monoxide. Slight traces of water and ammonia are given off from 182° to 300°. The horizontal of Fe_2O_3 begins at 300°. We recommend this excellent method for the automatic determination of iron.

20. Ferrous oxalate

Kornienko[35] and Doremieux and Boullé[36] also studied ferrous oxalate FeC_2O_4. They observed a minimal temperature of 195° at which the decomposition invariably occurred in two stages.

21. Basic ferric acetate

We have reproduced the method of determination by means of the basic acetate, which is a well known procedure for separating iron and aluminium, following the directions of Borck[38]. The precipitate, which filters well, loses water rapidly up to 155°; then the acetic radical is decomposed from 155° to 242°. The strictly horizontal level of the sesquioxide Fe_2O_3 begins beyond 242°. We also suggest this method for the automatic determination of iron.

22. Ferric cupferronate

The cupferronates of iron and copper are the most stable of all the known salts of this class, but the various workers such as Ferrari[39] and Lundell[40] have not taken this fact into account and they direct that the cupferronates be decomposed by heat and thus converted into oxides. The curve given by iron cupferronate has a perfectly horizontal level up to 100°. In the light of this finding, we suggest, on one hand, a rapid method of separating iron and titanium (the thermolysis curve of titanium cupferronate descends sharply from 40° on) and, on the other hand, an automatic determination of trivalent iron. The iron cupferronate decomposes rapidly beyond 100° and the horizontal of the sesquioxide Fe_2O_3 begins at 610°.

23. Ferric neocupferronate

Likewise, neocupferron gives a precipitate with trivalent iron, by the Bamberger and Baudisch[41] method. The product is perfectly stable up to 111–113°. The corresponding horizontal may be applied for the automatic determination of iron. The organic matter decomposes from 111° to 750° and disappears. The horizontal of Fe_2O_3 starts beyond 750°.

24. Ferric-N-benzoylphenylhydroxylamine

The brown-red precipitate produced with N-benzoylphenylhydroxylamine was prepared as directed by Chandra Shome[42] taking the numerous precautions indicated by this worker. The product loses water continuously up to 434°. In particular, at 110°, which this worker prescribed for drying, the curve slopes rapidly downward, and there is a sudden release of gas at 126°. The horizontal due to Fe_2O_3 begins above 434°.

25. Ferric α-nitroso-β-naphthol

The precipitate with α-nitroso-β-naphthol, first described by Ilinski and von Knorre[43], was prepared as directed by the Treadwell text[9]. This complex is stable up to 47°; it then decomposes slowly up to 164° where, as usual, there is a sudden evolution of nitric oxide NO. The organic matter then burns up to 630°, the temperature marking the start of the Fe_2O_3 horizontal. This finding shows that much too high temperatures are employed when this method is ordinarily used for determining iron.

26. Ferric oxinate

A number of workers have described the precipitation of $Fe(C_9H_6ON)_3$ by the action of oxine in acetic acid solution buffered with sodium acetate. These workers include Berg[44], Tsinberg[45], Zanko and Butenko[46], Raskin[47], and Borrel and Pâris[48]. The black product is stable up to 284° at least, which is remarkable for an organic chelate. The horizontal corresponding to this product is well suited to the automatic determination of iron. After the organic portion of the molecule has been destroyed, the oxide Fe_2O_3 is obtained near 900°.

27. Ferric 5,7-dichlorooxinate

The precipitation of the 5,7-dichlorooxinate, as described by Fresenius[49], and its filtration are slow and tedious. Ishimaru[50] found a horizontal extending from 112° to 198° for the chelate $Fe(C_9H_4Cl_2ON)_3$, and showed that Fe_2O_3 is obtained beyond 478°. We have been unable to reproduce these findings. Our curves always slope downward until 555° which marks the beginning of the level due to Fe_2O_3. We suggest that this method for the gravimetric determination of iron be discarded.

28. Ferric 5,7-dibromooxinate

Berg [44, 51] reported that the precipitate obtained with 5,7-dibromooxine filters well and that it can be immediately collected on a No. 4 crucible. However, it must be carefully washed with 10% acetone to remove any excess reagent. The chief disadvantage of this method is the necessity of working with very large volumes. On the other hand, it is the method of choice for microanalysis. Berg directs that the precipitate be dried at 120–140°, while Ishimaru[50] prescribes 118–140°. However, if the precipitate is washed as just described, it is perfectly dry from 72° on. The corresponding horizontal is suitable for the automatic determination of iron.

29. Ferric 5,7-diiodooxinate

Wendlandt[52] studied the precipitate produced by the action of alleged 5,7-diiodooxine. The resulting material is stable below 100°, if 100–150 mg are heated at the rate of 300° per hour. The oxide Fe_2O_3 appears above 410°.

30. Ferric benzenesulphinate

We did not use benzenesulphinic acid, which produces only incomplete precipitation, but its sodium salt gives quantitative results and a very filterable product. However, the curve shows that the latter oxidizes very quickly starting at 50° and yields a ferric sulphonate by taking up atmospheric oxygen. Accordingly, if this procedure is to be employed, it is necessary to wash the precipitate with water and ether, and then dry the product in the cold under reduced pressure. See Hecht and Donau[53].

The destruction of the sulphinate–sulphonate mixture occurs abruptly at 164° and the organic matter burns up to 875°, where the horizontal of the Fe_2O_3 begins.

31. Ferric dimine

Dimine, or the dithiocarbamate of cyclohexylethylamine, or Vulkacite 774, yields a voluminous precipitate with iron. Its discoverer Herrmann-Gurfinkel[54] decomposes this precipitate thermally and converts it into Fe_2O_3. In contrast to the curve given by the copper compound, in the present case there is a horizontal up to 115° which may be used for the automatic determination of iron. The molecular weight on the basis of 56 for iron is 1422, and the gravimetric factor is accordingly 0.039. The complex is completely decomposed at 640° yielding metallic iron. The latter slowly reoxidizes and the oxide Fe_2O_3 is obtained quantitatively from 960°.

32. Ferric benzothiazolecarboxylate

Yoshikawa and Shinra[55] found that the monohydrated iron chelate with benzothiazolecarboxylic acid is stable up to 150° and loses water between 150° and 160°.

33. Ferric picolinate

The iron salt of picolinic acid (pyridine-2-carboxylic acid) crystallizes with 4 molecules of water, which it loses between 75° and 145°.

34. Ferric quinaldate

The quinaldate (quinoline-2-carboxylate) crystallizes with 2 molecules of water; one of these is lost at 50° and the resulting monohydrate is stable up to 220°. (Thomas[57]).

35. Ferric p-butylphenylarsonate

We followed the directions of Craig and Chandlee[56] when preparing the iron salt of p-butylphenylarsonic acid. The precipitate tenaciously retains the reagent which does not wet readily, and consequently it is necessary to subject the product to a prolonged washing with hot water. After the customary loss of water, there is a level that is not horizontal between 80° and 242°, and consequently it is not feasible to weigh the ferric butylphenylarsonate. The destruction of the organic matter proceeds rapidly and the horizontal of the Fe_2O_3 is reached from 434° on. The method may serve as a separation procedure at most.

REFERENCES

1 C. Duval and Ng Dat Xuong, *Anal. Chim. Acta*, 5 (1951) 160.
2 J. N. Fuchs, *J. prakt. Chem.*, 17 (1839) 160.
3 S. Krishna and H. Singh, *J. Am. Chem. Soc.*, 50 (1928) 792.
4 J. V. Dubsky, E. Gravec and A. Langer, *Chem. Obzor.*, 12 (1937) 415; *C.A.*, 31 (1937) 5712.
5 F. Trombe, *Compt. rend.*, 215 (1942) 539; 216 (1943) 888; 225 (1947) 1156.
6 J. Bénard and J. Talbot, *Compt. rend.*, 226 (1948) 912.
7 F. de Carli and N. Collari, *Ann. Chim. (Rome)*, 40 (1950) 117.
8 E. Hubner-Kossan, *Thesis*, Sarrebruck, 1957.
9 F. P. Treadwell, *Manuel de chimie analytique*, French Ed., Dunod, Paris, 1922, Vol. 2, p. 87.
10 F. Paulik and L. Erdey, *Acta Chim. Acad. Sci. Hung.*, 13 (1957) 117.
11 H. Kato, H. Hosimija and S. Nakazima, *J. Chem. Soc. Japan*, 60 (1939) 1115.
12 A. Davidson, *Analyst*, 69 (1944) 374.
13 E. Schirm, *Chemiker-Ztg.*, 33 (1909) 877.
14 E. Schirm, *Chemiker-Ztg.*, 35 (1911) 897.
15 E. Moody, *Z. anal. Chem.*, 46 (1907) 247.
16 B. Glassmann, *Ber.*, 39 (1906) 3368.
17 B. Solaja, *Chemiker-Ztg.*, 49 (1925) 337.
18 J. V. Mendez and R. A. Pinto, *C.A.*, 34 (1940) 6190.
19 C. Cabannes-Ott, *Thesis*, Paris, 1958.
20 J. Loisel, *Diploma of Higher Studies*, Paris, 1955.
21 F. Claisse, F. East and F. Abesque, *Quebec Dept. Mines Prelim. Repts.* No. 305, 1954.
22 Shen-Chia-Shu and Chang-Miu-Shiao, *Acta chim. Sinica*, 26 (1960) 124.
23 H. Saito, *Sci. Repts. Tôhoku Imp. Univ.*, 16 (1927) 49.
24 A. F. Gill, *Can. J. Research*, 10 (1934) 703.
25 P. Truchot, *Rev. chim. pur. e appl.*, 10 (1907) 2.
26 C. Duval, *Anal. Chim. Acta*, 20 (1959) 23.
27 A. G. Ostroff and R. T. Sanderson, *J. Inorg. & Nuclear Chem.*, 9 (1958) 45.
28 W. W. Wendlandt and G. W. Ewing, *Anal. Chim. Acta*, 22 (1960) 497.
29 M. Capestan, *Ann. Chim. (Paris)*, 5 (1960) 213.
30 C. Duval, *Anal. Chim. Acta*, 20 (1959) 263.
31 W. W. Wendlandt, *Texas J. Sci.*, 10 (1958) 392.
32 F. Mohr, *Z. anal. Chem.*, 2 (1863) 250.
33 Y. Schwob, *Rev. matériaux construct. et trav. publ.*, No. 413 (1950) 33, 85.
34 S. Caillère and T. Pobeguin, *Bull. soc. franç minéral.*, 83 (1960) 36.
35 V. P. Kornienko, *Referat. Zhur. Khim.*, (1956), *Abstr.* No. 38984.
36 J. L. Doremieux and A. Boullé, *Compt. rend.*, 250 (1960) 3184.
37 W. Funk, *Z. anal. Chem.*, 45 (1906) 503.
38 H. Borck, *Z. anal. Chem.*, 51 (1912) 674.
39 F. Ferrari, *Ann. chim. appl.*, 4 (1915) 341.
40 G. E. F. Lundell, *J. Am. Chem. Soc.*, 43 (1921) 847.

41 E. BAMBERGER AND O. BAUDISCH, *Ber.*, 42 (1909) 3568.
42 S. CHANDRA SHOME, *Analyst*, 75 (1950) 27.
43 M. ILINSKI AND G. VON KNORRE. *Ber.*, 18 (1885) 2728.
44 R. BERG, *Z. anal. Chem.*, 71 (1927) 369; 76 (1929) 191.
45 S. L. TSINBERG, *Zavodskaya Lab.*, 4 (1935) 735.
46 A. M. ZANKO AND G. A. BUTENKO, *Zavodskaya Lab.*, 5 (1936) 415.
47 L. D. RASKIN, *Zavodskaya Lab.*, 5 (1936) 1129.
48 M. BORREL AND R. PÂRIS, *Anal. Chim. Acta*, 4 (1950) 267.
49 W. FRESENIUS, *Z. anal. Chem.*, 96 (1934) 433.
50 S. ISHIMARU, *J. Chem. Soc. Japan*, 55 (1934) 201.
51 R. BERG AND H. KÜSTENMACHER, *Z. anorg. Chem.*, 204 (1932) 21.
52 W. W. WENDLANDT, *Anal. Chim. Acta*, 15 (1956) 533.
53 F. HECHT AND J. DONAU, *Anorganische Mikrogewichtsanalyse*, Springer, Vienna, 1940, p. 49.
54 M. HERRMANN-GURFINKEL, *Bull. Soc. chim. Belg.*, 48 (1939) 94.
55 K. YOSHIKAWA AND K. SHINRA, *Nippon Kagaku Zasshi*, 77 (1956) 1418; *C.A.*, (1956) 2642.
56 K. A. CRAIG AND G. C. CHANDLEE, *J. Am. Chem. Soc.*, 56 (1934) 1278.
57 G. THOMAS, *Thesis*, Lyon, 1960.

CHAPTER 27

Cobalt

We have made a critical and thermogravimetric study of all of the procedures for determining cobalt gravimetrically. There are now about thirty such methods and to them we have added the investigation of a certain number of soluble salts such as the chloride and nitrate. See Duval and Duval[1].

However, we have not been successful in reproducing the precipitation of metallic cobalt by means of zinc as directed by Davies[2] nor the precipitation by isonitrosothiocamphor as given by Sen[3]. In this latter instance, we have not even been able to reproduce the reagent by following the procedure given in the paper nor have we been able to get into communication with this worker. Along another line, we have omitted the precipitation by thiosulphate as published by Gibbs[4], by sodium rubeanate as given by Rây and Rây[5] and by Voznesenki[6], and by the thioamide of oxanilic acid according to Majumdar[7] even though the latter has kindly provided us with some of the reagent. The reason for the omissions is that all of these methods terminate by weighing the cobalt as sulphate.

1. Metallic cobalt

A platinum crucible, which can be fitted into the silica ring of the thermobalance, serves as cathode in the electrolytic deposition of the cobalt which is weighed where it has been deposited on the inside of the crucible. The electrolyte is a 0.1 N solution of cobalt(II) sulphate to which ammonium acetate and ammonium hydroxide have been added. The current is applied at 4 V and 0.3 A. The deposit of metallic cobalt is washed with water, alcohol, and ether. It starts to gain weight at 190° due to the action of the oxygen of the air.

2. Cobalt(II) hydroxide

(i) Precipitation with sodium or potassium hydroxide If sodium or

potassium hydroxide is added to a solution of a cobalt(II) salt, the resulting pink precipitate progressively loses adsorbed water up to around 90°. However, the level which almost corresponds to the hydroxide $CoO \cdot H_2O$ was not the same for the various preparations studied, for instance, 99–112°, 81–91°. A rapid loss of water of constitution is then observed up to about 160°, followed by a slower loss up to 475°. A second horizontal, whose position is more constant, extends from 475° to 800°. The residue beyond 800° accords with the formula Co_3O_4. In the vicinity of 900° there is either a decrease or an increase in weight, which is quite small in either case. This fact has previously been reported by various workers; there is no simple explanation for it. However, if the temperature is raised beyond 900° the change in weight is sufficient to vitiate the determination. See Duval and Duval[1] and Kato, Hosimija and Nakazima[8].

(ii) Precipitation by yellow mercuric oxide We have followed the procedure described by Zimmermann[9] in which the solution containing the cobalt to be determined is treated with sodium chloride and boiled with an aqueous suspension of yellow mercuric oxide. The precipitate settles readily and filters remarkably well. The curve of the substance obtained shows constant weight up to around 130°; then the residual mercury derivatives sublime and decompose. A rather good horizontal due to Co_3O_4 starts at 475°. This level can be suggested for the automatic determination of cobalt.

3. Cobalt(III) oxide hydrate

(i) Precipitation by potassium persulphate The hydrous oxide $Co_2O_3 \cdot nH_2O$ was precipitated by the method of Dede[10] with potassium persulphate as the reagent. However, the reduction of the product by hydrogen to obtain metallic cobalt was not included. We found, in agreement with Hüttig and Kassler[11], that it is difficult to find any evidence of a definite hydrate on the curve. The break in the curve at 72° conforms essentially with the formula $Co_2O_3 \cdot 1H_2O$. The material then loses water and changes into the oxide Co_3O_4 up to 371°; the latter does not break down below 950°. This result demonstrates once again that the reduction in a Rose crucible is entirely unnecessary.

(ii) Precipitation by sodium hypochlorite The colloidal precipitate, which has no definite composition, was prepared by the method of Gibbs[12]

employing sodium hypochlorite. Here again, the curve shows a change of direction near 72°, and then another change of direction at around 180°, which agrees closely with the formula Co_2O_3 but without yielding a horizontal. The important difference between this and the preceding method resides in the fact that the decrease in weight accompanying the change from Co_2O_3 to Co_3O_4 occurs imperceptibly up to around 700°, and in addition the level is not strictly horizontal after passing this temperature. The method does not appear to be feasible for gravimetry; the precipitate cannot be collected on a No. 4 filtering crucible; it is necessary to use paper.

4. Cobalt aluminate

With respect to the transformation into aluminate, Salvetat[13] has pointed out that if hydrous cobalt(II) oxide, more or less oxidized, is strongly ignited with an excess of alumina, the cobalt will be found in the final product in the form of Thenard's blue, Al_2O_3 + CoO or $Co(AlO_2)_2$, and that the increase in weight of the alumina is due, in effect, to the oxide CoO. This unduly complicated procedure has led to a curve which slopes downward until 500°; the subsequent horizontal continues up to 890°. Beyond this temperature, the curve rises for some reason that has no chemical explanation. This method should be rejected in modern analytical practice.

5. Cobalt chloride

Cobalt(II) chloride $CoCl_2 \cdot 6H_2O$ has been heated by various workers, including Shen-Chia-Shu and Chang-Miu-Shiao[14] and Gvozdov and Erumova[15]. See also[16]. The salt has been used for making temperature-indicating crayons[15]. Duval[17] has offered an explanation of the color changes exhibited by sympathetic ink.

When heated by itself, the hexahydrate of cobalt(II) chloride begins to show some blue-violet dots at 35°. The salt loses water from 50° to 120° producing the dihydrate $CoCl_2 \cdot 2H_2O$; the latter is the cobalt salt of the blue acid which the author discovered:

$$3\ CoCl_2 \cdot 6H_2O \rightarrow Co[CoCl_3(H_2O)_3]_2 + 12\ H_2O$$

It is impossible to achieve a quantitative transformation into the anhydrous chloride unless the operation is carried out in the presence of hydrogen chloride gas. The water of this latter complex salt permits its hydrolysis and the resulting slate-blue mixture is a non-stoichiometrical mixture of $CoCl_2$ + CoO formed near 155°. Therefore, this pyrolysis is analogous to that of magnesium chloride.

The tetrahydrate $CoCl_2 \cdot 4H_2O$ and the monohydrate whose preparation Duval has described[17] cannot be produced by pyrolysis of the hexahydrate.

6. Cobalt bromate

The bromate $Co(BrO_3)_2 \cdot 6H_2O$ was heated by Rocchiccioli[18] at the rate of 150° per hour; it decomposed around 60°, losing water and bromine. Nothing was left in the crucible except the oxide CoO and no bromide. If the heating is less rapid (50° or 100° per hour), a dihydrate may be detected between 114° and 135°, and on raising the temperature a loss of bromine and oxygen is observed prior to the complete elimination of the two molecules of water. However, the horizontal corresponding to the anhydrous bromate is never obtained. Rocchiccioli has shown that the hexahydrate must be kept at 50–60° in the presence of phosphoric anhydride to obtain the anhydrous bromate.

7. Cobalt iodate

Rocchiccioli[18] heated the iodate $Co(IO_3)_2 \cdot 2H_2O$ at 150° per hour. She found it to be stable up to around 142°. The anhydrous salt was produced without loss of iodine from 220° on.

8. Cobalt sulphide

The precipitation of cobalt sulphide was conducted as described by Fresenius[19] by means of ammonium sulphide in the presence of ammonium hydroxide and ammonium chloride. The product does not satisfy the formula CoS. When heated, it loses water and sulphur by volatilization and combustion, while the sulphide is slowly oxidized to sulphate. The latter yields a horizontal extending from 700° to 820°, and then the compound partially decomposes giving a mixture of the violet anhydrous sulphate and the black oxide Co_3O_4. This method of determining cobalt is of little interest; we suggest that it be discarded.

9. Cobalt sulphate

(i) *Anhydrous sulphate*   The weighing of cobalt as the anhydrous

References p. 353

sulphate $CoSO_4$ is a current method of determination and other proce- dures come down to it in the end. They include: the precipitation by a thiosulphate as developed by Gibbs[12], by a sulphide as described by Gauhe[20], by rubeanic acid as given by Rây[5], the solution of the ignition residue of an organic material in sulphuric acid, etc. The method is said to be very accurate. The workers however are not in agreement as to the highest temperature at which anhydrous cobalt sulphate can exist, and we have seen figures ranging from 550° to 880°.

(ii) Heptahydrate   Duval and Wadier[21] found that the heptahydrate $CoSO_4 \cdot 7H_2O$, crystallized once from water, is not stable and loses weight from 36° on. See also Demassieux and Malard[22]. The heptahydrate level is not very distinct when the heating rate is 150° per hour. A slightly oblique level extends from 187° to 256°; it corresponds to a mixture consisting chiefly of monohydrate. A loss of 6 mg was registered by 248 mg of the anhydrous salt between the extremities of this level. Conse- quently, there is no need to consider weighing cobalt as the sulphate monohydrate. The dehydration is complete at 380° and the level be- longing to the anhydrous sulphate extends horizontally as far as 730°, and then slopes very slightly downward until 820°. Kato, Hosimija and Nakazima[8] give 500–600° as the range of highest precision. See also[16]. Ostroff and Sanderson[23] report 708° as the temperature at which de- composition sets in.

(iii) Moist sulphate   Duval[1] has also heated a moist sulphate such as is encountered in analytical practice; its composition is not far from $CoSO_4 \cdot 8H_2O$. The decrease in weight along the level of the anhydrous salt was 2.5 mg per 248 mg. The drying must be carried out not far from 600°. Loss of sulphur occurs beyond 820°. We recorded 2 mg at 893° and 8 mg at 950°. The residue is made up of the violet sulphate and the black oxide Co_3O_4. It is hardly realistic to expect an accuracy above 1/100 from this method of determining cobalt.

(iv) Schönites   Demassieux and Malard[22] have likewise heated the schönites produced from cobalt sulphate and the sulphates of potassium, rubidium, and cesium, respectively. See Chapter Nickel.

10. Cobalt sulphamate

Capestan[24] prepared the sulphamate (amidosulphonate) $Co(NH_2-$

$SO_3)_2$. When crystallized from its aqueous solution at 60°, it yields the pink dihydrate. One-fourth of its water is lost from 60° to 95°, and the resulting $Co(NH_2-SO_3)_2 \cdot \frac{3}{2} H_2O$ loses all of the remaining water from 165° to 195°. The blue anhydrous sulphamate is transformed into the sulphate from 350° to 460°.

11. Cobalt selenite

Cobalt(II) selenite $CoSeO_3 \cdot 2H_2O$ was studied by Rocchiccioli[18]. Its water is given up between 100° and 150° and the anhydrous salt decomposes near 600° yielding the oxide Co_3O_4.

12. Amminecobaltic salts

Various amminecobaltic salts, chosen from among the most simple of this class, have been subjected to pyrolysis.

(i) Luteocobaltic chloride First of all, Kékedy and his associates[25] investigated the luteocobaltic chloride $[Co(NH_3)_6]Cl_3$ by differential thermal analysis and thermogravimetry. This complex is stable up to 240° when heated in an atmosphere of ammonia. The compound $CoCl_2 \cdot NH_3$ NH_4Cl remains at about 285°. The last one is transformed into $CoCl_2 \cdot$ NH_3 at 420° Finally, cobalt(II) chloride is obtained at 450°. Only metallic cobalt is left near 600°. These workers have heated the intermediate products for control purposes. Furthermore, Kékedy, Kröbl, Szurkos and Kékedy[26] investigated several hexammine salts by the method of Erdey and Paulik.

In addition, Wendlandt[27] heated this complex also in air. The loss of weight begins to be apparent at 160° and there is a break in the curve at 360° where cobalt(II) chloride is formed. The latter gradually decomposes and yields the oxide Co_3O_4 at 655°.

(ii) Purpureocobaltic chloride It should be pointed out that Kékedy and his collaborators likewise studied purpureocobaltic chloride both by thermal analysis and[25] and thermogravimetry in an atmosphere of ammonia. This compound $[Co(NH_3)_5Cl]Cl_2$ starts to decompose at 200° and measurement becomes impossible around 250°. The following intermediate products are traversed: $CoCl_2 \cdot NH_3 \cdot NH_4Cl$ at 315°, $CoCl_2 \cdot NH_3$ at 460°, and $CoCl_2$ at 500°. The final product is metallic cobalt at 700°.

(iii) Various amminecobaltic complexes Wendlandt also heated vari-

ous complexes; we have summarized his findings in Table 9. The first column shows the maximum stability temperature; the second column gives the lowest temperature at which the oxide Co_3O_4 appears. All of these heatings were conducted on 80–100 mg samples at the rate of 5.4° per minute.

TABLE 9

THERMOLYSIS OF AMMINECOBALTIC SALTS

	Max. stability temp. (°C)	Lowest temp. at which Co_3O_4 appears (°C)
$[Co(NH_3)_4H_2OCl]Cl_2$	125	660
$[Co(NH_3)_5NO_2]Cl_2$	120	530
$[Co(NH_3)_5H_2O]Cl_3$	150	620
$[Co(NH_3)_5Cl]Cl_2$	170	645
$[Co(NH_3)_4CO_3]NO_3 \cdot \frac{1}{2}H_2O$	40	225
$[Co(NH_3)_5H_2O]Br_3$	150	540
$[Co(NH_3)_5F](NO_3)_2 \cdot 1(?)H_2O$	75	650
$[Co(NH_3)_5Br](NO_3)_2$	150	205
cis $[Coen_2Cl_2]Cl \cdot H_2O$	85	640
trans $[Coen_2Cl_2]Cl$	45	650
$[Coen_3]Cl_3 \cdot 3H_2O$	20	620
$[Co(NH_3)_5Br]Br_2$	130	550

Gibbons[28] investigated some amminecobaltic salts from the standpoint of gravimetric analysis. The *sulphate* of *dicobalti-μ-amino-μ-nitrooctammine* shows a constant weight between 25° and 122°, and the *antimonio-hexachloride* of *cobaltidichloro-bis-ethylenediamine* yields a horizontal between 75° and 225°.

Other amminecobaltic salts are discussed in 22 and 25.

13. Cobalt nitrate

Cobalt(II) nitrate $Co(NO_3)_2 \cdot nH_2O$ has been studied by Duval and Wadier[29]. It is extremely deliquescent and ordinarily contains more than 6 molecules of water. It is practically impossible to prepare a standard solution of this salt by direct weighing because it loses water continuously and in considerable amount from 34° on. The decomposition and the dehydration could not be distinguished from each other at the rate of heating we employed but the change slows down appreciably as 290° is neared. A slightly inclined level extends to around 700°, where it becomes almost horizontal. See also Wendlandt[30], who noted the appearance of the oxide Co_3O_4 only at 290° with our rate of heating. We have been unable to confirm this fact even with 85 mg of the product.

14. Sodium cobaltinitrite

Sodium cobaltinitrite $Na_3[Co(NO_2)_6]$ was heated by Wendlandt[27] at the rate of 5.4° per minute. He used 80–100 mg of the material. We used 148–150 mg and a heating rate of 300° per hour. The retained water was driven off at 40°, and then came the level of the anhydrous salt from 63° to 123°. Beyond this, the complex rapidly loses 3 NO_2 and a mixture of $CoO + 3 NaNO_2$ is present around 300°. The weight of this mixture then slowly increases up to 560° where the mixture consists of $CoO + 3 NaNO_3$, the cobalt oxide serving as acceptor for oxygen. The decomposition of the sodium nitrate then begins and the level due to the oxide mixture $2 CoO + 3 Na_2O$ starts at 860°.

15. Potassium cobaltinitrite

Potassium cobaltinitrite $K_3[Co(NO_2)_6]$ is a weighing form of cobalt; it is obtained by the classic procedure of Brauner[31], who did not recommend drying at 100°. This complex yields a horizontal extending from 43° to 160° which is suitable for the automatic determination of cobalt. As in the case of cesium cobaltinitrite, there is then a loss of 3 NO_2 per gram-molecule and formation of the mixture of $CoO + 3 KNO_2$, whose weight slowly increases from 245° to 813°. The disintegration of the potassium nitrate then sets in. The mixture of oxides $2 CoO + 3 K_2O$ has not yet reached its quantitative development at 950°.

16. Cobalt pyrophosphate

When cobalt is to be weighed as pyrophosphate, Dakin[32] starts with cobalt ammonium phosphate $CoNH_4PO_4 \cdot H_2O$ which he recommends drying at 100–105°. Strebinger and Pollak[33] have generalized the method for microgravimetric purposes. Actually, the curve rises slowly up to 98° (4 mg gain per 139 mg). Then a sudden and deep-seated decomposition occurs at this temperature. The anhydrous salt is not an intermediate product since water and ammonia are given off concurrently. The pyrophosphate $Co_2P_2O_7$ appears in the pure condition from 580° on and gives a horizontal. This is a good method for determining cobalt. Kato, Hosimija and Nakazima[8] reported 600° as the upper limit of the level due to the pyrophosphate, and 105–170° for that of the monohydrate; however, this salt is undoubtedly destroyed at 170°.

17. Cobalt arsenates

Masson, Charles-Messance, Bastick and Guérin[34] made an extended study of the cobalt(II) arsenates.

The starting point was monocobaltous binarsenate $CoO \cdot 2As_2O_5 \cdot 5H_2O$, which is prepared from the action, at 60°, of 65–70% arsenic acid on cobalt(II) hydroxide. It loses 4 molecules of water between 100° and 200° but it was not possible to detect the intermediate formation of the monohydrate. The anhydrous salt is obtained between 300° and 400°. The pink metarsenate is produced in the neighbourhood of 500° and along a short inclined level. This salt $CoO \cdot As_2O_5$ rapidly goes over to the violet pyroarsenate $2 CoO.As_2O_5$, which itself is decomposable from 750° on with production of the red orthoarsenate $3CoO \cdot As_2O_5$, whose pyrolysis cannot be detected below 1000°. These various compounds have been characterized by chemical analysis and by their X-ray spectra.

18. Silver cobalticyanide

Silver cobalticyanide $Ag_3[Co(CN)_6]$ was prepared by the method of Nenadkevich and Saltykova[35], and also by Saltykova[36]. It presents some interest for the automatic determination of cobalt; the gravimetric factor is 0.1146. The complex is stable from 96° to 252°. These workers have recommended 130° as the drying temperature; the corresponding level is perfectly horizontal. Cyanogen is given off from 450° to 745° and a mixture $Co_3O_4 + 9 Ag$ is left. Another decomposition is observed from 745° to 920° (doubtless due to the presence of the reducing silver) and the mixture $CoO + 3 Ag$ results.

19. Cobalt hydrazine thiocyanate

Cobalt dihydrazine thiocyanate $[Co(N_2H_4)_2](SCN)_2$ is precipitated by the Sarkar and Datta-Rây[37] method. This salt is stable only up to 79° and below this temperature it yields a good horizontal which would lend itself well to the automatic determination of cobalt. The decomposition is not so distinct as that of the corresponding cadmium compound because the hydrazine decomposes at the same time as the SCN groups, at 148° notably. The residue is not homogeneous.

20. Cobalt pyridine thiocyanate

The directions given by Spacu and Dick[38] were followed in the preparation of cobalt pyridine thiocyanate $Co(C_5H_5N)(SCN)_2$. This precipitate is stable only up to 60° with a slight gain in weight around 45°. Therefore, it should be dried in a desiccator as prescribed by these workers. The molecule of pyridine comes off at 125°. A short horizontal is observed; it is related to the existence of the thiocyanate $Co(SCN)_2$, which in turn decomposes rapidly.

21. Basic cobalt carbonate

The thermogravimetric curve of the basic carbonate $CoCO_3 \cdot 2Co(OH)_2$ has been traced[16]; Caillère and Pobeguin[39] have heated rhodochrosite in nitrogen. It starts to decompose at 340°, whereas, by differential thermal analysis in argon, they obtained 370°. This research was supplemented by an investigation of the infra-red absorption of the natural carbonates.

22. Pentamminecobaltiacido nitrates

Figlarz[40] has published thermolysis curves of the following compounds, which were heated in 5 mg portions, in nitrogen, on the Eyraud thermobalance, at the rate of 120° per hour: nitrates of carbonato-, formato-, aceto-, and propionatopentammine, namely $[Co(NH_3)_5CO_3]NO_3 \cdot \frac{1}{2}H_2O$, $[Co(NH_3)_5CO_2H](NO_3)_2$, $[Co(NH_3)_5CH_3CO_2](NO_3)_2$, $[Co(NH_3)_5C_2H_5CO_2](NO_3)_2$. The curves slope downward and the products explode below 250°, leaving a mixture of Co_3O_4 and CoO. However, if the nitrates are heated in hydrogen, flowing at the rate of 0.4 l per hour, the explosion occurs between 185° and 205°. Cobalt metal is obtained around 300°; it retains up to 10% of its weight of hydrogen.

23. Cobalt mercurithiocyanate

Cobalt mercurithiocyanate $Co[Hg(SCN)_4]$ was prepared by the method of Lamure[41]. The curve shows that, after the moisture has been removed, the complex salt is stable from 50° to 200°. The corresponding horizontal may be employed for the automatic determination of cobalt. Above 200°, there is rapid departure of cyanogen and mercury; the experiment was stopped at 245°. Consequently, the temperature of 90° suggested by Lamure is satisfactory for drying this salt.

References p. 353

24. Hexamminecobaltous mercuriiodide

The precipitate of hexamminecobalt(II) mercuriiodide $[Co(NH_3)_6]$ $(HgI_3)_2$, produced by the method of Taurinš[42], is unstable and can be dried only in a desiccator. It loses weight from 35° on; after loss of ammonia, iodine, and mercury, there remains, at 420°, a mixture of the oxides CoO and Co_3O_4. A slow oxidation proceeds beyond 688° and only Co_3O_4 is present around 960°. The method has some interest because of the analytical factor 0.079 for cobalt, but the precipitate is fragile and the manipulation is complicated.

25. Chloropurpureocobaltic molybdate

The method proposed by Carnot[43] employs chloropentamminecobalt-(III) molybdate (purpureo molybdate). After a critical study by Congdon and Chen[44], they reported that this is the least satisfactory method for the gravimetric determination of cobalt. They do not give the formula of the initial precipitate, but the residue at 110° is $Co_2O_3 \cdot 10NH_3 \cdot 6MoO_3$ according to Carnot, or $2CoO \cdot 7MoO_3$ according to Congdon and Chen. Our own findings can be summarized: the transformation of the bivalent cobalt into the purpureocobalti chloride is never entirely quantitative, and therefore the determination cannot be accurate no matter what formula is accepted.

We prepared the precipitate from ammonium molybdate and purpureocobaltic chloride. The pink product has a formula which accords as well as possible with $[Co(NH_3)_5Cl]Mo_3O_{10}$ or with the equivalent formulations $[Co(NH_3)_5Cl](Mo_2O_7) \cdot MoO_3$ or $[Co(NH_3)_5Cl](MoO_4) \cdot 2MoO_3$. A precipitate of this kind loses weight, because of the elimination of water and chlorine, up to around 167°; accordingly, drying at 110° can offer no surety. A fairly horizontal level extends from 167° to 206°, and in agreement with Carnot the corresponding composition is $3MoO_3 \cdot \frac{1}{2}Co_2O_3 \cdot 5NH_3 = 600$ with the analytical factor 0.098 for cobalt. The ammonia is then evolved and a new level extends from 338° to 836°; it is more definite than the preceding horizontal and represents the mixture $3MoO_3 \cdot 1/3Co_3O_4 = 512.12$ with the analytical factor 0.115 for cobalt. Beyond 836°, molybdic anhydride is driven off rapidly. From these findings we conclude: the formula given by Carnot is valid, that of Congdon and Chen is incorrect. The selected temperature is too low for proper drying of the precipitate. Since the method is not quantitative, it should be discarded for the determination of cobalt(II).

26. Cobalt formate

If cobalt formate $Co(CO_2H)_2 \cdot 2H_2O$ is heated at the rate of 150° per hour, it progressively becomes anhydrous between 152° and 209°. The level is not perfectly horizontal. Decomposition sets in definitely around 260° and becomes very rapid in the vicinity of 300°. Since the furnace atmosphere is reducing in character, Co_3O_4 is not obtained at once; a slow oxidation is observed between 372° and 400°. See also Zapletal et al.[45], Kornienko[46], Doremieux and Boullé[47].

27. Cobalt oxalate

The oxalate $CoC_2O_4 \cdot 2H_2O$ was precipitated as described by Classen[48] by adding a solution of potassium oxalate to a solution of a cobalt(II) salt. The crystalline precipitate is stable up to 108°. There is a slight gain in weight (1 mg per 260 mg) between 55° and 103°. The dehydration proceeds rapidly up to 192° and then more slowly from 192 to 256°. Weighing the anhydrous salt is out of the question; it decomposes as soon as it is formed with simultaneous loss of carbon monoxide and carbon dioxide. The perfectly horizontal level corresponding to the oxide Co_3O_4 begins at 285°. See also Kawagaki[49].

Potassium cobaltioxalate Wendlandt[27] has pyrolyzed potassium cobaltioxalate $K_3[Co(C_2O_4)_3] \cdot 3H_2O$ (which he designates as a cobaltiammine). If heated at the rate of 5.4° per minute, it loses moisture from room temperature on; the trihydrate is obtained at 60° but loses its water immediately. The curve descends continually as far as 340°. The author believes that he probably is dealing here with a mixture of Co_3O_4, K_2CO_3, and C. (It is likely that the carbon will burn at this temperature.)

28. Cobalt anthranilate

The anthranilate $Co(NH_2-C_6H_4-COO)_2$ was obtained by the procedure of Wenger, Cimerman and Corbaz[50] who recommend drying the product at 120–130°. The thermolysis curve of this anthranilate shows that moisture is given off up to 108°. A perfectly horizontal stretch begins there and continues to 290°. (Not even as little as 0.1 mg is lost from 200 mg.) Therefore, the temperatures given by these workers are correct. The destruction of the organic matter is indicated by a change in the curve at 403°. The equally fine horizontal due to the oxide Co_3O_4 starts at 639°.

References p. 353

In view of the excellent filtering qualities of this material, which can be collected and washed on a No. 4 fritted glass crucible, and also because of the fine horizontal given by this anthranilate, we recommend this precipitate for the automatic determination of cobalt.

Ishimaru[51], who likewise studied this precipitate, reports 107–202° as the limits of the level. If ignited with oxalic acid, this anthranilate yields the oxide Co_3O_4 above 1005°.

29. Cobalt 5-bromoanthranilate

Shennan[52] has given directions for bringing down cobalt with 5-bromo-anthranilic acid. The bright red precipitate filters rapidly, does not adhere to the walls of the vessels, and yields a thermolysis curve which includes a good horizontal for $Co(NH_2-C_6H_3BrCO_2)_2$ extending up to 194°. We suggest this horizontal likewise for the automatic determination of cobalt. The destruction of the organic matter is accompanied by abrupt changes in direction at 315°, 330°, and 826°. Contrary to all expectations, the residue at 946° consists of the oxide CoO.

30. Cobalt 2-aminonaphthoate

The 2-aminonaphthoate $Co(NH_2-C_{10}H_6-COO)_2$ was prepared as described by Shennan, Smith and Ward[53]. It is readily filtered on a No. 4 glass crucible, and does not stick to the walls of the vessels. Its weight remains constant up to 176°. These workers have prescribed drying the precipitate at 176°. When pyrolyzed, this product yields a mixture of the violet cobalt carbonate and the black oxide Co_3O_4 above 850°. This residue is much different from that given by the pyrolysis of the corresponding anthranilate.

31. Cobalt α-nitro-β-naphthol

There has been some controversy regarding the determination of cobalt by means of α-nitro-β-naphthol, especially on the part of Mayr and Feigl[54], Mayr[55], Mayr and Prodinger[56]. The reagent employed melts with decomposition at 103° but the point is not sharp. The curve obtained with the cobalt compound, supposed to be $Co(NO_2-C_{10}H_6-O)_3$, descends continuously from room temperature to 276° where an explosion occurs which empties the crucible. This method of determining cobalt should be abandoned.

32. Cobalt α-nitroso- β-naphthol

With α-nitroso-β-naphthol as reagent, it is necessary to distinguish two cases depending on the valency of the cobalt involved.

(i) Cobalt(II) Ilinsky and von Knorre[57] produced the brown-red precipitate $Co(NO-C_{10}H_6-O)_2$ by treating a neutral solution of a cobalt-(II) salt with the reagent in not too great excess. The product may be weighed as such. The level corresponding to it is not horizontal; it extends from 147° to 254°. The material decomposes abruptly at 254°, and the horizontal corresponding to the oxide Co_3O_4 starts at 589°.

(ii) Cobalt(III) The method of Mayr and Feigl[54] and Prodinger[58] was employed with cobalt(III). The curve shows, after removal of the wash water, an essentially horizontal level extending from 81° to 140° and agreeing with the formula $Co(NO-C_{10}H_6-O)_3$. A sudden explosion ensues at 270°; the residual carbon burns as far as 744°; the mixture of cobalt oxides undergoes a slow oxidation leading to Co_3O_4 around 960°. The procedure with the trivalent cobalt is more accurate than that with Co(II), but it does not possess the precision of the anthranilate method, for example.

33. Cobalt dinitrosoresorcinolate

We produced the precipitate $Co(C_6H_3O_4N_2)_2$ by means of the reagent prepared by the nitrosation of resorcinol[59]. The thermolysis curve slopes downward continuously. An explosion at 149° empties the crucible. We advise that this method be discarded. According to Orndorff and Nichols[60], this material should be dried for 2–3 hours at 105° in an atmosphere of carbon dioxide. They found then that decomposition occurred at 160–161°. We could not confirm this result. The decomposition temperature of explosive compounds depends greatly on the rate of heating.

34. Cobalt dinitrosoorcinolate

The reagent was prepared by nitrosation of orcinol, and was used to prepare the cobalt compound, whose composition, according to Guha-Sircar and Bhattacharjee[59] is $Co(C_7H_5O_4N_2)_2$. They suggest that the compound be dried for 2 or 3 hours at 115°. Our experience has been that this compound explodes around 112° even though it is heated very slowly. Up to this temperature, the curve slopes steeply downward. Since the precipitate is not stable, we reject the method.

35. Cobalt α-nitroso-β-naphthylamine

According to Guha-Sircar and Bhattacharjee[59], the precipitate produced with α-nitroso-β-naphthylamine should be dried at 110°. The curve obtained with this material shows a good horizontal going from 80° to 179° and corresponds to the formula $Co(C_{10}H_7ON_2)_3$. This level may be employed for the automatic determination of cobalt; the reagent is readily accessible. The compound decomposes abruptly at 247°; the carbon slowly burns up to 750°, the temperature which marks the start of the good horizontal corresponding to the oxide CoO. From 920° on, the curve rises due to the surface conversion of the CoO into Co_3O_4.

36. Cobalt β-nitroso-α-naphthylamine

Using this same procedure[59] but substituting the isomer of the preceding precipitant, i.e. β-nitroso-α-naphthylamine, we obtained a cherry red precipitate. The latter was more stable; it filtered well, and its curve included a horizontal stretch extending from 72° to 218°. An abrupt explosion occurred at about 250°, and the oxide Co_3O_4 was present above 609°. This method lends itself nicely to the automatic determination of cobalt; unfortunately, the reagent is difficult to prepare.

37. Cobalt diliturate

Cobalt diliturate $Co[C_4H_2O_3N_2(NO_2)_2] \cdot 8H_2O$ has been heated by Mihai[61] and Mihai and Roch[62]. On the Erdey thermobalance, the octahydrate is stable up to 80°, and the anhydrous compound is obtained at 240° but the latter does not yield a horizontal. The decomposition is vigorous beyond this temperature and the oxide Co_3O_4 appears above 340°. It has been suggested that this salt be used for the microgravimetric determination of cobalt.

38. Cobalt picolinate

Thomas[63] has studied cobalt picolinate (or pyridine-2-carboxylate). It carries 4 molecules of water which are removed between 75° and 145°. See also Lumme[64].

39. Cobalt quinaldate

The quinaldate (or quinoline-2-carboxylate) appears in two forms. One, prepared in neutral surroundings, is cream-colored; the other, which is prepared in dilute acetic acid medium, is reddish pink. If heated at 180° per hour, they yield different curves. The first, which is the dihydrate, loses water between 160° and 325°, whereas the second, the monohydrate, gives up its water between 160° and 325°. The level due to the anhydrous salt begins at 325°. See also Lumme[64].

40. Cobalt oxinate

Cobalt oxinate $Co(C_9H_6ON)_2 \cdot 2H_2O$ was prepared by the method of Berg[65] which was repeated by Tsinberg[66]. According to Borrel and Pâris[67], it loses water and wash alcohol up to 115°. The anhydrous oxinate is relatively sturdy and is stable up to 295°; the corresponding horizontal is suitable for the automatic determination of cobalt. The destruction of the oxine gives rise to the ordinary curve which differs but little from a straight line. Duval and Duval[1] found that at 960° the Co_3O_4 is still mixed with carbonaceous products.

41. Cobalt 2-methyloxinate

The 2-methyloxinate of Borrel and Pâris[68] crystallizes with one molecule of water which it retains up to 152°. The anhydrous compound then yields a stability level from 200° to 280°. Decomposition ensues and finely divided cobalt is left around 670°. Reoxidation subsequently leads to a mixture of the oxides CoO and Co_3O_4.

42. Cobalt 2-phenyloxinate

The 2-phenyloxinate, prepared by Bocquet and Pâris[69], has the formula $Co(C_6H_5-C_9H_5ON)_2 \cdot 1H_2O$. This hydrate has a variable yellow-brown color and it is stable up to 95°. The horizontal of the anhydrous chelate then extends up to 195° for a quantity of material ranging from 300 to 400 mg and heated up to 700° at the rate of 180° per hour and kept at this temperature for 2 hours.

43. Cobalt 2-o-hydroxyphenylbenzoxazole

Wendlandt[70] studied the chelate produced by the action of 2-o-hydroxyphenylbenzoxazole, which was introduced by Walter and Freiser[71]. The chelate proved to be stable below 210°; the oxide Co_3O_4 is obtained above 500°.

44. Cobalt phenylthiohydantoate

The directions published by Willard and Hall[72] were followed in the preparation of the cobalt salt of phenylthiohydantoic acid. The thermolysis curve shows the removal of wash water up to 87°; at this temperature the material consists of the complex salt $Co[C_6H_5-N=C(NH_2)-S-CH_2-CO_2]_2$, which is stable over an interval of only 2 or 3 degrees. Its decomposition is marked by two singular points at 205° and 510° respectively. It is quite remarkable that the residue above this temperature consists of the oxide Co_2O_3 whose level remains horizontal up to at least 946°. This method should be regarded as a means of separation rather than as a method for determining cobalt. See also Ishibashi[73].

45. Cobalt diphenylthiohydantoin

The blue precipitate $Co[NH-C(C_6H_5)_2-CO-N=CS]\cdot 2H_2O$ is obtained with diphenylthiohydantoin by the method of Garrido[74]. It slowly gains weight (10 mg per 170 mg) up to 247°. Decomposition occurs rather rapidly at this temperature, and at 560° the level due to the oxide Co_2O_3 appears, *i.e.* the same oxide is obtained as in the preceding case. The method apparently has some interest, but it is not capable of delivering highly accurate results.

46. Cobalt picrolonate

The picrolonate is anhydrous and stable up to 160–180°. It decomposes especially between 180° and 280°. According to Becherescu[75], the oxide Co_3O_4 is obtained between 310° and 390°.

47. Cobalt 8-quinaldate

Lumme[64] reports that orange or red quinoline-8-carboxylate $Co(C_{10}H_6O_2N)_2\cdot 2H_2O$, which is isomeric with the quinaldate, loses its water of crystallization between 80° and 164°. The anhydrous salt is stable up to 336°; it suddenly breaks down up to 400°, and then yields metallic cobalt at 676°. The reoxidation to Co_3O_4 begins at 700°.

REFERENCES

[1] R. Duval and C. Duval, *Anal. Chim. Acta*, 5 (1951) 84.

[2] J. L. Davies, *J. Chem. Soc.*, 28 (1875) 311; *Z. anal. Chem.*, 14 (1875) 343.

[3] D. C. Sen, *J. Indian Chem. Soc.*, 15 (1938) 473.

[4] W. Gibbs, *Z. anal. Chem.*, 3 (1864) 390.

[5] P. Rây and M. Rây, *J. Indian. Chem. Soc.*, 3 (1926) 1186.

[6] S. A. Voznesenki, *C.A.*, 34 (1940) 4057.

[7] A. K. Majumdar, *J. Indian Chem. Soc.*, 18 (1941) 415.

[8] H. Kato, H. Hosimija and S. Nakazima, *J. Chem. Soc. Japan*, 60 (1939) 1115.

[9] C. Zimmermann, *Ann.*, 232 (1886) 335.

[10] L. Dede, *Chemiker-Ztg.*, 35 (1911) 1077.

[11] G. F. Hüttig and R. Kassler, *Z. anorg. Chem.*, 184 (1929) 279.

[12] W. Gibbs, *Z. anal. Chem.*, 4 (1865) 426; 7 (1868) 259.

[13] Salvetat, *Compt. rend.*, 59 (1864) 292.

[14] Shen-Chia-Shu and Chang-Miu-Shiao, *Acta chim. Sinica*, 26 (1960) 124.

[15] S. P. Gvozdov and A. A. Erumova, *Izvest. Vysshykh Ucheb. Zavedenii Khim. i Khim. Tekhnol.*, 5 (1958) 154.

[16] *1st Colloq. Thermography, Kazan*, 1953, p. 78.

[17] C. Duval, *Compt. rend.*, 200 (1935) 934; *Rev. sci.*, 15 Oct. 1938.

[18] C. Rocchiccioli, *Thesis, Paris*, 1960.

[19] R. Fresenius, *Quantitative Analyse*, 6th Ed., 1875, Vol. 1, p. 264.

[20] F. Gauhe, *Z. anal. Chem.*, 4 (1863) 55.

[21] C. Duval and C. Wadier, *Anal. Chim. Acta*, 23 (1960) 541.

[22] N. Demassieux and C. Malard, *Compt. rend.*, 245 (1957) 1429.

[23] A. G. Ostroff and R. T. Sanderson, *J. Inorg. & Nuclear Chem.*, 9 (1958) 45.

[24] M. Capestan, *Ann. Chim. (Paris)*, 5 (1960) 222.

[25] L. Kékedy, A. Szurkos, R. Kröbl and E. Kékedy, *Acad. rep. populare Romîne, Filiala Cluj*, 9 (1958) 79; *C.A.*, (1959) 18721.

[26] L. Kékedy, P. Kröbl, A. Szurkos and E. Kékedy, *Studia Univ. Victor Babes Bolyai*, 3 (1958) 99.

[27] W. W. Wendlandt, *Texas J. Sci.*, 10 (1958) 392.

[28] D. Gibbons, *J. Chem. Soc.*, (1953) 1641.

[29] C. Duval and C. Wadier, *Anal. Chim. Acta*, 23 (1960) 257.

[30] W. W. Wendlandt, *Texas J. Sci.*, 10 (1958) 271.

[31] B. Brauner. *Z. anal. Chem.*, 16 (1877) 195.

[32] D. Dakin, *Z. anal. Chem.*, 39 (1900) 789.

[33] R. Strebinger and J. Pollak, *Mikrochemie*, 4 (1926) 16.

[34] J. Masson, B. Charles-Messance, M. Bastick and H. Guérin, *Compt. rend.*, 250 (1960) 859.

[35] K. A. Nenadkevich and V. S. Saltykova, *Zhur. Anal. Khim.*, 2 (1946) 123; *C.A.*, 41 (1947) 662.

[36] V. S. Saltykova, *Compt. rend. Acad. Sci. U.R.S.S.*, 49 (1945) 34; *C.A.*, 40 (1946) 3998.

[37] P. B. Sarkar and B. K. Datta-Rây, *J. Indian Chem. Soc.*, 7 (1930) 251.

[38] G. Spacu and J. Dick, *Z. anal. Chem.*, 71 (1927) 97.

[39] S. Caillère and S. Pobeguin, *Bull. Soc. franç. minéral.*, 83 (1960) 36.

[40] M. Figlarz, *Compt. rend.*, 250 (1960) 3844; 249 (1959) 2780.

[41] J. Lamure, *Bull. Soc. chim. France*, (1946) 661.

[42] A. Taurinš, *Z. anal. Chem.*, 101 (1935) 357.

[43] A. Carnot, *Compt. rend.*, 108 (1889) 741; 109 (1889) 109; *Ann. chim. anal.* 22 (1917) 121; *Bull. Soc. chim. France*, 21 (1917) 211.

[44] L. A. Congdon and T. H. Chen, *Chem. News*, 128 (1924) 132.

[45] V. Zapletal, J. Jedlička and V. Růžička, *Chem. Listy*, 50 (1956) 1406; *C.A.*, (1957) 2438.

[46] V. P. Kornienko, *Referat. Zhur Khim.* (1956), *Abstr. No.* 38984.

[47] J. L. DOREMIEUX AND A. BOULLÉ, *Compt. rend.*, 250 (1960) 3184.
[48] A. CLASSEN, *Ber.*, 10 (1877) 1315.
[49] K. KAWAGAKI, *J. Chem. Soc. Japan.*, 72 (1951) 1079.
[50] P. E. WENGER, C. CIMERMAN AND A. CORBAZ, *Mikrochim. Acta*, 2 (1938) 314.
[51] S. ISHIMARU, *J. Chem. Soc. Japan*, 55 (1934) 288.
[52] R. J. SHENNAN, *J. Soc. Chem. Ind.*, 61 (1942) 164.
[53] R. J. SHENNAN, J. H. F. SMITH AND A. M. WARD, *Analyst*, 61 (1936) 395.
[54] C. MAYR AND F. FEIGL, *Z. anal. Chem.*, 90 (1932) 15.
[55] C. MAYR, *Z. anal. Chem.*, 98 (1934) 402.
[56] C. MAYR AND W. PRODINGER, *Z. anal. Chem.*, 117 (1937) 334.
[57] M. ILINSKI AND G. VON KNORRE, *Ber.*, 18 (1885) 702.
[58] W. PRODINGER, *Organic Reagents used in Quantitative Inorganic Analysis*, Elsevier, New York, 1940.
[59] S. S. GUHA-SIRCAR AND S. C. BHATTACHARJEE, *J. Indian Chem. Soc.*, 18 (1941) 155, 161.
[60] W. R. ORNDORFF AND M. L. NICHOLS, *J. Am. Chem. Soc.*, 45 (1923) 1439.
[61] F. MIHAI, *Acad. rep. populare Romîne*, 4 (1957) 129; *C.A.*, (1960) 2000.
[62] F. MIHAI AND B. ROCH, *Stud. cerc. stiinte Ser. Timisoara*, 5 (1958) 113.
[63] G. THOMAS, *Thesis*, Lyon, 1960.
[64] P. LUMME, *Suomen Kemistilehti*, 32 (1959) 237, 241.
[65] R. BERG, *J. prakt. Chem.*, 115 (1927) 178.
[66] S. L. TSINBERG, *Zavodskaya Lab.*, 6 (1937) 1009.
[67] M. BORREL AND R. PÂRIS, *Anal. Chim. Acta*, 4 (1950) 267.
[68] M. BORREL AND R. PÂRIS, *Anal. Chim. Acta*, 5 (1951) 573.
[69] G. BOCQUET AND R. PÂRIS, *Anal. Chim. Acta*, 14 (1956) 201.
[70] W. W. WENDLANDT, *Anal. Chim. Acta*, 18 (1958) 638.
[71] J. L. WALTER AND H. FREISER, *Anal. Chem.*, 24 (1952) 984.
[72] H. H. WILLARD AND D. HALL, *J. Am. Chem. Soc.*, 44 (1922) 2219.
[73] S. ISHIBASHI, *J. Chem. Soc. Japan*, 61 (1940) 125.
[74] J. GARRIDO, *Anales soc. españ. fís. y quím.*, 43 (1947) 1195.
[75] D. BECHERESCU, *Acad. rep. populare Romîne, Timisoara*, 6 (1959) 115

CHAPTER 28

Nickel

In addition to the precipitates which can be employed for the gravimetric determination of nickel, and whose thermolysis curves have been discussed by Duval and Duval[1], there are available published reports of interesting studies of the structures of the hydrous nickel oxides, and papers dealing with the dehydration of nickel chloride, bromate, iodate, sulphamate, selenates, nitrate, formate, etc. Along with copper and cobalt, nickel is the metal which has benefitted most from the thermogravimetric techniques.

We have not included here the Davies[2] procedure in which nickel is precipitated by zinc from a solution of the sulphate or chloride supersaturated with ammonia; it is not possible to obtain a pure and quantitative deposit by this means. We have been graciously provided with a supply of the thioamide of oxanilic acid by its proponent Majumdar[3], but, as he too discovered, the precipitate is far from pure and it is necessary (as in the case of copper and cobalt) to finish the determination by some other method. Consequently, we believe this reagent will be applied only for separation purposes. Neither have we made recordings with the precipitates produced by phenylglyoxime or diacetylmonoxime, as suggested by Mironoff[4]; these precipitates are unstable and the procedure ends with a titration.

1. Metallic nickel

Metallic nickel has been deposited by the method of Fresenius and Bergmann[5] from an ammoniacal solution of the double nickel ammonium sulphate. The deposition was made in a small platinum crucible, which fitted into the carrier ring of the thermobalance and served as the cathode. After the deposit had been washed with water, alcohol, and ether, the oxidation of the cathodic nickel started at about 93°. The gain in oxygen at 283°, the upper limit of the experiment, was 2 mg per 50 mg. Therefore, it is not good practice to pass the cathode through a flame after washing the nickel with ether. The deposited metal should be dried below 93° in an oven or in a current of air.

Saito[6] made a study of the reduction of NiO (500 mg) in hydrogen (300 ml per minute) at a heating rate of 2° per minute. The reduction begins slowly at 230°, becomes very rapid at 300°, slows down at 350°, and is complete at 450°. If the reduced nickel is cooled to 100° in hydrogen, and then exposed to the air, it oxidizes instantly with almost explosive violence. The starting oxide was also reduced for 2 hours at 400°, 600°, 700°, 800°, and 1000°. The nickel reduced at 400° explodes at 150° in the air. The initial oxidation temperatures for the other forms are respectively 200°, 250°, 400°, and 500°. These findings are of interest because of their bearing on the catalytic properties of this metal.

2. Nickel oxide hydrate

(i) Precipitation by electrolysis Nicol[7] investigated the product of the electrolysis of a solution of nickel sulphate. Using the Guichard thermo-balance, he confirmed the formula $Ni(OH)_2$ at 96°. The conversion into the anhydrous oxide begins at 225°. When the precipitate is obtained from 0.1 N nickel sulphate solution, the product contains basic nickel sulphate which must be heated above 700° before it decomposes.

(ii) Precipitation by sodium (or potassium) hydroxide The Gibbs method[8] employing sodium hydroxide yields a true hydroxide $Ni(OH)_2$ between 115° and 260°.

Cabannes-Ott[9] reports that the infra-red spectrum furthermore shows OH bands, a small amount of carbonate, and some retained ions, especially if sulphate is the starting material. This true hydroxide takes up water with avidity. After complete dehydration, which may be a more or less lengthy operation, the oxide NiO appears, yielding another level extending from 600° to 1000°. The entrained SO_4^{2-} ions are eliminated only beyond 700°. There is never a difference between the two levels corresponding to one molecule of water because the retained impurities are removed between these two levels. The resulting NiO is never green but more or less blackish because of the presence of traces of oxide held in the interstitial spaces of the lattice.

Essentially the same curve is given by the material brought down with potassium hydroxide. We used the procedure published by Meineke[10]. If potassium hydroxide is added to the ammoniacal solution of nickel hydroxide, the resulting precipitate is gelatinous and of course readily absorbs carbon dioxide, but it contains less adsorbed contaminants. Cabannes-Ott[9] reports that the level of the true hydroxide is observed between 100° and 250°, and that of the oxide from 530° on.

(iii) Precipitation in the presence of persulphate If sodium (or potassium) hydroxide is added to a nickel solution containing an excess of persulphate, as directed by Dede[11] and Dede and Zieriacks[12], it is not certain whether the product is a hydrated Ni(III) oxide as stated by Moser and Maxymowicz[13]. We can, however, verify the statement of Vaubel[14] who reported that the ignition product is not NiO_2 as had been claimed. The curve obtained is identical with that given when potassium hydroxide alone is the precipitant. A break corresponding to the approximate composition $Ni(OH)_2$ occurs near 131° but there is no horizontal. The horizontal due to the oxide NiO begins at 440°. Assuredly, there is no appearance of the peroxide up to 900°.

(iv) Precipitation by iodide–iodate mixture The Moody[15] procedure employing a mixture of iodide and iodate has been carried out as described by Glassmann[16]. This provides a simple means of preparing very pure nickel hydroxide by collecting the product formed between 100° and 260°. On the other hand, its dehydration is difficult, and it is necessary to heat to at least 960° to arrive at the horizontal of the oxide NiO.

3. Nickel chloride

Nickel chloride $NiCl_2 \cdot 6H_2O$ has been studied by Shen-Chia-Shu and Chang-Miu-Shiao[17]. They report the passage from the hexahydrate to the tetrahydrate at 50°, and give 116° for the change of the tetrahydrate to the dihydrate, and 138° for the conversion of the dihydrate to the monohydrate, and finally 166° for the conversion of the monohydrate to the anhydrous salt. This investigation was supplemented by a differential thermal analysis.

4. Nickel bromate

Rocchiccioli[18] heated nickel bromate $Ni(BrO_3)_2 \cdot 6H_2O$ at the rate of 150° per hour. It starts to lose water at 54°. A dihydrate is obtained between 161° $\pm$ 4° and 200°. It is difficult to produce a pure anhydrous bromate near 280°; it has already begun to decompose.

5. Nickel iodate

Nickel iodate $Ni(IO_3)_2 \cdot H_2O$ was heated by Rocchiccioli[18]; she found that it progressively loses water from 90° on. The anhydrous salt is obtained at about 250°.

6. Nickel sulphide

(i) Precipitation with ammonium sulphide By means of ammonium sulphide in accord with the method of Fresenius[19]. The black precipitate of nickel sulphide is far from pure, even though it filters perfectly. This lack of purity is immediately reflected on the curve which is not reproducible. Water is given off rapidly up to 105°; from 105° to 360°, the mixture loses sulphur and ammonia. A mixture of nickel sulphide and sulphate of variable composition is reached at 520°. It takes up oxygen and between 701° and 838° the composition is approximately $NiSO_4$. It then loses the elements of sulphur trioxide, and nickel oxide NiO appears from 940° on. The method should be discarded for the determination of nickel.

(ii) Precipitation with sodium thiosulphate The method of Gibbs[20] employing sodium thiosulphate yields a precipitate of sulphur and nickel sulphide which adheres tenaciously to the walls of glass vessels. This mixture loses sulphur up to 300–350°. Between 350° and 800°, there is partial oxidation of the sulphide to sulphate and the latter is changed in part to the oxide NiO. The conversion into sulphate is furthermore far from being complete. Only nickel oxide NiO is really left beyond 800°. We suggest that this method likewise be discarded for the determination of nickel.

7. Nickel sulphate

The anhydrous sulphate $NiSO_4$ has been heated by Ostroff and Sanderson[21]; whereas Fruchart and Michel[22] and Demassieux and Malard[23] heated the heptahydrate. The former used the Mauer balance and heated the hexahydrate $NiSO_4 \cdot 6H_2O$ for 4 hours at 370° in the air. The resulting anhydrous sulphate began to decompose at 675°. (Other earlier values were 700°, 848°, 600°.) The residue consisted of NiO.

Fruchart and Michel[22] heated nickel sulphate heptahydrate at the rate of 0.6° per minute and made a differential thermal analysis at the same time. They observed that the hexahydrate is formed at 60°, where it undergoes a reversible allotropic transformation and then melts in its own water at 72°. The following hydrates appear transitorily: the tetra-hydrate at 104°, the dihydrate at 117°, and then the monohydrate, which is stable from 132° to 331°. The decomposition of the anhydrous sulphate begins at 730° and yields the oxide NiO.

Demassieux and Malard[23] found that 0.368 g of the heptahydrate, heated at the rate of 150° per hour, yields the monohydrate from 200° to 350°, the anhydrous salt from 460° to 750°, and the oxide NiO from 880° on.

8. Basic nickel sulphate

The basic sulphate $NiSO_4 \cdot 16H_2O$ was prepared by adhering closely to the directions given by Marshall[24] and Corminboeuf[25]. This material gives a good horizontal extending from 91° to 172°. Rapid dehydration sets in and all of the water has been driven off when the temperature has reached 200°. Up to 815°, the upper limit of our experiment, we were dealing with the anhydrous compound $NiSO_4 \cdot 8NiO$, but no decomposition of the sulphate was observed.

9. Nickel (potassium, rubidium, cesium) sulphate

Demassieux and Malard[23] investigated the thermolysis of the double sulphates of nickel with potassium, rubidium, and cesium. The samples weighed approximately 544 mg and were heated at the rate of 150° per hour. These three schönites yield similar curves. Four molecules of water are eliminated between 100–110° and around 180°. With regard to the last two molecules, the curve shows a distinct retardation between 180° and approximately 300° but no level is observed. It may be assumed that the anhydrous salts are formed from 300° upward. No loss was observed from 300° to 900°; volatilization and spattering occur above 900°.

10. Nickel sulphamate

Nickel amidosulphonate (sulphamate) $Ni(NH_2-SO_3)_2 \cdot 4H_2O$ was prepared at 60° in accord with the directions given by Capestan[26]. A partial dehydration starts at 100° and ends at 150° with a total loss of 2.5 molecules of water. The resulting $Ni(NH_2-SO_3)_2 \cdot 1.5H_2O$ is unstable and begins to decompose as soon as it is formed and in fact is manifested only through a change in the direction of the curve. A product corresponding to approximately $^2/_3$ H_2O per molecule is obtained at 210°. The loss of weight slows down at 210° but becomes rapid again near 360° and ends around 480° in a level corresponding to the sulphate. Analysis reveals a conversion of the hydrated amidosulphonate into ammonium sulphate and nickel sulphate.

11. Nickel ammonium sulphate

In this same paper Capestan[26] reported on the double nickel ammonium sulphate $NiSO_4 \cdot (NH_4)_2SO_4 \cdot 6H_2O$, which he prepared by crystallization at 15° from an equimolecular solution of the two sulphates. It loses all of its water in a single step from 130 to 225°. The resulting anhydrous product begins to decompose slowly at this latter temperature to yield the double sulphate 2 $NiSO_4 \cdot (NH_4)_2SO_4$. The latter, which is obtained around 430°, then decomposes further and leaves a residue of nickel sulphate at 500°.

12. Nickel nitrate

Nickel nitrate $Ni(NO_3)_2 \cdot 6H_2O$ + 0.13 H_2O is emerald green. It was studied at length by Weigel[27]. If heated rapidly (100° or 180° per hour), it melts at 57° to yield a dark green liquid; the levels are poorly defined. A yellow-green solid is obtained at 280°; it is a mixture. The black oxide NiO containing 3% of occluded gas appears at 340°. If heated slowly (10° or 5° per day), the green tetrahydrate $Ni(NO_3)_2 \cdot 4H_2O$ appears at 50°, the pale green dihydrate at 90°, a greyish green solid mixture at 180°, and the black oxide NiO containing 4% of occluded gas at 240°. See also Wendlandt[28] who reported 50°, 205°, and 505° as the respective temperatures at which the loss of weight begins, and the anhydrous salt and the oxide appear. He used 80 mg samples and a heating rate of 5.4° per minute.

13. Nickel arsenates

The arsenate was studied by Bastick and Guérin[29], first in vacuo, and then with the Chevenard thermobalance in nitrogen flowing at the rate of 500 ml per minute. The sample weighed 200 mg and the heating rate was 150° per hour. The starting material corresponded to the formula $NiO \cdot 2As_2O_5 \cdot 5H_2O$, which was prepared in a very acidic medium and always contained an excess of arsenic acid. There was no evidence of the presence of the binarsenate $NiO \cdot 2As_2O_5$. The metarsenate $NiO \cdot As_2O_5$ appears around 800° and gives an inflection at 900°, where the pyro-arsenate $2NiO \cdot As_2O_5$ furnishes a short level. The trinickelous ortho-arsenate $3NiO \cdot As_2O_5$ then forms progressively following a new loss of arsenic anhydride above 1000°.

14. Basic nickel carbonate

If, repeating the procedure of Gibbs[8], a nickel solution is treated with sodium carbonate, the resulting precipitate is not the carbonate $NiCO_3$, as he believed, but the so-called basic salt $NiCO_3 \cdot NiO$, which probably should be written as the orthocarbonate Ni_2CO_4, though this structure is still to be verified. This material is stable from 100° to 220° and yields an almost horizontal level. It then decomposes very slowly and the loss of the molecule of carbon dioxide continues up to 850°, where the horizontal of the oxide NiO starts.

François-Rosseti, Charton and Imelik[30] have furthermore made simultaneous studies of the poorly defined carbonates by thermal analysis and with the Eyraud balance at the rates 1.6°, 8°, and 10° per minute. The quasi-complete transformation of these carbonates into the oxide occurs between 300° and 350° (at the planned temperature gradients). Water and carbon dioxide are lost. The loss below 300° may be due to the elimination of adsorption products, notably water.

15. Nickel pyridine thiocyanate

Spacu[31] described the preparation of the nickel pyridine thiocyanate $[Ni(C_5H_5N)_4](SCN)_2$. The precipitate filters well. It is stable only up to 63° where it starts to lose one of the pyridine molecules. A horizontal between 110° and 130° therefore corresponds to $[Ni(C_5H_5N)_3](SCN)_2$. These two complexes may of course be recommended for the automatic determination of nickel. The interpretation of the rest of the thermogram is difficult since the results are only qualitatively reproducible. Although the oxide NiO is finally obtained above 920°, oxide contaminated with nickel sulphide and sulphate is already present from 350° on (as a minimum). Only oxide and sulphate are still present between 673° and 826°, but the amount of sulphate has no direct relation to the quantity of sulphur initially present in the complex. Summing up, the material should be dried in a desiccator below 63°; the weighing form is $[Ni(C_5H_5N)_4]$

$(SCN)_2$, whereas between $110°$ and $130°$ the formula is $[Ni(C_5H_5N)_3]$ $(SCN)_2$. Above $920°$ the material weighed is the oxide NiO.

16. Nickel molybdates

Corbet, Stefani, Merlin and Eyraud[32] and Corbet (Thesis) have studied the nickel molybdates. The latter used the Chevenard balance and a heating rate of $150°$ per hour in the air, and the Eyraud balance under reduced pressure ($5 \cdot 10^{-3}$ mm Hg) with the programmed rates of $180°$, $90°$, $45°$, and $22.5°$ per hour. This study was followed by a differential thermal analysis and by an X-ray investigation. Two products are obtained, depending on the amount of ammonia present during the preparation. The yellow product has the composition $NiO.MoO_3 \cdot xH_2O$ where x is between 1 and 2, and, apart from the water, the composition is constant, only the water content varying from one preparation to another. This water is eliminated up to around $500°$ under atmospheric pressure; subsequently, a horizontal is obtained for the anhydrous product designated as α. The second product is green; it does not correspond to a constant composition. Its formula is of the type $MoO_3 \cdot yNiO \cdot xH_2O$ where $y > 1$ and x has a variable value. If heated under atmospheric pressure, this material loses water up to around $605°$ without yielding a level; it apparently forms a mixture of the normal salt $NiMoO_4$ and the oxide NiO.

17. Nickel formate

Nickel formate $Ni(CO_2H)_2 \cdot 2H_2O$ is stable up to $150°$ when heated at the rate of $150°$ per hour. It then progressively loses water up to $240-245°$, and the level of the anhydrous salt extends to $270-275°$. An abrupt loss of water and carbon monoxide follows and the horizontal due to the oxide NiO begins at $310°$, but invariably there is a transitory production of metallic nickel which reoxidizes immediately. This salt has been studied also by Doremieux and Boullé[33], Kornienko[34], and Zapletal, Jedlička and Rüžička[35].

18. Nickel oxalate

The precipitate of nickel oxalate, produced in accord with the directions given by Gibbs[8] from oxalic acid and nickel sulphate in an alcoholic medium, has the formula $NiC_2O_4 \cdot 2H_2O$. It is not stable and loses water up to 232°, which marks the start of an oblique level, whose central portion corresponds to the anhydrous oxalate NiC_2O_4. The latter is unstable and loses carbon monoxide and most of the carbon dioxide abruptly at 272°. The nickel oxide is not obtained at once but via a basic carbonate, which is stable over a range of about 100°, and conforms very closely to the ensemble $NiCO_3 \cdot 20NiO$. The horizontal of the pure nickel oxide begins at 633°.

Wiedemann and Nehring[36] repeated this thermolysis. They also studied the mixed crystals with magnesium oxalate $MgC_2O_4 \cdot 2H_2O$, in air, hydrogen, nitrogen, and in vacuo.

19. Nickel anthranilate

The precipitated anthranilate $Ni(NH_2-C_6H_4-CO_2)_2$ yields a horizontal up to 310°. The drying temperature proposed by Funk and Ditt[37] and by Ishimaru[38] is suitable but too exclusive. This level is quite good for the automatic determination of nickel. Beyond 310°, the complex begins to decompose very rapidly and finely divided pure metal results at 640°. The latter quickly reoxidizes so that the oxide NiO is obtained quantitatively at 950°. If precipitated nickel anthranilate is ignited with oxalic acid, as was done by Ishimaru[38], the oxide NiO appears above 830°.

20. Nickel 5-bromoanthranilate

The excellent method of precipitation with 5-bromoanthranilic acid, suggested by Shennan[39], yields a compound $Ni(NH_2-C_6H_3Br-CO_2)_2$ which washes well. Its thermolysis curve includes a horizontal which is suitable for the automatic determination of nickel. This horizontal extends from 60° to 218°. The bromine and the organic materials are given off then but, in contrast to the preceding case, there is no reduction. The oxide NiO is obtained directly from 745° on.

21. Nickel 3-aminonaphthoate

The procedure with 3-aminonaphthoic acid, carried out as described by Shennan, Smith and Ward[40], results in a product whose curve is wholly identical with that of nickel anthranilate. However, the initial level corresponding to the formula $Ni(NH_2-C_{10}H_6-CO_2)_2$ remains horizontal only up to 154°. The loss of weight as far as 205° is quite minimal. Metallic nickel appears between 790° and 800°. It then quickly oxidizes up to 960°. The level up to 154° may be recommended for the automatic determination.

22. Nickel salicylaldehyde

In the method described by Duke[41], the nickel is precipitated by salicylaldehyde in the presence of an alkali tartrate and ammonia. In our opinion, this precipitate is not very usable; even though it does not begin to decompose below 230°; it shows a continuous gain in weight and the effect is rather large, amounting to 8 mg per 304 mg at 105°. In addition, this worker gives 0.1963 for the analytical factor for nickel; we find this too high and we believe it to be only 0.1940. During the pyrolysis of this compound, a very toxic material is released along with a very tenacious odor. The nickel oxide does not appear below 950°. We suggest that this method be discarded from all points of view.

23. Nickel oxinate

A precipitate whose formula is $Ni(C_9H_6ON)_2 \cdot 2H_2O$ is obtained with oxine in a solution buffered with acetic acid–acetate. In view of the fate of this precipitate when subjected to the action of heat, Berg[42], Kroupa[43], Fleck[44], and Fleck and Ward[45], all of whom have worked with this oxinate, dealt with it especially from the standpoint of its use in titrimetric determinations of nickel. Borrel and Pâris[46] found from the thermolysis curve that the dihydrate is stable up to 120°. A perfectly horizontal stretch from 160° to 232° corresponds to the anhydrous oxinate $Ni(C_9H_6ON)_2$; this level is well suited to the automatic determination of nickel with the analytical factor 0.1691. The decomposition then proceeds smoothly up to 690° where pure nickel is obtained, but starting at 770° the metal is reoxidized superficially and a level due to a mixture of $Ni + NiO$ appears. This level is a function of the size of the crucible and the atmosphere of the furnace. See also Sekido[47].

24. Nickel diiodooxinate

Wendlandt[48] prepared a material which he believed to be the diiodooxinate, but apparently it was not analyzed and it does not accord with the reported formula $Ni(C_9H_4I_2ON)_2$. A preparation of this kind was dried for 24 hours in the cold, and then heated in air at the rate of 4.5° per minute. It is stable only up to 80°; iodine is released and the organic matter burns. The level of the oxide NiO is reached at 505° when 100–150 mg of this product are taken for the thermolysis.

25. Nickel 2-methyloxinate

The 2-methyloxinate was studied by Borrel and Pâris[46]; it crystallizes in the form $Ni(CH_3-C_9H_5ON)_2 \cdot 1.5H_2O$. The stable monohydrate is formed near 65° and persists until around 215°, where it too begins to lose water and yields the anhydrous compound at 245°. The latter starts to decompose near 300°, slowly at first and then faster and faster. The pyrolysis slows down near 390° and resumes its normal course around 500°. The decomposition residue, which probably is a mixture of metal and oxide, reoxidizes at 600° as soon as the reducing materials derived from the decomposition of the organic matter have disappeared; the final product is the oxide NiO.

26. Nickel 2-phenyloxinate

The 2-phenyloxinate, prepared by Bocquet and Pâris[49], $Ni(C_6H_5-C_9H_5ON)_2 \cdot H_2O$ is dark brown and contains one molecule of water of crystallization which it starts to release at 95°. The anhydrous compound is stable up to 310°. The organic matter then burns off almost linearly as far as 700° where the horizontal of the oxide NiO starts.

27. Nickel picolinate

The picolinate or 2-pyridinecarboxylate $Ni(C_6H_4O_2N)_2 \cdot 4H_2O$ has been studied by Yoshikawa and Shinra[50], Thomas[51], and Lumme[52]. It loses its water between 75° and 145° and the level of the anhydrous compound extends as far as 230°.

28. Nickel quinaldate

Thomas[51] found that the bluish-green quinaldate or quinoline-2-carboxylate $Ni(C_{10}H_6O_2N)_2 \cdot 2H_2O$ is stable up to 165°. The anhydrous compound starts to decompose at 330°. See also Lumme[52].

29. Nickel benzothiazolecarboxylate

The benzothiazolecarboxylate was studied by Yoshikawa and Shinra[50]. It gives a level between 120° and 190°.

References p. 369

30. Nickel *o*-hydroxyphenylbenzoxazole

Wendlandt[53] investigated the thermal behaviour of the chelates produced by the *o*-hydroxyphenylbenzoxazole of Walter and Freiser[54]. The nickel compound is stable up to 230°; the resulting nickel oxide appears beyond 525°.

31. Nickel diliturate

Nickel diliturate, prepared by Mihai[55], has the formula $Ni[C_4H_2O_3N_2$ $(NO_2)_2] \cdot 8H_2O$. It, like the cobalt compound, is stable up to 80°, and then dehydration begins and yields the anhydrous salt whose horizontal extends from 180° to 240°. Complete disintegration beyond 380° leaves the oxide Ni_2O_3. It has been suggested that this salt be used for the microdetermination of nickel.

32. Nickel 8-quinaldate

The quinoline-8-carboxylate $Ni(C_{10}H_6O_2N)_2 \cdot 2H_2O$ is pale green. It is stable up to 100° and loses its water between 100° and 184°. The anhydrous salt is stable up to 350° and decomposes explosively beyond this temperature. The metallic nickel produced at 639° was found by Lumme[52] to be completely reoxidized at 700°.

33. Nickel α-nitroso-β-naphthylamine

The precipitation of nickel by α-nitroso-β-naphthylamine was carried out as described by Guha-Sircar and Bhattacharjee[56] to obtain the precipitate whose formula is $Ni(C_{10}H_7ON_2)_2$ and which these workers direct should be dried at 105°. Actually, the level corresponding to this compound is very short, going from 90° to 110°, but already at 103° there is loss of material and a not very violent explosion occurs near 345°. The nickel oxide NiO yields a horizontal above 462°. The method has little to recommend it.

34. Nickel β-nitroso-α-naphthylamine

The same workers[56] have described the procedure with the β-α-isomer, which seems to be more suitable; the horizontal given by the complex extends from 80° to 220° and lends itself well to the automatic determination of nickel. Unfortunately, the reagent is more difficult to prepare than the preceding one. Explosion occurs at 245° and the nickel oxide appears from 552° on.

35. Nickel dicyanodiamidine

The common reagent dicyanodiamidine sulphate yields a precipitate $Ni(C_2H_5N_4O)_2 \cdot 2H_2O$ which filters well. Grossmann and Schück[57], who first prepared this compound, advise that it be dried at 115°, or that it be converted into sulphate and electrolyzed. The recorded curve reveals that direct weighing of the precipitated material is possible. The level of the anhydrous complex extends from 123° to 240°. Accordingly, there is danger that the weight at 115° may be a little high. The method is suitable for the automatic determination of nickel. The NiO obtained near 720° is not entirely pure; it is contaminated with an apple-green basic carbonate, and consequently may not be used for a check determination.

36. Nickel nitroaminoguanidine

Phillips and Williams[58] recommend that the precipitate $NiO[HN=C(NHNO_2)NH-NH_2]_2$, which they prepared with nitroaminoguanidine, be dried at 110°. This temperature seems a little too high since the recorded curve reveals a loss of weight at 118° when the heating is very slow. The level is perfectly horizontal between 57° and 118° and meets the demands of automatic determination. An extremely violent explosion at 162–165° empties the crucible. The method is excellent but it requires very special care on the part of the operator.

37. Nickel dimethylglyoxime

We used the Bruncke[59] method for precipitating nickel with dimethylglyoxime. Ordinarily, the directions call for drying the red precipitate at 110–120°. However, this chelate is stable from 80° to 172° (or 180° in some recordings). The upper temperature is lowered when the heating is more rapid. The destruction of the organic matter is slow at first but it accelerates suddenly between 250° and 257°. Contrary to the common belief, there is no sublimation (except in vacuo) but rather two NO groups are lost per molecule of the oxime. The residue consists of carbon and metallic nickel. The level extending from 80° to 172° is suitable for the automatic determination of nickel.

38. Nickel diaminoglyoxime

The precipitation of the complex $Ni(C_2H_5O_2N_4)_2 \cdot 2H_2O$ by means of oxalene-diamineoxime or diaminoglyoxime or Niccolox was conducted along the lines laid down by Chatterjee[60] and Kuraš[61]. They prescribe drying at 120°, but in our opinion this temperature is too high. In fact, the curve includes a horizontal conforming to the above formula between 58° and 78° if the heating rate is low. The two molecules of water have disappeared at 120° and the salt begins to decompose. Strictly speaking, the procedure might serve for the automatic determination of nickel but the preceding method is just as good.

39. Nickel dicarbamidoglyoxime

The internal complex, obtained according to Feigl and Cristiani-Kornwald[62] with dicarbamidoglyoxime yields a curve which slants downward until 540° is reached. Nickel oxide appears beyond this temperature. We suggest that the method be discarded.

40. Nickel diphenylglyoxime

Atack[63] advises that the precipitate obtained from a nickel salt and diphenylglyoxime (benzildioxime) be dried at 110°. However, as in the preceding case and with very slow heating, it is not possible to find a horizontal in the thermolysis curve, which in fact declines without interruption as far as 960°, the temperature at which the nickel oxide NiO appears. The method can therefore not deliver accurate results in gravimetric analysis.

41. Nickel nioxime

The directions given by Voter, Banks and Diehl[64] were used with cyclohexanedionedioxime (nioxime). They suggest that the precipitate be dried at 110°. The curve discloses the existence of a horizontal as far as 115° where a rather violent explosion ensues. The method is excellently suited to the automatic determination of nickel but we advise using only that part of the level from 60° to 85°.

42. Nickel α-furildioxime

The precipitation of nickel with α-furildioxime, as suggested by Soule[65], has not been satisfactory in our laboratory. The chelate yields a curve which is inclined downward as far as 950° where nickel oxide appears. Admittedly, there is a break in the curve near 100° which corresponds closely to the anhydrous complex, but the latter decomposes at once without giving a level. At most, the procedure may serve for the separation of nickel, as outlined in a publication by Reed and Banks[66] of the U.S.A. Atomic Energy Commission, but less expensive reagents which deliver the same results are available.

REFERENCES

[1] R. DUVAL AND C. DUVAL, *Anal. Chim. Acta*, 5 (1951) 71.
[2] J. L. DAVIES, *Z. anal. Chem.*, 14 (1875) 343.
[3] A. K. MAJUMDAR, *J. Indian Chem. Soc.*, 18 (1941) 415.
[4] J. MIRONOFF, *Bull. soc. chim. Belg.*, 45 (1936) 1.
[5] H. FRESENIUS AND F. BERGMANN, *Z. anal. Chem.*, 19 (1880) 320.
[6] H. SAITO, *Sci. Repts. Tôhoku Imp. Univ.*, 16 (1927) 105.
[7] A. NICOL, *Compt. rend.*, 222 (1946) 1034.
[8] W. GIBBS, *Am. J. Sci.*, 4 (1867) 213; *Z. anal. Chem.*, 7 (1868) 259.
[9] C. CABANNES-OTT, *Compt. rend.*, 240 (1955) 68; *Thesis*, Paris, 1958.
[10] R. MEINEKE, *Lehrbuch der Chemische Analyse*, 1897, Vol. 1, p. 544.
[11] L. DEDE, *Chemiker-Ztg.*, 35 (1911) 1077.
[12] L. DEDE AND H. ZIERIACKS, *Z. anal. Chem.*, 124 (1942) 25.
[13] L. MOSER AND W. MAXYMOWICZ, *Z. anal. Chem.*, 67 (1925) 140.
[14] W. VAUBEL, *Chemiker-Ztg.*, 46 (1922) 978.
[15] E. MOODY, *Z. anal. Chem.*, 46 (1907) 247.
[16] B. GLASSMANN, *Ber.*, 39 (1906) 3368.
[17] SHEN-CHIA-SHU AND CHANG-MIU-SHIAO, *Acta chimica Sinica*, 26 (1960) 124.
[18] C. ROCCHICCIOLI, *Thesis*, Paris, 1960.
[19] R. FRESENIUS, *Quantitative Analyse*. 6th Ed., 1875, Vol. 1, p. 264.
[20] W. GIBBS, *Z. anal. Chem.*, 3 (1864) 389.
[21] A. G. OSTROFF AND R. T. SANDERSON, *J. Inorg. & Nuclear Chem.*, 9 (1958) 45.
[22] R. FRUCHART AND A. MICHEL, *Compt. rend.*, 246 (1958) 1222.
[23] N. DEMASSIEUX AND C. MALARD, *Compt. rend.*, 245 (1957) 1544.
[24] A. MARSHALL, *Analyst*, 24 (1899) 202.
[25] H. CORMINBOEUF, *Ann. chim. anal. et appl.*, 11 (1906) 6.
[26] M. CAPESTAN, *Ann. Chim. (Paris)*, 5 (1960) 225.
[27] D. WEIGEL, *Thesis*, Paris, 1960.
[28] W. W. WENDLANDT, *Texas J. Sci.*, 10 (1958) 392.
[29] M. BASTICK AND H. GUÉRIN, *La chimie des hautes températures, 2e Colloq. nat. C.N.R.S.*, 1957, p. 137.
[30] J. FRANÇOIS-ROSSETI, M. T. CHARTON AND B. IMELIK, *Bull. Soc. Chim. France*, (1957) 614.
[31] G. SPACU, *Bul. soc. stiinte Cluj*, 1 (1922) 314.
[32] F. CORBET, R. STEFANI, J. C. MERLIN AND C. EYRAUD, *Compt. rend.*, 246 (1958). 1696; F. CORBET, *Thesis*, Lyons, 1959
[33] J. L. DOREMIEUX AND A. BOULLÉ, *Compt. rend.*, 250 (1960) 3184.
[34] N. P. KORNIENKO, *Referat. Zhur. Khim.*, (1956), *Abstr. No.* 38984.
[35] V. ZAPLETAL, J. JEDLIČKA AND V. RŮŽIČKA, *Chem. Listy*, 50 (1956) 1406; *C.A.*, (1957) 2438.
[36] H. G. WIEDEMANN AND D. NEHRING, *Z. anorg. Chem.*, 304 (1960) 137.
[37] H. FUNK AND M. DITT, *Z. anal. Chem.*, 93 (1933) 241.

[38] S. Ishimaru, *J. Chem. Soc. Japan,* 55 (1934) 288.

[39] R. J. Shennan, *J. Soc. Chem. Ind.,* 61 (1942) 164.

[40] R. J. Shennan, J. H. F. Smith and A. M. Ward, *Analyst,* 61 (1936) 395.

[41] F. R. Duke, *Ind. Eng. Chem., Anal. Ed.,* 16 (1944) 750.

[42] R. Berg, *Z. anal. Chem.,* 71 (1927) 369.

[43] E. Kroupa, *Mikrochim. Acta,* 3 (1938) 306.

[44] H. R. Fleck, *Analyst,* 62 (1937) 378.

[45] H. R. Fleck and A. M. Ward, *Analyst,* 58 (1933) 388.

[46] M. Borrel and R. Pâris, *Anal. Chim. Acta,* 4 (1950) 267; 5 (1951) 573

[47] E. Sekido, *Nippon Kagaku Zasshi,* 80 (1959) 80; *C.A.,* (1960) 13950.

[48] W. W. Wendlandt, *Anal. Chim. Acta,* 15 (1956) 533.

[49] G. Bocquet and R. A. Pâris, *Anal. Chim. Acta,* 14 (1956) 201.

[50] K. Yoshikawa and K. Shinra, *Nippon Kagaku Zasshi,* 77 (1956) 1418; *C.A.,* (1958) 2642.

[51] G. Thomas, *Thesis,* Lyon, 1960.

[52] P. Lumme, *Suomen Kemistilehti,* B32 (1959) 237, 241.

[53] W. W. Wendlandt, *Anal. Chim. Acta,* 18 (1958) 638.

[54] J. L. Walter and H. Freiser, *Anal. Chem.,* 24 (1952) 984.

[55] F. Mihai, *Acad. rep. populare Romîne,* 4 (1957) 129; *C.A.,* (1960) 2000. F. Mihai and B. Roch, *Stud. cerc. stiinte. Ser. Timisoara,* 5 (1958) 113.

[56] S. S. Guha-Sircar and S. C. Bhattacharjee, *J. Indian Chem. Soc.,* 18 (1941) 155.

[57] H. Grossmann and B. Schück, *Chemiker-Ztg.,* 31 (1907) 535.

[58] R. Phillips and J. F. Williams, *J. Am. Chem. Soc.,* 50 (1928) 2469.

[59] O. Bruncke, *Z. angew. Chem.,* 20 (1907) 834, 1844.

[60] R. Chatterjee, *J. Indian Chem. Soc.,* 15 (1938) 608.

[61] M. Kuraš, *Collection Czechoslov. Chem. Communs.,* 12 (1947) 198.

[62] F. Feigl and A. Cristiani-Kornwald, *Z. anal. Chem.,* 65 (1925) 341.

[63] F. W. Atack, *Analyst,* 38 (1913) 316.

[64] R. C. Voter, C. V. Banks and H. Diehl, *Anal. Chem.,* 20 (1948) 459.

[65] B. A. Soule, *J. Am. Chem. Soc.,* 47 (1925) 981.

[66] S. A. Reed and C. V. Banks, *United States Energy Commission* A. E. C. D.,1819, Oak Ridge Tennesee, 9 March 1948

CHAPTER 29

Copper

Together with Marin[1,2] Duval made a thermogravimetric study of the precipitates containing copper and used for its gravimetric determination. In the course of this investigation, we discovered many new facts in inorganic chemistry, we have developed automatic procedures, and also rectified some rather serious errors. At the same time we have made a critical study of the best procedures.

There are far too many methods for determining copper gravimetrically. We have tested close to seventy such procedures. Our thermogravimetric and critical investigation has revealed that only about a dozen of these methods are acceptable and valid in all the cases met in actual practice. Nonetheless, we have not hesitated to lengthen this list because we found ourselves confronted by new interesting facts. We were the first to show that the internal complex produced by copper and cupferron is one of the most stable–if not the most stable–of the known materials of this category. Therefore, this cupferronate may be weighed as such, and without destroying it to obtain the copper oxide, and consequently the analytical factor is much improved; we can say the same regarding the use of neocupferron. We suggest that many of the published methods be discarded; they impress us as being incorrect, or they are not quantitative, or the formulas assigned to the precipitates by their authors are not exact, or finally the thermolysis curves given by the precipitates do not include any horizontal.

About ten of the methods are actually merely means of separating copper from other metals. The precipitates which they bring down must be converted into one of the forms discussed below. However, the authors of these methods have not always made the necessary complete investigations. They have been too hasty in assuming that the excellent results obtained with cobalt and especially nickel can be generalized and applied to copper.

We have uncovered a certain number of new facts regarding the oxidation of metallic copper. If precipitated by electrolysis or by displacement with iron, it oxidizes directly to cupric oxide when heated. But if the reduction is due to hydrazine, the copper yields cuprous oxide first. We have also found very notable differences in the temperature at which the various kinds of copper begin to oxidize. These findings are shown in the Table 10.

TABLE 10

OXIDATION TEMPERATURES OF VARIOUS KINDS OF COPPER

Copper deposited by	Temperature at which oxidation sets in (°C)
Electrolysis	67
Iron	103
Formaldehyde	140
Hydrazine	97
Hypophosphorous acid	194

These temperatures indicate that the varieties are in different states of division, since naturally the methods of heating were the same in the various trials.

Not less interesting were our findings relative to the action of various basic materials on a copper salt. Cabannes-Ott[3-5] employed infra-red spectrophotometry along with thermogravimetry to get an accurate picture of the structure of the various "hydroxides".

The study was supplemented by investigating the pyrolysis of various minerals, standard materials, catalyzing products, or in short by pyrolyzing copper derivatives which lend themselves to kinetic measurements.

We made a special study of the fate of cuprous sulphide when heated. We were the first to note that it then produces cupric oxide having the same percentage composition of metal following the formation and destruction of the intermediate copper sulphate. The purely fortuitous error is a consequence of the fact that the atomic weight of sulphur is twice that of oxygen. Cuprous iodide is converted into cupric oxide at a lower temperature than that reported in the literature, and accordingly we believe the physical constants of this iodide to be erroneous.

Several precipitates based on cupric thiocyanate have been reported to be free of organic matter, doubtless because of incomplete analyses. However, a good method of preparing cupric thiocyanate has been evolved from these errors, a material that hitherto has been a laboratory curiosity.

1. Metallic copper

(i) Precipitation by electrolysis With respect to electrolytic copper, it is common practice to dry the platinum gauze cathode in a gas flame and also to flame it after immersion in ethanol. It is quite proper to ask whether such a procedure is not likely to result in the oxidation of the newly deposited copper. We employed as cathode a small platinum dish which could be fitted into the carrier ring of the thermobalance, and coated the interior walls of this dish with a layer of electrolytic copper derived from a solution of copper sulphate acidified with sulphuric acid, as directed by Hecht and Donau[6], and deposited the metal by means of a 4-V current at 0.3 A. After rinsing the small dish with water and alcohol, it was heated gradually in the furnace. The weight remained quite constant up to 67°; beyond this temperature there was a gradual increase because of oxidation, while the colour changed from salmon-pink to black. In practice, therefore, after the cathode is immersed in alcohol and drained it should be kept in an oven below 67°. The gain in weight observed in our recording amounted to 20 mg per 100 mg at 255°.

(ii) Precipitation by iron The copper displaced by iron, as directed by Fuchs[7], begins to oxidize at 103°. The curve is identical with the previous one.

(iii) Precipitation by formaldehyde and Seignette salt When copper is precipitated by means of formaldehyde and sodium potassium tartrate as suggested by Hartwagner and Hersch[8], the reaction is carried out in a platinum dish which fits into the ring of the thermobalance. All useless manipulation is thus avoided. The copper obtained is relatively stable and does not begin to oxidize until around 140°.

(iv) Precipitation by hydrazine We followed the procedure of Jannasch and Biedermann[9] for precipitating copper by means of hydrazine. This method should not be confused with the one employing hydrazinium sulphate (see below). The resulting precipitate is brownish-red and retains but little wash water, which is rapidly removed up to 74°. However, the metal in this form is particularly sensitive to oxidation, which is true in general of metals produced by reduction with hydrazine. Determination by the automatic method can be recommended without hesitation. It is commonly believed that the copper is converted into CuO on heating. We have been able to uncover a new fact in this connection. The curve

shows that the oxidation proceeds until the copper has become univalent at 97°; in other words, the precipitated metal is converted into the red Cu_2O. However, since the latter is not within its stability zone, it gradually takes up oxygen and the weight increases slowly up to 787°. Above this temperature, the level of the cupric oxide CuO is reached. If the heating is continued above 1050°, cuprous oxide Cu_2O is found to reappear. We believe that the formation of this latter oxide in the vicinity of 100° may be due to the presence of a slight amount of hydrogen occluded in the reduced metal. Though not revealed by X-ray analysis, this hydrogen is still capable of reducing the layer of CuO formed initially. When the hydrogen has been consumed in the formation of water and eliminated as water vapour, the oxidation of the copper can then continue. The copper which has been precipitated by hydrazine and washed does not contain hydrazine, at least in an amount detectable by a spot test.

(v) Precipitation by hypophosphorous acid   Dallimore[10] has given a procedure for obtaining copper by the action of a 25% solution of hypophosphorous acid. The corresponding curve, analogous to the one given by electrolytic copper, remains horizontal up 194°, and then oxidation sets in rapidly. Even at 478°, 150 mg of this copper had taken up 16.5 mg of oxygen (theoretical 41 mg).

(vi) Precipitation by aluminium   We have been unable to reproduce the findings reported by Perkins[11], who used aluminium foils of diverse purities, unless an auxiliary metal, such as iron was added. The complete discharge of the colour of the solution under analysis was never observed. If the heating is too prolonged, the metal becomes contaminated with hydrous copper oxide.

(vii) Precipitation by zinc   Saito[12] studied copper powder, 0.074 mm in diameter, obtained by treating a solution of copper sulphate with zinc. He heated the specimens for 50 minutes at a series of constant temperatures: 250°, 300°, 400°, 500°, 600°, 700°. He found from these isotherms that the copper(II) oxide is formed especially at 165–170°, but the copper(I) oxide may be formed in small amount at lower temperatures. Saito has also oxidized the copper freshly obtained by the action of hydrogen on copper oxide.

2. Oxidation of copper

De Carli and Collari[13] also made a thermoponderal study of the oxidation of copper on the balance which they built. Their findings can be presented in several sections:

1. The isotherms at 500°, 600°, 650°, 690°, 750°, 800°, and 900° show a regular pattern with lengthening time:

$t°$	$Cu_2O\%$	$CuO\%$
500	70.00	30.0
600	87.5	12.5
650	95.0	5.0
690	95.5	4.5
750	96.0	4.0
800	98.5	1.5
900	98.5	1.5

2. At 400°, the increase of the layer of oxide is related to the time t in accord with an expression of the type:

$$1/P = A - B \log t$$

3. From 500° to 900°, the parabolic law prevails:

$$p^2 = Kt$$

4. From 500° to 900°, the law:

$$\log K = f(1/T)$$

divides into segments intersecting at 650°.

3. Oxidation of copper-tin alloys

De Carli and Collari[14] have also followed the oxidation of copper-tin alloys with the photographing thermobalance.

4. Copper(I) oxide

(i) Precipitation by hydroxylammonium chloride The method of Bayer[15] was employed to obtain cuprous oxide Cu_2O through the action of hydroxylammonium chloride in a basic medium. The brown-red precipitate does not adhere to the walls of the glass vessel, it filters well, and yields a very simple thermolysis curve. The latter discloses that the cuprous oxide formed in this way is stable up to 145° and may then be weighed after being dried below this temperature. The corresponding horizontal may also be employed for the automatic determination of copper. The oxidation proceeds slowly at first beyond 145° and becomes

rapid at around 285°, and near 350° the change into CuO is practically complete. The horizontal due to this oxide is observed from this temperature on.

(ii) Precipitation by sodium arsenite   The light green precipitate obtained from the action of sodium arsenite by the method described by Reichard[16] is stable up to 110° and yields Cu_2O which forms a horizontal extending from 150° to 288°. This product oxidizes slowly if the heating is continued and a new horizontal corresponding to the oxide CuO begins at 547°. The method appears to have less interest than the preceding one.

5. Copper hydroxide

Cabannes-Ott[3-5] has employed thermogravimetry, infra-red absorption, and X-ray spectrography to gain an exact picture of the nature of the products resulting from the action of a basic material on a cupric salt. The nature of these products changes with the anion of the salt, the nature of the base, the temperature, and the order in which the compounds enter the reaction. The hydroxide $Cu(OH)_2$ really exists. The best way to obtain it is to follow the procedure suggested by Peligot, namely to add potassium hydroxide to Schweitzer's reagent. Quite satisfactory results are obtained by adding the cupric solution to the potassium hydroxide solution (thus avoiding an excess of the cupric solution which can lead to the production of a basic salt but which provides a less favorable condition for the determination), or also by adding ammonia to a solution of copper acetate.

But the addition of a base, such as potassium hydroxide, to the solution of cupric sulphate, nitrate, or chloride results in the formation of basic salts, more or less clearly defined, possessing the approximate formula $CuX \cdot 3Cu(OH)_2 \cdot xH_2O$ in which X denotes SO_4, $(NO_3)_2$, and Cl_2, respectively.

The form of the precipitation is of minor importance from the analytical standpoint since CuO is finally weighed, but Cabannes-Ott found that the copper is not always quantitatively brought down when anions are absorbed.

(i) Precipitation by aqueous ammonia   Vaubel[17] proposed that the copper sulphate solution under analysis be boiled with aqueous ammonia; the resulting blue precipitate adheres to glass. Its thermogram includes three levels. The first appears after the extraneous moisture has been

driven off and extends up to about 200° (green product); its formula is not constant but approximates $CuSO_4 \cdot 3Cu(OH)_2 \cdot \frac{1}{2}H_2O$. The second level extends from 500° to about 650° (brown product); the composition is close to $CuSO_4 \cdot 3CuO$. The third level is due to the black oxide CuO; it begins near 865°. It should be noted that these findings are no longer valid if the initial solution is copper nitrate, chloride, or especially acetate rather than copper sulphate.

Among the papers dealing with this question, special note should be taken of the studies by Ishii[18] who reported the discovery of the hydrated hydroxide $Cu(OH)_2 \cdot 2H_2O$, which yields the oxide at 900°, and the investigations by Kato, Hosimija and Nakazima[19].

(ii) Precipitation by potassium hydroxide The well known method employing potassium hydroxide was taken from the Treadwell text on analytical chemistry. A 1 N solution of KOH is used in the cold. The form of the dehydration curve indicates that the retained water is not merely extraneous moisture but that part of it is bound water. However, Cabannes-Ott[5] found that a mixed formula such as $Cu(OH)_2 \cdot CuO \cdot H_2O$ is not very probable. The horizontal due to the oxide CuO begins near 439°.

(iii) Precipitation by sodium carbonate If, in accord with the Gibbs[20] procedure, a solution of copper sulphate is boiled with sodium carbonate, the black oxide CuO results. The thermolysis curve shows that the product is dry from 344° on, the temperature which marks the start of the horizontal extending beyond 947°. Accordingly, it is not possible to prepare a copper hydroxide by means of sodium carbonate or any other alkali carbonate.

(iv) Precipitation by morpholine The procedure, due to Malowan[21], in which a solution of copper sulphate is treated with an aqueous solution of morpholine yields a pale blue precipitate agreeing with the formula $Cu(OH)_2 \cdot H_2O$. This compound, which had been known previously, is quite stable up to 158°; it then slowly loses water up to 447° producing a new hydrate of the approximate composition 5 $Cu(OH)_2 \cdot H_2O$, which is stable from 447° to 762° and gives a horizontal between these two temperatures. A new loss of water is observed from 762° to 920° where the black oxide CuO is obtained.

(v) Precipitation by hexamine Hemmeler[22] described the reaction with hexamethylenetetramine. The resulting blue precipitate, whose for-

mula is approximately $Cu(OH)_2 \cdot H_2O$, gives a horizontal up to 90°. We suggest this method for the automatic determination of copper. About one molecule of water is lost up to 400° and a new hydrous oxide with a molecular weight of 100 appears, which accordingly is a little richer in water than the hydroxide $Cu(OH)_2$ whose molecular weight is 97.57. Finally, this latter product loses the remainder of its water from 563° to 780°, where the excellent horizontal of the oxide CuO starts.

(vi) Precipitation by hydrogen peroxide and saccharose   Jannasch and Routala[23] reported on the reaction of hydrogen peroxide plus saccharose which yields an emerald green precipitate with a cupric solution. The resulting hypothetical hydrous peroxide decomposes at room temperature and upward, yielding cuprous oxide starting at 220°. The latter then oxidizes slowly and cupric oxide is gradually obtained if the crucible is kept at about 960°.

6. Cupric chloride

Kamecki and Trau[24] studied the fate of the chloride $CuCl_2 \cdot 2H_2O$ on heating. We too have investigated this thermolysis several times. The dehydration sets in around 72°; it is complete at approximately 100° higher and the resulting yellow anhydrous chloride yields an inclined level. The salt is almost stable up to around 210°. No basic product appears. A double phenomenon then commences: loss of chlorine and gain of oxygen. The final copper oxide yields a horizontal from 730° on.

7. Cupric bromate

The hydrated copper bromate has been studied by Rocchiccioli[25]. When heated at the rate of 150° per hour, there is no observable difference between the hexahydrate and the anhydrous compound. Starting at 181° there is a violent loss of oxygen, and a horizontal is reached near 250° due to cuprous oxide and an oxybromide.

The deutero-hydrate $Cu(BrO_3)_2 \cdot 6D_2O$ loses 4 D_2O up to 160°. The same is true if the hexahydrate is heated at the rate of 50° per hour; it retains only 2 H_2O at 160°.

8. Cuprous iodide

Cuprous iodide was precipitated by treating the solution of the copper salt being determined with sulphur dioxide gas and adding potassium iodide. Pisani[26], the author of this procedure, recommended drying the precipitate at 110–120°. The curve shows that this iodide is stable up to 296°; it then rapidly loses iodine and takes up oxygen. The horizontal corresponding to the oxide CuO appears above 482°.

The temperatures given in the literature for the melting point (628°), the boiling point (772°), and the allotropic transformation (402°) appear doubtful to us. The curve obtained when cuprous iodide is heated in the air shows that this salt no longer exists at these temperatures.

9. Cupric iodate

Riegler's[27] method was employed for the precipitation of cupric iodate $Cu(IO_3)_2 \cdot H_2O$; it employs iodic acid and ethanol. He advised keeping the salt in a desiccator for 24 hours. The curve recorded with this material by Marin and Duval[1], Rocchiccioli[25], and Ishii[18] reveals a slight but very distinct gain of oxygen from room temperature up to 267°. The level corresponding to the anhydrous iodate extends from 287° to 440° and we suggest that this horizontal be employed for the automatic determination of copper. There is abrupt loss of iodine and oxygen beyond this point and the curve sharply changes direction at 497°. The residue slowly tends toward the composition CuO, namely the oxide which may be weighed from 650° on.

10. Copper periodates

Näsanen et al.[28,29] have constructed the thermolysis curves of two periodates, namely $Cu_2(OH)(H_2IO_6) \cdot H_2O$ and $Cu_5(IO_6)_2 \cdot 5H_2O$. The former loses its water irreversibly at 160°, producing $Cu_2(OH)(H_2IO_6)$ with no change in the shape of the crystals. A new loss of water occurs between 200° and 300° yielding a very fine brown powder whose composition is $Cu_4I_4O_{11}$. The latter then loses oxygen and iodine and produces copper oxide.

The second periodate yields an oblique level around 200° after losing 2.7 molecules of water. The hydrate $Cu_5(IO_6)_2 \cdot H_2O$ is present near 350°. Above 500° the residue consists of the oxide CuO while iodine, oxygen, and water are driven off.

11. Copper sulphide

(i) Precipitation by hydrogen sulphide   It is usually recommended that

the precipitate of cupric sulphide brought down by a stream of hydrogen sulphide gas be treated with a stream of hydrogen in a Rose crucible. We have carried out the experiment in air. The curve shows that under these conditions cupric sulphide loses combined sulphur, particularly starting at 100°; the cuprous sulphide, already partially oxidized, gives a visible minimum at about 300°. The oxidation to sulphate then begins, and all of the sulphate is still not decomposed at 950°. This fact had been pointed out in the classic treatise by Treadwell. High results are always obtained if the final weighing form is the oxide derived from sulphate.

(ii) Precipitation by thioacetic acid   The method of Schiff and Tarugi[30] was employed for the precipitation of the sulphide by means of thioacetic acid. Care must be taken to filter on paper and not on a fritted glass crucible, and to wash finally with two portions of alcohol to remove as well as possible the excess of thioacetic acid, whose odour is so persistent and unpleasant. The heating curve is very similar to the preceding one. It is not feasible to weigh the copper as cuprous sulphide or cupric sulphide. There is some chance of obtaining a weight around 1012° that is fairly correct for the oxide CuO resulting from the decomposition of the sulphate formed transitorily. However, the method cannot be recommended.

(iii) Precipitation by thioacetamide   Although Flaschka and Jakobljevich[31] apparently do not recommend the use of thioacetamide for the gravimetric determination of copper, we investigated the thermolysis of the sulphide produced by this reagent and copper sulphate. The curve differs from the two preceding ones in that it contains a horizontal up to 195°; the rest of the curve is essentially identical with the other two. In calculating on the basis of the copper oxide formed above 952°, the composition of the initial level then comes out close to CuS containing only 1/75th of Cu_2S. The minimum at 300° corresponds to essentially pure cuprous sulphide, *i.e.* not yet oxidized, and for a reason which will appear later. The use of thioacetamide is much less unpleasant than that of thioacetic acid, and while, in agreement with Flaschka and Jakobljevich, we cannot recommend it for the direct gravimetric determination of copper, it must be remembered that thioacetamide can serve for the preparation of the two copper sulphides in a relatively high state of purity.

(iv) Precipitation by thiosulphate   When thiosulphate is employed as the precipitant, Girard[32] directs that a mixture of the copper salt under

analysis and an excess of sodium thiosulphate be heated above 80°. Under these conditions, the precipitate consists of cuprous sulphide $Cu_2S = 159$ and some sulphur. The curve shown in Fig. 78 indicates that this sulphur is removed by heating, and pure cuprous sulphide is obtained at 260° (sometimes with a faint beginning of oxidation). In other words, the minimum is slightly above the dotted line. If the heating is continued, a more complete oxidation ensues, the product becomes greyish since

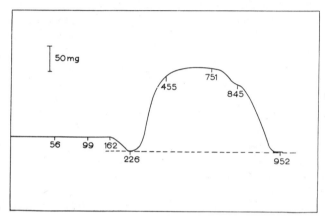

Fig. 78. Bell-shaped curve of cuprous sulphide.

it is a mixture of black cupric oxide and colourless anhydrous cupric sulphate:

$$2\ Cu_2S + 5\ O_2 \rightarrow 2\ CuO + 2\ CuSO_4$$

followed by the decomposition of the sulphate:

$$CuSO_4 \rightarrow SO_3 + CuO$$

Consequently, at 952°, the residue is an ensemble of CuO + CuO = 159, which has the same molecular weight as the initial cuprous sulphide. It is this fact that has led to the declaration that the method was very accurate; however, it is not cuprous sulphide that is weighed, for the residue contains absolutely no sulphur. It is necessary to ignite the residue at 950° and to make the calculations on the basis of CuO.

Ishii[18] prescribes a temperature of at least 850° for the complete conversion of cupric sulphide CuS into cupric oxide CuO.

(v) Minerals Gasco Sanchez *et al.*[33] have studied a new mineral whose formula is $Cu_{1.67}S$. In his general study, Saito[12] gives 8 curves for

chalcocite Cu_2S, 28 for chalcopyrite $CuFeS_2$, 7 for enargite Cu_3AsS_4, 2 for copper matte, etc.

12. Copper sulphate

We[1,34] have studied copper sulphate $CuSO_4 \cdot 5H_2O$ on various occasions, either from the analytical point of view since the copper contained in numerous inorganic and organic materials can be converted into sulphate, or because we considered this salt as a standard, or because we have studied at length the isotrimorphic system which it yields with zinc sulphate, or because we have replaced its hydrogen partially or wholly with deuterium. Assumed to be crystalline, and heated at the rate of 150° per hour, this salt, in powdered form, gradually loses 4 molecules of water from 67° to 153°; this is water that was coordinated to the copper atom. Then comes a horizontal extending as far as 250°; it has been attributed to the "monohydrate" but in truth the residue contains no water but only OH groups linked to the sulphur as SO_5H_2. The level due to the anhydrous salt begins at 303° and extends to 624°. We have never been able to observe any irregularity in the curve that might be interpreted as signalling the semihydrate. Sulphur dioxide is given off and the golden yellow basic sulphate $2CuO \cdot SO_3$ appears around 747°. Finally, the level of the oxide CuO begins above 910°. Many investigators, and likewise many students in laboratory courses, have pyrolyzed copper sulphate. Galimberti[35] reports that the transition from 5 to 3 molecules of water occurs from 30° to 38°, and from 3 H_2O to 1 H_2O between 68° and 74°, and from the monohydrate to the anhydrous salt from 200° to 220°. He used the Collari balance.

Ishii[18] obtained the oxide CuO at 800°. Orosco[36] also reports a trihydrate at 87°, and a monohydrate at 106°, and the anhydrous salt at 203°. Terem and Tugtepe[37] used isothermal and continuous heating. They too report the formation of the trihydrate by heating the pentahydrate for 1 hour. They prepared the semihydrate by rehydration of the anhydrous salt and they obtained the trihydrate in the same way. See also[38], Rigault[39] and Gordon and Denisov[40] who give 598° as the decomposition temperature of $CuSO_4$.

13. Copper sulphamate

Copper amidosulphonate $Cu(NH_2-SO_3)_2 \cdot 2H_2O$ was prepared by Capestan[41]. This blue salt loses one-fourth of its water from 70° to 80°. The green hydrate $Cu(NH_2-SO_3)_2 \cdot 1.5H_2O$ is completely dehydrated between

160° and 175° and produces the light yellow anhydrous amidosulphonate. The transformation into the imidodisulphonate $CuNH(SO_3)_2$ occurs between 230° and 250°. The level of the anhydrous sulphate then begins near 500°.

14. Copper hydrazinium sulphate

The compound $CuSO_4 \cdot (N_2H_5)_2SO_4$ was precipitated by the procedure of Reichinstein[42] in which a solution of copper sulphate is treated with hydrazinium sulphate. The colour of the copper solution is completely discharged. The precipitate settles well and is readily filtered. Its thermogram starts with a good horizontal which extends up to 132–133° and which has been recommended for the automatic determination of copper. Beyond this temperature, the hydrazine decomposes abruptly and reduces a portion of the copper sulphate. In fact, the bell-shape characteristic of the sulphide curves appears beyond 200°, followed by the decomposition of the copper sulphate and the appearance of the oxide, at 870°, as in the case of the sulphate. This method is remarkable in all respects; however, the copper salt being determined must be neutral and in the form of the sulphate.

15. Copper selenite

The copper selenite $CuSeO_3 \cdot H_2O$ or better $Cu(H_2SeO_4)$ was studied by Rocchiccioli[25]. Its molecule of water of constitution is not released below about 200°. There is no conversion to selenate but a basic selenite is formed whose level slopes downward between 378° and 520°. Copper oxide CuO is completely formed from 691° on.

16. Copper selenate

Malard[43] has studied the thermal behaviour of copper selenate $CuSeO_4 \cdot 5H_2O$. It starts to lose water at 80° and yields the monohydrate between 150° and 220°. The anhydrous salt is present from 265° on. A curve corresponding to the selenite begins at 480°, and the copper oxide CuO is formed around 700°.

The present writer has herself heated the double selenates of the schönite type $Cu(SeO_4) \cdot M_2SeO_4 \cdot 6H_2O$ in which M denotes potassium, rubidium, and cesium. At high temperatures, these compounds leave a mixture $M_2SeO_4 + CuO$; the respective temperatures being around 620°

for the potassium compound, 600° in the case of rubidium, and 800° for cesium.

17. Copper nitrate

Copper nitrate $Cu(NO_3)_2 \cdot 6H_2O$ was included by Peltier and Duval[44] in their study of the automatic determination of silver-copper alloys. This salt evolves water and nitrogen oxides from 58° on. A horizontal is observed from 148° to 200°, which corresponds to the existence zone of a crystalline green basic nitrate. Its formula (both calculated and found) is $Cu(NO_3)_2 \cdot 2Cu(OH)_2$ which gives the molecular weight 382. This compound decomposes from 200° to 607°, the decomposition is abrupt up to 252° and not so rapid thereafter. The residue consists of copper oxide CuO when heated up to 944°. Wendlandt[45] has repeated the pyrolysis of the trihydrate of this nitrate and reports that it starts to decompose at 70°, yielding the basic nitrate at 265°, and copper oxide CuO at 325°.

18. Copper dithiocarbamate

The ammonium dithiocarbamate method cannot be recommended. The Vogtherr[46] procedure for preparing the reagent $(NH_4)_2S_2CO$ results in a mixture. Klein[47] reports that it likewise leads to a mixture when it reacts with a copper salt, and when he heated the product in the air, with no special precautions, it gave cuprous sulphide around 245°, and the latter then was partially oxidized to sulphate. Copper oxide CuO was obtained above 970°.

19. Cuprous thiocyanate

The method of Kolthoff and Van der Meene[48] was employed for the precipitation, in the cold, of cuprous thiocyanate $CuSCN$. In general, authors do not appear to recommend the direct weighing of this thiocyanate. They prefer to convert it into CuO and into $CuSO_4$. The corresponding thermolysis curve indicates that cuprous thiocyanate is stable up to 300°; the level which it yields is not perfectly horizontal and slopes slightly upward. The gain in oxygen is less than 1/150th. Decomposition with evolution of sulphur and cyanogen sets in above 300°, and a minimum is found at 440° which satisfies the formula Cu_2S. The curve then takes on the bell shape given by thiosulphate. The black oxide CuO begins to appear near 950°. Nothing in this experiment indicates the presence of cupric thiocyanate. However, a simple procedure given below will deal with this salt.

20. Cupric nitroprusside

The nitroprusside $Cu[FeNO(CN)_5]$, prepared and washed as directed by Votocek and Pazourek[49], should be dried according to them at 110°. However, the thermolysis curve shows that the salt loses gaseous products even at 100°. We recommend a temperature not above 90°. Although the corresponding level is horizontal, we are reluctant to recommend this method for the automatic determination of copper because the precipitate passes through the filters. The remainder of the curve has little interest to the analyst; it refers to a mixture of iron and copper carbides.

21. Cuprous reineckate

The separation by means of Reinecke's salt was carried out as described by Mahr[50]; the necessary potassium chlorostannate was prepared by the method of Voyatzakis[51]. We found 16.64% of copper in the precipitate and not 16.36% as demanded by the formula $Cu[Cr(SCN)_4(NH_3)_2]$ = 381.5. Copper reineckate yields a horizontal up to 157°. Mahr prescribes drying at 110°. Constant weight can also be obtained very nicely after washing with alcohol and ether.

As shown by the curve, the decomposition of this complex is of course not simple. The residue consists of green copper chromite $CuO·Cr_2O_3$ and copper sulphide CuS. The calculated "molecular weight" of this mixture is 327, and the reading on the graph was 328. Half of the copper is converted into sulphate between 546° and 952°, with a maximum at 678°.

22. Cupric formate

Copper formate $Cu(COOH)_2·4H_2O$ has been studied by a number of workers: Zapletal, Jedlička, Rüžička[52], Kornienko[53], and ourselves. When heated at the rate of 150° per hour, it loses water continuously up to 120°, and is anhydrous from 120° to 200–205°. The salt then decomposes yielding water and carbon monoxide:

$$Cu(COOH)_2 \rightarrow H_2O + 2\ CO + CuO$$

The level of the CuO starts at 240°.

23. Cupric acetate

Copper acetate $Cu(CH_3COO)_2 \cdot H_2O$ begins to lose water at 107°. The dehydration continues gradually up to 167°. The anhydrous salt is not stable at this temperature and immediately begins to lose carbon monoxide. An explosion occurs around 270° and yields a mixture of copper and cuprous oxide, which gradually reoxidizes from 296° on. Duval[54] found that the resulting black copper oxide CuO maintains a constant weight beyond 510°.

24. Cupric oxalate

We used the procedure of Bornemann[55] for preparing copper oxalate CuC_2O_4; oxalic acid was added to a boiling solution of a copper salt. The precipitation is not complete and the precipitate is difficult to wash. The curve obtained with this product shows that after dehydration the resulting anhydrous oxalate is stable from 100° to 270°. It decomposes abruptly at 288°, but the reaction does not proceed according to the equation:

$$CuC_2O_4 \rightarrow CO + CO_2 + CuO$$

Cuprous oxide Cu_2O is doubtless obtained because of the presence of carbon monoxide which exercises a reducing action, and hence the reaction should be written as follows to fit the facts more closely:

$$2\ CuC_2O_4 \rightarrow CO + 3\ CO_2 + Cu_2O$$

However, this cuprous oxide is outside its stability zone and reoxidizes rapidly. The black oxide CuO is present from 494°, but as a whole its level is not strictly linear. There seems to be a scarcely perceptible maximum weight near 655°. The method cannot be recommended for the determination of copper. Ishii[18] reported that the limits of the level of the anhydrous oxalate are 150° to 210°.

25. Copper ethyl acetonedioxalate

Jilek and Lukas[56] suggest 105° for drying the precipitate $CuC_{11}H_{12}O_7$ produced by the action of ethyl acetonedioxalate, or they advise decomposing the material and weighing the resulting CuO. The curve obtained with this precipitate indicates that it does not maintain the preceding formula beyond 84°. It then gradually loses weight and decomposes suddenly near 240°–250°. The residue at 283° consists of pure cupric oxide CuO. The method therefore cannot be recommended; furthermore, the reagent is difficult to prepare.

26. Copper salicylaldehyde

The action of salicylaldehyde, employed in the Duke[57] method, yields a green chelated precipitate, which is identical with the product obtained by Pariaud and Labeille[58] with salicylhydramide. The compound is stable up to 105° (300° per hour); it is not of the imine type. It yields the level of the copper oxide CuO above 780°. This method need not be given serious attention. If the precipitation is conducted in accord with Pariaud and Labeille at pH 8.5 with $M/15$ salicylaldehyde and $M/30$ copper sulphate in 34% water–alcohol solution, the thermolysis curve of the precipitate has a different shape and copper oxide is obtained from 350° on.

27. Copper hexamine benzoate

The precipitate of the formula $3Cu(C_6H_5CO_2)_2 \cdot (CH_2)_6N_4$ obtained with benzoic acid and hexamethylenetetramine, as suggested by Ripan[59] and later by Dick[60], yields a level up to 60°. Beyond this, there is a sudden decomposition and the ignition residue is a mixture of black copper oxide and light green basic carbonate. Consequently, this residue cannot be considered as suitable for analytical purposes. We suggest that the method be abandoned.

28. Copper anthranilate

The excellent method suggested by Funk and Ditt[61] for the macro- and microdetermination of copper is equally good for its automatic determination. Sodium anthranilate is the precipitant. The level corresponding to the precipitate extends horizontally up to 225° where sudden decomposition occurs and continues up to 234°; from 234° to 335° there is additional but slower decomposition yielding metallic copper and a carbonate. Ishimaru[62], on the other hand, has reported 106–187° as the limits of the level of the dry anthranilate; he gives 473° as the lower temperature for the production of the oxide on ignition with oxalic acid. We worked in the open air and obtained a mixture of the black oxide and green basic carbonate which cannot be utilized gravimetrically.

29. Copper 5-bromoanthranilate

However, if 5-bromoanthranilic acid is used in place of the parent acid, then, according to Shennan[63], the precipitate has the formula $Cu(NH_2-C_6H_3Br-CO_2)_2$, and should be dried at 110°. The thermolysis curve confirms this fact because there is a good horizontal extending up to 204°. It is permissible to suggest this material for the automatic determination of copper. Furthermore, the reagent is readily prepared by brominating anthranilic acid. The decomposition of the salt is stormy up to 530°. This temperature marks the beginning of a horizontal due to a greenish mixture of oxide and carbonate, whose "mean molecular weight" is 118 per gram atom of copper.

30. Copper 3-aminonaphthoate

Shennan, Smith and Ward[64] have prepared the copper salt of 3-aminonaphthoic acid, which is a homologue of anthranilic acid. They report that the precipitate is very fine and hard to filter. We found this to be true, but in contrast to their finding, we found that the thermolysis curve descends continuously to 210°. Consequently, we are dubious of the propriety of drying the salt at 130° and then weighing it on the basis that it has attained constant weight, and also on the assumption that its formula is $Cu[C_{10}H_6(NH_2)CO_2]_2$. After the organic matter has been destroyed, the residue again is a greenish mixture of copper oxide and basic carbonate. Hence this method of determination should be discarded on all counts.

31. Copper cupferronate

The precipitation of copper by cupferron is conducted in an acetic acid milieu as suggested by Belluci and Chiucini[65]. Fresenius[66] suggested that the precipitate be heated with sulphur in a stream of hydrogen, a procedure which thus far has been impossible to follow with the thermobalance, and furthermore it strikes us as being entirely superfluous. Most of the writers suggest that the precipitate be subjected to a lengthy heating over a blast lamp with subsequent weighing of the resulting copper oxide CuO. The curve recorded with this precipitate revealed to us a fact which seemingly had escaped the notice of other chemists, namely, that of all the cupferronates studied with the thermobalance, those of iron and copper are by far the most stable and yield a quite horizontal level up to 107°. This finding leads us to suggest copper cupferronate for the automatic determination of this metal. It is hoped that this fact may be utilized in the future in separation operations since the other complexes of the same type decompose around 50°. Copper cupferronate decomposes rapidly above 118° and there is danger of losing copper oxide during the explosion. A mixture of oxides is left at 182°; the weight increases slowly but regularly. The constant weight obtained at 951° is that of copper oxide CuO.

32. Copper neocupferronate

Neocupferron, which is the ammonium salt of α-nitrosonaphthylhydroxylamine, was discovered by Baudisch and Holmes[67] who apparently did not suggest that it be employed for the gravimetric determination of copper. Likewise, there is no such proposal in the well known monograph "*Cupferron and Neocupferron*" by Smith published in 1940. We have therefore precipitated copper with a water solution of this reagent, at around 5° and in a medium acidified with hydrochloric acid. The voluminous greenish product filters well immediately, and should be washed with ice-water. There is marked shrinkage during the drying. The curve includes a good horizontal up to 89°, which may be used for the automatic determination of copper. This horizontal represents the existence zone of the internal complex, whose formula $Cu[C_{10}H_7N_2O_2]_2$ corresponds to a molecular weight of 437.88 and an analytical factor of 0.1451 for copper. The decomposition is rapid near 130°, not so fast after 156°, and the quantitative production of copper oxide CuO begins at 650°.

33. Copper N-benzoylphenylhydroxylamine

The technique developed by Shome[68] in which metals are precipitated by means of N-benzoylphenylhydroxylamine is more interesting for copper than for iron. The curve includes a level that is almost horizontal as far as 155°, where decomposition sets in abruptly. The constant weight due to copper oxide CuO is obtained from 642° on. However, the difficulty of filtering this precipitate, and the slight rise in the curve up to 155° prevent the employment of this method for the automatic determination of copper even though it is accurate in other respects.

34. Copper α-nitroso-β-naphthol

We used the procedure of Belluci and Chiucini[65] and of Wenger, Monnier and Besso[69] with α-nitroso-β-naphthol. These workers suggest drying the precipitate at 100–110°. The material whose formula is

Cu $(C_{10}H_6O-NO)_2$ yields a good horizontal from 62° to 172°; then an explosion empties the crucible. It should be pointed out that the copper compound is incomparably more stable than the cobalt or iron derivatives; it may be employed for the automatic determination of copper.

35. Copper α-nitroso-β-naphthylamine

By following the directions given by Guha-Sircar and Bhattacharjee[70] we have obtained with α-nitroso-β-naphthylamine a precipitate which loses weight from ordinary temperature on. We find little justification for retaining this method, which has been insufficiently studied, and which furthermore is not suitable for nickel or cobalt. Moreover, the product cannot be dried at 105°.

36. Copper β-nitroso-α-naphthylamine

We were unable to reproduce the findings of Guha-Sircar and Bhattacharjee[70] with β-nitroso-α-naphthylamine. Although this reagent is perfectly suitable for precipitating nickel or cobalt, we have not succeeded in precipitating copper with it, either in the cold or on boiling, no matter whether tartaric acid was present or not. No better success was achieved by raising the pH from 1 to 10.

37. Copper salicylaldoxime

We adopted the method of Jean[71] and Ducret[72] for the precipitation with salicylaldoxime. Furthermore, this form of precipitation had previously been employed in microanalysis. The internal complex $Cu[C_7-H_6O_2N]_2$ remains stable above 105–110°, $i.e.$ at temperatures suggested by these workers for drying the material, and even as high as 140°. The corresponding level may be proposed for the automatic determination of copper. This is an excellent method.

38. Copper benzoinoxime

The classic procedure with benzoinoxime has been repeated along the lines prescribed by Feigl[73], Strebinger[74] and in the text of Hecht and Donau[75]. After losing the adhering wash liquid, the complex, whose formula is $CuC_{14}H_{11}O_2N$, is stable from 60° to 143°. Ishibashi[76] reports 105° to 140°. The corresponding horizontal may be proposed for the automatic determination of copper. The compound decomposes abruptly between 143° and 156°, and then more slowly from 156° to 308° where all of the copper of the complex is obtained quantitatively in the metallic state within a very short time. Of course, this reduced copper oxidizes at once and the oxide CuO is obtained at 959°. Consequently, we are in agreement with the above workers who prescribed 105–110° as the drying zone.

39. Copper piperonaldoxime

As directed by Hovorka and Sykora[77], we worked in a solution buffered with sodium acetate and hexamethylenetetramine when precipitating copper with piperonaldoxime. However, we do not agree with them regarding the formula of the precipitate, which they give as $Cu(OH) \cdot C_8H_6NO_3$. Our product yields a horizontal extending as far as 100° and it corresponds to a molecular weight of only 126 per gram atom of copper. There is a second excellent horizontal between 422° and 611°. Cupric oxide is obtained from 774° on. We have been unable to get anything of much value from this method of precipitation.

40. Copper hydrazides

Duval, Buu-Hoï, Dat Xuong and Jacquinot[78] studied the internal complexes produced by copper and 5-chlorosalicylhydrazide, salicylhydrazide, 3,5-dichlorosalicylhydrazide, and nicotinic hydrazide. These products have been heated at the rate of 300° per hour. Their monohydrates are stable up to 75–80°; the molecule of water is lost between 135° and 140°. The anhydrous compounds exist up to about 200°. The organic matter decomposes in a complicated fashion and the level due to the copper oxide CuO is reached at temperatures which vary between 650° and 750° from one instance to another.

From these findings we developed a procedure for the microdetermination of copper in which 5-chlorosalicylhydrazide is the reagent. The limit of the method is 0.2 mg Cu per ml.

41. Copper diliturate

The diliturate $Cu[C_4H_2O_3N_2(NO_2)_2] \cdot 8H_2O$ was studied by Mihai[79]. It begins to lose water at 70° and is completely anhydrous at 230°. Suddenly it explodes; a new level beginning at 250° does not correspond precisely

to anything. The level of the copper oxide CuO starts at 380°. This diliturate, like those of cadmium, cobalt, and nickel are suggested for
microgravimetry.

42. Copper oxinate

The Berg[80] procedure was used with oxine. The precipitate is stated by
Borrel and Pâris[81] to have the composition $Cu(C_9H_6ON)_2 \cdot 2H_2O$; it loses
water rapidly beginning at 60°. The anhydrous oxinate is stable up to 269°
(or 300° according to Borrel and Pâris). It does not exhibit the peculiarities of cadmium oxinate. We suggest this level for the automatic determination of copper. The decomposition of the organic matter proceeds
up to a temperature which varies as always with the weight of material
and the velocity of the current of air passing through the furnace. When
300 mg of the oxinate was used, we found it was necessary to reach 834°
before the level due to the oxide CuO began. Borrel and Pâris obtained
constant weight by lengthy heating at 500°. See also Sekido[82].

43. Copper 5,7-dichlorooxinate

The precipitation with 5,7-dichlorooxine, as described by Fresenius[83],
was made with a reagent prepared in the laboratory from oxine. The
greenish precipitate $Cu(C_9H_4Cl_2ON)_2$ is stable from 63° to 117°. The
level extending between these temperatures is suitable for the automatic
determination of copper. Ishimaru[84] reported 122° and 192° as the
limits of this level; he found, as we did, that the copper oxide began to
appear at 525°.

44. Copper 5,7-dibromooxinate

Using the procedure of Berg[80] and Berg and Küstenmacher[85], we
obtained with 5,7-dibromooxine a precipitate $Cu(C_9H_4Br_2ON)_2$ which
yields a horizontal agreeing with this formula and extending from 170°
to 200°. Berg and Küstenmacher suggest drying the precipitate between
120° and 140° whereas Zan'ko and Bursuk[86] prescribe 160° to 190°. The
level can be suggested for the automatic determination of copper. The
compound decomposes and slowly sublimes at temperatures beyond 190°,
and there is nothing left in the crucible at 833°.

45. Copper 5,7-diiodooxinate

Wendlandt[87] studied the diiodooxinate whose formula is claimed to be $Cu(C_9H_4I_2ON)_2$. The precipitate begins to lose weight at 165°, and the level due to the oxide CuO begins at 455°. Mukherjee and Bannerjee[88] suggested that the compound be dried at 110–120°. Unfortunately, careful analysis of the precipitate reveals that it does not contain two atoms of iodine per molecule of oxine.

46. Copper 5-sulphooxinate

The reaction with 5-sulphooxine, studied by Lampitt, Hughes, Bilham and Fuller[89], seemingly does not yield a definite compound but instead an absorption product whose composition varies from one copper salt to the next. The thermolysis curve, besides being of no particular use, relates to a product containing less than one molecule of 5-sulphooxine per atom of copper, and which has a "molecular weight" of 161. We furthermore have reservations concerning the constitutional formula of the reagent as given in their paper. Our starting material certainly contained the SO_3H group in the 5-position, and its melting point was 310°. Examination of the precipitate under the microscope revealed that the material was not homogeneous. The mixture does not decompose under 295°, but the level is not strictly horizontal up to this temperature. The residual copper sulphate is completely converted into cupric oxide at 745°. We suggest that the method be discarded.

47. Copper 2-methyloxinate

The brownish red copper 2-methyloxinate $Cu(CH_3-C_9H_5ON)_2 \cdot H_2O$, prepared by Borrel and Pâris[90], is stable up to 110°. The molecule of water comes off between 110° and 140°; the horizontal of the anhydrous complex is present between 140° and 210°, and the colour turns greenish yellow. The methyloxinate decomposes rapidly beyond 210° and then more slowly between 265° and 400°. The decomposition again accelerates beyond 400° and produces copper oxide CuO.

48. Copper 2-phenyloxinate

The greenish black 2-phenyloxinate $Cu(C_6H_5-C_9H_5ON)_2 \cdot 2H_2O$ was prepared and studied by Bocquet and Pâris[91]. When they heated 300–

400 mg of this product to 700° at the rate of 3° per minute, they found the dehydration to begin at 45°. The anhydrous compound starts to decompose at 215°.

49. Copper quinaldate

Rây and Bose[92] prepared copper quinaldate or quinoline-2-carboxylate $Cu(C_{10}H_6O_2N)_2 \cdot H_2O$ and recommended that it be dried at 125°. The blue-green hydrate is stable up to 155°. The corresponding level may be recommended for the automatic determination of copper. The dehydration proceeds up to around 170° and the level of the anhydrous compound then extends as far as 295°; it too is suitable for a determination of this kind. At higher temperatures, the decomposition of the organic matter is more or less sudden because it forms much tar. The basic carbonate $CuCO_3 \cdot CuO$ or the orthocarbonate Cu_2CO_4 is produced above 800°; we have not clarified this matter. See also Yoshikawa and Shinra[93], Lumme[94] and Thomas[95].

50. Copper 5,6-benzoquinaldate

We followed the technique developed by Mallik and Majumdar[96] for preparing the copper salt of 5,6-benzoquinaldic acid. The precipitate filters well and these workers suggest that it be dried between 110° and 120°. The curve reveals a rapid loss of water up to 70°; doubtless this is moisture since the loss proceeds from room temperature on. The salt $Cu(C_{14}H_8O_2N)_2$ then yields a horizontal extending from 70° to 143° and we suggest that it be used for the automatic determination of copper. The organic matter decomposes abruptly at 720° but it is likely that the residue contains metallic copper or univalent copper because a slow re-oxidation occurs up to 1011°, where only pure CuO is left. This method is of interest also because of its analytical factor.

51. Copper 8-quinaldate

Quinoline-8-carboxylic acid, discovered by Majumdar[97], yields a blue precipitate $Cu(C_{10}H_6O_2N)_2$, which may be filtered after standing for 3 hours. Following removal of the wash water, the curve shows a strictly horizontal level extending from 110° to 221° and this agrees closely with the formula given above. Majumdar directs that the compound be dried at 150°, while Gilbreath and Haendler[98] suggest 110–120°. Accordingly these workers are in accord with the findings taken from the curve. We suggest this level for the automatic determination of copper. There is an abrupt loss of carbon dioxide at 226°, and up to 752° we find the ordinary curve of the combustion of carbon and tarry materials. The horizontal of cupric oxide CuO begins beyond 752°. Lumme[94] reports that the anhydrous salt gives a level extending from 191° to 287° with explosive decomposition near 305°.

52. Copper picolinate

The picolinate or pyridine-2-carboxylate was studied by Yoshikawa and Shinra[93], Thomas[95], Pâris and Thomas[99], and Lumme[94]. The first workers give 110–130° as the limits of the level for drying the compound. Thomas reports 2 molecules of water which begin to depart at 50°. Lumme found that the monohydrate $Cu(C_6H_4O_2N)_2.H_2O$ occurs as violet plates and is stable up 73°, while the anhydrous form remains stable from 120° to 300°. The oxide CuO is obtained from 441° upward. We likewise obtained a monohydrate but no dihydrate.

53. Copper biguanide sulphate

The precipitation of copper by means of biguanide sulphate was conducted as described by Rây and Roy-Chowdhury[100]. The pink precipitate $[Cu(C_2H_7N_5)_2]SO_4.3H_2O$ should be dried, according to them, at 50–70°. Actually, the curve given by this salt shows that it is not stable in the form of the trihydrate and that there is definite dehydration already at 40°, and that this loss of water is complete at 94°. The horizontal of the anhydrous complex extends from 94° to 146°, and we suggest this level for the automatic determination of copper, with the analytical factor 0.1745 for the metal. (We do not consider the temperatures 50–70° suitable for drying since they are situated on a downward slope of the curve.) Decomposition sets in at higher temperatures, apparently by a complicated mechanism, with production of cupric oxide CuO from 833° on. The minimum, and then the ascent of the curve at 565°, may be accounted for by the passage to copper sulphide arising from the reduction of the sulphate.

54. Copper thionalide

The thionalide method grew out of the researches of Berg and Roebling[101] who advised drying the product at 105°. The curve shows that the moist product loses its extraneous wash water up to 81°, the temperature at which the excellent horizontal conforming to the hydrate $Cu(C_{12}H_{10}ONS)_2 \cdot H_2O$ begins. This level extends as far as 121°. Consequently, the temperatures prescribed by these workers are correct; however, Umemura[102] in 1940 reported 140–200° as the limits of the anhydrous product. This level, as well as the following one, may serve for the automatic determination of copper. We observed the loss of the molecule of water of crystallization from 121° to 148°. Then a short level appears, extending up to 167° and agreeing with the anhydrous complex. The decomposition of the organic matter then sets in progressively with formation of a new level from 450° to 611°; it corresponds to a mixture. The end of the curve relates to the usual decomposition of copper sulphate which will have been formed via the sulphur of the complex. The resulting copper oxide does not appear below 950°.

55. Copper mercaptobenzothiazole

The directions given by Kuraš and Spacu[103,104] were followed for the precipitation of the copper derivative of mercaptobenzothiazole. The decomposition curve discloses that the complex, whose formula is $Cu(C_7H_4NS_2)_2$, is stable up to 145°. Therefore, the temperatures of 110–120° suggested by various workers are correct, and consequently the method may be suggested for the automatic determination of copper. From then on the destruction of the complex follows a complicated course and does not lead to pure cupric oxide, a finding which is in agreement with the earlier reports of Prodinger and Springer. The determination of copper by the mercaptobenzothiazole method is excellent, and furthermore the reagent is readily available.

56. Copper benzothiazolecarboxylate

Yoshikawa and Shinra[93] report that the monohydrated complex obtained by the action of benzothiazolecarboxylic acid is stable from 125° to 175°.

57. Copper 2-*o*-hydroxyphenylbenzoxazole

Wendlandt[105] studied the chelates produced with the 2-*o*-hydroxyphenylbenzoxazole of Walter and Freiser[106]. The copper chelate is stable up to 165° and the resultant copper oxide CuO appears above 525°.

58. Copper dithizonate

Copper dithizonate prepared by Pariaud and Archinard[107] gives cupric oxide with an explosion at 150°. The theoretical weight of the oxide is obtained at 717°.

59. Copper benzotriazole

For the preparation of the complex with the benzotriazole of Curtis[108] we have used laboratory preparations of the reagent, or the material purchased from the Eastman Kodak Company carrying the highest guaranties possible as to the nitrogen content, melting point, and ultraviolet absorption spectrum. The precipitate was thrown down in the presence of a tartrate. However, our results differed considerably from those reported by Curtis. The precipitate does not have the presumed formula $Cu(C_6H_4N_3)_2$ apparently confirmed by the excellent analyses published by Curtis. The molecular weight in one of our trials was 214; this precipitate loses weight from ordinary temperatures on. After the organic matter has been burned away, a mixture of copper and cuprous oxide is left at 932° and gradually takes up oxygen again. The theoretical weight of cupric oxide is obtained at 1017°. Until more information is available, we must regard this method as doubtful for the determination of copper.

60. Copper 6-nitrobenzimidazole

The green precipitate obtained with 6-nitrobenzimidazole, described by Vejdilek and Vorisek[109], appears to contain no organic matter. Our qualitative analysis revealed none, and the thermolysis curve is entirely identical with that given with ammonia, for instance. There is a horizontal up to around 158°, corresponding to a basic oxide with the molecular weight of 110, and a second level between 475° and 750° which corresponds to a basic oxide whose molecular weight is 100, but there is no assurance that these hydrates of cupric oxide are definite compounds. Consequently, we have rejected this method for the determination of copper.

61. Copper dicyandiamidine

The red inner complex $Cu(C_2H_5ON_4)$ produced with dicyandiamidine or guanylurea was prepared as directed by Grossmann and Mannheim[110], who recommended that the precipitate be dried at 120–140°. These temperatures strike us as too high. It is better not to exceed 100°, but in any event the curve does not include a very good level between 70° and 120°. The decomposition is rather rapid above 130°, and a mixture containing chiefly CuO is obtained from 425° on. We suggest that this method be discarded; in addition it is not strictly quantitative.

62. Copper dimine

In line with the procedure of Herrmann-Gurfinkel[111], we employed an 0.8% solution of dimine (cyclohexylethylaminedithiocarbamate) using a 50% excess of the reagent. The curve shows no level corresponding to the anhydrous complex; it continues to fall as far as 283°. The residual copper then slowly reoxidizes but the level corresponding to the oxide CuO is not horizontal enough. Although this method has some interest with respect to the determination of iron, it is less satisfactory for copper, and in any case it would be better if a titrimetric or a colorimetric procedure constituted the final step.

63. Copper pyridine bichromate

The complex salt $[Cu(C_5H_5N)_4]Cr_2O_7$ was precipitated with potassium bichromate and pyridine as described by Spacu[112]. He directed that the product be washed with acetone and ether and then dried in a desiccator at room temperature. Actually, the curve shows that this complex is stable up to 64°. From this temperature to 170°, and especially beginning at 80°, there is a loss of exactly 4 molecules of pyridine, leaving copper bichromate $CuCr_2O_7$ which yields a very short slightly inclined level. Prolonged heating above 223° probably will result in explosion. The level extending up to 64° may be suitable for the automatic determination of copper.

64. Copper bisethylenediamine iodomercurate

Copper(II) bisethylenediamine iodomercurate(II) $[Cuen_2][HgI_4]$ was prepared as described by Spacu and Suciu[113], who advise drying the product in a desiccator under reduced pressure. We found this procedure to be justified. Actually, the pyrolysis curve discloses that the complex gains a little weight (1 mg per 350 mg) up to 82°. True decomposition sets in at 99°; it is very rapid around 343° where the mercury and the iodine are driven off. The residue produced beyond 850° consists of the black oxide CuO.

65. Copper propylenediamine iodomercurate

The blue-violet precipitate [Cupn$_2$][HgI$_4$] is readily prepared with pro-pylenediamine by the method of Spacu and Spacu[114]. The curve is completely identical with that given by the ethylenediamine compound just described, and the precipitate is stable up to 157°. Accordingly, drying in a desiccator is not essential. Because of the favorable analytical factor, we suggest the corresponding level for the automatic determination of copper. Subsequently the propylenediamine, the mercury, and the iodine are driven off, and there is an abrupt change in the direction of the curve near 425°. The horizontal due to the copper oxide CuO begins at 720°.

66. Copper o-phenylenediamine iodomercurate

An analogous compound was obtained with o-phenylenediamine by Tarasevich[115]; its formula is [Cu(phen)$_2$][HgI$_4$]. This product yields a curve of the same type as those just described and the level corresponding to the above formula extends horizontally from 63° to 90°. At higher temperatures there is rapid escape of the mercury, iodine, and organic matter. The horizontal due to the oxide CuO starts at 560°. The method can be highly recommended since the precipitate forms readily, settles well, and may be filtered off immediately. In addition, the analytical factor for copper is excellent.

67. Copper pyridine thiocyanate

The procedure reported by Spacu and Dick[116] with potassium thio-cyanate and pyridine yields an unstable precipitate [Cu(C$_5$H$_5$N)$_2$](SCN)$_2$ which these workers weighed after it had been kept in a desiccator in the cold. Actually, the curve shows that this complex, differing in this respect from the others of the same type, loses 2 molecules of pyridine simultaneously between 46° and 134°, and then the resulting cupric thiocyanate decomposes in its turn. This latter compound will be dis-cussed later.

68. Copper isoquinoline thiocyanate

The complex [Cu(C$_9$H$_7$N)$_2$](SCN)$_2$, prepared in accord with the procedure described by Spakowski and Freiser[117] from thiocyanate and isoquinoline, shows the phenomenon of thixotropy to a marked extent while it is being washed. It is stable only up to 103° and is already decomposed to a considerable degree at 143°. At 345° the residue consists of practically pure cuprous sulphide. The remainder of the curve is already known to us; it shows the conversion of this sulphide to the sulphate and then to cupric oxide. This method is fairly satisfactory, but the level extending as far as 103° is not strictly horizontal so that it cannot be recommended for the automatic determination of copper.

69. Cupric thiocyanate

(i) Precipitation in the presence of benzidine   We have followed, step by step, the procedure given by Spacu and Macarovici[118] who used thiocyanate and benzidine for precipitating copper. We obtained a blue material to which they assigned the formula [Cubzd]$(SCN)_2$, but to our great astonishment our trials never yielded a product with this composition. In fact, its qualitative analysis and the calculations based on the curve showed that the material is free of organic matter. Burriel-Marti[119] modified the procedure and claimed to have obtained the above complex. Our product is stable up to around 168° and its composition agrees rather with $Cu(SCN)_2 \cdot \frac{1}{2}H_2O$, in other words with cupric thiocyanate, which admittedly is only slightly known in inorganic chemistry. It had previously been reported in the anhydrous condition. When heated, it yields cuprous sulphide and cupric oxide. The former oxidizes as usual to give copper sulphate, which subsequently loses sulphuric anhydride and only cupric oxide is present at 894°. We have been able to demonstrate that of 5 molecules of the final oxide, 3 are present prior to 500° and that the others come from the destruction of the transitory sulphate. Summing up, this method should be discarded as a means of determining copper. The benzidine, which we have been unable to detect by spot test reactions, serves only to facilitate the preparation of the cupric thiocyanate, which however cannot of itself be used for the gravimetric determination of copper.

(ii) Precipitation in the presence of tolidine   We came to identical conclusions with regard to the complex "containing tolidine" which Spacu and Macarovici[118] formulated as [Cutol]$(SCN)_2$. The curve obtained was perfectly identical with the preceding one, even though this should not have been the case since tolidine is the homolog of benzidine. After the

precipitate was washed, no tolidine could be detected in it by spot test reactions with gold(III) salts. Here again we are dealing with a product which consists of the monohydrate of cupric thiocyanate, and which is stable between 53° and 168°. Hence this procedure should be discarded as a means of determining copper. However, it does provide a good method for preparing pure cupric thiocyanate.

(iii) Precipitation in the presence of o,o'-dianisidine The precipitate obtained with thiocyanate and *o,o'*-dianisidine, by Spacu and Macarovici[118] and more recently by Buscarons and Loriente[120], consists of black cupric thiocyanate free of organic matter. When heated it yields cupric oxide from 885° on. The method should be discarded.

(iv) Precipitation in the presence of α-naphthylamine What has been stated in the preceding sections applies equally to the product obtained by Buscarons–Ubeda and Perez[121], who claimed that they had prepared $[Cu(C_{10}H_7NH_2)_2](SCN)_2$ by means of thiocyanate and α-naphthylamine. We repeated the procedure and found that it results indubitably in the precipitation of blackish cupric thiocyanate and contains no organic matter. The thermogram again shows a horizontal extending as far as 225°, followed by formation of cuprous sulphide near 348°, partial oxidation to sulphate, and eventual appearance of cupric oxide at 917°. Once more, we believe that the method should be discarded.

70. Copper 2-chloro-7-methoxy-5-thiolacridine

The precipitate produced with 2-chloro-7-methoxy-5-thiolacridine as directed by Das-Gupta[122] loses weight from room temperature on. It appears to be difficult to dry this material to constant weight at 100°. Furthermore, we have some doubts about the structure of the reagent when prepared by the method used by this worker. The residual cupric oxide begins to form at 554°. We recommend the abandonment of this method.

71. Copper picrolonate

Becherescu[123] reports that the picrolonate with 2.5 H_2O begins to lose water at 130–150°. The decomposition of the organic matter sets in above 200°, and yields cupric oxide at 280–300°.

References p. 402

72. Copper amine halogenides

Wendlandt and Whealy[124] prepared complexes of the general formula $(amine+H)_nCuX_{2+n}$ where the amine may be aniline, pyridine, ethylamine, methylamine, p-toluidine, trimethylamine, or the tetramethylammonium radical, and X denotes chlorine or bromine. The chlorine complexes decompose between 100° and 235°, while those containing bromine decompose from 75° to 110°. The levels due to the copper oxide CuO appear from 420° to 605°.

REFERENCES

1 Y. MARIN AND C. DUVAL, Anal. Chim. Acta, 6 (1952) 47.
2 Y. MARIN, Diploma of Higher Studies, Paris, 1949.
3 C. OTT, Compt. rend., 236 (1953) 2224.
4 C. OTT, Compt. rend., 238 (1954) 1113.
5 C. CABANNES-OTT, Thesis, Paris, 1958.
6 F. HECHT AND J. DONAU, Anorganische Mikrogewichtsanalyse, Springer, Vienna, 1940, p. 161.
7 DR. FUCHS, J. prakt. Chem., 17 (1839) 160.
8 F. HARTWAGNER, Z. anal. Chem., 52 (1912) 17; P. HERSCH, Analyst, 63 (1938) 486.
9 P. JANNASCH AND T. SEIDEL, J. prakt. Chem., 91 (1915) 152; P. JANNASCH AND K. BIEDERMANN, Ber., 33 (1900) 631.
10 P. B. DALLIMORE, Pharm. J., 29 (1908) 271.
11 G. E. PERKINS, J. Am. Chem. Soc., 24 (1902) 478.
12 H. SAITO, Sci. Rep. Tôhoku Imp. Univ., 16 (1927) 50.
13 F. DE CARLI AND N. COLLARI, Metallurgia ital., 1 (1952) 1.
14 F. DE CARLI AND N. COLLARI, Chim. & Ind. (Milan), 33 (1951) 77.
15 A. BAYER, Z. anal. Chem., 51 (1912) 733.
16 C. REICHARD, Ber., 31 (1898) 2166.
17 W. VAUBEL, Z. angew. Chem., 22 (1909) 1716; Z. anal. Chem., 49 (1910) 314.
18 K. ISHII, J. Chem. Soc. Japan, 52 (1931) 167.
19 H. KATO, H. HOSIMIJA AND S. NAKAZIMA, J. Chem. Soc. Japan, 60 (1939) 1115.
20 W. GIBBS, Z. anal. Chem., 7 (1868) 258.
21 L. S. MALOWAN, Mikrochemie, 26 (1939) 319.
22 A. HEMMELER, Ann. chim. appl., 26 (1936) 237.
23 P. JANNASCH AND O. ROUTALA, Ber., 45 (1912) 598.
24 J. KAMECKI AND J. TRAU, Bull. acad. polon. sci. Classe (III), 3 (1955) 111.
25 C. ROCCHICCIOLI, Thesis, Paris, 1960.
26 F. PISANI, Compt. rend., 47 (1858) 294.
27 E. RIEGLER, Z. anal. Chem., 43 (1904) 214.
28 R. NÄSANEN, R. UGGLA AND Y. IRVONEN, Suomen Kemistilehti, B, 30 (1957) 33.
29 R. NÄSANEN, R. UGGLA AND K. A. HELIN, Suomen Kemistilehti, 31 (1958) 163.
30 R. SCHIFF AND N. TARUGI, Ber., 27 (1898) 3437.
31 H. FLASCHKA AND H. JAKOBLJEVICH, Anal. Chim. Acta, 4 (1950) 485.
32 J. GIRARD, Ann. chim. anal., 4 (1899) 382.
33 L. GASCO SANCHEZ, T. BATUECAS RODRIGUEZ, H. SANG GARCIA AND F. FERNANDEZ-CELLINI, Anales real. soc. españ. fís. y quím., B, 55 (1959) 315.
34 C. DUVAL, Anal. Chim. Acta, 16 (1957) 223.
35 L. GALIMBERTI, Boll. sci. fac. chim. ind. Bologna, (1940) 272.
36 E. OROSCO, Ministério trabalho ind. e com. Inst. nac. tecnocol., (Rio de Janeiro), (1940); C.A., 35 (1941) 3485.

37 H. N. TEREM AND M. TUGTEPE, *Rev. fac. sci. univ. Istanbul*, *Sér. C*, 22 (1957) 70.
38 *1st Colloq. Thermography, Kazan*, 1953, p. 46.
39 C. RIGAULT, *Atti accad. sci. Torino*, 93 (1958-59) 489.
40 B. E. GORDON AND A. M. DENISOV, *Ukrain. Khim. Zhur.*, 19 (1953) 368: *C.A.*, (1955) 5062.
41 M. CAPESTAN, *Ann. Chim.* (Paris), 5 (1960) 215.
42 G. REICHINSTEIN, *Zavodskaya Lab.*, 15 (1949) 1407.
43 C. MALARD, *Compt. rend.*, 248 (1959) 2761.
44 S. PELTIER AND C. DUVAL, *Compt. rend.*, 226 (1948) 1727.
45 W. W. WENDLANDT, *Texas J. Sci.*, 10 (1958) 392.
46 M. VOGTHERR, *Ber. deut. pharm. Ges.*, 8 (1898) 228.
47 J. KLEIN, *Rep. Anal. Chem.*, 7 (1887) 629.
48 I. M. KOLTHOFF AND G. H. P. VAN DER MEENE, *Z. anal. Chem.*, 72 (1927) 337.
49 E. VOTOCEK AND J. PAZOUREK, *Chemiker-Ztg.*, 42 (1918) 475; *J. Soc. Chem. Ind.*, 37 (1918) 751.
50 C. MAHR, *Z. anorg. Chem.*, 225 (1935) 386.
51 E. VOYATZAKIS, *Bull. Soc. chim. France*, 1 (1934) 1356.
52 V. ZAPLETAL, J. JEDLIČKA AND V. RŮŽIČKA, *Chem. Listy*, 50 (1956) 1406.
53 P. KORNIENKO, *Ref. Zhur. Khim.*, (1956), *Abstr.* No. 38984; *C. A.*, (1959) 957.
54 C. DUVAL, *Anal. Chim. Acta*, 20 (1959) 264.
55 G. BORNEMANN, *Chemiker-Ztg.*, 23 (1899) 565.
56 A. JILEK AND J. LUKAS, *Chem. Listy*, 19 (1925) 275.
57 F. R. DUKE, *Ind. Eng. Chem., Anal. Ed.*, 16 (1944) 750.
58 J. C. PARIAUD AND P. LABEILLE, *Bull. Soc. chim. France*, (1956) 429.
59 R. RIPAN, *Bul. soc. stiinte Cluj*, 6 (1931) 286.
60 J. DICK, *Z. anal. Chem.*, 111 (1938) 260.
61 H. FUNK AND M. DITT, *Z. anal. Chem.*, 93 (1933) 241.
62 S. ISHIMARU, *J. Chem. Soc. Japan*, 55 (1934) 288.
63 R. J. SHENNAN, *J. Soc. Chem. Ind.*, 61 (1942) 164.
64 R. J. SHENNAN, I. H. F. SMITH AND A. M. WARD, *Analyst*, 61 (1936) 395.
65 I. BELLUCI AND A. CHIUCINI, *Gazz. chim. ital.*, 49 (1919) 187.
66 R. FRESENIUS, *Z. anal. Chem.*, 50 (1911) 42.
67 O. BAUDISCH AND S. HOLMES, *Z. anal. Chem.*, 119 (1940) 241.
68 SUDHIR CHANDRA SHOME, *Analyst*, 75 (1950) 27.
69 P. E. WENGER, D. MONNIER AND Z. BESSO, *Anal. Chim. Acta*, 3 (1949) 660.
70 S. S. GUHA-SIRCAR AND S. C. BHATTACHARJEE, *J. Indian Chem. Soc.*, 18 (1941) 155.
71 M. JEAN, *Bull. Soc. chim. France*, 10 (1943) 201.
72 L. DUCRET, *Bull. Soc. chim. France*, 12 (1945) 88
73 F. FEIGL, *Ber.*, 56 (1923) 2083.
74 R. STREBINGER, *Mikrochemie*, 1 (1923) 72.
75 F. HECHT AND J. DONAU, *Anorganische Mikrogewichtsanalyse*, Springer, Vienna, 1940, p. 155.
76 S. ISHIBASHI, *J. Chem. Soc. Japan*, 61 (1940) 125.
77 V. HOVORKA AND V. SYKORA, *Chem. Listy*, 33 (1939) 275.
78 C. DUVAL, NG BUU-HOÏ, NG DAT XUONG AND M. JACQUINOT, *Mikrochim. Acta*, (1953) 212.
79 F. MIHAI, *Acad. rep. populare Romîne*, 4 (1957) 129.
80 R. BERG, *Z. anal. Chem.*, 204 (1932) 208.
81 M. BORREL AND R. PÂRIS, *Anal. Chim. Acta*, 4 (1950) 267.
82 E. SEKIDO, *Nippon Kagaku Zasshi*, 80 (1959) 80; *C.A.*, (1960) 13950.
83 W. FRESENIUS, *Z. anal. Chem.*, 96 (1934) 433.
84 S. ISHIMARU, *J. Chem. Soc. Japan*, 55 (1934) 201.
85 R. BERG AND H. KÜSTENMACHER, *Z. anorg. Chem.*, 204 (1932) 215.
86 A. M. ZAN'KO AND A. K. BURSUK, *Zavodskaya Lab.*, 6 (1937) 675.
87 W. W. WENDLANDT, *Anal. Chim. Acta*, 15 (1956) 533.

88 A. K. MUKHERJEE AND B. BANERJEE, *Naturwissenschaften*, 42 (1955) 416.
89 L. H. LAMPITT, E. B. HUGHES, P. BILHAM AND C. H. F. FULLER, *Analyst*, 51 (1926) 327.
90 M. BORREL AND R. PÂRIS, *Anal. Chim. Acta*, 5 (1951) 573.
91 G. BOCQUET AND R. A. PÂRIS, *Anal. Chim. Acta*, 14 (1956) 201.
92 P. R. RÂY AND M. K. BOSE, *Z. anal. Chem.*, 95 (1933) 400; *Mikrochemie*, 17 (1935) 11.
93 K. YOSHIKAWA AND K. SHINRA, *Nippon Kagaku Zasshi*, 77 (1956) 1418; *C.A.*, (1958) 2642.
94 P. LUMME, *Suomen Kemistilehti*, 32 (1959) 237, 241.
95 G. THOMAS, *Thesis*, Lyon, 1960.
96 A. K. MALLIK AND A. K. MAJUMDAR, *Sci. and Culture (Calcutta)*, 14 (1949) 477.
97 A. K. MAJUMDAR, *J. Indian Chem. Soc.*, 18 (1941) 419.
98 J. R. GILBREATH AND H. M. HAENDLER, *Ind. Eng. Chem., Anal. Ed.*, 14 (1942) 866.
99 R. A. PÂRIS AND G. THOMAS, *15th Intern. Congr. Pure and Appl. Chem.*, Lisboa, 1956, *Actas do Congresso* II Vol. VI p. 25, Lisboa, 1958.
100 P. RÂY AND J. ROY-CHOWDHURY, *J. Indian Chem. Soc.*, 18 (1941) 149.
101 R. BERG AND W. ROEBLING, *Angew. Chem.*, 48 (1935) 597.
102 T. UMEMURA, *J. Chem. Soc. Japan*, 61 (1940) 25.
103 M. KURAŠ, *Collection Czechoslov. Chem. Communs.*, 11 (1939) 367.
104 G. SPACU AND M. KURAŠ, *Z. anal. Chem.*, 102 (1935) 24.
105 W. W. WENDLANDT, *Anal. Chim. Acta*, 18 (1958) 638.
106 J. L. WALTER AND H. FREISER, *Anal. Chem.*, 24 (1952) 984.
107 J. C. PARIAUD AND P. ARCHINARD, *Rec. trav. chim.*, 71 (1952) 634.
108 J. A. CURTIS, *Ind. Eng. Chem., Anal. Ed.*, 13 (1941) 349.
109 Z. VEJDILEK AND J. VORISEK, *Chem. Obzor.*, 20 (1945) 138.
110 H. GROSSMANN AND I. MANNHEIM, *Chemiker-Ztg.*, 42 (1918) 17.
111 M. HERRMANN-GURFINKEL, *Bull. Soc. chim. Belg.* 48 (1939) 94.
112 P. SPACU, *Z. anal. Chem.*, 115 (1939) 423.
113 G. SPACU AND G. SUCIU, *Z. anal. Chem.*, 78 (1929) 329.
114 G. SPACU AND P. SPACU, *Z. anal. Chem.*, 89 (1932) 187.
115 N. I. TARASEVICH, *Zhur. Anal. Khim.*, 4 (1949) 108.
116 G. SPACU AND J. DICK, *Z. anal. Chem.*, 78 (1929) 241.
117 A. E. SPAKOWSKI AND H. FREISER, *Anal. Chem.*, 21 (1949) 486.
118 G. SPACU AND C. G. MACAROVICI, *Z. anal. Chem.*, 102 (1935) 350.
119 F. BURRIEL-MARTI AND J. BARCIA-GOYANES, *Anales soc. españ. fís. y quím.*, B, 50 (1954) 281, 285.
120 F. BUSCARONS AND E. LORIENTE, *Anales soc. españ. fís y quím*, B, 44 (1948) 215.
121 F. BUSCARONS-UBEDA AND R. A. PEREZ, *Anales soc. españ. fís. y quim.* B, 38 (1942) 159.
122 DAS-GUPTA, *J. Indian Chem. Soc.*, 18 (1941) 43.
123 D. BECHERESCU, *Acad. rep. populare Romîne*, Timisoara, 6 (1959) 115.
124 W. W. WENDLANDT AND R. WHEALY, *Texas J. Sci.*, 11 (1959) 475.

CHAPTER 30

Zinc

Duval and De Clercq[1] have made a critical study of the thermolysis curves of the precipitates employed for the determination of zinc. We have omitted the precipitation of calcium zincate, proposed by Bertrand and Javillier[2], which eventually becomes a determination as sulphate. We also omitted the precipitation as borneol glucuronate, suggested by Quick[3]; the reagent is difficult to prepare, and furthermore we have been unable to find the slightest trace of zinc borneol glucuronate in 1500 ml of urine from dogs that had been fed for several days with borneol, either in paste form or by tube. We think that any reagent as capricious as this should be discarded. On the other hand, we have included in our study several soluble zinc salts: bromate, sulphate, sulphamate, formate, etc.

1. Metallic zinc

We have deposited zinc electrolytically in accord with the technique of Wenger, Cimerman and Tschanun[4]. The zinc does not maintain a constant weight beyond 54°; the weight then increases linearly due to an uptake of oxygen as far as 1025° (gain of oxygen: 6.0 mg per 117.9 mg). Therefore the cathode should always be dried at a low temperature in an oven.

2. Zinc hydroxide

(i) Precipitation by aqueous ammonia   The method introduced by Vaubel[5] was used for the precipitation of zinc hydroxide with ammonium hydroxide. The curve shows an abrupt change of curvature near 100° during the loss of water, and then the hydroxide appears, though it is more or less carbonated. The loss then slows down and a basic carbonate is obtained around 200°, but is does not release all of its carbon dioxide until a temperature of about 1000° is reached. We advise heating to at least 1010° to obtain a constant weight of zinc oxide ZnO.

(ii) Precipitation by mercuric oxide The Volhard[6] method, in which the hydrous oxide is brought down by mercuric oxide, yields a product whose curve is analogous to the preceding one. However, the intermediate basic carbonate is stable enough to produce a horizontal extending from 272° to 520°. In this case also, it is essential to carry the temperature to at least 970° to break down this carbonate and obtain the oxide ZnO quantitatively.

(iii) Precipitation by dimethylglyoxime The technique of Herz[7] employing dimethylglyoxime as precipitant yields a hydroxide and carbonate more stable than those discussed above, and which produce an approximately horizontal level between 100° and 240°. Only the basic carbonate remains around 280° and it slowly decomposes from 945° on. The material must be heated above 1000° to obtain the pure oxide.

(iv) Precipitation by morpholine The product obtained with morpholine, as suggested by Malowan[8], yields a curve identical with those just discussed but it is much sharper. In fact, there is a good horizontal between 60° and 95°; the carbonate remains quite stable between 200° and 388°. Then it slowly loses carbon dioxide and the shape of the curve changes near 566°. The weight of the oxide does not become constant unless the heating is carried to at least 910°.

(v) Precipitation by piperidine When piperidine is used as precipitant in accord with the directions of Herz[7], there is no level due to the hydrous oxide, but a very stable basic carbonate is obtained between 470° and 758°. The release of its carbon dioxide stops suddenly around 920°, the temperature which marks the start of the level due to the oxide.

3. Zinc bromate

Zinc bromate $Zn(BrO_3)_2 \cdot 6H_2O$ was studied by Rocchiccioli[9] who found that it loses 5 molecules of water from 56° to 153°. The anhydrous salt was not observed. The rest of the water and the oxygen depart concurrently and zinc bromide $ZnBr_2$ alone remains at about 280°.

4. Zinc sulphide

The various workers who have investigated the transformation of zinc sulphide into the oxide give very different figures for the passage from one form to the other. The curve which we constructed with a sulphide precipitated by hydrogen sulphide and heated in the air did not include any strictly horizontal level between 48° and 165°, and consequently it is difficult to use this as the weighing form as has been recommended by some workers. The departure of the sulphur accelerates around 230° and stops abruptly at 945° where the horizontal due to the oxide begins. We have been unable to detect the transient formation of the sulphate which certain chemists have reported.

Burriel-Marti, Barcia-Goyanes and Quiertant-Alvarez[10] repeated this work and have heated the zinc sulphide in the presence of barium sulphate (lithopone) hoping to discover an industrial method of analysis.

Zinc blende Zinc blende was heated by Saito[11] in air, and in sulphur dioxide, at temperatures increasing at the rate of 2° per minute; he constructed isotherms at 600°, 700°, 800° over periods of 50, 90, and 40 minutes. The fineness of the grains had a great influence on the reaction:

$$2ZnS + 3O_2 \rightarrow 2ZnO + 2SO_2$$

since it began at 647° with 0.10 mm grains as compared with 810° with 0.20 mm grains. The presence of iron oxide likewise exerts a decided effect. When the grains are coarse, decrepitation occurs at 350°.

If zinc sulphate is to be prepared by roasting, it is necessary to heat the finely powdered ore as slowly as possible, and of course at a temperature below the decomposition temperature of the sulphate. It may be assumed that:

$$ZnSO_4 \; (^{720°}) \rightarrow 3 \; ZnO \cdot 2SO_3 \; (^{765°}) \rightarrow ZnO$$

5. Zinc sulphate

A number of workers have studied the thermolysis of zinc sulphate. We used the method of Gutbier and Staib[12] which provided us with a sulphate containing a little more than 7 molecules of water. The curve revealed the removal of the water of crystallization and the excess sulphuric acid up to 300°. The level of the anhydrous sulphate extends from 300° to 788°. Friedrich[13] reported 675° as marking the start of the decomposition. This break-down proceeds actively from 788° and is complete around 950° with a quantitative yield of the oxide. The curve did not reveal the formation of either the hydrate or the basic sulphate.

Through the simultaneous use of differential thermal analysis, Bui-Nam[14] was able to distinguish $3ZnO \cdot 2SO_3$, previously reported by Saito, and also the pentahydrated sulphate.

Noshida[15] found that the heptahydrate yields the oxide at 950°. See also Paulik and Erdey[16] and Saito[11].

6. Zinc selenite

Zinc selenite $ZnSeO_3 \cdot 4H_2O$ was studied by Rocchiccioli[9]. If heated at the rate of 300° per hour, it loses its water from 90° on. The relatively robust anhydrous salt yields a level extending from 300° to 500°. The oxide ZnO gives a level which starts near 690° and continues as far as 950°. See also Korneeva and Novoselova[17].

7. Zinc sulphamate

Zinc amidosulphonate (sulphamate) $Zn(NH_2-SO_3)_2 \cdot 2H_2O$ was prepared by Capestan[18]. It loses all of its water in a single step from 105° to 155°. The resulting anhydrous salt decomposes from 335° to 450° yielding zinc sulphate.

8. Zinc nitrate

Noshida[15] reported that zinc nitrate $Zn(NO_3)_2 \cdot 6H_2O$ loses nitrogen oxides at 350°. See also Wendlandt[19]. We traced the curve at various heating rates and found a loss of weight from ordinary temperatures upward. The dehydration is observable as low as 40°; we have never observed any intermediate hydrate. A level is reached at about 350° but it is not strictly horizontal. The non-carbonated oxide is definitely present only beyond approximately 880°.

9. Zinc ammonium phosphate

We followed the directions given by Tamm[20] for the precipitation of zinc ammonium phosphate $ZnNH_4PO_4 \cdot H_2O$. The curve first presents a good horizontal extending from 50° to 167°; it corresponds to the anhydrous $ZnNH_4PO_4$. This level is suitable for the automatic determination of zinc. At higher temperatures, both water and ammonia are given off and there is quantitative production of the pyrophosphate $Zn_2P_2O_7$ only at 610°. However, the loss of weight between 410° and 610° is inconsiderable. We do not agree with Luff[21] who claims that the decomposition has not yet begun at 180°. Noshida[15], on the other hand, gives 550° as the temperature for the conversion into pyrophosphate. However, he does not clearly state the temperature at which the anhydrous salt is produced; he merely states that this occurs below 200°.

10. Zinc pyrophosphate

If the zinc solution being determined is treated directly with sodium pyrophosphate, the curve given by the precipitate shows that the product must be heated likewise to 610° to obtain the horizontal level of zinc pyrophosphate. See Chapter Phosphorus.

11. Smithsonite

Smithsonite from Tha Nguyen (Tonkin) was studied by Caillère and Pobeguin[22]. They found that it begins to decompose at 325° in nitrogen, and at 310° in air. These findings are in accord with the results given by differential thermal analysis.

12. Basic zinc carbonate

(i) Precipitation by ammonium carbonate The precipitation of the basic carbonate was conducted as described by Schirm[23] by means of ammonium carbonate. The precipitate gradually releases retained water up to 370°, where a perfectly horizontal stretch begins; it corresponds to the formula $5ZnO \cdot 2CO_2$. This latter compound is very stable and persists up to 880°; decomposition occurs beyond this temperature. Carbon dioxide is given off and the oxide is produced quantitatively near 1000°. Doubtless, Noshida[15] had mistaken this basic carbonate for the oxide itself.

(ii) Precipitation by guanidinium carbonate According to Grossmann and Schück[24], guanidinium carbonate yields a basic carbonate. The curve slants downward as far as 425° where the level of the oxide begins. However, this level is not perfectly horizontal and we do not recommend this method.

(iii) Precipitation by trimethylphenylammonium carbonate The basic carbonate was precipitated as suggested by Schirm[23] by means of tri-methylphenylammonium carbonate. The product yields a curve greatly resembling the preceding one. The level due to the zinc oxide begins at 250°.

13. Zinc cyanamide

The procedure described by Marckwald and Gebhardt[25] was employed for the precipitation of zinc cyanamide $ZnCN_2$. The curve obtained on heating this material reveals a rapid loss of water up to around 105°, followed by a horizontal level extending from 105° to 152°. This level corresponds to the formula $ZnCN_2$ and it may be used for the automatic determination of zinc. The decomposition is almost complete at 640°, but it results in the formation of the basic carbonate. The latter does not yield the oxide ZnO until the temperature has gone above 810°.

14. Zinc pyridine thiocyanate

Zinc pyridine thiocyanate $Zn(C_5H_5N)_2(SCN)_2$, prepared as described by Spacu[26], yields a curve which shows that the complex is stable only up to 71°. However, it does undergo a slight increase in weight (1 mg per 213 mg) from 20° to 71°. Hence it is better to dry the compound at low temperatures in a desiccator, as recommended by Spacu. Decomposition sets in around 70° and continues up to 780° where the horizontal of the oxide ZnO begins.

15. Zinc mercurithiocyanate

The mercurithiocyanate was prepared as described by Lundell and Bee[27], who suggested that the precipitate, whose formula is Zn[Hg (SCN)₄], be dried between 102° and 108°. This suggestion is well founded since the curve reveals that the precipitate is stable up to 270°. The corresponding horizontal is suitable for the automatic determination of zinc. However, there is a slight gain in weight between 120° and 270° (0.8 mg per 245 mg). The complex decomposes above 270°, and the horizontal of the zinc oxide starts around 820°.

16. Zinc formate

Zinc formate $Zn(CO_2H)_2·2H_2O$ is stable up to approximately 114°. It gradually loses water up to near 162°. The anhydrous salt is not stable; it begins to decompose at 178° and the level of the zinc oxide begins in an indecisive manner between 340° and 400°. This oxide seems not to be carbonated. See also Zapletal, Jedlička and Růžička[28], and Doremieux and Boullé[29].

17. Zinc oxalate

The oxalate dihydrate $ZnC_2O_4 \cdot 2H_2O$, prepared as directed by Classen[30], is stable up to 75°; then it loses its water of crystallization up to 165° but the anhydrous oxalate does not yield a horizontal level. In other words, it is not stable; it suddenly decomposes near 280° with evolution of carbon monoxide and carbon dioxide. There is no transient carbonate, and the resulting zinc oxide gives a horizontal which begins at 590°. The level of the dihydrate may be employed for the automatic determination of zinc. See also Erdey and Paulik[31], Doremieux and Boullé[29], Kawagaki[32]. The latter reports that the oxide appears at 450°.

18. Zinc anthranilate

We adopted the procedure of Funk and Ditt[33] for the preparation of the anthranilate. This method has been generalized for use in micro-analysis by Pregl and Roth[34] and Cimerman and Wenger[35], who recommend drying the salt between 105° and 110° for 30 minutes. Ishi-maru[36] reported the existence of a horizontal between 105° and 141°. When the product is ignited with oxalic acid it produces zinc oxide above 645°. We have not been able to verify this finding. Our curves are inclined downward between 105° and 110° and the calculations based on the weights of oxide formed above 500° do not agree with the formula $Zn(C_7H_6O_2N)_2$.

19. Zinc 5-bromoanthranilate

For the case of the 5-bromoanthranilate, we used the procedure given by Shennan[37], who suggests drying for 1 hour at 110°. The formula of the precipitate is $Zn(NH_2-C_6H_3Br-CO_2)_2$. The curve discloses that this material is stable up to 123°; the corresponding level may be suggested for the automatic determination of zinc. A slow loss of weight occurs from 123° to 230° where a more rapid decomposition occurs. This stops near 490° and yields a basic carbonate. The latter is stable as far as 875°; it then loses carbon dioxide and produces zinc oxide beyond 955°.

20. Zinc oxinate

Zinc oxinate, prepared as directed by Berg[38] in slightly acid solution, is the dihydrate $Zn(C_9H_6ON)_2 \cdot 2H_2O$, whereas if prepared according to the method of Wenger, Cimerman and Frank[39] it accords more closely with the formula $Zn(C_9H_6ON)_2 \cdot 1.5H_2O$. Borrel and Pâris[40] report that the precipitate dried at the ambient temperature is however the dihydrate which is stable up to 72°; it loses water slowly and regularly until the anhydrous oxinate is obtained at 170°. This complete dehydration may be accomplished likewise by heating the material for a sufficiently long time at 130–140°. On the other hand, De Clercq and Duval[1] found that the hydrate with 1.5 H_2O is stable only up to 65°, and that the anhydrous oxinate yields a horizontal extending from 127° to 284°, a level which we suggest be used for the automatic determination of zinc. Beyond this temperature, we found the customary pyrolysis curve of oxine, and a mixture of zinc oxide and zinc carbonate appeared at 960° and upward. See also Sekido[41]. Borrel and Pâris report that the oxide ZnO is obtained free of carbonate by prolonged heating at 700°.

21. Zinc methyloxinate

It is very difficult to remove the excess precipitant from zinc methyloxinate $Zn(CH_3-C_9H_5ON)_2 \cdot H_2O$, which was prepared as directed by Merritt and Walker[42]. They recommend that the resulting material be kept between 130° and 140°. Borrel and Pâris [43] found however that it is stable up to 130°, that it then loses the molecule of water up to 180° and yields the anhydrous salt between 180° and 255°. The chelate then decomposes, very slowly at first up to around 400°, and the abatement observed between 350° and 400° appears difficult to interpret. The decomposition becomes regular beyond 400° and continues thus until the residue of zinc oxide ZnO is obtained.

22. Zinc phenyloxinate

The bright yellow precipitate of anhydrous zinc phenyloxinate $Zn(C_6H_5-C_9H_5ON)_2$, prepared by Bocquet and Pâris[44], begins to decompose at 330°. The level due to the resulting zinc oxide seems to commence near 820°.

23. Zinc monosalicylaldoxime

The complex with monosalicylaldoxime $Zn(C_7H_6O_2N)_2$, prepared as directed by Flagg and Furman[45], has been studied by Rynasiewicz and Flagg[46] with the Chevenard thermobalance. We had suggested that the method be abandoned since we had been unable to obtain a horizontal. The preceding workers have shown that the correct drying temperature lies between 25° and 285°, and the level begins between 25° and 135° depending on the initial humidity of the product. The final product is zinc oxide which appears between 500° and 1000°. The very considerable instantaneous release of gas between 290° and 310° is dangerous and produces an intolerable odour.

24. Zinc dithizonate

The precipitate of zinc dithizonate, obtained by the procedure of Wassermann and Ssupronowitsch[47], is best dried in a desiccator as they recommend. The curve shows that the drying can be conducted at as high as 68° without significant error. Furthermore, Pariaud and Archinard[48] report that this material explodes at 225° leaving zinc oxide. We regard this procedure as a secondary method, at least as regards the gravimetric determination of zinc.

25. Zinc 2,7-diaminofluorene

The method employing 2,7-diaminofluorene is due to Nino and Calvet[49] who recommend drying the precipitate at 90° and applying the analytical factor 0.197. The curve obtained with this precipitate includes a horizontal section extending from 98–100° to 207–210°; we find this corresponds to a molecular weight of 220.5, and accordingly an analytical factor of 0.299 for zinc. This level gives us the impression that the material is not homogeneous. The destruction of the organic matter sets in at 207° and is complete at 970°. The latter temperature marks the start of the level due to the zinc oxide containing no carbonate.

26. Zinc picolinate

The pyridine-2-carboxylate $Zn(C_6H_4O_2N)_2$ has been heated by Lumme[50] and by Thomas[51]. The former reports 2 molecules of water, the latter found 4 molecules which begin to come off at 70°.

27. Zinc quinaldate

The precipitation of zinc quinaldate was due to Râ y and his associates [52-55] who recommend drying the white precipitate at 125°. The formula is

$Zn(C_{10}H_6O_2N)_2 \cdot H_2O$. The compound yields a horizontal extending as far as 240° and this level can serve for the automatic determination of zinc. The decomposition of the anhydrous compound then begins around 340° and the level due to zinc oxide contaminated with carbonate starts near 770°. The study was repeated by Kiba and Sato[56] who give 110–170° as the limits of the level due to the monohydrate, and by Lumme[57], and Thomas[51].

28. Zinc picrolonate

Zinc picrolonate, which was studied by Becherescu[58], undergoes a continuous loss of weight, notably at 160–280° and at 430–490°. Zinc oxide is obtained beyond this latter temperature.

REFERENCES

[1] M. DE CLERCQ AND C. DUVAL, *Anal. Chim. Acta*, 5 (1951) 282.
[2] G. BERTRAND AND M. JAVILLIER, *Compt. rend.*, 143 (1906) 900.
[3] A. J. QUICK, *Ind. Eng. Chem., Anal. Ed.*, 5 (1933) 26.
[4] P. WENGER, C. CIMERMAN AND G. TSCHANUN, *Mikrochim. Acta*, 1 (1937) 51.
[5] W. VAUBEL, *Angew. Chem.*, 22 (1909) 1716.
[6] J. VOLHARD, *Liebigs Ann.*, 198 (1879) 331; 199 (1879) 6.
[7] W. HERZ, *Z. anorg. Chem.*, 26 (1901) 90.
[8] L. S. MALOWAN, *Mikrochemie*, 26 (1939) 319.
[9] C. ROCCHICCIOLI, *Thesis*, Paris, 1960.
[10] F. BURRIEL-MARTI, C. BARCIA-GOYANES AND M. QUIERTANT-ALVAREZ, *Las Ciencias*, 24 (1959) 501.
[11] H. SAITO, *Sci. Repts. Tôhoku Imp. Univ.*, 16 (1927) 52.
[12] A. GUTBIER AND K. STAIB, *Z. anal. Chem.*, 61 (1922) 97.
[13] K. FRIEDRICH, *Metallurgie*, 6 (1909) 175.
[14] BUI-NAM, *Compt. rend.*, 249 (1959) 1108.
[15] I. NOSHIDA, *J. Chem. Soc. Japan*, 48 (1927) 520.
[16] F. PAULIK AND L. ERDEY, *Acta Chim. Acad. Sci. Hung.*, 13 (1957) 117.
[17] I. V. KORNEEVA AND A. V. NOVOSELOVA, *Zhur. Neorg. Khim.*, 4 (1959) 2220.
[18] M. CAPESTAN, *Ann. Chim. (Paris)*, 5 (1960) 222.
[19] W. W. WENDLANDT, *Texas J. Sci.*, 10 (1958) 392.
[20] H. TAMM, *Chem. News*, 24 (1871) 148; *Z. anal. Chem.*, 13 (1874) 320.
[21] G. LUFF, *Chemiker-Ztg.*, 45 (1921) 613.
[22] S. CAILLÈRE AND T. POBEGUIN, *Bull. soc. franç. minéral.*, 83 (1960) 36.
[23] E. SCHIRM, *Chemiker-Ztg.*, 35 (1911) 1177, 1193.
[24] H. GROSSMANN AND B. SCHÜCK, *Chemiker-Ztg.*, 30 (1906) 1205.
[25] W. MARCKWALD AND H. GEBHARDT, *Z. anorg. Chem.*, 147 (1925) 42.
[26] G. SPACU, *Bull. soc. stiinte Cluj*, 1 (1922) 361.
[27] G. E. F. LUNDELL AND H. BEE, *Eng. Mining J.*, 99 (1915) 701.
[28] V. ZAPLETAL, J. JEDLIČKA AND V. RŮŽIČKA, *Chem. Listy*, 50 (1956) 1406; *C.A.*, (1957) 2438.
[29] J. L. DOREMIEUX AND A. BOULLÉ, *Compt. rend.*, 250 (1960) 3184.
[30] A. CLASSEN, *Z. anal. Chem.*, 18 (1879) 189, 373.
[31] L. ERDEY AND F. PAULIK, *C.A.*, (1956) 3952.
[32] K. KAWAGAKI, *J. Chem. Soc. Japan*, 72 (1951) 1079.

33 H. Funk and M. Ditt, *Z. anal. Chem.*, 91 (1932) 332.
34 F. Pregl and H. Roth, *Organische Mikroanalyse*, 4th Ed., Springer, Berlin, 1935, p. 117.
35 C. Cimerman and P. E. Wenger, *Mikrochemie*, 18 (1935) 53.
36 S. Ishimaru, *J. Chem. Soc. Japan*, 55 (1934) 288.
37 R. J. Shennan, *J. Soc. Chem. Ind.*, 61 (1942) 164.
38 R. Berg, *Z. anal. Chem.*, 71 (1927) 171.
39 P. E. Wenger, C. Cimerman and D. Frank, *Compt. rend. soc. phys. et hist. nat. Genève*, 53 (1936) 57; *Mikrochemie*, 24 (1938) 148, 153, 162.
40 M. Borrel and R. Pâris, *Anal. Chim. Acta*, 4 (1950) 267.
41 E. Sekidos, *Nippon Kagaku Zasshi*, 80 (1959) 80; *C.A.*, (1960) 13950.
42 L. L. Merritt Jr. and J. K. Walker, *Ind. Eng. Chem., Anal. Ed.*, 16 (1944) 387.
43 M. Borrel and R. Pâris, *Anal. Chim. Acta*, 5 (1951) 573.
44 G. Bocquet and R. A. Pâris, *Anal. Chim. Acta*, 14 (1956) 201.
45 J. L. Flagg and N. H. Furman, *Ind. Eng. Chem., Anal. Ed.*, 12 (1940) 663.
46 J. Rynasiewicz and J. L. Flagg, *Anal. Chem.*, 26 (1954) 1506.
47 I. S. Wassermann and I. B. Ssuprunowitsch, *Chem. Zentr.*, 106 (1935 II) 3268.
48 J. C. Pariaud and P. Archinard, *Rec. trav. chim.*, 71 (1952) 634.
49 E. L. Nino and F. Calvet, *Anales soc. españ. fís. y quím.*, 32 (1934) 698.
50 P. Lumme, *Suomen Kemistilehti*, 32 (1959) 241.
51 G. Thomas, *Thesis*, Lyon, 1960.
52 P. R. Rây and M. K. Bose, *Z. anal. Chem.*, 95 (1933) 400.
53 P. R. Rây and N. K. Dutt, *Z. anal. Chem.*, 115 (1939) 265.
54 P. R. Rây and A. K. Majumdar, *Z. anal. Chem.*, 100 (1935) 324.
55 P. R. Rây and P. C. Sarkar, *Mikrochemie*, 27 (1939) 64.
56 T. Kiba and S. Sato, *J. Chem. Soc. Japan*, 61 (1940) 133.
57 P. Lumme, *Suomen Kemistilehti*, 32 (1959) 237.
58 D. Becherescu, *Acad. rep. populare Romîne Timisoara*, 6 (1959) 115.

CHAPTER 31

Gallium

The gravimetric procedures for determining gallium all eventually come down to the determination of the oxide Ga_2O_3. Weighing as the oxinate leaves much to be desired since this compound is sublimable and decomposes readily. We have omitted from this discussion the precipitation as basic acetate since it is not quantitative, and likewise the precipitation as oxide by means of azide, which has proved to be dangerous and difficult to carry out. We also have excluded the precipitation by a bromate-bromide mixture since the product cannot be filtered satisfactorily. On the other hand, and as in the case of aluminium, the precipitation of the hydrous oxide or hydroxide by ammonia gas is entirely satisfactory. It may be pointed out that the determination of gallium by means of 5,7-dibromooxine is much superior to that of aluminium with the same reagent. We[1] have proposed two new methods for determining gallium that have been suggested by the use of the thermobalance. We do not understand why the ferrocyanide has been proposed for the gravimetric determination of gallium.

In Chapter 4 (p. 59) we have in addition explained why the use of thermogravimetry has suggested to us a very simple method for the separation of gallium and iron.

1. Gallium hydroxide

(i) Precipitation by aqueous ammonia We adopted the procedure of Hillebrand and Lundell[2] for the precipitation by ammonia, and used methyl red for following the course of the reaction. The curve obtained from this product, after it has been washed, shows a loss of moisture up to 110° where there is a distinct break corresponding to the hydroxide $Ga(OH)_3$. The latter then slowly loses its water of constitution up to 408° where the level of constant weight for the oxide Ga_2O_3 begins. The porcelain filtering crucible is not attacked at this temperature. It is obvious that our automatic method of determination makes it possible to

avoid the use of temperatures in the vicinity of 1200–1300° required to reach constant weight, nor is there need for are such devices as a tared vessel or a double porcelain crucible, etc. Moreover, since we avoid the use of filter paper, there is no risk that the oxide Ga_2O_3 will be reduced, and consequently no danger of volatilizing the oxide Ga_2O.

(ii) Precipitation by gaseous ammonia If a very slow stream of air charged with ammonia is passed into the solution of gallium nitrate being analyzed, and the temperature is kept low, the precipitate begins to appear as soon as the pH reaches 3.4, and it disappears from pH 9.7 on. The color change of the methyl red, as used in the preceding section, can serve as a guide for keeping the pH between these limits. The suspension is then brought to 70° for 2–3 minutes; the precipitate collects on the surface of the liquid, and may immediately be gathered on a porcelain filtering crucible. It does not stick to the walls. This modification of the method discovered by Trombe[3] leads to a curve which is identical with the preceding, but the hydroxide appears here at 85° and the oxide at 400°. This method cannot be applied for the separation of gallium and iron.

(iii) Precipitation by urea The method developed by Willard and Fogg[4] was employed for the precipitation by urea. The pyrolysis curve reveals formation of the hydroxide beginning at 65°, and there is a slight change in direction at 283° when 2/3rds of the water of constitution are given off. The constant weight corresponding to the oxide starts at 475°. The method is not rigorously quantitative and furthermore it is not capable of separating gallium from iron.

(iv) Precipitation by tannin According to Moser and Brukl[5], the water suspension of tannin acts like a negative hydrosol towards gallium hydroxide. The curve given by the resulting flocculent precipitate shows two changes of direction, one, at 65°, signalling the end of the removal of the extraneous moisture, the other, at 283°, where gallium oxide is obtained and the accompanying organic matter starts to burn. The mixture should be heated to not less than 520° to reach the horizontal of the oxide Ga_2O_3. The tannin method will not separate either iron or aluminium from gallium.

(v) Precipitation by thiosulphate and aniline Aniline completes the reaction of sodium thiosulphate which is employed in the Moser and Brukl[5] procedure for removing iron. The dry hydroxide appears only

after 158° is reached, and a temperature of 546° is needed to arrive at the horizontal due to the oxide. We regard the present method as less satisfactory than all those which have been suggested for the elimination of iron from gallium. If traces of iron are present, the curve B (Fig. 59) shows a distinct rise.

(vi) Precipitation by sulphite or bisulphite Solid sulphite $Na_2SO_3 \cdot 7H_2O$ or a solution of bisulphite yields a product which is extremely easy to filter, but it is necessary to heat the precipitate to at least 813° to accomplish complete dehydration or to obtain the theoretical weight of the oxide Ga_2O_3. This method was first proposed by Dennis and Bridgman[6]. If known amounts of ferric nitrate are added to the solution being determined, the curve A (Fig. 59) does not rise near 800° as all the others do. If the ignition residue is taken up in hydrochloric acid (free from iron), the familiar test for iron with hydroxylammonium chloride and α,α'-dipyridyl gives a negative response. Since the dilution limit of this reaction is pD 6.18, it is obvious that the sulphite method may be seriously considered for determining gallium in the presence of iron, or it may be employed for removing the final traces of iron from gallium oxide which is already quite pure. It should be pointed out that our trials have dealt with amounts of iron not exceeding 1%.

2. Gallium ferrocyanide

The precipitation of gallium as the ferrocyanide $Ga_4[Fe(CN)_6]_3$ as described by Lecoq de Boisbaudran[7] has not found much favour in gravimetry because the precipitate is very difficult to filter (sometimes 2 days), and furthermore it changes its composition spontaneously since it is necessary to work in a very acidic medium, and it becomes contaminated with Prussian blue. The mixture of oxides finally obtained does not contain gallium and iron in the expected ratio of 4 : 3. The method may serve for separation purposes, but much better methods are now generally known. Under such conditions, the interpretation of the thermolysis curve has but slight interest; it slopes downward until 800° is reached, where constant weight is attained due to the oxides and the iron carbide FeC_2.

3. Gallium cupferronate

We used the method of Moser and Brukl[5] to obtain the unstable cupferronate. The product is already partially decomposed at 45°; beyond this temperature the curve descends slowly, though in an irregular fashion. Constant weight requires heating to 745° and this temperature is also necessary to obtain a quantitative conversion into the oxide Ga_2O_3. Any iron which may be present with the gallium will of course be coprecipitated.

4. Gallium camphorate

Ato[8] discovered the camphorate method and we followed his directions. Information furnished by the curve enabled us to modify the end of the determination in an interesting way. Actually, if the precipitate of the anhydrous camphorate $Ga_2[C_8H_{14}(CO_2)_2]_3$ is washed with ethanol, acetone, and ether it dries quickly and retains this formula up to 125°. It is merely necessary to keep the material between 94° and 110° for 30 minutes and to multiply its weight by 0.189 to obtain the weight of gallium present. The organic matter decomposes above 125° and is entirely removed at 478°, the temperature at which the level corresponding to the oxide Ga_2O_3 begins. It should be noted that this excellent method for separating and determining gallium cannot be employed if iron and/or indium are present.

5. Gallium oxinate

Geilmann and Wrigge[9] developed the method used for the precipitation of gallium oxinate. They recommended drying the product between 110° and 150°, but the curve shows that gallium oxinate loses weight from room temperature on. No level is obtained below 180° where the crucible is found to be coated with carbon. The compound under consideration is obviously sublimable and the region from 110° to 150° corresponds precisely to the highest rate of decomposition. Consequently, we have grave doubts about determining gallium as the oxinate.

6. Gallium 5,7-dibromooxine

In contrast, the determination with 5,7-dibromooxine as suggested by Gastinger[10] is excellent. The product remains anhydrous and does not decompose between 100° and 224°. It may be weighed as $Ga(C_9H_4Br_2 ON)_3$, which contains 7.14% of gallium.

References p. 420

7. Gallium ammonia and ethylenediamine complexes

Finally, it should be mentioned that Kochetkova and Tronev[11], who were not pursuing an analytical purpose, studied the behaviour of complexes of gallium with ammonia and ethylenediamine.

REFERENCES

[1] T. DUPUIS AND C. DUVAL, *Anal. Chim. Acta*, 3 (1949) 324.
[2] W. F. HILLEBRAND AND G. E. F. LUNDELL, *Applied Inorganic Analysis*, 1929, p. 387.
[3] F. TROMBE, *Compt. rend.*, 215 (1942) 539; 216 (1943) 888.
[4] H. H. WILLARD AND H. C. FOGG, *J. Am. Chem. Soc.*, 59 (1937) 1179, 2422.
[5] L. MOSER AND A. BRUKL, *Monatsh.*, 51 (1929) 325.
[6] L. M. DENNIS AND J. A. BRIDGMAN, *J. Am. Chem. Soc.*, 40 (1918) 1544.
[7] A. LECOQ DE BOISBAUDRAN, *Ann. Chim. et Phys*, 10 (1877) 124.
[8] S. ATO, *Sci. Papers Inst. Phys. Chem. Research Tokyo*, 12 (1929/30) 225; 15 (1931) 289.
[9] W. GEILMANN AND F. W. WRIGGE, *Z. anorg. Chem.*, 209 (1932) 135.
[10] E. GASTINGER, *Z. anal. Chem.*, 126 (1944) 373.
[11] A. P. KOCHETKOVA AND V. G. TRONEV, *Zhur. Neorg. Khim.*, 2 (1957) 2043.

CHAPTER 32

Germanium

The following precipitates have been examined by Dupuis and Duval[1]: hydroxide, sulphide, magnesium germanate, β-naphthoquinoline germanooxalate, hexamethylenetetramine molybdogermanate, pyridine molybdogermanate, cinchonine molybdogermanate, and oxine molybdogermanate. Furthermore, Dupuis[2] employed the thermobalance (along with the infra-red spectrometer) to settle the question as to whether germanous hydroxide is amphoteric and whether or not it should be regarded as germanoformic acid.

1. Germanium hydroxide

The tannin method, devised by Schoeller[3] and improved by Holness[4], is rated as excellent by Davies and Morgan[5] in their critical study. We followed the directions of Holness. The adsorbed water is given off up to 114° and the curve descends without interruption until the temperature reaches 900°. The tannin is destroyed from 114° to 180° and then the carbon gradually burns away. The crucible must be heated to between 900° and 950° to obtain satisfactory constant figures for GeO_2.

Piétri, Haladjian, Périnet and Carpéni[6] report that specimens of hydrated GeO_2, dried in vacuo over calcium chloride or potassium hydroxide, yield an initial level which extends to around 200–220°. The remaining water is held still more firmly and does not entirely leave the oxides until the temperature is brought to 700–800°. The water of hydration is divided into two distinct parts. The first portion, which is variable, does not exceed 3.4% as a maximum, and corresponds to 1 H_2O per 5 GeO_2, bound irreversibly. The metastable hydrate may correspond to the isopolyacid $H_2Ge_5O_{11}$ previously postulated in solution. In this interesting study, the authors also include a dehydration–rehydration curve of an oxide GeO_2 carrying 5% of water. The potassium salt of the presumed acid $K_2Ge_5O_{11}$ is amorphous; it too has been followed by the method of thermogravimetric analysis[7] and more exactly by recycling.

2. Germanium sulphide

The sulphide GeS_2 was precipitated as directed by Winkler[8], Johnson and Dennis[9], and Geilmann and Brünger[10]. This method gives a product which includes a little precipitated sulphur and the latter burns and volatilizes up to 410°, while the sulphide loses some weight due to conversion into the dioxide GeO_2. The weight of this oxide stays constant from 410° to 946°, which was the upper limit of our experiment.

We did not investigate the Willard and Zuehlke[11] method in which the sulphide is brought down by the action of hydrogen sulphide or potassium sulphide, since the procedure ends with a volumetric operation.

3. Magnesium orthogermanate

The determination as magnesium orthogermanate as described by Müller[12] is not entirely satisfactory according to Davies and Morgan[5] because the precipitate retains magnesia. The curve given by the precipitate shows first the removal of the moisture, followed by a sudden break near 60°. The absorbed water (or bound to the magnesia) departs slowly up to 280°. Strictly constant weight is observed from 280° to 814°.

By infra-red spectrography, we have proved that this compound spontaneously decomposes to yield two products:

$$Mg_2GeO_4 \rightarrow GeO_2 + 2\ MgO$$

4. β-Naphthoquinoline germanioxalate

The method of Willard and Zuehlke[11] was employed for the precipitation of β-naphthoquinoline germanioxalate. There is an inflection in the curve between 103° and 120° and doubtless corresponds to the appearance of the anhydrous product. It is difficult to assign to the anhydrous compound a simple formula which agrees with the molecular weight calculated on the basis of a single $Ge(C_2O_4)_3$ group. The destruction of the organic molecule occurs abruptly between 177° and 321°; the residual carbon burns slowly up to 800°. The oxide GeO_2 remains beyond this temperature. It must be stressed that this method requires a reagent which, as well as its decomposition products, gives rise to a particularly clinging and disagreeable odour, which in addition is harmful to the health of the operator (headache, nose-bleed, etc.).

5. Hexamine germanimolybdate

We have hoped to extend to germanium the method given[13] for the determination of silicon, namely precipitation through the reaction of an aqueous 10% solution of hexamethylenetetramine with a solution of germanimolybdic acid, which is buffered at pH 3. The curve obtained with the resulting precipitate includes a horizontal extending from $70°$ to $90°$ and this corresponds to a mixture of germanimolybdate bearing an excess of the organic base. The horizontal from $440°$ to $813°$ corresponds to the mixture $GeO_2 + 12\ MoO_3$, with a molecular weight of 1832; this is the only certain portion of the curve. The molybdic anhydride MoO_3 sublimes rapidly above $813°$. The method is less recommendable than the oxine procedure.

6. Pyridine germanimolybdate

The procedure of Geilmann and Brünger[10] was used for the precipitation with pyridine, and the corresponding factor for germanium is given as 0.0326. On the other hand, the method of Illingworth and Keggin[14] gives a factor of 0.033. The only horizontal included in the curve of this germanimolybdate extends from $429°$ to $813°$; it corresponds to the sum of $GeO_2 + 12\ MoO_3$. Geilmann and Brünger suggested that the precipitate be dried at $160°$ but the curve shows a particularly rapid descent here and the results obtained can only be approximate.

7. Cinchonine germanimolybdate

We repeated the precipitation with cinchonine described by Davies and Morgan[5] who claimed that they obtained a product with the formula $H_4[Ge(Mo_{12}O_{40})](C_{19}H_{22}O_2N_4)$. We traced the curve corresponding to 2 mg of the metal and then another with 9 mg because the above workers report that the method gives low results when applied to amounts greater than 5 mg. We must admit that the curves are not reproducible. Although an indecisive level is obtained between $93°$ and $121°$ it does not correspond to the above formula nor to the more rational formula $H_4[Ge(Mo_2O_7)_6](C_{19}H_{22}ON_2)_4$. The level which appeared between $450°$ and $900°$ does not conform to the usual mixture $GeO_2 + 12\ MoO_3$. It contains an excess of molybdenum. Consequently, we do not recommend this method of determining germanium.

8. Oxine germanimolybdate

The procedure of Alimarin and Alekseeva[15] with oxine yields a precipitate $GeO_2 \cdot 12MoO_3 \cdot (C_9H_7ON)_4$ but this formula only applies between 50° and 115°. The residue $GeO_2 + 12\ MoO_3$ is stable from 496° to 920°. The two corresponding levels are suitable for the automatic determination of germanium; obviously the former is the more interesting. The above workers advise that the precipitate be dried at 110°. Though the curve is almost horizontal from 93° to 343°, the most accurate results will be obtained below 115°.

REFERENCES

[1] T. DUPUIS AND C. DUVAL, *Anal. Chim. Acta*, 4 (1950) 186.
[2] T. DUPUIS, *Rec. trav. chim.*, special number in honour of C. J. VAN NIEUWENBURG 79 (1960) 519.
[3] W. R. SCHOELLER, *Analyst*, 57 (1932) 57; 66 (1936) 589.
[4] H. HOLNESS, *Anal. Chim. Acta*, 2 (1948) 254.
[5] G. R. DAVIES AND G. MORGAN, *Analyst*, 63 (1938) 388.
[6] A. PIÉTRI, J. HALADJIAN, G. PÉRINET, and G. CARPÉNI, *Bull. Soc. chim. France*, (1960) 1909.
[7] G. CARPÉNI, Y. HAMANN, J. HALADJIAN AND G. PÉRINET, *Bull. Soc. chim. France*, (1960) 1903.
[8] C. WINKLER, *J. prakt. Chem.*, 34 (1886) 177; *Z. anal. Chem.*, 26 (1887) 363.
[9] E. B. JOHNSON AND L. M. DENNIS, *J. Am. Chem. Soc.*, 47 (1925) 790.
[10] W. GEILMANN AND K. BRÜNGER, *Z. anorg. Chem.*, 196 (1931) 312.
[11] H. H. WILLARD AND C. W. ZUEHLKE, *Ind. Eng. Chem., Anal. Ed.*, 16 (1944) 322.
[12] J. H. MÜLLER, *J. Am. Chem. Soc.*, 44 (1922) 2496.
[13] C. DUVAL, *Compt. rend.*, 218 (1944) 119, 198; *Anal. Chim. Acta*, 1 (1947) 33.
[14] J. W. ILLINGWORTH AND J. F. KEGGIN, *J. Chem. Soc.*, (1935) 575.
[15] I. P. ALIMARIN AND O. A. ALEKSEEVA, *J. Appl. Chem. (S.S.S.R.)*, 12 (1939) 1900; *C.A.*, 34 (1940) 5623, 7777.

CHAPTER 33

Arsenic

In addition to the usual weighing forms (pentoxide, sulphides, arsenates, arsenimolybdates) which have been investigated by Dupuis and Duval[1], there have also been studies of arsenious oxide with regard to its use as a standard, and of the natural sulphides, realgar and orpiment.

1. Arsenious oxide

Arsenious oxide As_4O_6 taken from a fresh bottle yields a thermolysis curve which is horizontal up to 200°. Then sublimation begins abruptly and continues as far as 534° where the crucible will be empty. There is a very marked inflection between 350° and 450°. Consequently standard solutions can be prepared by direct weighing of this compound. (Duval[23])

2. Arsenic oxide

The weighing form arsenic oxide As_2O_5 is obtained if arsenic or one of its derivatives is dissolved in nitric acid, in other words, if orthoarsenic acid is heated. Although Bäckström[2] gives 435–450° as the limiting temperatures for the existence of As_2O_5, and though Auger[3] on the contrary reports 400° as the upper limit, the thermolysis curve of a material whose composition approximated $As_2O_5 \cdot ^3/_2H_2O$ showed that the anhydrous oxide As_2O_5 starts to appear at 193° and that decomposition into arsenious oxide sets in at 246°. Guérin and Masson[4] and also Guérin and Boulitrop[5] do not agree with this finding and believe that the arsenic oxide, which they themselves prepared by oxidizing arsenious oxide with nitric acid, does not decompose below 400° either in the air or in vacuo.

3. Arsenic trisulphide

(i) Precipitation by hydrogen sulphide The sulphide As_2S_3 has been

prepared in various ways. An alkali arsenite, in a solution containing hydrochloric acid, was treated at length with a stream of hydrogen sulphide as directed by Berzelius[6], Wenger and Cimerman[7], and Schwarz-Bergkampf[8]. The resulting precipitate has been found to be almost pure around 200° and it is stable up to 265°. Beyond this temperature, it sublimes and is superficially oxidized. The crucible is completely empty by the time the temperature has reached 670°. Yosida[9] advises that 200° be not exceeded for the drying.

(ii)Precipitation as xanthogenate   The method involving a xanthogenate is due to Tarugi and Sorbini[10]. It appears that the resulting product is not homogeneous and its composition does not conform strictly to the formula $As(SCS-OC_2H_5)_3$, though it is stable up to 51°. The level extending from 223° to 273° is not perfectly horizontal and agrees essentially with the formula As_2S_3. As was true of the preceding product, this material starts to sublime at 273°. We suggest that this method be discarded for gravimetric purposes.

4. Arsenic pentasulphide

The pentasulphide As_2S_5 precipitated from an alkali arsenate, in a solution containing hydrochloric acid, by means of hydrogen sulphide as directed by Bunsen[11] and Wenger and Cimerman[7], yields a horizontal extending from 78° to 245°. The sulphide sublimes from this temperature on. On the other hand, Taimni and Tandon[12] claim that the pentasulphide precipitated by means of sodium sulphide is stable up to 270°. It then begins to sublime (on the Stanton balance at the heating rate of 5.3° per minute) and the crucible is empty at 550°. Yosida[9] warns against drying at temperatures above 130°.

5. Realgar

Saito[13] investigated the thermal behaviour of a sample of realgar from Osoreyama. Its composition was: As, 60.20%; S, 25.88%; Fe, 0.88%; insolubles, 13.00%. The curve showed a continuous descent to around 215°. Then oxidation set in producing arsenious oxide and sulphur dioxide. The crucible suffered notable loss of weight only between 550° and 750°.

6. Orpiment

Orpiment was also heated by Saito[13]; the specimen came from Jozankei and contained As, 58.32%; S, 37.40%; Fe, 0.56%; insolubles, 3.70%. As in the case of realgar, the sample weighed 0.150 g and was ground to pass a 200-mesh sieve, the air current flowed at 100 ml per minute and the heating rate was 2° per minute. The mineral began to decompose at 200° and gave off sulphur dioxide and arsenious oxide up to 450°. From then on to 700°, the loss became much slower.

7. Calcium arsenate

The precipitation of calcium arsenate is discussed in Chapter Calcium. It should be noted that calcium pyroarsenate may be weighed after heating between 350° and 950°.

8. Ammonium magnesium arsenate

Similarly, the precipitation as ammonium magnesium arsenate is described in Chapter Magnesium. Magnesium pyroarsenate may be weighed after heating between 415° and 884°.

9. Silver arsenate

Silver arsenate Ag_3AsO_4 was prepared as directed by Desbourdeaux[14] and Eschweiler and Röhrs[15]. We do not regard it as a proper means of determining arsenic. The thermolysis curve is a slanting straight line. A dry sample, weighing 347.5 mg at 20°, showed a gain of 6 mg at 950°. When cooled, this same sample did not regain its initial weight, there was a negative difference of 1.3 mg. We have found no explanation for this finding which is well reproducible. Yosida[9] gives 550–780° as the limits of the level he obtained.

10. Silver thallium arsenate

By applying the method of Spacu and Dima[16], we obtained silver thallium arsenate Ag_2TlAsO_4; they suggest that it be dried for 15 minutes in vacuo. Actually, this compound may be weighed at any temperature between 20° and 846°; the curve is a straight line at whose extremities we recorded: initial weight: 360.73 mg; final weight: 360.05 mg. This is an excellent method for the automatic determination of arsenic.

11. Lead hydrogen arsenate

The precipitate obtained by treating disodium arsenate with lead nitrate yields a good horizontal between 81° and 269° agreeing with the formula $PbHAsO_4$. This level is suitable for the automatic determination of arsenic. Two molecules share the loss of 1 molecule of water and the pyroarsenate $Pb_2As_2O_7$ is obtained above 320°. The conversion is quantitative and the pyrophosphate is stable up to 950° at least.

12. Bismuth arsenate

Bismuth arsenate $BiAsO_4 \cdot \frac{1}{2}H_2O$ was prepared by the method given by Carnot[17], who suggests drying the product at 110°. However, the curve reveals that water is lost up to 645°; first a rapid loss due to removal of wash water and acid, then a slower removal of a half molecule of water of crystallization from 100° to 645°. A horizontal conforming to the formula of the anhydrous arsenate extends from 645° to 951°.

13. Iron arsenate

The precipitation of iron arsenate as proposed by Fresenius[18] gave no results of value for gravimetric purposes.

14. Uranium arsenate

If ammonium salts are absent, the Puller[19] procedure in which an arsenate is treated with a uranium(VI) salt yields a precipitate which is of no value for gravimetric analysis.

Puller also reported on the precipitate obtained by this reaction in the presence of ammonium salts but gave no directions. We have found that if the solution of the arsenate is treated with ammonium acetate and uranium(VI) acetate, the precipitate settles readily and filters well. The curve shows a continuous loss until 500° is reached; then a horizontal extends from 500° to 800° and perhaps corresponds to the pyroarsenate. There is a further descent beyond 800°. We have not obtained information of value to gravimetric analysis from this method.

15. Ammonium arsenimolybdate

Ammonium arsenimolybdate was prepared by the method given in the

Carnot textbook[20]. The composition ranges from $As_2O_5 \cdot 17MoO_3$ to $As_2O_5 \cdot 23MoO_3$, a fact which of itself presages unsatisfactory results. The curve given by one of these precipitates greatly resembled that of ammonium phosphomolybdate. The level relating to the mixture of the two oxides extends from 690° to 735° (sometimes 765°). The method is useless for determination purposes, but the temperature at which the molybdic anhydride is lost is always above 730°. See Dupuis[21].

16. Strychnine arsenimolybdate

Embden[22] reports that strychnine arsenimolybdate may be employed for the determination of arsenic in the same manner as the corresponding phosphomolybdate. However, the composition of this arsenimolybdate is not known and the curve obtained from the mixted precipitate yields erratic results. The level corresponding to the mixture $As_2O_5 \cdot xMoO_3$ extends from 424° to 778°. Dupuis[21] suggests that the method be rejected as regards the determination of arsenic.

REFERENCES

[1] T. DUPUIS AND C. DUVAL, Anal. Chim. Acta, 4 (1950) 262.
[2] H. BÄCKSTRÖM, Z. anal. Chem., 31 (1892) 663.
[3] V. AUGER, Compt. rend., 134 (1902) 1059.
[4] H. GUÉRIN, Bull. Soc. chim. France, (1955) 1536; H. GUÉRIN AND J. MASSON, Compt. rend., 241 (1955) 415.
[5] H. GUÉRIN AND R. BOULITROP, Compt. rend., 236 (1953) 83.
[6] J. BERZELIUS, Poggendorfs Ann., 7 (1826) 2.
[7] P. E. WENGER AND C. CIMERMAN, Helv. Chim. Acta, 14 (1931) 718.
[8] E. SCHWARZ-BERGKAMPF, Z. anal. Chem., 69 (1926) 341.
[9] Y. YOSIDA, J. Chem. Soc. Japan, 60 (1940) 915.
[10] N. TARUGI AND F. SORBINI, Boll. chim. farm., 51 (1912) 361.
[11] R. BUNSEN, Ann., 192 (1878) 319.
[12] I. K. TAIMNI AND S. N. TANDON, Anal. Chim. Acta, 22 (1960) 34.
[13] H. SAITO, Sci. Repts. Tôhoku Imp. Univ., 16 (1927) 129.
[14] L. DESBOURDEAUX, Bull. sci. pharmacol., 27 (1920) 225; 28 (1921) 289.
[15] W. ESCHWEILER AND W. RÖHRS, Z. angew. Chem., 36 (1923) 465.
[16] G. SPACU AND L. DIMA, Z. anal. Chem., 120 (1940) 317.
[17] A. CARNOT, Compt. rend., 121 (1895) 20.
[18] R. FRESENIUS, Manuel de chimie analytique, 6th French Ed., Dunod, Paris, 1912, p. 167
[19] R. E. O. PULLER, Z. anal. Chem., 10 (1871) 72.
[20] A. CARNOT, Traité d'analyse des substances minérales, Dunod, Paris, Vol. 2, p. 569.
[21] T. DUPUIS, Thesis, Paris, 1954, p. 23.
[22] G. EMBDEN, Z. physiol. Chem., 113 (1921) 138.
[23] C. DUVAL, Anal. Chim. Acta, 13 (1955) 429.

CHAPTER 34

Selenium

There are only a few methods for determining selenium gravimetrically Usually its compounds are reduced by hydrazine, sulphur dioxide, or stannous chloride (the best method). If the element is tetravalent, it may be drastically oxidized with hydrogen peroxide and then precipitated as lead selenate. Duval and Doan[1], who made the present study, did not include the methods involving the use of permanganate, bromate, iodate, and titanium chloride since these procedures have been found to yield unsatisfactory results.

1. Elementary selenium

(i) Precipitation by hydrazine The precipitation by means of hydrazine was conducted as outlined by Hovorka[2] and Meyer[3]. The resulting elementary selenium is violet and readily oxidizes when heated in the air. The curve reveals a maximum gain of 2 mg per 36 mg at 350°, where the sublimation becomes noticeable. Matsuura[4] reported 200°. The crucible is empty at 540°.

(ii) Precipitation by sulphur dioxide in acetone An acetone solution of sulphur dioxide is used in the method given by Hovorka[5]. The red variety of selenium is precipitated and it must be heated until it becomes violet before it can be filtered. The thermogram is again slightly ascending and includes a level from 152° to 217°. The initial weight is regained at 330° which marks the beginning of the departure of the selenium.

(iii) Precipitation by ammonium sulphite When ammonium sulphite is used, as proposed by Treadwell[6], the curve of the resulting elementary selenium is not ascending. In this case, sublimation starts only after the temperature reaches 370°.

(iv) Precipitation by stannous chloride The stannous chloride method,

which was originated by Taboury and Gray[7], may be applied for the determination of either selenium or tin. The elementary selenium does not oxidize but it is much more sublimable than the preceding varieties; the selenium begins to be lost at 272°.

2. Selenium sulphide

Taimni and Tandon[8] report that sodium sulphide produces a selenium sulphide SeS_2. It is stable up to 210° where it begins to sublime.

3. Lead selenate

Before adding lead nitrate, the solution containing tetravalent selenium should be oxidized by means of 100 volume (30%) hydrogen peroxide solution. The resulting selenate yields a precipitate of lead selenate as described by Ripan-Tilici[9] and Spacu[10]. This salt yields a good horizontal up to 330° and consequently it is not essential to dry it in a desiccator. Decomposition is slow at temperatures above 330° and is not complete until about 900° is reached. Lead oxide PbO results. For details see Duval and Doan[1].

REFERENCES

[1] C. DUVAL AND U. M. DOAN, *Anal. Chim. Acta*, 5 (1951) 566.
[2] V. HOVORKA, *Collection Czechoslov. Chem. Communs.*, 4 (1932) 300.
[3] J. MEYER, *Z. anal. Chem.*, 53 (1914) 145.
[4] K. MATSUURA, *J. Chem. Soc. Japan*, 52 (1931) 730.
[5] V. HOVORKA, *Collection Czechoslov. Chem. Communs.*, 7 (1935) 125.
[6] F. P. TREADWELL, *Manuel de chimie analytique*, 3rd French Ed., Dunod, Paris, 1920, Vol. 2, p. 256.
[7] M. F. TABOURY AND E. GRAY, *Compt. rend.*, 213 (1941) 481.
[8] I. K. TAIMNI AND S. N. TANDON, *Anal. Chim. Acta*, 22 (1960)553.
[9] R. RIPAN-TILICI, *Z. anal. Chem.*, 102 (1935) 343.
[10] P. SPACU, *Bull. Soc. chim. France*, 3 (1936) 159.

CHAPTER 35

Bromine

1. Bromide ion Br⁻

Silver bromide Silver bromide may be dried, away from the light, between 70° and 946°. The corresponding straight line may serve for the automatic determination of bromine or silver. See Dupuis and Duval[1].

A precipitate is often seen to turn violet or black on its surface while being weighed. It has been recommended in such cases to treat the precipitate with a drop of nitric acid in order, it is said, to redissolve any silver that may have been produced. It is also asserted that there is a loss of halogen at the moment a photographic plate is exposed; in fact, it has been even claimed that a sub-halide of silver is formed. However, the thermobalance does not disclose any loss of weight between 30° and 135° if a layer of silver bromide, either with or without gelatin, is irradiated with an ultra-violet lamp.

When silver bromide (250 mg) is heated (350° per hour) in hydrogen, the reduction starts at room temperature. The loss of weight does not become important until the temperature reaches 300° and is complete at 480° where the crucible contains only metallic silver. Duval[2], who carried out this experiment, found no indication of the formation of the compound Ag_2Br.

2. Hypobromite ion BrO⁻

We have found no record of trials with respect to this ion.

3. Bromate ion BrO₃⁻

We have found no record of a special gravimetric method for this ion. However, we have traced the thermolysis curves of some bromates. See Chapter Potassium.

REFERENCES

[1] T. DUPUIS AND C. DUVAL, *Anal. Chim. Acta*, 4 (1950) 615.
[2] C. DUVAL, *Mikrochemie*, (1956) 1430.

CHAPTER 36

Rubidium

This chapter deals with the following compounds of rubidium, which have been retained for its gravimetric determination: chloride, perchlorate, sulphate, hexanitrocobaltate (cobaltinitrite), hexachlorostannate (stannichloride), hexachloroplatinate (platinichloride), 6-chloro-5-nitro-*m*-toluenesulphonate, tetraphenylboron compound.

Some studies of the bisulphate, carbonate, ferrocyanide, and the chlorotitanate(IV) are also known.

1. Rubidium chloride

We prepared the chloride by dissolving rubidium carbonate in hydrochloric acid. The moist acidic product does not acquire constant weight below 88°; the curve shows a change of slope at 58°. No loss by vapourization is observed around 500°; the permissible drying range extends from 88° to 605°. See Duval and Duval[1]. Kitajima[2] gives 740° as the limit.

2. Rubidium perchlorate

Rubidium perchlorate $RbClO_4$ is obtained by subjecting rubidium chloride to the action of dilute perchloric acid and washing the residue with ethanol. Its curve descends as far as 95° which marks the departure of the water and the remaining acid. The anhydrous perchlorate yields a horizontal which extends up to 343°. The loss is still insignificant at 408° (about 1/300th). Explosive decomposition sets in near 473° and is finished at 530°. The equally horizontal level of the anhydrous chloride begins at this point. The perchlorate therefore may be dried between 100° and 400° before weighing.

3. Rubidium sulphate

Rubidium sulphate, prepared by the action of dilute sulphuric acid on the chloride, loses moisture up to 76°. The level of the anhydrous sulphate then extends horizontally as far as 877°. Kitajima[2] gives 490°. We did not observe any loss by volatilization at 877°.

4. Rubidium hydrogen sulphate

Tomkova, Jirů and Rosický[3] found that the acid sulphate $RbHSO_4$ is transformed into the pyrosulphate $Rb_2S_2O_7$ at 265°; the latter yields the sulphate Rb_2SO_4 at 445°.

5. Rubidium nitrate

Rubidium nitrate $RbNO_3$ is anhydrous and very stable. It gives a horizontal up to at least 430°. The formation of the oxide is complete at 880°.

6. Rubidium sodium cobaltinitrite

The precipitate $(Rb,Na)_3[Co(NO_2)_6] \cdot xH_2O$, formed by the action of rubidium chloride or sulphate on an aqueous solution of sodium cobaltinitrite, has no fixed formula, either with regard to the water content or the Rb/Na ratio. The curve descends from room temperature as far as 283°. There is loss of water of crystallization in this range and also hydrolysis of the complex, which loses 3 NO_2 groups. There is no discontinuity indicating the production of the anhydrous compound. A horizontal extends from 283° to 550°, but it corresponds to a mixture $CoO + RbNO_3 + NaNO_3$, which is constant within the same experiment but variable from one lot of precipitate to another. Consequently, this compound cannot be used for the gravimetric determination of rubidium, but only for the separation of the sum Rb + Na. A regular loss of weight sets in beyond 550°; it is the resultant of several phenomena: loss of NO_2 from the sodium and rubidium nitrates, conversion of these metals into their oxides and peroxides, further oxidation of the cobalt to Co_3O_4 followed by redecomposition into CoO. The residue is brown-black; its complete analysis has no significance with regard to the objective of the present study.

7. Rubidium carbonate

Rubidium carbonate $4Rb_2CO_3 \cdot 7H_2O$ was heated on an Ainsworth balance by Reisman[4].

8. Rubidium ferrocyanide

Cappellina and Babani[5] report that rubidium ferrocyanide $Rb_4[Fe(CN)_6] \cdot 2H_2O$ loses its water between 85° and 130°.

9. Rubidium chlorostannate

Fresenius[6] found that the hexachlorostannate $Rb_2[SnCl_6]$ can be prepared by mixing an alcohol solution of stannic chloride with a concentrated hydrochloric solution of rubidium chloride. After 12 hours, the precipitate is filtered off and washed with absolute alcohol. The pyrolysis curve shows a regular increase in weight as far as 341°; at this temperature the increase has amounted to 4 mg per an initial weight of 224 mg. The corresponding cesium salt produces the same effect but in a less clear-cut fashion. This increase must be due to an uptake of oxygen; up till now the production of perstannates has been reported only through the action of hydrogen peroxide. The decomposition of the chlorostannate begins sharply near 407° and is complete near 650°; chlorine and stannic-chloride are given off. The residue consists of rubidium chloride and stannic oxide; the mixture yields a horizontal from 650° to 812°. Because of the many uncertainties in the interpretation of this curve, we do not think it advisable to recommend the chlorostannate for an accurate determination of rubidium. The quantity of the residual stannic oxide is a function of the hydrolysis occurring during the analytical operations, and consequently the residue between 650° and 812° does not have a constant composition.

10. Rubidium chlorotitanate

Morozov and Toptygin[7] made a thermal stability study of rubidium hexachlorotitanate $Rb_2[TiCl_6]$.

11. Rubidium chloroplatinate

The chloroplatinate $Rb_2[PtCl_6]$ is obtained by the action of chloroplatinic acid on rubidium chloride; the product is washed with ethanol. It is stable up to 674°. Kitajima[2] reports 420°. Four of the six atoms of chlorine are lost slowly up to around 1000°. The residue consists of platinum metal and rubidium chloride. The latter sublimes at higher temperatures.

12. Rubidium 6-chloro-5-nitro-*m*-toluenesulphonate

As is true in the case of potassium, the method of Benedetti-Pichler and Bryant[8], employing the 6-chloro-5-nitro-*m*-toluenesulphonate, leads in the case of rubidium to a voluminous precipitate which yields a level that extends to about 220° and is suitable for the automatic determination of rubidium.

13. Rubidium tetraphenylboron compound

Wendlandt[9] studied the precipitate of rubidium tetraphenylboron compound, which was prepared in accord with the directions of Raff and Brotz[10], and dried in the air for 24 hours. The specimen weighed 100–150 mg and was heated at the rate of 4.5° per minute. The complex is stable up to 240°. It then decomposes and above 730° produces the metaborate $RbBO_2$:

$$Rb[B(C_6H_5)_4] + 30\ O_2 \rightarrow RbBO_2 + 24\ CO_2 + 10\ H_2O$$

REFERENCES

[1] T. DUVAL AND C. DUVAL, *Anal. Chim. Acta*, 2 (1948) 110.
[2] I. KITAJIMA, *J. Chem. Soc. Japan*, 55 (1934) 199.
[3] D. TOMKOVA, P. JIRŮ AND J. ROSICKÝ, *Collection Czechoslov. Chem. Communs.*, 25 (1960) 957.
[4] A. REISMAN, *Anal. Chem.*, 32 (1960) 1566.
[5] F. CAPPELLINA AND B. BABANI, *Atti accad. sci. ist. Bologna*, 11 (1957) 76.
[6] L. FRESENIUS, *Z. anal. Chem.*, 86 (1931) 187.
[7] I. S. MOROZOV AND D. J. A. TOPTYGIN, *Zhur. Neorg. Khim.*, 5 (1960) 88.
[8] A. A. BENEDETTI-PICHLER AND J. T. BRYANT, *Ind. Eng. Chem., Anal. Ed.*, 10 (1938) 107.
[9] W. W. WENDLANDT, *Chemist Analyst*, 26 (1957) 38; *Anal. Chem.*, 28 (1956) 1001.
[10] P. RAFF AND W. BROTZ, *Z. anal. Chem.*, 133 (1951) 241.

CHAPTER 37

Strontium

This chapter will deal with thermolysis curves given by the hydroxide, fluoride, bromide, chlorate, iodate, sulphate, thiosulphate, sulphamate, nitrate, arsenate, niobate, carbonate, aluminate, formate, oxalate, and tartrate.

1. Strontium hydroxide

Strontium hydroxide has been studied by Burriel-Marti and Garcia-Clavel[1] with continuous heatings at 5°, 1.8°, and 0.6° per minute, and then with isothermal heatings at 40°, 50°, 80°, 95°, 100°, 105°, and 110°. It is difficult to distinguish between adsorbed water and water of hydration. Although the starting material contains close to 8 molecules of water, it may be said that the monohydrate is produced in the vicinity of 95°. We found the true hydroxide $Sr(OH)_2$ between 160° and 475°, and the zone of the oxide SrO ranges from 650° to 800° (or higher). The study was checked by differential thermal analysis.

2. Strontium fluoride

Peltier and Duval[2] pointed out that strontium is not often weighed as the fluoride since the titrimetric method employing ferric chloride is preferable. The fluoride is precipitated by the action of potassium fluoride on strontium chloride, and the product is washed with water. The curve, or rather the straight line, given by this precipitate up to 950° gives the operator a wide choice with regard to the ignition temperature. The weight of the porcelain crucible did not change 0.1 mg (per 583.4 mg) during the operation.

3. Strontium bromide

Collari and Galimberti[3] used their balance to trace the dehydration curve of the bromide $SrBr_2 \cdot 6H_2O$. The sample weighed 0.1847 g. They did not interpret the curve.

4. Strontium chlorate

The chlorate $Sr(ClO_3)_2$ crystallizes in the anhydrous form. It has been heated to 120° by infra-red radiation by Philibert[4] on the Gallia balance.

5. Strontium iodate

The iodate $Sr(IO_3)_2$, produced by the action of iodic acid on strontium chloride, is a mixture of the anhydrous salt and the monohydrate. See Peltier and Duval[2], Rocchiccioli[5], and Bestougeff[6]. The curve shows that the product is completely free of water at 157°. It may be weighed in this form up to 600°. The Rammelsberg reaction sets in at this temperature, *i.e.* the transformation into paraperiodate with noticeable evolution of iodine vapours around 680°. The reaction:

$$5\ Sr(IO_3)_2 \rightarrow Sr_5(IO_6)_2 + 4\ I_2 + 9\ O_2$$

is complete at 750°; the paraperiodate is yellow with a reddish tinge. Obviously, the weighing can be made in this form; the transformation is strictly quantitative. The relative error is 1 : 256 in the case of this curve for a residual weight of 174 mg.

6. Strontium sulphate

The Fresenius method was used for the precipitation of the sulphate $SrSO_4$: strontium chloride is treated with sulphuric acid and ethanol and the suspension is allowed to settle for 12 hours. The weighing can be made after the precipitate has been kept between 100° and 300°. A slight but measurable loss of weight is found above 300°; it is due to decomposition and the release of sulphur trioxide fumes. At 884°, the upper limit of our experiment, the loss of weight was 5.3 mg per 476 mg, and this was checked also by direct weighing of the crucible, before and after the trial, using a balance sensitive to 0.01 mg. Accordingly, the decomposition begins well below the conventional "white hot" temperature.

7. Strontium thiosulphate

When the strontium thiosulphate, which served as our starting material, was crystallized above 50° it retained 1 molecule of water which was lost slowly between 70° and 180° and more rapidly at higher temperatures. See Duval[7]. However, it is necessary to assume that the removal of water is not complete below 240° but the salt begins to decompose then. Sulphur is given off and the essentially horizontal level obtained between 350° and 507° corresponds to a mixture of sulphate, sulphite, and evolving sulphide. The uptake of oxygen becomes considerable above 507° and strontium sulphate free of sulphide is obtained from 700° on. When it is desired to prepare standard solutions of strontium thiosulphate, the salt should not be heated above 170° before it is weighed out.

8. Strontium sulphamate

Capestan[8] reports that, like the corresponding salts of the alkali metals, anhydrous strontium amidosulphonate $Sr(NH_2-SO_3)_2$ is converted into the imidodisulphonate between 230° and 250°, and then into the sulphate $SrSO_4$ between 300° and 350°.

9. Strontium nitrate

Strontium nitrate is stable up to 280° and yields a perfectly horizontal level which starts at room temperature. Decomposition then sets in and proceeds very slowly but becomes explosive above 600°. This disintegration appears to be complete near 820°. The succeeding horizontal is due to the oxide SrO. See also Wendlandt[9].

10. Strontium hydrogen arsenate

Monosodium arsenate and strontium chloride yield a precipitate whose formula is $SrHAsO_4.H_2O$. This product loses moisture and the water of crystallization with no discontinuity between 35° and 132°. The curve changes its slope because of the onset of the decomposition reaction:
$$2\ SrHAsO_4 \rightarrow H_2O + Sr_2As_2O_7$$
which is complete at 412°. Strontium pyroarsenate is stable from this point up to at least 950°, the highest temperature of the study conducted by Peltier and Duval[2].

11. Strontium niobate

The niobate $Sr_7Nb_{12}O_{37}\cdot 33H_2O$, like all the salts of this type, was studied by Pehelkin et al.[10]. By successive dehydrations it yielded the following hydrates: $6H_2O$ at $120°$, $2H_2O$ at $240°$, $1.5H_2O$ at $300°$, $1H_2O$ at $360°$, $0.6H_2O$ at $380°$. The metaniobate appears at $690°$.

12. Strontium carbonate

The carbonate $SrCO_3$, precipitated by the action of ammonium carbonate on strontium chloride, does not acquire constant weight until around $410°$. It is then stable up to $1100°$. Because of the way in which the furnace was constructed we did not obtain quantitative conversion into the oxide SrO. See also[11].

The mineral *strontianite* yielded different results. Pobeguin and Caillère[12] found that the decomposition starts at $865°$ in air and at $880°$ in nitrogen. The study was also followed by differential thermal analysis.

13. Strontium aluminate

Maekawa[13] made a study of strontium aluminate.

14. Strontium chromate

Strontium chromate was prepared, in the cold, by the action of potassium bichromate and strontium chloride in the presence of 50% alcohol. It, like barium chromate, decomposes slightly when heated. The loss, which is barely noticeable at $400°$, reaches 4 mg per 338 mg at $950°$. A small amount of chromite is produced. The colour of the product is not changed by cooling. We advise that the drying be carried out between $100°$ and $400°$.

15. Strontium ferrocyanide

Strontium ferrocyanide $Sr_2[Fe(CN)_6] \cdot 13H_2O$ has been studied by Cappellina and Babani[14]. They found two levels; the first from $20°$ to $110°$ due to the nonahydrate, the other from $110°$ to $140°$ corresponding to the tetrahydrate.

16. Strontium formate

The formate $Sr(CO_2H)_2$ was heated by Zapletal *et al.*[15]

17. Strontium oxalate

The monohydrate of the oxalate $SrC_2O_4 \cdot H_2O$ is prepared, in a solution containing acetic acid, by adding a 10% aqueous solution of potassium oxalate drop by drop to a boiling solution of strontium chloride. The curve does not agree with the data reported by Winkler[16] who gave 132° as the dehydration temperature. See also Erdey and Paulik[17]. The monohydrate cannot be recommended at any temperature. It begins to dehydrate at 43°, and water is being freely given off at 131°. The molecule of water has disappeared completely at 177°. The anhydrous oxalate may be weighed after it has been kept between 177° and 400°. The decomposition into carbon monoxide and carbon dioxide is not instantaneous as in the case of calcium oxalate. It occurs between 400° and 520°. The strictly horizontal stretch due to the carbonate extends from 520° to 1100°.

18. Strontium tartrate

Although the tetrahydrated tartrate exists, the salt brought down in the presence of acetic acid by tartaric acid and strontium chloride is the monohydrate $SrC_4H_4O_6 \cdot H_2O$. The pyrolysis curve discloses that the water of crystallization is not removed completely until the temperature reaches 200°, where the tartrate radical begins to disintegrate. Constant weight is attained at 810° through conversion into carbonate. Accordingly, this method of determination has but little interest in the case of strontium.

REFERENCES

[1] F. BURRIEL-MARTI AND M. E. GARCIA-CLAVEL, 16th Intern. Congr. Pure and Applied. Chem., Paris, 1957, p. 861.
[2] S. PELTIER AND C. DUVAL, Anal. Chim. Acta, 1 (1947) 355.
[3] N. COLLARI AND L. GALIMBERTI, Boll. scient. fac. chim. ind. Bologna, 18 (1940) 9.
[4] P. PHILIBERT, Thesis, Clermont-Ferrand, 1958.
[5] C. ROCCHICCIOLI, Thesis, Paris, 1960.
[6] N. BESTOUGEFF, Diploma of Higher Studies, Paris, 1960.
[7] C. DUVAL, Anal. Chim. Acta, 20 (1959) 20.
[8] M. CAPESTAN, Ann. Chim. (Paris), 5 (1960) 222.
[9] W. W. WENDLANDT, Texas J. Sci. 10 (1958) 392.
[10] V. A. PEHELKIN, A. V. LAPITSKII, V. I. SPITSYN AND Y. P. SIMANOV, Zhur. Neorg, Khim., 1 (1956) 1784; C.A., (1957) 2445.
[11] 1st Colloq. Thermography, Kazan, 1953, p. 7.
[12] T. POBEGUIN AND S. CAILLÈRE, Bull. Soc. franç. minéral., 83 (1960) 36.
[13] G. MAEKAWA, J. Soc. Chem. Ind. Japan, 46 (1943) 751.
[14] F. CAPPELLINA AND B. BABANI, Atti accad. sci. ist. Bologna, 11 (1957) 76.
[15] V. ZAPLETAL, J. JEDLIČKA AND V. RŮŽIČKA, Chem. Listy, 50 (1956) 1406; C.A., (1957) 2438.
[16] C. WINKLER, Angew. Chem., 31 (1918) 80.
[17] L. ERDEY AND F. PAULIK, C.A., 50 (1956) 3952.

CHAPTER 38

Yttrium

The thermolysis studies dealing with the compounds of yttrium have included the hydroxide, fluoride, chloride, sulphate, nitrate ,oxalate, cupferronate, neocupferronate, oxinate, and quinaldinate (2-methyloxinate).

1. Yttrium hydroxide

The hydroxide is obtained by gradual addition of dilute ammonium hydroxide to an essentially neutral solution of yttrium nitrate (spectroscopically pure). After the gelatinous colourless precipitate has stood overnight it is filtered on paper. The thermogram obtained by linear heating of the still moist product at 300° per hour related to 81.4 mg of the oxide Y_2O_3 (less than 1/1000 holmium oxide). The curve slants downward until the temperature is near 856° where the horizontal of the oxide Y_2O_3 begins. No discontinuity indicative of the formation of the hydroxide $Y(OH)_3$ was observed. Calculations based on the original curve show that this compound should be present near 365°. See Duval[1].

2. Yttrium fluoride

Wendlandt and Love[2] used the method of Popov and Knudsen[3] to prepare the fluoride $YF_3 \cdot H_2O$. The water is lost between 40° and 60°. Conversion into the oxyfluoride YOF begins between 600° and 690° but the change is not complete at 900°.

3. Yttrium chloride

Wendlandt[4] reports that the chloride $YCl_3 \cdot 6H_2O$ loses its water above 110° and becomes anhydrous at 275°. Rapid decomposition occurs beyond 375° leading to the oxychloride YOCl. The latter is stable above 450°.

4. Yttrium sulphate

Wendlandt[5] heated 85–100 mg of the sulphate $Y_2(SO_4)_3 \cdot 8H_2O$ at the rate of 5.4° per minute. The loss of water begins at 70° and the dehydration is complete at 195°.

5. Yttrium nitrate

Yttrium nitrate $Y(NO_3)_3 \cdot 6H_2O$ was studied by Wendlandt[6]; it starts to lose water at room temperature. He did not obtain levels for the intermediate hydrates or the anhydrous compound, but he did find a short level between 440° and 475° corresponding to the oxynitrate $YONO_3$. The level of the oxide begins at 660°.

6. Yttrium oxalate

Duval[1] prepared the oxalate by treating the nitrate with oxalic acid in a slightly acidic solution. The system was heated on the water bath near 60° and produced a very fine white precipitate which was allowed to settle overnight and then filtered on paper. The thermolysis curve shows that the starting material cannot be the nonahydrate which Wendlandt[7] reported as being present up to 45°. We found the level corresponding to the dihydrate between 180° and 300° (Wendlandt gives 180–260°). The formation of the oxide Y_2O_3 is complete at 680° according to our observations, whereas Wendlandt reports 735°. Both of us used essentially the same rate of heating. We, personally, worked in aluminite crucibles and our product showed only the slightest trace of holmium lines.

7. Yttrium cupferronate

Yttrium cupferronate, prepared by Wendlandt[8], cannot be used in the same manner as the iron and copper salts. It yields the oxide Y_2O_3 above 300°.

8. Yttrium neocupferronate

Wendlandt and Bryant[9] prepared yttrium neocupferronate. It loses weight from ordinary temperatures on and yields the oxide Y_2O_3 beyond 523°.

References p. 444

9. Yttrium oxinate

According to Wendlandt[10], yttrium oxinate $Y(C_9H_6ON)_3$ gives an almost horizontal level between 250° and 300°. The curve then reveals decomposition of the organic matter from 350° to 525°. The level due to the oxide appears rather indistinctly between 700° and 800°. However, the determination of yttrium by means of oxine is not quantitative.

10. Yttrium methyloxinate

Wendlandt[11] reports that the methyloxinate $Y(C_{10}H_8ON)_3$ is stable up to 65°. When heated linearly at 4.5° per minute, the level due to the oxide Y_2O_3 starts at 565°.

REFERENCES

[1] C. DUVAL, *Anal. Chim. Acta*, 10 (1954) 321. .
[2] W. W. WENDLANDT AND B. LOVE, *Science*, 129 (1959) 842.
[3] A. I. POPOV AND G. E. KNUDSEN, *J. Am. Chem. Soc.*, 76 (1954) 3921.
[4] W. W. WENDLANDT, *J. Inorg. & Nuclear Chem.*, 5 (1957) 118.
[5] W. W. WENDLANDT, *J. Inorg. & Nuclear Chem.*, 7 (1958) 51.
[6] W. W. WENDLANDT, *Anal. Chim. Acta*, 15 (1956) 435.
[7] W. W. WENDLANDT, *Anal. Chem.*, 30 (1958) 58; 31 (1959) 408.
[8] W. W. WENDLANDT, *Anal. Chem.*, 27 (1955) 1277.
[9] W. W. WENDLANDT AND J. M. BRYANT, *Anal. Chim. Acta*, 13 (1955) 550.
[10] W. W. WENDLANDT, *Anal. Chim. Acta*, 15 (1956) 109.
[11] W. W. WENDLANDT, *Anal. Chim. Acta*, 17 (1957) 277.

CHAPTER 39

Zirconium

This chapter will deal with the nitrate and with the precipitates which may serve for determining zirconium. Among the organic acids which have been proposed recently as precipitants for zirconium, mandelic acid seems to be the best. In fact, we found that the mandelate may be weighed in place of the customary dioxide. Our study was made with hafnium-free zirconium whereas other workers probably used the commercial grades. However, the shape of the curves is not changed significantly.

The compounds discussed in this chapter are: the hydroxide or rather the hydrous oxide precipitated by: ammonium hydroxide, gaseous ammonia, aniline, dimethylaniline, diethylaniline, benzylamine, xylidine, pyridine, piperidine, quinoline, phenylhydrazine, and tannin; the "peroxide", iodate, sulphate, selenite and basic selenite, nitrate, the phosphate precipitated by diammonium phosphate and metaphosphoric acid, methyl zirconyl phosphate, arsenate, benzoate, m-nitrobenzoate, phenoxyacetate, m-cresoxyacetate, phthalate, cinnamate, benzilate, salicylate, cupferronate, mandelate, p-bromomandelate, diphenate, oxinate, the complexes with arrhenal and atoxyl, propylarsonate, hydroxyphenylarsonate.

Among all the precipitants of the hydrous oxide, aniline yields a product whose composition most closely approaches the composition $Zr(OH)_4$. However, Cabannes-Ott[1] was unable to detect any OH groups when she examined this material by infra-red absorption. Only water molecules were visible, which perhaps were a little less firmly bound in the case of the hydrous oxide resulting from the action of aniline as precipitant.

Speter[2] reported that he obtained a precipitate with picric acid. We have been unable to duplicate his findings.

1. Zirconyl oxide hydrate

(i) Precipitation by aqueous ammonia When aqueous ammonia is

added to a solution of zirconyl chloride, the familiar gelatinous precipitate appears. Its thermolysis curve reveals the rapid elimination of the imbibed water up to 120°; there is nothing to indicate the composition $ZrO_2.2H_2O$, but around 120° a distinct break corresponds to the composition $ZrO_2.H_2O$. The loss of water of constitution then proceeds from 120° to 800°, but no great error is incurred if it is assumed that the portion of the curve from 400° to 800° is almost horizontal. Consequently, the branch of the curve beginning at 400° may be considered for the automatic determination as the oxide ZrO_2. See Stachtchenko and Duval[3].

(ii) Precipitation by gaseous ammonia   If, however, the precipitation is made not with aqueous ammonia, but by means of a slow stream of air charged with ammonia vapours, as described by Trombe[4], the precipitate yields a curve which is identical with the preceding one. The break related to the hypothetical compound $ZrO_2.H_2O$ appears near 143°. The level related to the ZrO_2 becomes practically horizontal at 467° and perfectly horizontal beginning at 637°.

(iii) Precipitation by aniline   If aniline is used instead of ammonia, as suggested by Jefferson[5], the curve will be found to have an entirely different aspect. There is a sudden change of direction near 105°, and an essentially horizontal level extends from 105° to 150°. The height of this level agrees rather well with the composition $ZrO_2.2H_2O$. This hydrate then gradually loses water; a new break is found not far from 789° and agrees rather well with $ZrO_2.H_2O$, but the final dehydration is rather difficult to accomplish. It is necessary to keep the specimen at 972° for an hour to obtain the weight which corresponds to the ZrO_2.

(iv) Precipitation by dimethylaniline   Jefferson[5] likewise suggested the use of dimethylaniline. The resulting precipitate yields a thermolysis curve very similar to the one just discussed. However, the level due to the hydrate $ZrO_2 \cdot 2H_2O$ is less distinct, but the constant weight corresponding to the oxide ZrO_2 begins at 842°.

(v) Precipitation by diethylaniline   Well fractionated diethylaniline yields a colourless precipitate which becomes slightly brownish after a time. Its thermolysis curve shows an abrupt change at 60° during the dehydration but this has no special significance. The loss of water then proceeds more slowly and the constant weight due to the zirconium oxide is reached at only 223°. Therefore, this method of precipitation appears

more interesting than the others and we recommend its retention for the automatic determination.

(vi) Precipitation by cyanate Cabannes-Ott[1] studied the precipitate produced by potassium cyanate in the presence of ammonium chloride. It contained no chloride ions. The level due to the oxide began at 550°.

(vii) Precipitation by benzylamine The precipitate obtained with benzylamine, as suggested by Jefferson[5], yields a curve which slants downward until the temperature reaches 584°. There is an inflection at 373° which corresponds almost to the composition $ZrO_2.H_2O$.

(viii) Precipitation by xylidine When we used xylidine, as proposed by Jefferson[5], we obtained no horizontal level corresponding to $ZrO_2 \cdot 2H_2O$. There was only a change of direction near 732° but it did not correspond to any simple composition. The last traces of water were given off beyond 868° but the release was very slow and it was necessary to bring the temperature to at least 1028° before the constant weight due to ZrO_2 was obtained. Consequently the method cannot be recommended.

(ix) Precipitation by pyridine Precipitation of the hydrous oxide by means of pyridine as described by Jefferson[5] gives a product whose curve is similar in shape but which shows a rather lengthy inflection between 114° and 232°. The corresponding horizontal gives a molecular weight of 154.72 as compared with 159.22 calculated for $ZrO_2 \cdot 2H_2O$. Likewise, around 731°, our result is a little low as compared with the calculated value for $ZrO_2 \cdot H_2O$. The constant weight due to the oxide ZrO_2 is reached near 850°.

(x) Precipitation by piperidine When we employed piperidine as precipitant as suggested by Jefferson[5], the curve was always of the same type with constant weight around 473°. An inflection was observed between 40° and 100° corresponding rather well to the composition $ZrO_2 \cdot 2H_2O$.

(xi) Precipitation by quinoline Jefferson[5] likewise proposed quinoline as precipitant. The curve does not show constant weight until the temperature reaches 878°. Despite its appearance of being dry, the precipitate tenaciously retains quinoline and water.

(xii) Precipitation by phenylhydrazine The precipitation of hydrous

zirconium oxide by phenylhydrazine was conducted as described by Allen[6]. The product loses water up to around 792°. The dehydration slows down near 716° and the curve agrees with $ZrO_2 \cdot H_2O$. The oxide ZrO_2 appears above 792°.

(xiii) Precipitation by tannin Schoeller[7] suggested tannin for the precipitation of the hydrous oxide. The product shows a continuous loss of weight as far as 730°, with an abrupt change of direction and a very slow loss of weight from 635°. Because of the reducing action of the medium, this method involves a slight reduction of the zirconium. In fact the section of the curve extending from 730° to 880° does ascend very slightly and then becomes level. It is the only curve in this group which exhibits this peculiarity.

2. Zirconium "peroxide"

According to Bailey[8], if zirconyl sulphate is treated in very dilute sulphuric acid with 100 volume (30%) hydrogen peroxide, the precipitate obtained has the formula $Zr_2O_5 \cdot 4H_2O$. We repeated this experiment and found that the colourless precipitate does not yield any level related to this peroxide. The curve presents a striking resemblance to that given by the product brought down by ammonia; however, the break starts at 75° and the loss of weight due to the removal of the water is not noticeable below 955°, but between 300° and 955° the loss of weight does not amount to 1/70th. We do not regard this procedure with much interest and all the less so because the precipitate filters very poorly. If then we have obtained a peroxide, the active oxygen which it supposedly contains must have come off at the same time as the water without giving any evidence of a definite compound.

3. Basic zirconium iodate

When a zirconium salt was treated with potassium iodate in accord with the method of Chernikhov and Uspenskaya[9], we obtained a white precipitate which apparently had no definite composition. The corresponding pyrolysis curve descends regularly as far as 395° where an almost horizontal level begins and extends to around 480°. We have isolated the compound corresponding to this level. Determination of the iodine and zirconium led to a complicated formula of a basic zirconium iodate. The above workers recommend drying the precipitate at 45° but the curve is descending at this temperature. There is a rapid loss of iodine and oxygen above 480°; however, the zirconium oxide which appears above 950° is not pure, and the absorbed iodine gives it a yellowish tint. We believe this method should be abandoned.

4. Basic zirconium selenite

The method with selenious acid was proposed by Smith and James[10]; it has been investigated by Schumb and Pittman[11], Coppieters[12] and Simpson and Schumb[13]. The colourless precipitate is a so-called basic selenite whose formula depends intimately on the conditions of preparation. Its curve descends regularly or at least it does not yield any level near 150° that corresponds to constant figures. Consequently this basic selenite has been neglected by the previous workers; they destroyed it by heating and weighed the residue of zirconium dioxide, which does not present much of special interest from the analytical viewpoint. However, if in accord with the suggestion of Claassen[14], the basic selenite is digested between 80° and 100° with a large excess of selenious acid for 2–20 hours, the neutral selenite $Zr(SeO_3)_2$ is produced. The latter gives a horizontal that is correct up to 537° (heating rate 150° per hour and sample weight 200 mg). Decomposition then ensues, brown fumes are evolved, and finally zirconium oxide is obtained. We have applied this improved Claassen procedure to the automatic determination of zirconium and in microanalysis.

5. Zirconium sulphate

The sulphate $Zr(SO_4)_2 \cdot 4H_2O$ has been studied in various ways by Komissarova, Plyushchev and Yuranova[15]. The salt loses its water in two or three stages depending on the rate at which it is heated. Three molecules of water are removed between 100° and 160°. The last molecule is eliminated slowly: $\frac{1}{2} H_2O$ at 190–215° and the rest at 300–340°. The anhydrous salt then decomposes with loss of sulphur trioxide in the interval 450–800°. The above workers have come to the conclusion that the monohydrate has the formula $[H_2ZrO(SO_4)_2]$; we prefer to write it as $Zr(H_2SO_5)(SO_4)$.

6. Zirconyl nitrate

Wendlandt[16] heated the nitrate $ZrO(NO_3)_2 \cdot 5H_2O$ but could find no level for the anhydrous salt. The level due to the oxide ZrO_2 starts at 575°.

7. Zirconyl phosphate

(i) Precipitation by diammonium phosphate   The phosphate precipitated by Stumper and Mettelock[17] by means of diammonium phosphate is rich in water and dehydrates rapidly up to around 282°. At this temperature, the formula of the material agrees with that of a diacid zirconyl phosphate $ZrO(H_2PO_4)_2$ but the latter is not stable for long. It loses the elements of 2 molecules of water very regularly and produces zirconium pyrophosphate ZrP_2O_7 at a temperature which is difficult to determine but which may be situated between 850° and 880°.

(ii) Precipitation by metaphosphoric acid   If the precipitation is carried out as described by Willard and Hahn[18] by means of metaphosphoric acid, the curve descends regularly as far as 700°, the temperature at which the horizontal of zirconium pyrophosphate starts.

8. Methyl zirconyl phosphate

Willard and Hahn[18] described the precipitation as methyl zirconyl phosphate; a large excess of the reagent is used. The curve greatly resembles that of the phosphate. There is a change of direction between 98° and 121° and zirconium pyrophosphate begins to appear at 655°. We can find but little of interest in this method in view of the slowness of precipitation (2 days on the water bath) and the difficulty of preparing the reagent.

9. Zirconyl hydrogen arsenate

The precipitation of zirconium, in acid surroundings, by means of disodium arsenate as described by Axt[19] and Claassen and Visser[20], leads to an abundant precipitate, difficult to wash, and clinging to the walls of the glass container. It corresponds approximately to the formula $ZrOHAsO_4$ which is manifested on the curve by a break at 121°. The above workers advise redissolving this precipitate in an acid and reprecipitating the zirconium by ammonium hydroxide. This may be a happy solution. In fact, even though the thermolysis curve shows a good horizontal beginning at 637°, the latter does not correspond precisely to the formula $Zr_2As_2O_7$ for a pyroarsenate. The precipitate retains excessive amounts of the precipitant, especially ammonium arsenate, so that at present the method of determination appears to be of not much value.

10. Zirconyl benzoate

The method of Rao, Venkataramaniah and Raghava Rao[21] was used by Wendlandt[22] for preparing zirconium benzoate $ZrO(C_6H_5CO_2)_2$, whose thermolysis curve he then traced. This complex is stable up to 75° and then begins slowly to decompose. The decomposition rate increases above 200° and the level of the oxide starts at 500°. Although the separation of zirconium by means of benzoic acid appears to be quantitative in 0.15 N hydrochloric acid, we have not found the product to possess exactly the formula given above.

11. Zirconyl m-nitrobenzoate

The precipitation of zirconium as the m-nitrobenzoate was suggested by Osborn[23]. The white product, whose formation is quantitative, yields a thermolysis curve which exhibits the following features. It descends slowly until around 350° where a sudden explosion frees the nitro groups. The carbon then burns very slowly up to 827° and results in a partial reduction of the zirconium. Then the curve slowly ascends and it is necessary to bring the specimen up to at least 1000° to reach the horizontal corresponding to the zirconium oxide ZrO_2. This method, like those that follow (the mandelate excepted), is of interest only for separation purposes.

12. Zirconyl salicylate

The salicylate, whose formula is alleged to be $ZrO(HO-C_6H_4-CO_2)_2$, was prepared by Viswanadha Sastry and Raghava Rao[24]. Wendlandt[22] found it to be stable up to 100° where it begins to decompose slowly. After a number of decomposition stages, the level of the oxide starts at 790°. We[25] have discarded this method.

13. Zirconyl phenoxyacetate

These same workers[24] prepared the phenoxyacetate with the alleged composition $ZrO(C_6H_5-OCH_2-CO_2)_2$. Wendlandt[22] found that its curve included a horizontal stretch up to 80°, and the organic matter decomposed up to 400°. A tar was left, and the level of the zirconium oxide began at 565°.

14. Zirconyl m-cresoxyacetate

The m-cresoxyacetate, with the approximate formula $ZrO(CH_3-C_6H_4-OCH_2-CO_2)_2$, was prepared by Venkataramaniah and Raghava Rao[26]. Wendlandt[22] reports that it is stable up to 80°. The rate of decomposition increases above 250° and tar is produced at 375°. If the heating is carried farther, the level of the oxide begins at 585°.

15. Zirconium phthalate

The phthalate (impure) $Zr(CO_2-C_6H_4-CO_2)_2$ is likewise stable up to 80°, according to Wendlandt[22], and then starts to decompose slowly. The level due to the oxide begins at 525°. The composition of the precipitate is variable. See Duval[25].

16. Zirconyl cinnamate

The cinnamate $ZrO(C_6H_5-CH=CH-CO_2)_2$, prepared by Wenkateswarlu and Raghava Rao[27] should also be considered merely as a method of separation. The formula given above is not strictly applicable[25]. The precipitate is stable[22] up to 75°. The level of the zirconium oxide begins at 600°.

17. Zirconium diphenate

The diphenate, prepared by Banerjee[28], has the approximate formula $Zr[(C_6H_4-CO_2)_2]_2$. Wendlandt[22] found that it is stable only up to 50° where it begins slowly to decompose. The rapid decomposition of the organic matter commences at 400° and the level due to the zirconium oxide starts at 505°.

18. Zirconium benzilate

The benzilate precipitate is the worst of all. It was prepared by Venkataramaniah and Raghava Rao[29]. Wendlandt[22] found that the compound starts to decompose at 85°, and above 225° the decomposition is rapid with a break in the curve around 450°. The tar oxidizes beyond 450° and leaves free zirconium oxide from 500° on.

19. Zirconium cupferronate

When zirconium is precipitated by means of cupferron as described by Thornton and Hayden[30] and Brown[31], the product is very unstable and decomposes abruptly around 90°. The constant weight due to zirconium oxide is obtained from 745° on. The method holds little of interest.

20. Zirconium mandelate

By applying the excellent method of Kumins[32], we obtained with mandelic acid a quantitative precipitation of zirconium. The precipitate $Zr(C_6H_4-CHOH-CO_2)_4$ is stable as far as 188° so that there is no need to destroy it and weigh the resulting zirconium oxide. Of course we recommend the horizontal corresponding to the mandelate for the automatic determination of zirconium. As in the case of the m-nitrobenzoate, there is partial reduction of the zirconium during the decomposition of the organic matter, and in fact the curve ascends somewhat after 820°, and the attainment of constant weight due to the zirconium dioxide requires heating to 960° at least.

21. Zirconium p-bromomandelate

In view of the excellence of the mandelate method, it was quite logical to increase the molecular weight of the precipitant in the hope of securing even better results. This was done by Oesper and Klingenberg[33] who used p-bromomandelic acid. Curves traced with samples of the precipitate furnished to us by these workers showed a good horizontal from 64° to 170° but this level did not correspond to the reported formula $Zr(BrC_6H_4-CHOH-CO_2)_4$. Instead of the calculated value 1010, the curve in one run showed a molecular weight of 849, and in another 901. Accordingly the preparation did not have a constant composition. The results obtained with the p-chloro- and the p-fluoromandelic acids were no more satisfactory. Wendlandt[22] does not agree with our findings. He found that the precipitate agrees well with the normal salt $Zr(BrC_6H_4-CHOH-CO_2)_4$ and it gave him a value of 980 for the molecular weight. He found it to be stable up to 150° when heated at the rate of 4.5° per minute; it then disintegrates slowly, and the curve shows a break at 325°. The level due to the oxide ZrO_2 starts at 550°.

22. Zirconium p-fluoromandelate

Duval[25] found that the p-fluoromandelate yields results which are only 1% away from the calculated figure. See also Belcher, Sykes and Tatlow[34].

23. Zirconium oxinate

The very lengthy procedure of Balanescu[35] provided us with a precipitate of zirconium oxinate; it loses weight up to 820° when heated. This worker suggested that it be dried between 130° and 140°. Actually, the dehydration is practically complete around 100°, but the anhydrous oxinate is not stable and the molecular weight deduced from the curve is 705 as opposed to the calculated value 659.2. We therefore suggest that the method be discarded, and all the more because we found that the temperature must be raised to 1011° to obtain zirconium oxide quite free of carbon.

A direct determination after drying at 130°, and independent of the thermobalance, revealed to us that the formula of the precipitate is not $Zr(C_9H_6NO)_4$. In their general study of the oxinates, Borrel and Pâris[36] likewise came to the conclusion that the oxinate prepared by the Balanescu technique would be a mixture. Wendlandt[22] has repeated the experiment and believes that the formula of the chelate is in agreement with $2ZrO_2 \cdot 3C_9H_7ON$. It starts to lose weight at 45° and gives no level in the range 100–200°. The level of the oxide was reported to begin at 610°, a finding that astonished us greatly since the oxinates and, in a general manner, the derivatives of quinoline are difficult to burn completely. See also Kiba and Ikeda[37] and Süe and Wetroff[38].

24. Zirconium methylarsonate

The method due to Chandelle[39] in which zirconium is precipitated by means of arrhenal (methylarsonic acid) does not appear to yield a definite compound. The corresponding curve slants downward continuously as far as 520°, and in particular we have not observed any level at 50° which would permit a satisfactory weighing in the form of $Zr(CH_3-AsO_3)_2$. Consequently, the procedure should be considered solely as a separation method. We obtained zirconium oxide at 971°; there is a slow reoxidation of the reduced zirconium beginning at 521°, the reduction being due to the action of the arsenious oxide and the carbon monoxide in the furnace.

25. Zirconium p-aminophenylarsonate

The precipitation with atoxyl (sodium p-aminophenylarsonate) was conducted as described by Chandelle[40]. The curve greatly resembles that given by the material brought down by means of arrhenal. The reoxidation commences at 700° and a horizontal due to the zirconium oxide is reached near 959°.

26. Zirconium propylarsonate

The method of Arnold and Chandlee[41] was used for the preparation of the propylarsonate. The curve slants downward until the temperature reaches 942° and the resulting mixture slowly reoxidizes. The horizontal due to the zirconium oxide begins near 959°.

27. Zirconium phenylarsonate

What we have just said may be repeated word for word with respect the phenylarsonate prepared as outlined by Medvedeva[42]. The curve slants downward until 636° and the slow reoxidation seems to be complete at 969°. It is impossible to dry the phenylarsonate quantitatively at a low temperature.

28. Zirconium p-hydroxyphenylarsonate

Claassen[43] described the preparation of the p-hydroxyphenylarsonate. Its curve again has a form much like that of the preceding compounds but the minimum is less accentuated and the horizontal due to the zirconium oxide begins at around 900°.

Accordingly, all of these arsenical derivatives give identical results, and they should rather be considered as separation reagents. One rather than the other of them may be selected depending on the nature of the elements accompanying the zirconium.

The ascent of the decomposition curve in the case of the product obtained by means of arrhenal might suggest that the valence of zirconium has been brought below four. We therefore carefully collected the products which appeared at the minima of the curves and made a qualitative analysis of these materials. Only zirconium and oxygen were found; there was no arsenic present. Quantitative examination of these specimens showed a composition poorer in oxygen than that demanded by the formula ZrO_2. However, X-ray analysis by the Debye-Scherrer powder method delivered spectra completely identical with those of pure zirconium dioxide. Accordingly, we are still without a satisfactory explanation of our findings.

REFERENCES

1 C. CABANNES-OTT, Thesis, Paris, 1958.
2 M. SPETER, Continental Met. & Chem. Eng., 1 (1926) 83.
3 J. STACHTCHENKO AND C. DUVAL, Anal. Chim. Acta, 5 (1951) 410.
4 F. TROMBE, Compt. rend., 215 (1942) 539; 216 (1943) 888.
5 A. MacMICHAEL JEFFERSON, J. Am. Chem. Soc., 24 (1902) 540.
6 E. T. ALLEN, J. Am. Chem. Soc., 25 (1903) 421.
7 W. R. SCHOELLER, Analyst, 69 (1944) 259.
8 G. H. BAILEY, J. Chem. Soc., 49 (1886) 149, 481; Z. anal. Chem., 28 (1889) 699.
9 Y. A CHERNIKHOV AND T. A. USPENKAYA, Zavodskaya Lab., 10 (1941) 248.
10 M. M. SMITH AND C. JAMES, J. Am. Chem. Soc., 42 (1920) 1764.
11 W. G. SCHUMB AND F. E. PITTMAN, Ind. Eng. Chem., Anal. Ed., 14 (1942) 512.
12 V. COPPIETERS, Ing. chim. (Milan), 22 (1938) 179, 233.
13 G. SIMPSON AND W. C. SCHUMB, C.A., (1931) 2071; (1933) 681; (1935) 1359.
14 A. CLAASSEN, Z. anal. Chem., 117 (1939) 252.
15 L. N. KOMISSAROVA, V. E. PLYUSHCHEV AND L. I. YURANOVA, Izvest. Vysshykh. Ucheb. Zavedinii Khim. i Khim. Tekhnol., (1958) 37; C.A., (1958) 16965.
16 W. W. WENDLANDT, Anal. Chim. Acta, 15 (1956) 435.
17 R. STUMPER AND P. METTELOCK, Bull. Soc. chim. France, 8 (1947) 674.
18 H. H. WILLARD AND R. B. HAHN, Anal. Chem., 21 (1949) 293
19 M. AXT, Ing. chim. (Milan), 22 (1938) 26.
20 A. CLAASSEN AND J. VISSER, Rec. trav. chim., 62 (1943) 172.
21 C. L. RAO, M. VENKATARAMANIAH AND B. S. V. RAGHAVA RAO, J. Sci. Ind. Research. (India), 108 (1951) 152.
22 W. W. WENDLANDT, Anal. Chim. Acta, 16 (1957) 129.
23 G. H. OSBORN, Analyst, 73 (1948) 381.
24 T. VISWANADHA SASTRY AND B. S. V. RAGHAVA RAO, J. Indian Chem. Soc., 28 (1951) 354.
25 C. DUVAL, Traité de micro-analyse minérale, Presses scientifiques internationales, Paris, 1955, Vol. 2, p. 48.
26 M. VENKATARAMANIAH AND B. S. V. RAGHAVA RAO, Anal. Chem., 23 (1951) 539.
27 C. VENKATESWARLU AND B. S. V. RAGHAVA RAO, J. Indian Chem. Soc., 28 (1951) 354.
28 G. BANERJEE, Z. anal. Chem., 147 (1955) 409.
29 M. VENKATARAMANIAH AND B. S. V. RAGHAVA RAO, J. Indian Chem. Soc., 28 (1951) 257.
30 W. H. THORNTON AND E. M. HAYDEN, Am. J. Sci., 38 (1914) 137.
31 J. BROWN, J. Am. Chem. Soc., 39 (1917) 2358.
32 C. A. KUMINS, Anal. Chem., 19 (1947) 376.
33 R. E. OESPER AND J. KLINGENBERG, Anal. Chem., 21 (1949) 1509.
34 R. BELCHER, A. SYKES AND J. C. TATLOW, Anal. Chim. Acta, 10 (1954) 38.
35 G. BALANESCU, Z. anal. Chem., 101 (1935) 101.
36 M. BORREL AND R. PÂRIS, Anal. Chim. Acta, 14 (1950) 267.
37 T. KIBA AND T. IKEDA, J. Chem. Soc. Japan, 60 (1939) 911.
38 P. SÜE AND G. WETROFF, Compt. rend., 196 (1933) 1813.
39 R. CHANDELLE, Bull. soc. chim. Belg., 46 (1937) 283.

[40] R. CHANDELLE, *Bull. soc. chim. Belg.*, 48 (1939) 12.
[41] F. W. ARNOLD JR. AND G. C. CHANDLEE, *J. Am. Chem. Soc.*, 57 (1935) 8.
[42] A. MEDVEDEVA, *Zavodskaya Lab.*, 11 (1945) 254.
[43] A. CLAASSEN, *Rec. trav. chim.*, 61 (1942) 299.

CHAPTER 40

Niobium

Up to the present six methods have been suggested for the precipitation of niobium but all of them eventually require the element to be weighed as Nb_2O_5. None of these procedures seems to have been generally accepted and there is still need of a satisfactory method for the determination of niobium. This chapter will deal with the precipitation by means of tartaric acid, cupferron, hexamethylenetetramine + pyrogallol, tannin, oxine, and phenylarsonic acid. The primary material used by Doan and Duval[1] was spectrographically pure niobic oxide which was melted in a platinum crucible with potassium acid sulphate.

1. Niobic oxide

(i) Precipitation with tartaric acid   The precipitation with tartaric acid was conducted as directed by Schöller and Webb[2] and Schöller[3]. The curve given by the precipitate shows first a rapid elimination of water up to 74°; decomposition then sets in and the level due to the niobic oxide Nb_2O_5 starts at 731°. Tashiro[4] reported 800°.

(ii) Precipitation by cupferron   Cupferron was employed as described by Schöller and Webb[2] and Pied[5]. The precipitate appears rapidly and yields niobic oxide from 650° on.

(iii) Precipitation by hexamine and pyrogallol   Alimarin[6], who suggested the procedure with hexamethylenetetramine + pyrogallol, advised that the precipitate be dried between 900° and 1000°. The curve slants downward until the temperature reaches 987°, but the residue is not niobic oxide. Actually, the curve slowly ascends as far as 1026° because of the uptake of oxygen. Therefore, the material must be dried at this latter temperature, at least, if the precipitate remains in the furnace in contact with the gases evolved during the decomposition of the organic matter.

(iv) Precipitation by tannin The procedure with tannin is due to Schöller[3,7]; it yields a precipitate whose thermolysis curve is completely identical with that given by the precipitate obtained with tartaric acid. The level due to the oxide Nb_2O_5 commences at 822° in this case.

(v) Precipitation by oxine Süe[8] suggested that the precipitate obtained with oxine be dried for 2 hours at 115°, whereas Kiba and Ikeda[9] reported a level extending from 110° to 130°. In our opinion, we are not dealing here with a definite compound but rather with an absorption product which is brought down quantitatively, but retains variable amounts of oxine depending on the procedure employed. In one of our trials, we found approximately two molecules of oxine per molecule of niobic oxide. The curve reveals a rapid elimination of water up to 50°, followed by a poorly defined level from 50° to 70°. The precipitate then slowly but distinctly loses weight as far as 205°; the weight at 115° cannot be constant. The oxine has been completely destroyed when the temperature has reached 649°, and the level due to the niobic oxide Nb_2O_5 begins there.

(vi) Precipitation by phenylarsonic acid The precipitate obtained by the method of Alimarin and Frid[10] with phenylarsonic acid gives a curve which descends until 650° is reached. From then on the curve ascends because of the uptake of oxygen, and constant weight due to niobic oxide Nb_2O_5 is obtained from 1008° on. Therefore the method seems to possess little interest as a means of determining niobium, but X-ray spectroscopy should yield information concerning the nature of the product which appears at the minimum of the curve and whose calculated formula agrees approximately with Nb_2O.

2. Other niobium compounds

Finally, it should be pointed out that Spitsyn and Lapitskii[5] made a study with the thermobalance of the *solid state reaction* between niobic oxide and sodium carbonate, and that Tashiro[4] reported 110–180° as the existence region of the *nitrate* $Nb(NO_3)_5$, and that Pehelkin, Lapitskii, Spitsyn and Simanov[12] have made a general study of the *niobates* of the $M_7Nb_{12}O_{37}$ type of salts, where M denotes Be, Mg, Ca, Sr, Ba, Pb. The respective chapters dealing with these cations include a discussion of these niobates.

REFERENCES

1 U. M. DOAN AND C. DUVAL, *Anal. Chim. Acta*, 6 (1952) 81.
2 W. R. SCHÖLLER AND H. W. WEBB, *Analyst*, 54 (1929) 704.
3 W. R. SCHÖLLER, *Analyst*, 54 (1929) 456.
4 T. TASHIRO, *J. Chem. Soc. Japan*, 52 (1931) 727.
5 H. PIED, *Compt. rend.*, 179 (1924) 1897.
6 I. P. ALIMARIN, *Zavodskaya Lab.*, 13 (1947) 547.
7 W. R. SCHÖLLER, *Analyst*, 61 (1936) 806.
8 P. SÜE, *Compt. rend.*, 196 (1933) 1022.
9 T. KIBA AND T. IKEDA, *J. Chem. Soc. Japan*, 60 (1939) 911.
10 I. P. ALIMARIN AND B. I. FRID, *Zavodskaya Lab.*, 7 (1938) 913.
11 V. I. SPITSYN AND A. V. LAPITSKII, *Zhur Neorg. Khim.*, 1 (1956) 1771; *C.A.*, (1957) 2441.
12 V. A. PEHELKIN, A. V. LAPITSKII, V. I. SPITSYN AND Y. P. SIMANOV, *Zhur Neorg. Khim.*, 1 (1956) 1784; *C.A.*, (1957) 2445.

CHAPTER 41

Molybdenum

Dupuis and Duval[1] have published an over-all survey of the thermo-gravimetry of the precipitates used for determining molybdenum. In addition, Dupuis[2] accurately determined the temperature at which the volatility of molybdic anhydride becomes appreciable, either when the oxide is alone or included in phosphomolybdates, silicomolybdates, germanimolybdates, arsenimolybdates, etc.

We will also investigate in this chapter the sulphide MoS_2, and the various simple and complex molybdates. We will omit the methods which prescribe weighing the element as the oxide MoO_2, or as the free metal, and also the procedure calling for the reduction of the trisulphide to the bisulphide by means of hydrogen.

1. Molybdic acids

Dupuis[2] successively employed three molybdic acids.

(i) White commercial acid   A white commercial acid, whose composition was found to be $MoO_3 \cdot 7H_2O$, showed a slight gain in weight with maximum flattening between 160° and 222° and corresponding to a gain of 2 mg per 500 mg. It loses water in a noticeable fashion above 277°, and then more slowly beginning at 394°. The level of the anhydride extends from 443° to 715° where volatilization sets in, proceeds very slowly and continues up to 775° (0.2%).

(ii) Yellow crystalline acid   A yellow crystalline acid whose composition was $MoO_3 \cdot 2H_2O$. It was very stable and came from the deposit produced by an old solution of molybdic reagent; it contained no ammonia. This compound lost its 2 molecules of water rather abruptly from 66° on, and then more slowly between 200° and 407°. The molybdic anhydride began to sublime near 767°.

(iii) Yellow acid   A yellow acid, prepared by the action of nitric acid

References p. 468

(sp. gr. 1.16) on ammonium molybdate; its composition was MoO_3· $1.8H_2O$. It did not deposit until several months had passed. This material gave a curve which starts with a horizontal extending up to 70°, followed by a regular loss of water up to 120°, remaining stationary from 120° to 153°, and again accelerating as far as 193° where it slows down again and terminates at 300°. The molybdic anhydride MoO_3 obtained is stable from 300° to 773°, and then sublimes rapidly.

2. Ammonium paramolybdate

The thermolysis curve of ammonium paramolybdate $3(NH_4)_2O·7MoO_3$ $·4H_2O$ was obtained by progressive heating of the material from 0° to 900° over a period of three hours. The curve shows that the molybdic anhydride coming from the decomposition of this salt has a horizontal stretch between approximately 400° and 750–760°. The following figures taken from one of the curves show the losses suffered by 239.8 mg of the molybdic anhydride at temperatures above this latter temperature range:

Temp. (°C)	751	796	809	821	835	850
Loss (mg)	0	0.98	2.94	4.90	8.82	17.64

Accordingly, if the anhydride is alone, its volatility does not appear below 751°. This finding puts a stop to all the controversies and exposes the non-validity of the figures published by Treadwell[3], Niederl[4], Péchard[5], Brinton and Stoppel[6], Wolf[7], Sterling and Spuhr[8], and especially Pavelka and Zucchelli[9]. In addition Fig. 57 shows the network of isotherms whose interpretation weakens the findings of these latter workers.

The paramolybdate is stable up to approximately 109°. It breaks down as far as 158°, and from 158° to 208° there is a second horizontal. Analysis of the corresponding product yields the global composition $2(NH_4)_2O·5MoO_3$ or that of an equimolecular mixture of the tri- and dimolybdate. This association would seem rather to be a mixture in view of the descending aspect of the level and the variability of the extreme temperatures of this level with changes in the heating rates. A renewed loss of ammonia and water occurs from 208° to 256°, and from 256° to 304° there is a third horizontal stretch whose limits are more constant; it is the existence region of ammonium tetramolybdate $(NH_4)_2O·4MoO_3$. The remainder of the ammonia comes off rapidly up to 356° and then more slowly up to 393°, which marks the start of the fourth horizontal level; it is due to molybdic anhydride. As stated above, the latter level extends to 751°. See also Hegedus, Suovari and Neugebauer[10].

3. Molybdenum trisulphide

(i) Precipitation by hydrogen sulphide The directions given by Moser and Beer[11] are followed if hydrogen sulphide is employed. The colloidal nature of the precipitate enables it to adsorb water, hydrogen sulphide, and sulphuric acid which continue to be released even after the temperature has passed 300°. Consequently, drying at 200° cannot yield the pure product. The trisulphide does not produce any level; therefore this weighing form can hardly be recommended. If purified at once, it yields the anhydride quantitatively beginning at 490°; beyond this temperature the anhydride gives a horizontal up to 695°, and at higher temperatures volatilization occurs as noted above. This method of determination is accurate but lengthy; the digestion in a pressure vessel and the washing require much care.

(ii) Precipitation by sodium sulphide Taimni and Tandon[12] claim to have obtained a definite trisulphide $MoS_3 \cdot 2H_2O$ by means of sodium sulphide. They heated their product on the Stanton balance. There was a loss of 1–2% near 80° and then a horizontal extended as far as 170°. Two molecules of water came off from 170° to 290°, and the anhydrous sulphide was formed from 290° to 390°. The level of the oxide MoO_3 extended from 430° to 800°. We have not been able to reproduce these findings when the material is heated at the rate of 5.3° per minute on the Chevenard thermobalance, and we consider the precipitation of the dihydrate as doubtful.

(iii) Precipitation by thioacetic acid The precipitation with thioacetic acid as suggested by Herstein[13] results in a product which filters readily, but washing does not remove all of the retained acid. The curve is the same as that of the product brought down by hydrogen sulphide. The trioxide MoO_3 is obtained from 450° on, and follows a horizontal level which extends as far as 780° where the oxide begins to sublime. Apart from the extremely persistent odour of the reagent, we believe that its use endangers the health of the operator, and this reagent cannot be recommended for series determinations.

4. Phosphomolybdic acid and salts

See Chapter Phosphorus regarding phosphomolybdic acid and the various phosphomolybdates.

References p. 467

5. Calcium molybdate

The thermolysis curve of calcium molybdate shows that constant weight of this salt begins at 230°.

6. Barium molybdate

Barium molybdate, prepared in accord with the procedure of Smith and Bradbury[14], does not give a constant weight corresponding to the formula $BaMoO_4$ until the temperature exceeds 320°. Satisfactory results were not obtained by drying at 110° or even at 250°. Precipitation with a strontium salt is not quantitative and consequently we have omitted strontium molybdate from this discussion.

7. Lead molybdate

Lead molybdate was prepared as directed by van Dyke-Cruser and Miller[15], who did not suggest any particular drying temperature, beyond merely heating the product over a bare flame. The salt does not reach constant weight until around 505°. The procedure conversely is not suitable for the determination of lead.

8. Cadmium molybdate

Cadmium molybdate is discussed in Chapter Cadmium. The horizontal extending from 82° to 250° may serve for the automatic determination of cadmium but not of molybdenum.

9. Silver molybdate

The precipitate of silver molybdate was produced as suggested by Mc Cay[16]. It is stable from 89° to 250°. At 946° the weight of the crucible and its contents is almost the same as at 200° but reactions involving gain and loss of weight occur between these temperatures, reactions which apparently are of complicated natures. Therefore, the temperature 110° recommended by McCay is suitable for drying this salt.

10. Mercurous molybdate

Mercury(I) molybdate was prepared by the method of Hillebrand[17],

Treadwell[18], and Yosida[19], who suggest that this salt be ignited and that the final weighing be in the form of MoO_3. The curve obtained with this molybdate is more complicated than that of mercurous chromate; in fact the level extending from 75° to 135° indicates that the composition of the precipitate is not Hg_2MoO_4. Instead, it is a mixture of the latter and basic mercury nitrate; the composition varies according to the pH of the surroundings. The second level is not so well marked; it extends to around 267° and no longer corresponds to a definite compound. The third level goes from 476° to 675° and agrees satisfactorily with a compound $Hg_2O \cdot 3MoO_3$, which loses its mercury up to 793°. The level of the molybdic anhydride then extends over about 100 degrees. However, Yosida suggests that the mercurous salt be kept below 600°. It would be advisable moreover not to go beyond 880° when igniting to constant weight. Therefore this determination must be conducted with considerable care, and since it takes no advantage of the high atomic weight of mercury the procedure is of little use. The molybdic anhydride obtained from the mercurous mixture is less volatile than that coming from ammonium paramolybdate.

11. Bismuth molybdate

Dupuis[2] heated bismuth molybdate and observed a level going from 410° to 805°; it is due to the mixture of bismuth oxide and molybdenum oxides derived from the decomposition of the salt. The method is better suited to the determination of bismuth than of molybdenum. The molybdic anhydride sublimes alone above 805°.

12. Purpureocobaltic molybdate

See Chapter Cobalt regarding the chloropentamminecobaltic (purpureo) molybdate.

13. Benzoinoxime molybdate

The complex with benzoinoxime, prepared as described by Knowles[20], should be kept for an hour on the water bath before filtering. He specifies 500° as the ignition temperature, whereas Ishibashi[21] gives 530–550°. The thermolysis curve shows that the anhydrous complex yields no level; it decomposes in continuous fashion up to 570°; the level of the molybdic anhydride MoO_3 begins there and the horizontal continues as far as 755°

Duval 30
References p. 468

before undergoing the usual volatilization. We cannot find much of interest in this method of determining molybdenum because of the waste of expensive organic material, which in addition emits a clinging disagreeable odour during its decomposition.

14. Cinchonine molybdate

The red precipitate $MoC_{22}H_{33}O_7N_5S_{13}$, obtained with cinchonine as directed by Johnson[22], yields a curve which constantly descends. The crucible becomes completely empty because the molybdic anhydride sublimes as soon as the carbon stops burning. Between 100° and 150° there is an inflection with an oblique tangent which roughly corresponds to the anhydrous complex. Despite the fact that the analytical factor is favourable for molybdenum, this method of determining molybdenum does not offer sufficient accuracy.

15. o,o'-Dianisidine molybdate

Ubeda and Gonzales[23] precipitated molybdenum by means of o,o'-dianisidine. They recommend that the precipitate be kept at 80°. The curve includes no horizontal levels except the one due to the molybdic anhydride; it extends from 600° to 780°. There is admittedly a change of direction at 52° but the succeeding linear portion is not horizontal. No great accuracy can be attained by weighing the anhydrous complex.

16. Oxine molybdate

Balanescu[24] found that oxine reacts in a solution buffered with acetic acid and ammonium acetate to give the complex $MoO_2(C_9H_6ON)_2$ which filters and washes very well. The thermolysis curve contains a good horizontal between 60° and 322° and, along with the quantitative character of the reaction, leads us to suggest this procedure for the automatic determination of molybdenum. The remainder of the curve reveals that the molybdic anhydride finally produced sublimes at the same time that the carbon burns. Balanescu and Ishimaru[25] respectively suggest 130–140° and 135–159° as the temperature limits for drying the product. These temperatures are suitable but they may be extended at either end.

The matter was complicated by Busev and Chzhan Fan'[26] when they described two other compounds obtained with oxine, namely $H_2Mo_4O_{11}$ $(C_9H_6ON)_7\cdot11H_2O$ which is yellow, and $Mo_2O_3(C_9H_6ON)_4\cdot H_2O$ which is black. These products give thermolysis curves which are scarcely legible.

Borrel and Pâris[27] included this oxinate in their general study and found a horizontal level up to 138° which conforms to the hydrate $MoO_2(C_9H_6ON)_2 \cdot \frac{1}{2}H_2O$. Beyond 138° this hydrate loses its water and yields the well-known oxinate $MoO_2(C_9H_6ON)_2$ which remains stable up to about 326°. It then undergoes vigorous decomposition, which accelerates and leads to the oxide MoO_3 which is slightly volatile at 700°.

17. 5,7-Dibromooxine molybdate

The dibromooxinate of Wendlandt and Romano Rao[28] was dried in the air and then pyrolyzed at the rate of 5.4° per minute. The precipitate corresponds to the formula $MoO_2(C_9H_4Br_2ON)_2 \cdot \sim H_2O$ but the water content may be fortuitous. Dehydration sets in at 65° and the horizontal extending from 105° to 165° corresponds to the anhydrous chelate. The compound starts to decompose rapidly beyond 260°, and the curve shows breaks at 355° and 395°. The oxide MoO_3 appears at 560°.

18. 2-Methyloxine molybdate

Borrel and Pâris[29] found that the 2-methyloxinate has the complex formula $7[MoO_2(CH_3-C_9H_5ON)_2] \cdot MoO_3 \cdot H_2O$ which holds as far as 170°. The molecule of water is lost up to 260° and the anhydrous compound decomposes slowly up to 370° and then more quickly up to 425°. Finally, the thermolysis yields a horizontal level between 430° and 470° which has not been interpreted. The last phase of the decomposition leads to the oxide MoO_3 which is slightly volatile at 700°.

19. 2-Phenyloxine molybdate

The yellow 2-phenyloxinate was prepared by Bocquet and Pâris[30]. Its formula is $MoO_2(C_6H_5-C_9H_5ON)_2$. This anhydrous compound is stable up to 305°; beyond this temperature the organic matter appears to burn uniformly up to around 700°.

20. Thiocarbohydrazide molybdate

Duval and Tran Ba Loc[31] studied the precipitate obtained with thiocarbohydrazide in the cold at pH 2.0. A water solution of the reagent was used. When the precipitate is heated at the rate of 150° per hour, there is a continuous loss while the wash liquids are being driven off and while the organic matter is decomposing; there is no transitory intermediate level. Consequently, it is necessary to weigh the residual oxide. We have done this at 500° after recording the level of constant weight. It should be noted that this determination can be carried out in the presence of tungsten and uranium.

21. Vanillidene benzidine molybdate

The complex with vanillidene–benzidine was obtained as directed by Hovorka[32] by treating the molybdate to be determined with an acetic acid solution of the reagent. The latter is readily prepared by heating vanillal (vanillin) with benzidine in the proportions prescribed in the published account. The precipitate is washed easily and filters well. The curve slants downward until the temperature reaches 580°, which marks the start of the horizontal of the molybdic anhydride. This horizontal continues as far as 780°. Consequently, there is a notable difference between this curve and that of the corresponding tungsten compound, which in the anhydrous condition is stable from 100° to 170°.

REFERENCES

[1] T. Dupuis and C. Duval, *Anal. Chim. Acta*, 4 (1950) 173.
[2] T. Dupuis, *Compt. rend.*, 228 (1949) 841; *Thesis*, Paris, 1954; *Mikrochemie*, 35 (1950) 449.
[3] W. F. Treadwell, *Z. Elektrochem.*, 19 (1913) 219.
[4] J. B. Niederl and E. P. Silbert, *J. Am. Chem. Soc.*, 51 (1929) 376.
[5] E. Péchard, *Compt. rend.*, 114 (1892) 173.
[6] P. Brinton and A. E. Stoppel, *J. Am. Chem. Soc.*, 46 (1925) 2448.
[7] K. Wolf, *Z. angew. Chem.*, 31 (1918) 140.
[8] C. Sterling and W. P. Spuhr, *Ind. Eng. Chem. Anal. Ed.*, 12 (1940) 33.
[9] F. Pavelka and A. Zuchelli, *Mikrochemie*, 31 (1943) 69.
[10] A. J. Hegedus, K. Suovari and J. Neugebauer, *Z. anorg. Chem.*, 293 (1958) 56.
[11] L. Moser and M. Beer, *Z. anorg. Chem.*, 134 (1924) 67.
[12] I. K. Taimni and S. N. Tandon, *Anal. Chim. Acta*, 22 (1960) 34.
[13] B. Herstein, *U.S. Dept. Agr. Bull.*, No. 150, p. 44; *C.A.*, (1913) 37.
[14] E. F. Smith and R. H. Bradbury, *Ber.*, 24 (1891) 2930.
[15] F. van Dyke-Cruser and E. H. Miller, *J. Am. Chem. Soc.*, 26 (1904) 676.
[16] Le Roy W. McCay, *J. Am. Chem. Soc.*, 56 (1934) 2548.
[17] W. F. Hillebrand, *Chem. News.*, 78 (1898) 218.
[18] F. P. Treadwell, *Manuel de chimie analytique*, 2nd French Ed., Dunod, Paris, 1920, p. 284.
[19] Y. Yosida, *J. Chem. Soc. Japan*, 61 (1940) 130.
[20] H. B. Knowles, *Bur. Standards J. Research*, 9 (1932) 1.
[21] S. Ishibashi, *J. Chem. Soc. Japan*, 61 (1940) 125.
[22] C. M. Johnson, *Iron Age*, 132 (1933) 16.
[23] F. Butcarons Ubeda and E. Loriente Gonzales, *Anales soc. españ. fís y quím.*, 40 (1944) 1312.

[24] G. Balanescu, *Z. anal. Chem.*, 83 (1931) 470.
[25] S. Ishimaru, *J. Chem. Soc. Japan*, 55 (1934) 201.
[26] A. I. Busev and Chzhan Fan', *Zhur. Anal. Khim.*, 15 (1960) 455.
[27] M. Borrel and R. Pâris, *Anal. Chim. Acta*, 4 (1950) 267.
[28] W. W. Wendlandt and D. V. Romana Rao, *Anal. Chim. Acta*, 17 (1957) 525.
[29] M. Borrel and R. Pâris, *Anal. Chim. Acta*, 5 (1951) 573.
[30] G. Bocquet and R. Pâris, *Anal. Chim. Acta*, 10 (1956) 201.
[31] C. Duval and Tran Ba Loc, *Compt. rend.*, 240 (1955) 1097; *Mikrochim. Acta*, (1956) 458.
[32] V. Hovorka, *Collection Czechoslov. Chem. Communs.*, 10 (1939) 527.

CHAPTER 42

Ruthenium

Because of the rarity of this element only a few studies have been devoted to its determination by gravimetric procedures.

1. Metallic ruthenium

The greater part of the workers who have determined this element as metallic ruthenium have decomposed the precipitate thermally and as often as possible they have worked in hydrogen and cooled the metal in carbon dioxide. We believe these precautions to be unnecessary because the metal apparently does not oxidize when it is in the form of ingots, wires or even as a fine powder. Goto[1] advises that the temperature not be allowed to go above 500°; he also points out that the oxide RuO_2 appears above 850°. The curves we have traced do not show an ascent before 950°.

2. Ruthenium sulphide

Weighing as the sulphide had already been tried by Gilchrist and Wichers[2] who did not find a definite product. Taimni and Salaria[3] claimed to have obtained the definite sulphide by means of a 2 N sodium sulphide solution and gave $Ru_2S_3 \cdot 2H_2O$ as its formula. We did not find any level in the vicinity of 85° where they suggested drying the compound, and later Taimni and Tandon[4] actually did find that the precipitate loses weight from 60° to 180°. The level extending from 180° to 240° does not correspond to the anhydrous form. The precipitate is converted completely into the metal at 500°.

3. Ruthenium thionalide

On the other hand, the precipitate which Rogers, Beamish and Russell[5] prepared with thionalide is perfectly definite and is stable up to 170–175°

(100 mg heated at the rate of 150° per hour). The decomposition becomes considerable from 235° on. After an abrupt change in direction at 516°, the decomposition becomes still more rapid and the level due to the ruthenium metal starts at 570–580°.

4. Ruthenium 2-phenylbenzothiazole

We made a thermal study of the precipitate formed by ruthenium and 2-phenylbenzothiazole. However, the curve slants downward from room temperature to 690° at least with a sudden fall at 516°, as in the preceding case. See Champ, Fauconnier and Duval[6].

REFERENCES

[1] H. GOTO, *J. Chem. Soc. Japan*, 55 (1934) 326,
[2] R. GILCHRIST AND E. WICHERS, *J. Am. Chem. Soc.*, 57 (1935) 2567.
[3] I. K. TAIMNI AND G. B. S. SALARIA, *Anal. Chim. Acta*, 11 (1954) 329.
[4] I. K. TAIMNI AND S. N. TANDON, *Anal. Chim. Acta*, 22 (1960) 553.
[5] W. J. ROGERS, F. E. BEAMISH AND D. S. RUSSELL, *Ind. Eng. Chem., Anal. Ed.*, 12 (1940) 561.
[6] P. CHAMP, P. FAUCONNIER AND C. DUVAL, unpublished results.

CHAPTER 43

Rhodium

Up to the present, it has been suggested that rhodium be determined gravimetrically with the metal as the weighing form and after it has been cooled in nitrogen. The initial precipitation can be accomplished with: thionalide, 2-mercaptobenzoxazole, or thiobarbituric acid.

1. Metallic rhodium

(i) Grains The metal obtained from the German firm Heraeus was in the form of silvery grains, less than 1 mm across. When 260 mg were heated linearly at 200° per hour, the curve remained horizontal up to 662° at least. Oxygen was taken up then, the amount depending on the temperature (and hence on the heating period). The gain amounted to 13 mg at 946° which was the upper limit of the experiment. The surface of the grains became dark probably, as was stated by Berzelius and later workers, because of a coating with the oxide Rh_2O_3. However, the oxidation is far from complete at this temperature and with this rate of heating; in fact, the final weight was 273.9 mg as opposed to the calculated 321.8 mg.

(ii) Produced by decomposition of complexes If now, a study is made of the rhodium produced by decomposing the inner complexes, *i.e.* the metal obtained from the precipitates formed with thionalide, etc., it is truly astounding to find that the metal resists oxidation up to 1000° at least. This discovery suggests that it would be well to discontinue the use of hydrogen and nitrogen during the reduction and cooling periods since this entirely unnecessary complication is not in accord with the thermograms traced by Duval, Champ and Fauconnier[1]. Furthermore, Goto[2] as early as 1934 showed that metallic rhodium may be weighed after it has been kept under 500°.

2. Rhodium thionalide

The method of Kienitz and Rombock[3] was used to precipitate the orange $Rh(C_{12}H_{10}ONS)_3$, free of rhodium chloride; the precipitant is thionalide. These workers used especially the procedure in which the excess of thionalide is determined iodometrically, but the method can be employed also for the gravimetric and microgravimetric determination of rhodium. In reality, the curve shows that, following removal of the moisture, the precipitate remains anhydrous and maintains a constant weight from 79° to 250° at the stated rate of heating. The organic matter decomposes above 250° and the curve shows a steep decline around 370°. Constant weight can be relied on, starting at 417°. As we have said, the level remains horizontal up to 1000°; the residual metal is grey and has a spongy appearance. We recommend the complex with thionalide for the automatic determination of rhodium; the molecular weight of the complex is 751.69 and the analytical factor is 0.1369. We have been able to accomplish this determination with the thermobalance and as little as 6.5 mg of rhodium.

3. Rhodium 2-mercaptobenzoxazole

The precipitate obtained with 2-mercaptobenzoxazole and rhodium chloride $RhCl_3$ has the formula $Rh(C_7H_4ONS)_3$. It is readily produced by following the directions given by Haines and Ryan[4]. The curve obtained with this precipitate greatly resembles the one given with thionalide, but it is advisable not to exceed 150° for the automatic determination of the inner complex, whose level, after the departure of the water, extends from 92° to 155°. The organic matter decomposes slowly and the pure metallic rhodium appears near 459°. The molecular weight of the complex is 553.39 with the analytical factor 0.1859. This method likewise has allowed us to go as low as 6.5 mg of rhodium. In our opinion, it is just as recommendable as the preceding method from the point of view of the insolubility and handling of the precipitate, but the analytical factor is a little higher in the present case.

4. Rhodium thiobarbiturate

Currah, McBryde, Cruikshank and Beamish[5] precipitated the choco-late-brown material by treating sodium chlororhodate(III) in acid solution with thiobarbituric acid. We think the precipitation is quantitative but the precipitate does not appear to have a definite composition. It contains 33.60% of rhodium. One experiment carried out with 41.27 mg of this dry precipitate yielded 13.87 mg of the metal, which does not agree with the proposed formulas. Consequently the method cannot be employed for the automatic determination of rhodium, but it may be used if the metal is the weighing form; the latter appears in the pure state around 480°. The precipitate itself, of unknown nature, remains stable between 105° and 220° on the average.

REFERENCES

[1] C. DUVAL, P. CHAMP AND P. FAUCONNIER, *Anal. Chim. Acta*, 12 (1955) 138.
[2] H. GOTO, *J. Chem. Soc. Japan*, 55 (1934) 326.
[3] H. KIENITZ AND L. ROMBOCK, *Z. anal. Chem.*, 117 (1939) 241.
[4] R. L. HAINES AND D. E. RYAN, *Can. J. Research*, 27 (1949) 72.
[5] J. E. CURRAH, W. A. E. MCBRYDE, A. J. CRUIKSHANK AND F. E. BEAMISH, *Ind. Eng Chem., Anal. Ed.*, 18 (1946) 120.

CHAPTER 44

Palladium

This chapter will deal largely with the results obtained by Champ, Fauconnier and Duval[1] in their study of the thermal behaviour of the precipitates used for the gravimetric determination of palladium. The heating and cooling operations were not conducted in hydrogen. They feel that urea, which was suggested by Drechsel[2], should be omitted from the list of precipitants, and they also were unable to prepare 5-methyl-8-hydroxyquinoline in a state of purity sufficient for the precipitation of palladium as proposed by Sa[3]. On the other hand, they have added to the list the reaction with thiobarbituric acid, a reagent proposed by Currah, McBryde, Cruikshank and Beamish[4]. It follows then that their investigation was limited to precipitation by the following reagents: carbon monoxide, hydrazine, mercury cyanide, ethylene, acetylene, sodium formate, potassium iodide, hydrogen sulphide, dimethylglyoxime, methylbenzoylglyoxime, salicylaldoxime, β-furfuraldoxime, cyclohexanedionedioxime, α-nitro-β-naphthol, α-nitroso-β-naphthol, the hydrazide of *m*-nitrobenzoic acid, 6-nitroquinoline, thiobarbituric acid, *o*-phenanthroline, thionalide, 2-mercaptobenzothiazole, oxine, α-picolic acid.

1. Metallic palladium

(i) Precipitation by carbon monoxide   Brunk's[5] method in which carbon monoxide is passed into a solution of palladium(II) chloride in the presence of sodium acetate, yields a grey deposit of the metal which may be weighed without using a stream of hydrogen. Actually, the curve shows that the metal remains unaltered up to 384°. Oxidation is distinctly discernible near 410° and the resulting oxide PdO yields a flattened maximum between 788° and 830°. This oxide decomposes at higher temperatures and according to Tashiro[6] the initial weight of metal is regained in air. This metal again yields the same curve when reheated under like conditions.

(ii) Precipitation by hydrazine The reduction with hydrazine in a solution containing hydrochloric acid is conducted at the boiling point as directed by Jannasch and Bettges[7]. The black precipitate is dry from 45° on and the curve furnished by this metallic palladium is identical with that just described.

(iii) Precipitation by ethylene Ogburn and Brastow[8] used ethylene as the reducing agent. The resulting precipitate of palladium behaves like the metal just discussed.

(iv) Precipitation by sodium formate Palladium is precipitated as a black powder by sodium formate in the procedure described by Treadwell and Hall[9]. Its curve has the same form as that given by the metal precipitated by means of carbon monoxide. However, it should be noted that in the present case the metal begins to oxidize at 242°.

2. Palladium iodide

The precipitate of PdI_2 obtained with potassium iodide is discussed in Chapter Iodine. Its curve includes a horizontal extending from 84° to 365° which can serve for the automatic determination of iodine or palladium. Tashiro[6] states that this level extends from 200° to 350°.

3. Palladium sulphide

The black sulphide, alleged to be PdS, is precipitated from an acidic solution. It loses water rapidly up to 150°; the level corresponding to the sulphide extends from 150° to 179°. It is not strictly horizontal, a finding which leads to the suspicion that the sulphide is not stable. The complete decomposition is finished at 450° and takes place in two stages: an abrupt release of gas corresponding to 2/3rds of the sulphur around 220°, and then a progressive loss of the final third between 220° and 450°. The residue consisting of palladium then oxidizes in accord with the same mechanism as in the case with carbon monoxide. Taimni and Salaria[10] prepared a sulphide $PdS \cdot 2H_2O$ by means of 2 N sodium sulphide in a solution buffered with acetic acid–ammonium acetate. The product was dried in a desiccator in the cold. We have been unable to duplicate this preparation. Taimni and Tandon[11] found that the product loses weight from 50° to 420°.

4. Palladium cyanide

The precipitation with mercury cyanide was carried out as directed by the Rose textbook[12]. The white precipitate is not stable and it is impossible to weigh the anhydrous cyanide $Pd(CN)_2$. Its thermolysis curve slants downward until the temperature reaches 502° and there is distinct release of cyanogen beginning at 330°, where the curve shows a break. The palladium then reoxidizes as in the case with carbon monoxide; the maximum weight of the oxide PdO is near 711°.

5. Palladium acetylide

The procedure of Erdmann and Makowka[13] with acetylene yields a curious red-brown compound which is not explosive. After losing the extraneous water, the material yields an essentially horizontal level between 71° and 109°. Accordingly, this acetylide is comparatively stable; it has the formula PdC_2 near the middle of this level. The infra-red absorption spectrum will sometime undoubtedly show that this compound contains triply-bound carbon atoms. The decomposition and the combustion of the resulting carbon is complete near 384° and the remainder of the curve once more is identical with the preceding thermograms.

6. Palladium dimethylglyoxime

The most classic precipitate of all is obtained by means of dimethylglyoxime. The procedure employed was that of Wunder and Thüringer[14], Holzer[15], Hecht and Donau[16]. The precipitate filters well and, following the elimination of the moisture, the curve includes a perfectly horizontal section extending from 45° to 171° and conforming closely to the expected formula $Pd(C_4H_7O_2N_2)_2$. Tashiro[6] gave 100–200° as these limits. We suggest this new weighing form; it obviously is more advantageous than metallic palladium ignited in a stream of hydrogen. It is equally indicated for the automatic determination of palladium. The dimethylglyoxime suddenly decomposes when the temperature exceeds 195° and especially as it approaches 228°; the residue consists of finely divided carbon mixed with palladium. The carbon burns away between 259° and 356°. The palladium then oxidizes in conformity with the usual pattern.

References p. 481

7. Palladium methylbenzoylglyoxime

The curve obtained with the precipitate produced by means of methyl-benzoylglyoxime closely resembles the thermogram of the dimethyl-glyoxime product. However, the precipitate is apparently less stable; it begins to lose weight at even 177°. We suggest that the horizontal extending from 52° to 150° be employed for the automatic determination of palladium. This complex was prepared by the method given by Holzer[15], who proposed 110° as the drying temperature. In passing, we wish to point out that it is essential to control the pH closely during the preparation of the reagent. Otherwise, the *amphi*-form will be obtained and it is incapable of forming a precipitate with palladium.

8. Palladium salicylaldoxime

The inner complex with salicylaldoxime was prepared by the method of Holzer[15] and Hecht and Donau[16]. It retains water more tenaciously than the two preceding products but the level corresponding to this compound can nonetheless be used for the automatic determination of palladium. It extends from 93° to 197° and includes the temperature 110° advocated by the above workers. The curve is identical with that of the dimethylglyoxime-palladium complex. The decomposition must not be conducted in a room that is not well ventilated; the odour of the disintegration products persists for days and furthermore the vapours are toxic when inhaled.

9. Palladium furfuraldoxime

Hayes and Chandler[17] recommend against using a temperature above 110° for drying the precipitate produced with furfuraldoxime. Its curve likewise is analogous to that obtained from the dimethylglyoxime product. The horizontal extending from 51° to 156° is suitable for the automatic determination of palladium. The precipitate retains but little water and dries quickly. Metallic palladium appears at 400° (minimum point of the curve). The metal then takes up oxygen and regains its initial weight at 974°.

10. Palladium nioxime

The yellow precipitate obtained with cyclohexanedionedioxime (nioxime), as directed by Voter, Banks and Diehl[18], dries immediately. It may be suitable for the automatic determination of palladium. When heated linearly at the rate of 400° per hour, it explodes at 169°. The level corresponding to the complex is perfectly horizontal up to this temperature.

11. Palladium α-nitro-β-naphthol

The yellow needles, $Pd(NO_2-C_{10}H_6-O)_2$, obtained with α-nitro-β-naphthol, as directed by Herfeld and Gerngross[19] and Mayr[20], yielded a curve that descended steadily as far as 836°; but the loss of weight up to 132° was very slight. Strictly speaking, it would be better to dry the precipitate below 92°. However, in various respects, the α-nitro-β-naphthol method has proved to be much inferior to various other procedures employing the preceding oximes.

12. Palladium α-nitroso-β-naphthol

In contrast, the precipitate obtained by means of α-nitroso-β-naphthol was found to be more suitable, and the level in its thermolysis curve extends to at least 245° and lends itself well to the automatic determination of palladium. The temperature 135° advocated by the above workers is suitable for producing a good weighing form but is too restrictive. Tashiro [6] reported 100–250° as the temperature limits of the level.

13. Palladium m-nitrobenzoic hydrazide

Vorisek and Vejdelek[21] used the hydrazide of m-nitrobenzoic acid as precipitant. The curve given by this complex shows a well-marked horizontal between 85° and 214°, but unfortunately the precipitate is hard to filter and the precipitation is far from quantitative (dimethylglyoxime reveals palladium in the filtrate). We have discarded this method.

14. Palladium 6-nitroquinoline

6-Nitroquinoline is interesting since it precipitates only palladium from among the six metals occurring in crude platinum. In addition, the yellow flocks, obtained as directed by Ogburn and Riesmayer[22] may be weighed as such; their formula is $Pd(C_9H_6O_2N_2)_2$. The curve shows a horizontal up to 198° at least, and this level may also be employed for the automatic determination of palladium. The organic matter is destroyed between 300° and 432°. Here again, the ignition in hydrogen may be avoided, and furthermore the analytical factor is lowered considerably.

15. Palladium thiobarbiturate

We employed the directions given by Currah, McBryde, Cruikshank and Beamish[4] for precipitating palladium with thiobarbituric acid. (The reagent was dissolved in 50% alcohol.) Although the precipitation has been found to be quantitative, the thermolysis curve shows that the precipitate is of little real use since there is a steady loss of weight until the temperature reaches 62° and then the palladium is set free. Consequently, this procedure provides no weighing form except the metal itself.

16. Palladium o-phenanthroline

Ryan and Fainer[23] introduced a method which is much more interesting from all points of view. The precipitant is o-phenanthroline and the precipitate, after drying at 110°, has the formula $PdClC_{12}H_8N_2$. The thermolysis curve starts with a horizontal which extends to the relatively high temperature of 389°. This level could be recommended for the automatic determination of palladium if it were not for the high cost of the reagent. The decomposition of the organic matter proceeds regularly from 400° to 766°, the temperature at which only palladium remains in the crucible. When applying this method, it should be kept in mind that the formula of the precipitate changes with the anion associated with the palladium before the precipitation.

17. Palladium oxinate

According to Kiba and Ikeda[24], palladium oxinate should be dried between 120° and 140°. It is not used in analysis.

18. Palladium thionalide

Umemura[25] reports that palladium thionalide is stable up to 195°.

19. Palladium 2-mercaptobenzothiazole

Majumdar[26] suggested the use of 2-mercaptobenzothiazole. When the corresponding chelate was heated on the Chevenard thermobalance the material proved to be stable up to 388°. Metallic palladium appeared beyond 500°.

20. Palladium picolinate

Majumdar[26] also found that picolinic acid yields a precipitate whose formula is $Pd(C_6H_4O_2N)_2$. If prepared between pH 3 and pH 7, the product is stable up to 380°. Palladium appears above 420°.

REFERENCES

1 P. CHAMP, P. FAUCONNIER AND C. DUVAL, Anal. Chim. Acta, 6 (1952) 250.
2 E. DRECHSEL, J. prakt. Chem., 20 (1879) 469.
3 A. M. SA, Rev. asoc. bioquim. arg., 16 (1944) 11.
4 J. E. CURRAH, W. A. E. McBRYDE, A. J. CRUIKSHANK AND F. E. BEAMISH, Ind. Eng. Chem. Anal. Ed., 18 (1946) 120.
5 O. BRUNCK, Z. angew. Chem., 25 (1912) 2479.
6 M. TASHIRO, J. Chem. Soc. Japan, 52 (1931) 232.
7 P. JANNASCH AND W. BETTGES., Ber., 37 (1904) 2210.
8 S. C. OGBURN AND W. C. BRASTOW, J. Am. Chem. Soc., 55 (1933) 1307.
9 F. P. TREADWELL AND W. T. HALL, Ind. Eng. Chem., Anal. Ed., 2 (1930) 255.
10 I. K. TAIMNI AND G. B. S. SALARIA, Anal. Chim. Acta, 11 (1954) 336.
11 I. K. TAIMNI AND S. N. TANDON, Anal. Chim. Acta, 22 (1960) 554.
12 H. ROSE, Handbuch der analytischen Chemie, 6th Ed. 1875, Vol. 2, p. 204.
13 H. ERDMANN AND O. MAKOWKA, Z. anal. Chem., 46 (1907) 143.
14 W. WUNDER AND V. THÜRINGER, Z. anal. Chem., 52 (1913) 101.
15 H. HOLZER, Z. anal. Chem., 95 (1933) 392.
16 F. HECHT AND J. DONAU, Anorganische Mikrogewichtsanalyse, Springer, Vienna, 1940, p. 184.
17 J. R. HAYES AND G. C. CHANDLER, Ind. Eng. Chem., Anal. Ed., 14 (1942) 491.
18 R. C. VOTER, C. V. BANKS AND H. DIEHL, Anal. Chem., 20 (1948) 652.
19 H. HERFELD AND O. GERNGROSS, Z. anal. Chem., 94 (1933) 7.
20 C. MAYR, Z. anal. Chem., 98 (1934) 402.
21 J. VORISEK AND Z. VEJDELEK, Chem. Listy, 37 (1943) 50, 65, 91.
22 S. C. OGBURN AND A. H. RIESMAYER, J. Am. Chem. Soc., 50 (1928) 3018.
23 D. E. RYAN AND P. FAINER, Can. J. Research, 27 (1949) 67.
24 T. KIBA AND T. IKEDA, J. Chem. Soc. Japan, 60 (1939) 911.
25 T. UMEMURA, J. Chem. Soc. Japan, 61 (1940) 25.
26 A. K. MAJUMDAR AND M. M. CHAKRABARTTY, Z. anal. Chem., 162 (1958) 96; 161 (1958) 100.

CHAPTER 45

Silver

The methods which have been proposed for the gravimetric determination of silver may be classified as follows: weighing as silver deposited electrolytically at the cathode, or precipitated by an ammoniacal solution of cuprous chloride, or by hypophosphorous acid, or by metals such as cadmium and aluminium, by hydroxylamine, by formaldehyde and ammonia, by vitamin C; weighing as chloride, bromide, iodide; or as sulphide precipitated by means of hydrogen sulphide, sodium sulphide, sodium thiosulphate; or weighing as selenite, as silver thallium arsenate; weighing as cyanide, thiocyanate, chromate, cupric propylenediamine argentoiodide, cobaltic diethylenediamine thiocyanate argentothiocyanate; as oxalate; as a complex with rhodanine, and as a complex with thionalide. See Marin and Duval[1].

Details will also be given regarding the oxide, sulphate, nitrate, permanganate, antimonate, oxinate, dithizonate, cobalticyanide, etc.

The precipitation of silver as the peroxide Ag_2O_2 at the anode, as proposed by Schucht[2], does not seem to be quantitative (around 95%), and since this procedure is difficult to carry out we do not intend to take it into account here.

1. Metallic silver

(i) Precipitation by electrolysis Metallic silver was deposited by electrolysis from a solution of silver nitrate containing sulphuric and tartaric acid. The electrolysis vessel was a small platinum crucible, fitted into the carrier ring of the thermobalance, and serving as the cathode. We used the current density and the voltage given by Friedrich and Rapoport[3]. After washing the metal in the inside of the crucible with water and alcohol, the deposit was heated to 950°. A horizontal straight line was obtained as soon as the alcohol had been removed rapidly. This metallic deposit differs from both copper and cobalt. There is therefore no objection to drying the cathode by heating prior to weighing the deposit.

(ii) Precipitation by ammoniacal cuprous chloride The procedure of Millon and Commaille[4] was used for the precipitation of silver by treating a solution of silver nitrate, saturated with ammonia, with an ammoniacal solution of copper(I) chloride. The thermogram consisted of a straight line up to 948° and hence was similar to the previous case.

(iii) Precipitation by hypophosphorous acid Mawrow and Mollow[5] described the procedure in which the reducing agent is hypophosphorous acid. After the metallic silver was washed with 95% alcohol, it yielded a horizontal extending to approximately 477°. A slight and progressive ascent was then observed because of fixation of oxygen; this gain in weight amounted to 4 mg per 109 mg at 960°. Consequently, it may be stated that silver is like gold in that, depending on the state of division, these metals are capable of taking oxygen from the air even below their melting points.

(iv) Precipitation by cadmium The metallic silver which has been deposited through the displacement action of a small cadmium rod, as directed by Classen[6], apparently is not pure. It retains some product which begins to oxidize at ordinary temperatures. The weight remains fairly constant thereafter and agrees with the theoretical from 167° to 530°. A slow and regular rise was subsequently observed, as in the preceding instance. The total gain in weight reached 5 mg per 263 mg.

(v) Precipitation by aluminium The use of thin aluminium foil as prescribed by Tarugi[7] yields a deposit of silver, which likewise is not pure. It increases in weight up to 232°, and then at 423° the weight becomes constant and equal to the initial weight.

(vi) Precipitation by hydroxylamine When silver nitrate is treated with hydroxylamine and potassium hydroxide, as suggested by Lainer[8], the resulting finely divided silver is readily oxidized from 600° on, but less distinctly than in the case where hypophosphorous acid is the reductant. The uptake of oxygen reaches a maximum of 2 mg per 200 mg at 948°.

(vii) Precipitation by formaldehyde When a solution of silver nitrate is boiled with a great excess of formaldehyde and 15% ammonia, the precipitate is not metallic silver. Taran[9] suggested that the product be ignited at 450–500° to obtain a constant weight of metallic silver. The

curve shows a very short level, then a slow descent from 40° to 160°, and then a sudden decomposition at this temperature. The constant weight due to metallic silver begins effectively at 500° and we advise that the drying be conducted above this temperature.

The initial precipitate is grey; it is known as "reduced silver". It adheres to glass vessels. It is known that NH groups situated between two carbon atoms are specific for the detection of silver. Under the conditions of the present experiment, one molecule of ammonia is combined with two molecules of formaldehyde, giving dihydroxymethylamine $HOCH_2-NH-CH_2OH$. The latter, on replacing the central hydrogen atom with silver, yields a precipitate whose molecular weight is 184, namely $HOCH_2-NAg-CH_2OH$. We analyzed this material and then thermally decomposed it.

(viii) Precipitation by vitamin C   The reducing properties of vitamin C have been utilized by Stathis[10] to precipitate metallic silver. The curve led us to the same conclusions as those derived in the case of the precipitation with hypophosphorous acid. The straight line ascends slowly, beginning at 624°; the uptake of oxygen amounts to 3 mg per 101 mg of metal at 882°.

2. Silver oxide

Saito[11] made a study of dry silver oxide Ag_2O. When 1 g was heated at the rate of 120° per hour it began to decompose at 350°. The reaction is finished around 450°.

Moist silver oxide does not appear to contain any hydroxide; the curve is the same after the water has departed. Furthermore, the infra-red absorption spectrum gives no indication of the presence of OH groups in the precipitate.

3. Silver chloride

The weighing of silver as the chloride AgCl is taken up in Chapter Chlorine. A perfectly horizontal level extends from 70° to 600° and is suitable for the automatic determination of silver or chlorine. Consult this chapter also regarding an investigation of the blackening of silver chloride. Saito[11] reported that the volatilization of silver chloride is not appreciable until the temperature reaches 800°.

4. Silver bromide

The weighing of silver as the bromide AgBr is taken up in Chapter Bromine. This salt may be weighed after it has been dried between 70° and 946°. Matsuura[12] gives only the range 100–500°. The corresponding straight line is suitable for the automatic determination of bromine or silver. See also in the cited chapter a study by Duval[13] of the darkening of silver bromide.

5. Silver iodide

The Chapter Iodine discusses the weighing of silver as the iodide AgI. This salt may be dried without change of weight between 60° and 900°. The corresponding straight line may be employed for the automatic determination of iodine or silver.

6. Silver sulphide

(i) Precipitation by sodium sulphide   Silver sulphide Ag_2S was prepared by the action of sodium sulphide as directed by Taimni and Salaria[14]. According to Taimni and Tandon[15], after this salt has been air-dried it gives an almost horizontal level up to 570° (Stanton balance). The increase in weight (1%) may be due to impurities.

(ii) Precipitation by hydrogen sulphide   The precipitation with hydrogen sulphide was made as directed by Fresenius[16]. He specified 100° as the drying temperature but this is not restrictive. As a matter of fact, after the wash water is removed, the curve shows a horizontal extending from 69° to 615° and corresponding to this sulphide. Ishii[17] recommended carrying the salt to above 170° in dry air. A transformation begins at 615° and gives rise to a mixture of metallic silver and silver sulphate in no simple stoichiometric relation; the minimum weight is reached at 774°. The horizontal of the sulphide may be used for the automatic determination of silver.

(iii) Precipitation by sodium thiosulphate   The reaction of sodium thiosulphate with silver nitrate was conducted by the method of Faktor[18]. After passing through a number of colour changes, a black precipitate of silver sulphide is obtained which is almost dry at 65° and completely dry near 129°. There is no further change up to 649°. The curve closely resembles the preceding one but the weighing range lies between 129° and 649°.

(iv) Preparation by the dry method   Silver sulphide prepared by the dry method gave Saito[11] characteristic curves showing the occurrence of the roasting reaction:

$$Ag_2S + 3 O_2 \rightarrow SO_2 + Ag_2SO_4$$

which starts around 450° if 1 g is heated at the rate of 120° per hour in a steady stream of air flowing at the rate of 100 ml per minute. The resulting sulphate begins to decompose near 800° and this action is complete in the vicinity of 920°. When Saito prepared the sulphide in an aqueous medium, the curves had the same form; they show a transient reduction of the sulphate and then reoxidation between 550° and 580°.

7. Silver sulphate

When 300 mg of silver sulphate Ag_2SO_4 were heated at the rate of 120° per hour, the curve included a horizontal up to 800°, precisely as when the salt came from the oxidation of the sulphide.

8. Silver selenite

Silver selenite Ag_2SeO_3 was precipitated, as described by Narui[19], by means of selenious acid in a solution buffered with sodium acetate (pH 4.3–10.2). The curve starts with a level which is essentially horizontal up to 690°. However, careful scrutiny discloses a slight superficial oxidation into selenate between 48° and 475°. The gain in weight amounts to nearly 2 mg per 136 mg and seems to be too great to permit the use of this method for the automatic determination of silver. Narui suggested that the selenite be dried at 105°; we advise keeping the precipitate below 48° under reduced pressure in a desiccator. Silver selenite starts to decompose at 690°, but the decomposition is not yet completed when the temperature has reached 960°. Rocchiccioli[20] found that the salt decomposes near 600° and that at 1000° the crucible contains a little selenium and a shiny deposit of silver.

9. Silver nitrate

Duval[21] has studied silver nitrate as a standard, and Peltier and Duval[22] investigated it from the standpoint of determining silver from the sum of Ag + Cu as described in Chapter 4. Silver nitrate maintains a constant weight up to 342°, then undergoes a slight loss of weight up to 473°, a sudden decomposition up to 608°, and a much slower loss from 608° to 811°. From this temperature on the weight remains constant again (pure silver).

10. Silver arsenate

Silver arsenate Ag_3AsO_4 is discussed in Chapter Arsenic. It is not suitable for determining silver and yields a slowly ascending straight line. However, Yosida[23] reports that this arsenate gives a horizontal from 550° to 780°.

11. Silver thallium arsenate

On the other hand, as we have pointed out, the curve or rather the straight line due to silver thallium arsenate Ag_2TlAsO_4 extending from 20° to 846° lends itself well to the automatic determination of arsenic, silver, and thallium.

12. Silver metaantimonate

Silver metaantimonate $AgSbO_3$ has been studied by Dupuis[24] who prepared it by treating a concentrated solution of silver nitrate with "potassium pyroantimonate". The thermolysis curve shows that the moist product starts to lose water at ordinary temperatures. At the end of the level, which extends from 100° to 125°, there remain only 2 molecules of water per molecule of the metaantimonate, namely $AgSbO_3.2H_2O$. This water departs slowly and steadily up to 560° where the horizontal corresponding to the anhydrous metaantimonate $AgSbO_3$ commences. A bend at 176° in the descent corresponds to the loss of the second molecule of water. The metaantimonate is stable up to 732°; then the curve rises indicating a partial oxidation of the silver oxide.

13. Silver cyanide

Silver cyanide was precipitated by the method of Fresenius[16], who recommended 100° as the drying temperature. It gives a very simple thermolysis curve with a horizontal extending from the ordinary temperature to 237°, where cyanogen begins to come off. After a jog in the curve at 425°, metallic silver is obtained beginning at 516°, the paracyanogen being decomposed in its turn. The horizontal due to the cyanide can be suggested for the automatic determination of silver.

14. Silver thiocyanate

Silver thiocyanate AgSCN was precipitated as recommended by Van Name[25] who dried the product at 115°. It yielded a curve with a horizontal extending from room temperature up to 224°. This horizontal lends itself to the automatic determination of silver and equally well of thiocyanate. Beyond 224°, 2 molecules share the sudden loss of 1 atom of sulphur; the residue consists of the double salt, silver thiocyanate cyanide AgSCN–AgCN. We believe this is a new salt; it yields an equally horizontal level extending from 225° to 370°. Beyond 370°, cyanogen is evolved and silver sulphide Ag_2S is left. The end of the curve of course shows the formation of silver sulphate and then metallic silver.

15. Silver chromate

Weighing silver as the chromate has been taken up in Chapter Chromium. It should be pointed out again that this salt maintains its weight constant from 92° to 812°; Ishii[17] gives only above 650°. We have reported that the green chromite is produced beyond 812°.

16. Silver permanganate

Satava and Körbl[26] have investigated the pyrolysis of silver permanganate. This salt begins to lose oxygen at 160°; the decomposition is complete at 480–500°. The residue is silver manganite $AgMnO_2$.

17. Silver molybdate

According to Kato, Hosimiya and Nakazima[27], silver molybdate Ag_2MoO_4 may be weighed after drying between 250° and 500°.

18. Cupric propylenediamine argentoiodide

The method of Spacu and Spacu[28] was used for the preparation of the cupric propylenediamine argentoiodide $[Cupn_2][AgI_2]_2$. The product is stable up to 155° and it is not necessary to dry the material in a vacuum desiccator as these workers suggested. The horizontal is suitable of course for the automatic determination of silver. Beyond 155°, decomposition sets in and the complex breaks down by a complicated mechanism. A mixture of silver and copper oxide is obtained at 960°.

19. Cobaltic diethylenediaminethiocyanate argentothiocyanate

The complex cobaltic diethylenediamine dithiocyanate argentothiocyanate $[Coen_2(SCN)_2][Ag(SCN)_2]$ was prepared as described by Spacu[29]. The reagent was prepared in our laboratory by the procedure of Groszmann and Schück[30]. The curve obtained with this complex has a horizontal extending as far as 144°; it may be used for the automatic determination of silver but not of cobalt. The subsequent pyrolysis follows a course which has proved to be complicated, and which eventually yields metallic silver and cobalt oxide CoO.

20. Silver (thallium) cobalticyanides

The mixed silver thallium cobalticyanides $Ag_2Tl[Co(CN)_6]$, $AgTl_2[Co(CN)_6]$, and the triargentic salt $Ag_3[Co(CN)_6]$ were heated to 600° by Dragulescu and Tribunescu[31].

21. Silver oxalate

The precipitate of silver oxalate $Ag_2C_2O_4$ obtained by the method of von Reis[32] is stable up to 101° and gives a horizontal that is satisfactory for the automatic determination of silver. This oxalate settles and filters well. It explodes suddenly at 140°; the level due to metallic silver starts at 156°.

22. Silver rhodanine

The internal complex $AgC_3H_2ONS_2$, produced as directed by Feigl and Pollak[33] with rhodanine, filters rather slowly. They recommend drying the product at 70–75° in dry air. Actually, the curve given by this material remains almost horizontal up to 158°. Then the compound decomposes quietly and leaves metallic silver at 600°. The latter oxidizes slightly and the gain in weight amounts to 2 mg per 100 mg.

23. Silver thionalide

The complex with thionalide has been recommended by Kiba[34] for the separation of silver, followed by its determination as the chloride. The thermolysis curve shows that this complex is more or less stable up to 105° and not up to 190° as reported by Umemura[35]. It then loses weight very slowly and yields a mixture of silver and silver sulphide around 720°. This sulphide then reoxidizes to silver sulphate up to 960°.

24. Silver dithizonate

Pariaud and Archinard[36] studied silver dithizonate by heating it at the rate of 100° or 150° per hour. The weight remains constant as far as 120°; then decomposition sets in and becomes violent, almost explosive, at 186°. This break-down continues slowly and steadily up to 465°, then accelerates appreciably up to 570°, where the residue consists only of a mixture of silver sulphate and a little sulphide gradually changing into sulphate, as revealed by a gentle rise of the curve. Finally, the sulphate decomposes at 800° and metallic silver is obtained from 950° on.

25. Silver oxinate

The yellow and the green form of silver oxinate have been investigated by Wendlandt and van Tassel[37]. These varieties were prepared by the method of Block, Bailar and Pearce[38] and heated at the rate of 5.4° per minute. The yellow form is the more stable and starts to lose weight slowly at 140°. The descent then becomes rapid with a break at 280°. However, it is not possible to obtain a level for the compound containing one atom of silver per molecule of oxine. A new decomposition begins above 370° and the level due to the metallic silver starts at 505°.

The green form commences to lose weight at 120° and there is a break in the curve at 225°. No level corresponding to a definite compound could be found. The loss becomes more rapid beyond 415° and metallic silver is obtained beginning at 600°.

REFERENCES

1 Y. MARIN AND C. DUVAL, *Anal. Chim. Acta*, 4 (1950) 393.
2 L. SCHUCHT, *Z. anal. Chem.*, 32 (1883) 485.
3 A. FRIEDRICH AND S. RAPOPORT, *Mikrochemie*, 18 (1935) 227.
4 E. MILLON AND C. COMMAILLE, *Compt. rend.*, 56 (1863) 309.
5 F. MAWROW AND G. MOLLOW, *Z. anorg. Chem.*, 61 (1909) 96.

SILVER491

6 A. CLASSEN, *Z. anal. Chem.*, 5 (1866) 402.
7 N. TARUGI, *Gazz. chim. ital.*, 33 (1903) 223.
8 A. LAINER, *Monatsh.* 9 (1888) 533.
9 E. N. TARAN, *J. Appl. Chem. (S.S.S.R.)*, 9 (1936) 520; *C.A.*, (1936) 7487.
10 C. STATHIS, *Anal. Chem.*, 20 (1948) 271.
11 H. SAITO, *Sci. Rep. Tôhoku Imp. Univ.*, 16 (1927) 131.
12 K. MATSUURA, *J. Chem. Soc. Japan*, 52 (1931) 730.
13 C. DUVAL, *Mikrochim. Acta*, (1956) 1430.
14 I. K. TAIMNI AND G. B. S. SALARIA, *Anal. Chim. Acta*, 12 (1955) 519.
15 I. K. TAIMNI AND S. N. TANDON, *Anal. Chim. Acta*, 22 (1960) 555.
16 R. FRESENIUS, *Quantitative Analyse*, 6th Ed., 1875, p. 301.
17 K. ISHII, *J. Chem. Soc. Japan*, 52 (1931) 167.
18 F. FAKTOR, *Pharm. Post*, 33 (1900) 169.
19 Y. NARUI, *J. Chem. Soc. Japan*, 63 (1942) 746.
20 C. ROCCHICCIOLI, *Thesis*, Paris, 1960.
21 C. DUVAL, *Anal. Chim. Acta*, 16 (1957) 222.
22 S. PELTIER AND C. DUVAL, *Compt. rend.*, 226 (1948) 1727.
23 Y. YOSIDA, *J. Chem. Soc. Japan*, 60 (1940) 915.
24 T. DUPUIS, *Thesis*, Paris, 1954.
25 G. VAN NAME, *Am. J. Sci.*, 10 (1901) 451.
26 V. SATAVA AND J. KÖRBL, *Chem. Listy*, 51 (1957) 27.
27 H. KATO, H. HOSIMIYA AND S. NAKAZIMA, *J. Chem. Soc. Japan*, 60 (1939) 1115.
28 G. SPACU AND P. SPACU, *Z. anal. Chem.*, 89 (1932) 190; 90 (1932) 182.
29 P. SPACU, *Bul. soc. stiinte Cluj.*, 7 (1934) 568.
30 H. GROSZMANN AND B. SCHÜCK, *Ber.*, 39 (1906) 1897.
31 C. DRAGULESCU AND P. TRIBUNESCU, *Stud. cerc. sti. chim. Jassy*, 6 (1959) 59.
32 M. A. VON REIS, *Ber.*, 14 (1881) 1172.
33 F. FEIGL AND J. POLLAK, *Mikrochemie*, 4 (1926) 185.
34 T. KIBA, *J. Chem. Soc. Japan*, 59 (1938) 577.
35 T. UMEMURA, *J. Chem. Soc. Japan*, 61 (1940) 25.
36 C. PARIAUD AND P. ARCHINARD, *Rec. trav. chim.*, 71 (1952) 634.
37 W. W. WENDLANDT AND J. H. VAN TASSEL, *Science*, 127 (1958) 242.
38 B. P. BLOCK, J. C. BAILAR AND D. W. PEARCE, *J. Am. Chem. Soc.*, 73 (1951) 2971.

CHAPTER 46

Cadmium

Duval[1] made a critical study of the methods of determining cadmium. This investigation has been supplemented by various recordings. He omitted the precipitation by means of the allyl iodide–hexamethylenetetramine mixture of Evrard[2]. Still other workers[3,4] had previously criticized this method for its non-quantitative character and because the precipitate occludes excessive amounts of impurities. However, Kiba and Sato[5] point out that the complex $CdI_2[(CH_2)_6N_4 \cdot C_3H_5I]_2$ may be maintained from 75° to 105°.

The precipitation of cadmium as the cobalticyanide, as suggested by Evans and Higgs[6], is predominantly a separation procedure, and they finish the determination by weighing cadmium sulphate. The same applies to the method of Pass and Ward[7] with β-naphthoquinoline, which employs an iodine titration as the final step. But here again Kiba and Sato[5] have reported that the level from 110° to 130° corresponds to the complex $H_2(C_{13}H_9N)_2[CdI_4]$.

1. Metallic cadmium

The cadmium deposited electrolytically as directed by Okač[8] should not be dried by heating; it begins to oxidize at 25° in the open air.

2. Cadmium hydroxide

The white precipitate of cadmium hydroxide $Cd(OH)_2$, prepared from cadmium chloride and potassium hydroxide, yields a perfectly horizontal level extending from 89° to 170°. The latter is suitable for the automatic determination of cadmium. Dehydration is complete at 371° and the oxide CdO is stable up to 880°. Its decomposition is considerable even at 950°.

A critical study by Cabannes-Ott[9,10] showed that a better method of obtaining cadmium hydroxide $Cd(OH)_2$ is the one suggested by de Schulten

CADMIUM 493

in which cadmium iodide is heated with potassium hydroxide in a nickel crucible. The product is assuredly $Cd(OH)_2$ and not $CdO.H_2O$.

3. Cadmium chloride

Cadmium chloride $CdCl_2.5H_2O$ is not stable at this stage of hydration. It loses water steadily from 35° up to 210°. The level of the anhydrous chloride goes as far as 600°. Decomposition accompanied by oxidation then ensues with formation of the oxide CdO, which in turn sublimes and the crucible is empty around 1012°. See Duval[11].

4. Cadmium sulphide

The yellow cadmium sulphide CdS may be precipitated by hydrogen sulphide, sodium sulphide, or a thiosulphate. Although Winkler[12] recommended drying at 130°, and though Sarudi[13] gave 110°, it may be pointed out that these temperatures depend on the precipitating agent. A slow and very slight loss (water and sulphur) is observed up to 400° (heating rate 300° per hour). Then the weight stays constant (cadmium sulphide) up to 650°. Part of the sulphide then undergoes oxidation to sulphate and another part sublimes. The maximum in the curve extends from 777° to 800°, and is followed by decomposition. At 1020° the residue consists of a mixture of cadmium oxide, sulphide, and sulphate.

Taimni and Salaria[14] were the first to report that the sulphide prepared by means of sodium sulphide loses weight up to 140°. Later, Taimni and Tandon[15] found that this weight remained constant up to 300°. They recognized that the temperatures from 105° to 115° recommended previously are too low and that the analytical factor is not correct.

5. Cadmium sulphate

Cadmium sulphate $3CdSO_4.8H_2O$ starts to lose water near 40°; at 74° it has been converted into the monohydrate $CdSO_4.H_2O$, which remains intact up to about 156°. The good horizontal of the monohydrate is suitable for the automatic determination of cadmium. The dehydration is complete from 320° on and the anhydrous sulphate remains unchanged up to 906°. No formation of a basic sulphate was observed near 700° or at 827° as had been reported. Cadmium sulphate differs from copper sulphate in this respect.

We have also used as a standard substance[16] the commercial grade of cadmium sulphate, which is usually obtained by crystallization between 80° and 100°; it corresponds to the monohydrate. As a matter of fact, it loses no water below 160°. If the sample is heated continuously at the rate of 300° per hour, the dehydration is complete at 261°. The horizontal of the neutral anhydrous sulphate begins beyond 261°. Noshida[17] gave 210–820° as the existence zone of the anhydrous sulphate.

6. Cadmium sulphamate

Capestan[18] found that the dihydrate of cadmium amidosulphonate $Cd(NH_2-SO_3)_2 \cdot 2H_2O$ loses its water of hydration from 80° to 125°. The resulting anhydrous salt passes directly into the sulphate from 335° to 440°.

7. Cadmium nitrate

We used cadmium nitrate tetrahydrate $Cd(NO_3)_2 \cdot 4H_2O$ in our study. Water began to be lost at 40° when 392 mg were heated at the rate of 150° per hour. An inflection near 91° almost accords with the dihydrate. The anhydrous nitrate is manifested by an essentially horizontal level between 200° and 294°. The curve scarcely descends at all as far as 344°, but then nitrogen oxides are suddenly evolved and the oxide CdO remains from about 505–510° on. Wendlandt[19] repeated this thermolysis and has given less precise temperatures. We wish to point out that the melting point of 360° reported for this nitrate strikes us as incorrect since the decomposition is already quite marked at this temperature.

8. Cadmium dihydrazine iodide

Cadmium dihydrazine iodide $[Cd(N_2H_4)_2]I_2$ is a white compound, which precipitates and filters well. It was prepared as directed by Jilek and Kohut[20]. The product is stable up to 166°, and above this temperature it loses all of its hydrazine up to 273°. The resulting cadmium iodide is not stable; it decomposes at once and cadmium oxide, still contaminated with particles of iodine, remains at 675°. This excellent method, which is fully quantitative and easily carried out, is proposed for the automatic determination of cadmium. It is sufficient to record the curve in the vicinity of 110°, which is the temperature also selected by Jilek and Kohut.

9. Cadmium ammonium phosphate

The well-known method of Carnot[21] for preparing cadmium ammonium phosphate was used as modified by Winkler[12] and by Dick[22]. The curve shows that the monohydrate is perfectly stable up to 100–103°, which has been suggested for drying. The monohydrate remains intact even up to 122°; the corresponding level is suitable for the automatic determination of cadmium. The decomposition is not sharp. The water and the ammonia are eliminated concurrently, as in the case of the corresponding magnesium and manganese phosphates. The transformation into pyrophosphate can be regarded as complete at 581°. Subsequently, slight changes in weight are observed, either up or down, but they do not exceed 1/150. In all respects it is preferable to weigh this material after it has been dried at a low temperature. Noshida[17] reported that the conversion into the pyrophosphate occurs at 550°.

10. Cadmium carbonate

The carbonate was prepared by the procedure given by Wendehorst[23]. The material was dry at 75°. However, it is not stable and its weight decreases until the temperature is near 488°. There is no need to heat the residual oxide to 1000° and furthermore there is considerable danger of loss by sublimation. It is quite sufficient to use 500° for the ignition temperature. In addition the method has little to recommend it as regards the determination of cadmium. See[17,24] for additional information about cadmium carbonate.

11. Cadmium molybdate

Cadmium molybdate $CdMoO_4$ maintains its weight at a constant level up to 250°; it then begins to turn brown and lose weight. The residue between 893° and 946° corresponds almost to the sum $CdO + MoO_2$, which indicates that there has been a loss of 1 atom of oxygen per molecule of $CdMoO_4$. Cadmium can be conveniently determined automatically by using the level which extends from 82° to 250°.

12. Cadmium dipyridine chloride

One of the best methods of determining cadmium is provided by cadmium dipyridine chloride $CdCl_2 \cdot 2C_5H_5N$. This compound was prepared by the procedure given by Malatesta and Germain[25] and Kragen[26]. The complex is stable up to 70°. One molecule of pyridine is given off up to 107°. A new horizontal which corresponds to $CdCl_2 \cdot C_5H_5N$ extends from 107° to 177°. The second molecule of pyridine is given off up to 270°. The horizontal due to cadmium chloride extends from 270° to 610°. The method is outstandingly suited to the automatic determination of cadmium.

13. Cadmium dipyridine thiocyanate

Less interesting than the preceding is the analogous cadmium dipyridine thiocyanate $Cd(SCN)_2 \cdot (C_5H_5N)_2$, which was investigated by Spacu and Dick[27] and Vornweg[28]. It is not very stable at ordinary temperatures and decomposes noticeably from 30° on. One molecule of pyridine comes off between 27° and 77° and leaves the complex $Cd(SCN)_2 \cdot (C_5H_5N)$ whose horizontal extending from 77° to 101° may be used for the automatic determination of cadmium. The other molecule of pyridine comes off at 149° and the decomposition then progresses rapidly.

14. Tetramminecadmium ferrocyanide

The tetrammine of cadmium ferrocyanide was precipitated by the procedure given by Luff[29] using cadmium chloride as the starting material. The product is highly insoluble, easy to filter, and Luff advises that it be dried at 100–110°. The curve reveals that the complex is stable up to 127°; it slowly decreases in weight until the temperature is around 267°. Then the decomposition becomes faster; the cyanogen decomposes violently up to 430°. The residue seems to be a mixture of the oxides of the two metals.

15. Tetramminecadmium triiodomercurate

Tetramminecadmium triiodomercurate $[Cd(NH_3)_4](HgI_3)_2$ precipitates remarkably well and is easily dried with alcohol and ether. Taurins[30] recommends keeping this precipitate in a vacuum desiccator; actually it is only stable up to 69° where it suddenly begins to decompose. The crucible is empty at 959°; nothing explicit can be said about the disintegration of the complex. Two levels extending from 152° to 162° and from 248° to 475°, respectively, do not correspond to definite compounds.

16. Cupric diethylenediamine tetraiodocadmiate

The complex known as copper diethylenediamine tetraiodocadmiate $[Cuen_2][CdI_4]$ is prepared very well by the technique of Spacu and Suciu[31]. It is readily washed, filters nicely, and is stable. Its thermolysis curve shows no variation in weight up to 79°, and then there is a slight increase as far as 173° (1 mg per 112 mg). The complex begins to disintegrate at 240°; the residual metal produced at 771° slowly reoxidizes and the oxide CdO is obtained at 950°. It should be pointed out that this compound may be dried in a desiccator as suggested by Spacu and Suciu or it may be kept in an oven below 79°.

17. Brucine tetrabromocadmiate

The method of Meurice[32], modified by Nikitina[33], yields a precipitate of brucine tetrabromocadmiate $[(CH_3O)_2C_{21}H_{21}O_2N_2]_2[CdBr_4]$, which washes and filters extremely well. The level extending from 120° to 250° is perfectly suited to the automatic determination of cadmium. These temperatures conform to those indicated by Meurice for this excellent method. The factor for cadmium is 0.092.

18. Cadmium dithiourea reineckate

Cadmium dithiourea reineckate $[Cd(SCN_2H_4)_2][Cr(SCN)_4(NH_3)_2]$ when precipitated according to Mahr[34] and Mahr and Ohle[35] can be filtered off readily as soon as it is formed, and does not adhere to the walls of the container. It yields a horizontal extending as far as 167° and this level is suitable in all respects for the automatic determination of cadmium. It is more difficult to interpret the remainder of the course of the curve since it includes the destruction of the organic matter and the reoxidation of the two metals present. The residue in the crucible corresponds at 946° to a mixture of cadmium chromite and oxide. Kiba and Sato[5] have reported 110–140° for the limits of the level of the reineckate.

19. Cadmium formate

Cadmium formate has been heated by Zapletal, Jedlička and Rüžička[36] and also in the writer's laboratory. The salt is anhydrous and begins to lose weight at room temperature. A slightly oblique level extends from 100° to 200–210° and corresponds to the appearance of Cd[Cd(OH)

$(HCO_2)_3$] *i.e.* a complex produced by the condensation of two molecules. If the heating is continued, the oxide CdO is obtained around 300–305°.

20. Cadmium acetate

Cadmium acetate, a standard substance, was found by Duval and Wadier[37] to lose its two molecules of water between 63° and 190°. A level due to the anhydrous acetate then extends as far as 286°. Decomposition sets in abruptly and the level of the oxide CdO starts at 344°.

21. Cadmium oxalate

The oxalate $CdC_2O_4 \cdot 3H_2O$ yields the oxide CdO quantitatively when ignited. If prepared by the method described by Dick[38] but without the use of paper pulp, it yields this oxide only if the temperature exceeds 771°. The trihydrate, the anhydrous oxalate, and the carbonate are not shown on the curve. However, after the carbon monoxide and carbon dioxide have disappeared, a horizontal begins at 296° and goes as far as 675° and this level may be attributed to a mixture of the oxide and basic carbonate. In one of our trials, the mixture weighed 138.2 mg as opposed to the calculated 136.28 mg. This latter weight was not reached until the temperature exceeded 771°. Kawagaki[39] gives 380° as the temperature marking the end of the decomposition, and Hagenmuller[40], who employed slow heating, discovered a level due to the anhydrous oxalate extending from 100° to 245°. Using isothermal heating, he noted that the decomposition was complete in 50 hours at 300° and in 2 hours at 350°. The titer of the metal reached 0.895 at 350° under reduced pressure (0.1 mm Hg).

22. Cadmium benzidine

Precipitation of the white complex [Cd(NH_2–C_6H_4–C_6H_4–NH_2)] is not quite quantitative and filtration is difficult even on a No. 4 sintered-glass crucible. The product is stable between 77° and 270°. See Barcelo[41] and Heller and Machek[42].

23. Cadmium anthranilate

Cadmium anthranilate was prepared by the procedure of Wenger and Masset[43] and Funk[44]. The drying temperature 105–110° given by them is correct. The thermolysis curve starts with a very slow rise up to 222° where decomposition begins. The following table, which was constructed from the results obtained with 118.25 mg of the dry anthranilate, shows that the increase is slight:

Temp. (°C)	51	87	150	222
Increase (mg)	0.6	0.9	1.8	2.0

Since the leaflets of cadmium anthranilate dry readily, particularly after they have been washed with ethanol, it is suggested that they be kept in a desiccator below 40° before being weighed.

Ishimaru[45] has given 105–152° for the limits of the level of the anthranilate. When ignited with oxalic acid, the salt yields the oxide beyond 453°.

24. Cadmium oxinate

Cadmium oxinate $Cd(C_9H_6ON)_2 \cdot 2H_2O$ is one of the most stable, if not the most stable of the oxinates; it does not begin to decompose and sublime until the temperature reaches 384°. The crucible is empty above 950°. A strange behaviour, which had already been noted by Berg[46], and by Wenger, Cimerman and Wyszewianska[47] who suggested the use of the oxinate in microanalysis, occurs from 116° to 384°. Near 130° there is a sudden change in the rate at which weight is being lost and this might give the impression that the complex is completely dry. However, the curve constructed with 65 mg of dry oxinate shows that the latter does not reach constant weight until the temperature is 280°. At 130° 2 mg of water and tenaciously held oxine still remain to be driven off. Accordingly the drying should be accomplished between 280° and 384° to obtain the correct weight. See also Sekido[48].

Borrel and Pâris[49] showed that the dihydrate is stable up to 116°, and that the anhydrous compound yields a level from 195° to 345°, and that the cadmium oxide is formed above 900°. However, the hydrate containing $^3/_2$ H_2O reported by Berg was not observed on our curves.

25. Cadmium 8-quinaldate

The product obtained by the action of 8-carboxyquinoline was prepared by the method of Majumdar[50]. Its formula is $Cd(C_{10}H_6O_2N)_2$, and the thermolysis curve of this white, voluminous precipitate reveals up to 89° a rapid elimination of the water retained during its preparation. A good horizontal agreeing with the formula just given extends from 89° to 263°, and we suggest this level for the automatic determination of cadmium. The decomposition of the organic matter becomes rather vigorous around 400° where nothing remains but a little carbon and the cadmium oxide. At 950° the residue consists of metallic cadmium mixed with a little of the oxide.

26. Cadmium 2-quinaldate

Cadmium quinaldate (quinoline-2-carboxylate) provides an excellent method of determining cadmium. This isomer of the complex just discussed, was prepared by Rây and Bose[51] and by Majumdar[52], who verified the formula $Cd(C_{10}H_6O_2N)_2$. The precipitate may be filtered off after 5 minutes; it is well suited for the automatic determination of cadmium. The pyrolysis curve shows in fact a strictly horizontal section going from 66° to 197° which accords with this formula. The decomposition of the organic matter then proceeds with singular points at 410°, 581°, 904°. At 950° the residue consists of cadmium oxide. Yoshikawa and Shinra[53] have also studied this chelate and they give only 110–130° as the limits of the level of the anhydrous compound.

Lumme[54] repeated our investigation as did Thomas[55]. The latter used a white monohydrate $Cd(C_{10}H_6O_2N)_2 \cdot H_2O$ which proved to be stable up to 130°, and the anhydrous salt began to decompose only at 340°.

27. Cadmium picolinate

The pyridine-2-carboxylate (picolinate) $Cd(C_6H_4O_2N)_2$, which is anhydrous, was heated by Thomas[55]. See also Thomas and Pâris[56].

28. Cadmium mercaptobenzothiazole

The method given by Spacu and Kuraš[57] was used to prepare the complex with mercaptobenzothiazole $Cd(NH_3)_2(C_7H_4NS_2)_2$. Its curve, like that of the anthranilate, rises slightly at first and reaches a maximum at 151° which represents a gain of 1.5 mg relative to a theoretical weight of 122.45 mg of the complex. At 946°, the residue consists of a mixture of cadmium oxide and sulphide. Here again, more accurate data are obtained if the material is dried below 45° in a desiccator.

29. Cadmium thionalidate

The thionalidate was prepared by Umemura[58] and proved to be stable up to 210°.

30. Cadmium mercaptobenzimidazole

The complex with mercaptobenzimidazole $CdOH \cdot NH_3 \cdot C_7H_5N_2S$, described by Kuraš[59], yields a thermolysis curve which steadily slants downward. The theoretical weight for the complex occurs on the curve somewhere between 80° and 125°, but the linear part of the graph between these temperatures is not perfectly horizontal. In this case also, it is necessary to dry the material in the desiccator. The residue at 946° is a mixture of cadmium sulphide and oxide.

31. Cadmium 2-phenyloxinate

The dark yellow 2-phenyloxinate $Cd(C_6H_5-C_9H_5ON)_2 \cdot H_2O$ was prepared by the method of Bocquet and Pâris[60]. A specimen weighing 200 mg was heated at the rate of 180° per hour. It lost its water from 45° on and the anhydrous salt was stable up to 335°. The organic matter then burned away almost linearly, and the level of the oxide CdO began at 800°.

32. Cadmium benzothiazolecarboxylate

The benzothiazolecarboxylate, prepared by Yoshikawa and Shinra[53], yields a horizontal going from 75° to 135°.

33. Cadmium 2-hydroxy-o-phenylbenzoxazole

Wendlandt[61] studied the chelate obtained with 2-hydroxy-o-phenylbenzoxazole, which was introduced by Walter and Freiser[62]. It is stable below 285°. The oxide CdO appears beyond 505°.

These findings show that cadmium is one of the elements which have benefitted most from the introduction of organic reagents into inorganic chemistry.

REFERENCES

[1] C. DUVAL, *Anal. Chim. Acta*, 4 (1950) 190.
[2] V. EVRARD, *Ann. chim. anal.*, 11 (1929) 322; *Natuurw. Tijdschr. (Ghent)*, 11 (1929) 191.
[3] L. C. HURD AND R. W. EVANS, *Ind. Eng. Chem., Anal. Ed.*, 5 (1933) 16.
[4] A. D. MITCHELL AND A. M. WARD, *Modern Methods in Quantitative Analysis*, 1932, p. 20.

[5] T. Kiba and S. Sato, *J. Chem. Soc. Japan*, 61 (1940) 133.
[6] B. S. Evans and D. G. Higgs, *Analyst*, 70 (1945) 158.
[7] A. Pass and A. M. Ward, *Analyst*, 58 (1933) 667.
[8] A. Okač, *Z. anal. Chem.*, 89 (1932) 109.
[9] C. Cabannes-Ott, *Thesis*, Paris, 1958.
[10] C. Cabannes-Ott, *Compt. rend.*, 240 (1955) 68.
[11] C. Duval, *Anal. Chim. Acta*, 20 (1959) 264.
[12] L. W. Winkler, *Z. angew. Chem.*, 34 (1921) 383.
[13] I. Sarudi, *Z. anal. Chem.*, 121 (1941) 348.
[14] I. K. Taimni and G. B. S. Salaria, *Anal. Chim. Acta*, 11 (1954) 54.
[15] I. K. Taimni and S. N. Tandon, *Anal.Chim. Acta*, 22 (1960) 555.
[16] C. Duval, *Anal. Chim. Acta*, 13 (1955) 429.
[17] I. Noshida, *J. Chem. Soc. Japan*, 48 (1927) 520.
[18] M. Capestan, *Ann. Chim. (Paris)*, 5 (1960) 213.
[19] W. W. Wendlandt, *Texas J. Sci.*, 10 (1958) 392.
[20] A. Jilek and B. Kohut, *Chem. Listy*, 33 (1939) 252.
[21] A. Carnot and P. M. Proromont, *Compt. rend.*, 101 (1885) 59.
[22] J. Dick, *Z. anorg. Chem.*, 82 (1930) 401.
[23] E. Wendehorst, *Z. angew. Chem.*, 41 (1928) 567.
[24] *1st Colloq. Thermography*, Kazan, 1953, p. 95.
[25] G. Malatesta and A. Germain, *Boll. chim. farm.*, 53 (1914) 225.
[26] S. Kragen, *Monatsh.*, 37 (1916) 391.
[27] G. Spacu and J. Dick, *Z. anal. Chem.*, 73 (1928) 279.
[28] G. Vornweg, *Z. anal. Chem.*, 120 (1940) 243.
[29] G. Luff, *Chemiker-Ztg.* 49 (1925) 513.
[30] A. Tauriņš, *Z. anal. Chem.*, 97 (1934) 27.
[31] G. Spacu and G. Suciu, *Z. anal. Chem.*, 77 (1929) 340.
[32] R. Meurice, *Ann. chim. anal.*, 8 (1926) 130.
[33] E. J. Nikitina, *Zavodskaya Lab.*, 7 (1938) 409.
[34] C. Mahr, *Angew. Chem.*, 53 (1940) 257.
[35] C. Mahr and H. Ohle, *Z. anal. Chem.*, 109 (1937) 1.
[36] V. Zapletal, J. Jedlička and V. Růžička, *Chem. Listy*, 50 (1956) 1406.
[37] C. Duval and C. Wadier, *Anal. Chim. Acta*, 23 (1960) 541.
[38] J. Dick, *Z. anal. Chem.*, 78 (1929) 414.
[39] K. Kawagaki, *J. Chem. Soc. Japan*, 72 (1951) 1079.
[40] P. Hagenmuller, *Compt. rend.*, 234 (1952) 1168.
[41] J. Barcelo, *Anales. soc. españ. fís, y quím.*, 32 (1934) 91.
[42] K. Heller and F. Machek, *Mikrochemie*, 19 (1936) 147.
[43] P. Wenger and E. Masset, *Helv. Chim. Acta*, 23 (1940) 34.
[44] H. Funk, *Z. anal. Chem.*, 123 (1942) 241.
[45] S. Ishimaru, *J. Chem. Soc. Japan*, 55 (1934) 288.
[46] R. Berg, *Z. anal. Chem.*, 71 (1927) 321.
[47] P. Wenger, C. Cimerman and M. Wyszewianska, *Mikrochemie*, 18 (1935) 182.
[48] E. Sekido, *Nippon Kagaku Zasshi*, 80 (1959) 80; *C.A.*, (1960) 13950.
[49] M. Borrel and R. Pâris, *Anal. Chim. Acta*, 4 (1950) 267.
[50] A. K. Majumdar, *J. Indian Chem. Soc.*, 22 (1945) 309.
[51] P. Rây and M. K. Bose, *Z. anal. Chem.*, 95 (1933) 400.
[52] A. K. Majumdar, *Analyst*, 64 (1939) 874.
[53] K. Yoshikawa and K. Shinra, *Nippon Kagaku Zasshi*, 77 (1956) 1418; *C.A.*, (1958) 2642.
[54] P. O. Lumme, *Suomen Kemistilehti*, B, 32, (1959) 237, 241.
[55] G. Thomas, *Thesis*, Lyon, 1960.
[56] G. Thomas and R. A. Pâris, *Bull. Soc. chim. France*, (1953) 472; *15th Intern. Congr. Pure and Applied Chem.*, Lisboa, 1958, Actas do Congresso, II, Vol. VI, p. 25.
[57] G. Spacu and M. Kuraš, *Z. anal. Chem.*, 102 (1935) 108.

[58] T. Umemura, *J. Chem. Soc. Japan*, 61 (1940) 133.
[59] M. Kuraš, *Chem. Obzor.*, 14 (1939) 51.
[60] G. Bocquet and R. A. Pâris, *Anal. Chim. Acta*, 18 (1958) 639.
[61] W. W. Wendlandt, *Anal. Chim. Acta*, 15 (1956) 533.
[62] J. L. Walter and H. Freiser, *Anal. Chem.*, 24 (1952) 984.

CHAPTER 47

Indium

Indium is determined gravimetrically as the hydroxide, sulphide, phosphate, luteocobaltic chloroindate(III), oxinate, and dithiocarbamate. See Dupuis and Duval[1]. Some ammoniated complexes have also been investigated by thermogravimetry.

1. Indium hydroxide

(i) Precipitation by aqueous ammonia The hydroxide may be prepared by the method of Moser and Siegmann[2] by treating a boiling solution of indium nitrate with dilute ammonium hydroxide. After filtering in the cold, and washing with water, the resulting flocculent product is found to release moisture up to 119°. This temperature corresponds to the appearance of the hydroxide In(OH)$_3$ which is not stable; there is no corresponding horizontal. Five-sixths of the water of constitution has been eliminated when the temperature has reached the neighbourhood of 235°, and then a new break is observed, which is much sharper than in the following case. The constant weight corresponding to the oxide starts at 345°. Takeno[3] reports the range 700–830° for this oxide, and in order to avoid the reacquisition of water by the ignited product, outside the thermobalance, the authors of textbooks of analytical chemistry have directed that the heating be carried to 850°, 1000°, and 1200°, respectively.

(ii) Precipitation by hexamine Use of hexamethylenetetramine instead of ammonium hydroxide was suggested by Moser and Siegmann[2]. The precipitate then loses the imbibed water up to 86°. The hydroxide decomposes at once, but a distinct horizontal is observed between 287° and 380° after five-sixths of the water of constitution has been eliminated. The compound corresponding to the formula In$_2$O$_3$·H$_2$O has not been reported as yet. It is observed in a more or less distinct but nevertheless constant fashion on all of our recordings. It loses its water at 546° where the horizontal of the oxide In$_2$O$_3$ begins.

(iii) Precipitation by potassium cyanate   Moser and Siegmann have also studied a method of precipitation based on the determination of aluminium by means of potassium cyanate. The hydroxide appears on the curve at 87° and the monohydrate at 250°. However, the latter does not give a well marked horizontal. The formation of the oxide is complete from 475° on. The cyanate method is excellent.

2. Indium sulphide

Indium sulphide In_2S_3 is precipitated from a solution containing acetic acid as directed by Thiel and Luckmann[4] by means of hydrogen sulphide dissolved in acetone. This sulphide may be weighed from 94° to 221°; the corresponding level is perfectly horizontal. Taimni and Tandon[5] report 90°–340° for the sulphide obtained by means of sodium sulphide. If the heating of the In_2S_3 is continued in the air, a double transformation ensues:

(1) A quite irregular interrupted loss of weight is observed from 221° to 320°; sulphur dioxide is evolved in spurts. The resulting product is the sulphide of indium(II) InS, which is stable between 320° and 544°. It is formed by the reaction:

$$In_2S_3 + O_2 \rightarrow 2\,InS + SO_2$$

and hence a new possibility is provided for the automatic determination of indium.

(2) On continuing the heating, a second oxidation occurs; oxygen is taken up from the air and sulphur dioxide is lost. The trioxide is obtained beginning at 690° through the reaction:

$$4\,InS + 7\,O_2 \rightarrow 2\,In_2O_3 + 4\,SO_2$$

The three levels corresponding respectively to In_2S_3, InS, and In_2O_3 are perfectly parallel.

3. Indium phosphate

According to Ensslein[6], indium phosphate is precipitated by reaction of diammonium phosphate with a solution of indium nitrate containing acetic acid. The curve obtained from this white material indicates that it reaches almost constant weight at 477°. The level corresponding to this phosphate is never exactly horizontal.

4. Luteocobaltic chloroindate

Ensslein[6] found that indium(III) chloride reacts in the presence of hydrochloric acid with luteocobaltic chloride to give a yellow precipitate $[Co(NH_3)_6][InCl_6]$. The latter slowly loses weight, even at 105°; at 130° the loss already amounts to 2 mg per 267 mg. The decomposition becomes rapid from then on: chlorine, ammonia, and hydrochloric acid are expelled and only the oxide CoO remains in the crucible at 800° after the departure of the chloride. The two distinct changes in direction of the curve at 345° and 450° do not correspond, however, to simple results and it appears to be difficult to represent this pyrolysis by a single equation. In view of the correct result it furnishes, we advise that the precipitate be washed with ethanol and ether, and then dried at a temperature below 95°.

5. Indium oxinate

The oxinate was prepared as directed by Geilmann and Wrigge[7] who recommend drying the product between 110° and 150°. These temperatures are suitable; in fact, the excellent horizontal of the anhydrous oxinate $In(C_9H_6ON)_3$ extends from 100° to 285°. The organic matter decomposes then and the indium oxide does not appear until the temperature exceeds 1000°.

6. Indium diethyldithiocarbamate

The method of Ensslein[6] for preparing the diethyldithiocarbamate yields a white precipitate $In[CS_2N(C_2H_5)_2]_3$, which gives an almost horizontal stretch between 100° and 210°. The temperature of 105°, recommended by Ensslein, seems proper. Strictly speaking, if more precise results are demanded, the temperature should not exceed 157° during the drying. The remainder of the curve greatly resembles that of the sulphide; the oxide In_2O_3 is obtained above 600°, after passage into the sulphide InS, which is not very stable between 320° and 480°.

7. Indium chloride complexes

Tronev and Kochetkova[8] have heated the complexes of the type $M_2[InCl_5 \cdot H_2O]$ where M denotes NH_4, Rb, Cs. They lose water and then $InCl_3$, leaving 2 MCl. The decomposition temperatures rise from NH_4 to Cs. These same workers[9] worked with the complex $[Inen_3]Cl_3$ and found that one molecule of ethylenediamine is lost at 167° with production of $[Inen_2Cl_2]Cl$. A new decomposition occurs at 305–318° with formation of $InCl_3$ and ethylenediamine. The hexammine $[In(NH_3)_6]Cl_3$

loses ammonia at 150°, 180°, and 286°, yielding respectively the tri-, di- and then the monoammine.

REFERENCES

[1] T. DUPUIS AND C. DUVAL, *Anal. Chim. Acta*, 3 (1949) 330.
[2] L. MOSER AND F. SIEGMANN, *Monatsh.*, 55 (1930) 18.
[3] R. TAKENO, *J. Chem. Soc. Japan*, 54 (1933) 741.
[4] A. THIEL AND H. LUCKMANN, *Z. anorg. Chem.*, 172 (1928) 359.
[5] I. K. TAIMNI AND S. N. TANDON, *Anal. Chim. Acta*, 22 (1960) 556.
[6] F. ENSSLEIN, *Metall u. Erz.* 38 (1941) 305.
[7] W. GEILMANN AND F. WRIGGE, *Z. anorg. Chem.*, 209 (1932) 129.
[8] V. G. TRONEV AND A. P. KOCHETKOVA, *Khim. Redkikh. Elementoo Akad. Nauk S.S.S.R.*, (1957) 89.
[9] A. P. KOCHETKOVA AND V. G. TRONEV, *Zhur. Neorg. Khim.*, 2 (1957) 2043.

CHAPTER 48

Tin

Seven gravimetric methods are known for the determination of tin, namely with selenious acid, ammonia, pyridine, tannin, cupferron, ammonium sulphide, phenylarsonic acid. Dupuis and Duval[1] not only examined these precipitates with respect to their thermolysis but also included an investigation of the thermal behaviour of stannous chloride as a standard.

1. Tin in alloys

Taboury and Gray[2] worked out a procedure for determining tin in its alloys with lead and antimony. The method was based on the reduction of stannous chloride with selenious acid and then weighing the elementary selenium. The curve can of course serve inversely for the determination of selenium. We wish only to state here that after the retained wash water was eliminated, the selenium maintained a constant weight up to 272°; it sublimed abruptly from 352° on.

2. Tin oxide hydrate

(i) Precipitation by aqueous ammonia The method given in the Treadwell textbook was followed for the precipitation of the hydrous oxide of tin(IV) from a hydrochloric acid solution of tin(IV) chloride by ammonium hydroxide. Of course the curve showed no level corresponding to the hydroxide $Sn(OH)_4$. The oxide SnO_2 begins to appear in a pure condition at 834°. Murayama[3] reported beyond 700°. The calculation shows that the composition $Sn(OH)_4$ is attained on the curve at 58°, while the composition $SnO_2.H_2O$ is present at 98°. Spinedi and Gauzzi[4] have also investigated the formation of the oxide.

(ii) Precipitation by pyridine The precipitation with 20% pyridine solution in the presence of 5–10 g of ammonium chloride per 100 ml of

solution was conducted as described by Ostroumov[5]. The curve given by this product was identical with that given by the precipitate obtained with ammonium hydroxide. The dioxide SnO_2 is formed above 650°. It should be noted that the washed precipitate still retains pyridine, and the latter is suddenly released at 77°. The composition agrees with the formula $SnO_2.2H_2O$ at 80°.

(iii) Precipitation by tannin   The precipitation by means of tannin was carried out as prescribed by Schoeller and Holness[6], but no paper pulp was used since its combustion interferes with the thermolysis curve. The latter slants downward until the temperature reaches 875°, but no horizontal related to the oxide was observed.

3. Tin cupferronate

The precipitate obtained by adding a 6% water solution of cupferron to an acidic tin solution[7-12] gave a level for the oxide at 747°. Since no better procedure is available at present, we suggest the use of this classic method for the automatic determination of tin, despite the high temperature required to obtain the level due to the oxide.

4. Tin sulphide

Treadwell, and also Hecht and Donau[13] report that the precipitate produced by ammonium sulphide retains considerable sulphur which burns and escapes up to 433° while the sulphide is changing into the dioxide SnO_2, whose level is fairly horizontal. Murayama[3] gives 830° for the conversion but this temperature strikes us as being very high despite the difference in the rates of heating.

5. Tin phenylarsonate

In accord with Knapper, Craig and Chandlee[14], we obtained a white precipitate with phenylarsonic acid. The thermolysis curve of this product is quite irregular. The starting material gives no horizontal; it decomposes between 430° and 591° and leads to an oxide, probably impure, whose weight does not remain constant up to 954°, the upper limit of our experiment.

6. Tin chloride

The chloride $SnCl_2 \cdot 2H_2O$ taken as a standard gave a thermolysis curve which revealed to Duval[15] that the two molecules of water are retained in their entirety only up to 28°. This salt dehydrates progressively up to 214° and the anhydrous salt obtained in this way sublimes almost immediately. Above 507° it leaves an oxychloride, but not in quantitative fashion.

REFERENCES

[1] T. DUPUIS AND C. DUVAL, Anal. Chim. Acta, 4 (1959) 201.
[2] M. F. TABOURY AND E. GRAY, Compt. rend., 213 (1941) 481.
[3] K. MURAYAMA, J. Chem. soc. Japan, 51 (1930) 786.
[4] P. SPINEDI AND F. GAUZZI, Ann. Chim. (Rome), 47 (1957) 1297.
[5] E. A. OSTROUMOV, C.A., 31 (1937) 4226, 4616.
[6] W. R. SCHOELLER AND H. HOLNESS, Analyst, 71 (1946) 70, 217.
[7] A. KLING AND A. LASSIEUR, Compt. rend., 170 (1920) 1112; Chim. & Ind. (Paris), 4 (1920) 155.
[8] E. GRAY, Compt. rend., 212 (1941) 904.
[9] V. MACK AND F. HECHT, Mikrochim. Acta, 2 (1937) 230.
[10] N. H. FURMAN, Ind. Eng. Chem., Anal. Ed., 15 (1923) 1071.
[11] A. PINKUS AND J. CLAESSENS, Bull. Soc. chim. Belg., 31 (1927) 413.
[12] W. D. MOGERMAN, J. Research Natl. Bur. Standards, 33 (1944) 307.
[13] F. HECHT AND J. DONAU, Anorganische Mikrogewichtsanalyse, Springer, Vienna, 1940, p. 178.
[14] J. S. KNAPPER, K. A. CRAIG AND G. C. CHANDLEE, J. Am. Chem. Soc., 55 (1933) 3945.
[15] C. DUVAL, Anal. Chim. Acta, 16 (1957) 223.

CHAPTER 49

Antimony

Morandat and Duval[1] made a thermolytic study of the pertinent compounds of antimony. In the course of this investigation they found an interesting new fact about the sulphide Sb_2S_3 which may simplify its determination.

They omitted from their study a number of items such as the reduction of the trisulphide with hydrogen in a Rose crucible[2], and heating of the trisulphide in carbon dioxide as suggested by Treadwell[3], and the precipitation of antimony by displacement with cadmium, tin, or iron, which is not quantitative and besides the deposited antimony is difficult to handle. Nevertheless, Goto[4] recommends drying the elementary antimony between 100° and 250°. At all events, the deposition by electrolysis is not practical, it is hard to render this operation quantitative, and so it too was omitted. The use of the pentasulphide was likewise not included, though Salaria[5] repeated the investigation of this compound. Since sodium antimonate is scarcely admissible as a weighing form for sodium, it likewise cannot be given serious consideration for the determination of antimony. Another omission was the indirect determination of antimony as given by Rose[6] and by Levol[7] in which the final step is the weighing of metallic gold.

1. Antimony oxides

   (i) *Weighing as* Sb_2O_4 The Bunsen[8-10] method was used when the oxide Sb_2O_4 was the desired weighing form, *i.e.* the trisulphide was oxidized by nitric acid. The curve of the deliquescent product slants steadily downward until 900° is reached; this temperature marks the start of the horizontal of the oxide Sb_2O_4. However, the loss of water is slight from 700° to 900° and is of the order of 1 mg per 66 mg of the anhydrous material. Goto[4] reported 700–950° as the limits of the level of this oxide.

(ii) Weighing as Sb_2O_5   To obtain Sb_2O_5 as the weighing form, Duval and Morandat carefully adhered to the directions of Vortmann and Metzl[11,12]. The orange sulphide is brought down by hydrogen sulphide in the usual manner and is then oxidized by heating with a mixture of 3 parts of ferric oxide and 1 part of ferric nitrate in such manner that the final mixture contains the oxide Sb_2O_5 and a calculable quantity of ferric oxide.

When ferric nitrate is heated by itself, it yields ferric oxide from 413° on, but if antimony sulphide is present constant weight is attained only at a much higher temperature, namely at 827° as shown by the curve traced with the same heating rate. Though the above workers recommend the use of a blast lamp, this is not necessary for a satisfactory ignition. The method yields excellent results. It likewise shows that antimonic oxide is stable up to around 950° or even higher under these conditions, whereas when heated alone it decomposes from 430° on. It is very likely that the two oxides are not simply juxtaposed in the final mixture. Goto[4] gives 800° as the lower limit for the existence of the pentoxide. See also Wanmaker and Verheijke[13].

2. Antimony trisulphide

(i) Preparation by hydrogen sulphide   The trisulphide was prepared by means of a stream of hydrogen sulphide as described by Treadwell[3]. The precipitate contains sulphur and also pentasulphide. The hydrogen sulphide was passed through the boiling solution for 45 minutes. The ignition was conducted, however, in the open air instead of in a stream of hydrogen sulphide followed by carbon dioxide in the familiar Treadwell oven. Under these conditions, the thermolysis curve of the product reveals an interesting fact which we believe is being reported for the first time; after the elimination of the water and sulphur, the curve includes a perfectly horizontal stretch from 176° to 275°, which agrees with the content of Sb_2S_3, as checked by analysis, and which may be used for the automatic determination of antimony. The product is black and homogeneous and its detection obviously simplifies the method employed up till now since it is sufficient to work in air with the furnace set at 176° and keeping the compound there for about 10 minutes. Beyond this temperature, this sulphide oxidizes and is converted first into the oxide Sb_2O_3, which slowly takes up more oxygen to form Sb_2O_4, which, as was seen, gives a horizontal level starting at about 700°.

(ii) Preparation by sodium sulphide The trisulphide prepared by means of sodium sulphide as directed by Taimni and Agarwal[14] was reported by Taimni and Tandon[15] to be stable up to 310°. When heated it yields the oxide Sb_2O_3 between 530° and 580°, and between 580° and 610° undergoes the gain in weight leading to the oxide Sb_2O_4.

(iii) Precipitation by ammonium thiocyanate In accord with the procedure given by Rây[16], we treated the antimony solution being analyzed with a boiling solution of ammonium thiocyanate. The sulphide Sb_2S_3 precipitated under these conditions is much purer than that produced by hydrogen sulphide. However, the curve has the same shape, particularly beginning at 170°. Up to this temperature the precipitate shows a slight gain in weight because of oxidation. The horizontal satisfying the trisulphide extends from 170° to 292°; temperatures in excess of 292° are not necessary for drying the material even though Rây suggested their use. Decomposition then sets in, there is formation of the oxide Sb_2O_3, followed by conversion to Sb_2O_4. This method is just as precise as the hydrogen sulphide procedure; it is acceptable for the automatic determination of antimony. Obviously it is more convenient since it avoids the use of hydrogen sulphide.

3. Antimony pentasulphide

The pentasulphide was prepared as directed by Salaria[5]; we have some doubts about its purity. Taimni and Tandon[15] report that it undergoes a continuous loss of weight between 190° and 410°. The first horizontal from 410° to 570° corresponds nearly to Sb_2O_4. A new loss of weight begins at 570° and a level starting at 590° is related to the oxide Sb_2O_4.

4. Stibnite

Natural stibnite was heated by Saito[17] during his general study; the grains were 0.074 mm across. The oxidation reaction:
$$2Sb_2S_3 + 9O_2 \rightarrow 2Sb_2O_3 + 6SO_2$$
begins at 290° and continues up to 500° if the rate of heating is 2° per minute. The oxidation into Sb_2O_4 occurs above 500° and notably from 520° on. This oxide remains stable up to at least 800°.

5. Sodium antimonate

The formation and the course of the pyrolysis of the antimonate

References p. 516

Na[Sb(OH)$_6$] have been discussed in Chapter Sodium. See also Dupuis[18]. It should be noted that this antimonate is stable up to 126° and on ignition it yields the metaantimonate NaSbO$_3$, which is stable from 600° to 950°. Goto[4] gives from 450° to 700°.

6. Chromic triethylenediamine thioantimonate

The reagent [Cren$_3$]Cl$_3$ required for the preparation of chromic triethyl-enediamine thioantimonate was prepared in our laboratory by heating completely dehydrated chrome alum for three hours on the water bath with ethylenediamine; the resulting sulphate was then treated with barium chloride. The precipitate of interest to us, [Cren$_3$]SbS$_4$ was then obtained and washed as directed by Spacu and Pop[19]. See also Hecht and Donau[20]. The curve furnished by this precipitate indicates that the complex undergoes a slight decrease in weight from room temperature on, so that logically it should be weighed after being kept in a vacuum desiccator. However, calculation of the error proved to us that at 65° it had lost only 1/300th of its weight; above this temperature the loss becomes more important and at 575° the material consists of a mixture of oxides and green chromite which does not acquire constant weight until 800° is reached.

7. Tartar emetic

Tartar emetic (potassium antimonyl tartrate), regarded as a standard, was heated by Duval[21]. The anhydrous material was found to be stable up to 149° and the semi-hydrate, used in medicine, loses water at 100°. One molecule of water of constitution departs slowly between 149° and 250° and the Bougault inner anhydride is represented by a horizontal extending as far as 334°. An explosion then occurs; the organic matter burns, and, as is well known, the so-called Serullas alloy is left above 540°. Consequently, it is possible to prepare a standard solution by direct weighing of the commercial semi-hydrate, or the anhydrous material, or the Bougault anhydride.

8. Antimonyl pyrogallate

The procedure described by Feigl[22] was followed exactly with pyro-gallol and it yielded a pyrogallate whose formula was C$_6$H$_3$O$_3$(SbOH). He advises drying the product at 110°. The curve obtained with this precipitate reveals the loss of moisture up to 74° and then comes a horizontal agreeing with the data of Feigl and showing that the pyrogallate is stable up to 140°. The organic matter has finished burning at 813° and leaves the anhydride Sb_2O_3, which slowly reoxidizes to Sb_2O_4. The method may be favourably considered for the automatic determination of antimony.

9. Antimonyl gallate

Gallic acid yielded a precipitate with the approximate formula SbOH $[C_6H_2(OH)_3CO_2]$ which corresponds to 39.74% of antimony. It was prepared by the procedure published by Gomez and Romero[23], who report that this precipitate contains 39.67% antimony. The corresponding thermolysis curve shows that after the water has been driven off there is a horizontal extending from 114° to 163°. The organic matter burns up to 610°, and then the oxide Sb_2O_3 takes up oxygen as in the preceding cases and the oxide Sb_2O_4 is obtained essentially in the vicinity of 856°.

10. Antimonyl tannin

According to Tamm[24], treatment of the antimony solution to be determined with tannin yields a precipitate whose formula is $SbO_3(C_7HO_3)_2\cdot 3H_2O$ or $SbO(C_7HO_4)_2\cdot 3H_2O$. Our own analysis appears rather to indicate that the precipitate contains one molecule of gallic acid per two SbO radicals, and the curve has a horizontal between 90° and 224°. According to this level, the precipitate contains 55.20% of antimony. The destruction of the organic matter is complete at 687° and beyond this temperature there is an essentially horizontal level corresponding to the oxide Sb_2O_4. Several writers have previously reported that the Tamm method does not give exact results.

11. Antimony thionalide

The complex between thionalide and an antimony salt was found by Umemura[25] to be stable below 200°.

12. Antimony oxinate

Much that is highly contradictory has been published on the subject of the precipitation of antimony by oxine. In his classic book, Prodinger[26] reports that the precipitation is not complete. Pirtea[27] announces that it is quantitative and that the yellow precipitate contains 21.97% of antimony, which is in accord with the formula $Sb(C_9H_6ON)_3$. Pirtea also prescribes a pH zone from 6.0 to 7.5 for the precipitation, and this seems to be impossible without contaminating the precipitate with antimony oxychloride. Furthermore, Goto[4] has reported that the precipitation is complete above pH 1.5. We have tried to use the Pirtea procedure; the corresponding curve has a good horizontal up to 111° which does not agree at all–as might have been expected–with the formula given above, but instead our analysis shows 54.03% of antimony, and hence the precipitate contains an excess of this metal. Consequently, the determination of antimony by means of oxine need not be taken into account.

13. Cobaltic *trans*-dichloroethylenediamine antimony(V) hexachloride

Belcher and Gibbons[28] suggested that antimony be determined as the antimony(V) hexachloride of cobaltic *trans*-dichloroethylenediamine $[Coen_2Cl_2][SbCl_6]$. The thermolysis curve obtained with the yellow precipitate reveals that the product is anhydrous and may be dried at 110° without loss. The horizontal then extends up to around 153°; a very slow loss of weight then occurs as far as 222°, and then becomes very considerable and finally stops near 672°. The crucible then contains nothing but a mixture of antimony oxides and cobalt oxide (antimonate?).

REFERENCES

1 J. MORANDAT AND C. DUVAL, *Anal. Chim. Acta*, 4 (1950) 498.
2 H. ROSE, *Handbuch der analytischen Chemie*, 6th Ed., 1871, Vol. 2, p. 293.
3 F. P. TREADWELL, *Manuel de chimie analytique*, 3rd French Ed., Dunod, Paris, 1920, Vol. 2, p. 205.
4 H. GOTO, *J. Chem. Soc. Japan*, 55 (1934) 326.
5 G. B. S. SALARIA, *Anal. Chim. Acta*, 17 (1957) 395.
6 H. ROSE, *Ann. Physik*, 77 (1849) 110.
7 A. LEVOL, *Ann. chim. et phys.*, [3] 1 (1841) 504.
8 R. BUNSEN, *Ann.*, 106 (1858) 3.
9 A. SIMON AND W. NETH, *Z. anal. Chem.*, 72 (1927) 308.
10 H. BAUBIGNY, *Compt. rend.*, 124 (1897) 499.
11 G. VORTMANN AND A. METZL, *Z. anal. Chem.*, 44 (1905) 525.
12 P. WENGER AND C. CIMERMAN, *Helv. Chim. Acta*, 14 (1931) 718.
13 L. WANMAKER AND M. L. VERHEIJKE, *Philips Research Repts.*, 11 (1956) 1.
14 I. K. TAIMNI AND R. P. AGARWAL, *Anal. Chim. Acta*, 10 (1954) 312.
15 I. K. TAIMNI AND S. N. TANDON, *Anal. Chim. Acta*, 22 (1960) 35.
16 H. N. RÂY, *J. Indian Chem. Soc.*, 17 (1940) 586.
17 H. SAITO, *Sci. Repts. Tôhoku. Imp. Univ.*, 16 (1927) 138.
18 T. DUPUIS, *Rec. trav. chim.*, 71 (1952) 111.
19 G. SPACU AND A. POP, *Z. anal. Chem.*, 111 (1937-38) 254.
20 F. HECHT AND J. DONAU, *Anorganische Mikrogewichtsanalyse*, Springer, 1940, p. 178.

21 C. Duval, *Anal. Chim. Acta*, 16 (1957) 222.

22 F. Feigl, *Z. anal. Chem.*, 64 (1924) 45.

23 J. O. Gomez and J. G. Romero, *Técnica mét. (Barcelona)*, 4 (1948) 461; *C.A.*, 43 (1949) 4972.

24 H. Tamm, *Chem. News*, 24 (1871) 207, 221.

25 T. Umemura, *J. Chem. Soc. Japan*, 61 (1940) 25.

26 W. Prodinger, *Organic Reagents used in quantitative Inorganic Analysis*, Elsevier, New York, 1940, p. 115.

27 T. I. Pirtea, *Z. anal. Chem.*, 118 (1939) 26.

28 R. Belcher and D. Gibbons, *J. Chem. Soc.*, (1952) 4775.

CHAPTER 50

Tellurium

There is no good gravimetric method for determining tellurium. It is much more oxidizable than selenium and hence it should be weighed at low temperatures when obtained in its reduced state. The best of the four known procedures is the one employing vanadyl sulphate. With regard to an automatic determination, it will be necessary to look into the possibility of weighing as tellurium dioxide (obtained preferably by means of hexamethylenetetramine) or as lead tellurate, whose use has not yet been worked out for this purpose. However, a good method is still to be discovered. See Doan and Duval[1].

1. Elementary tellurium

(i) Precipitation by hydrazine   The precipitation by means of hydrazine was conducted as directed by Meyer[2], Gutbier and Huber[3], and Drew and Porter[4], who recommend drying at 100–103° or 110°. Matsuura[5] suggests 120–300°, temperatures which are obviously too high. Since the curve rises continuously, we advise keeping the temperature below 40° during the drying because on being cooled the finely divided tellurium retains oxygen with which it combined when heated.

(ii) Precipitation by hypophosphite   The Clauder[6] procedure in which hypophosphite functions as the reductant also gives results that are too high. He recommends drying at 132° prior to weighing the precipitate. The increases obtained with an initial weight of 159.3 mg of tellurium on heating to the respective temperatures were taken from the curve:

Temp. (°C)	77	139	229	342	440	608	872	1011
Increase (mg)	1.2	2.0	2.9	3.1	3.6	3.8	4.0	4.4

(iii) Precipitation by potassium iodide and sulphur dioxide   The procedure of Ripan and Macarovici[7] was followed for the precipitation by potassium iodide and sulphur dioxide gas. They suggest a correction

factor of 8.5% for this determination; the precipitated tellurium is far from pure. The curve shows a constant weight up to about 100°, a descent as far as 350°, and then a reoxidation from 350° to 716°. All of the evidence indicates that the heated material is a mixture and that the residue of tellurium dioxide at 1000° contains firmly occluded iodine. Since the exact formula of the material weighed is not known, and since the thermolysis curve is not good, we think it would be well to abandon this method.

(iv) Precipitation by vanadyl sulphate Bilek[8] found that the results are too low when the precipitate of tellurium obtained with vanadyl sulphate was dried at 95°. The curve shows a constant weight up to 90°; then there is a slow rise. The uptake of oxygen amounts to 4.2 mg per 15 mg at 539°.

2. Tellurium dioxide

(i) Precipitation by pyridine Precipitation with pyridine yields tellurium dioxide (tellurous anhydride), which Jilek and Kot'a[9] and Macarovici[10] suggest be dried at 105–130°. The graph yields a straight line beyond 100°.

(ii) Precipitation by hexamine When hexamethylenetetramine is used as the precipitant as suggested by Hecht and John[11] and Clauder[6], the precipitate filters well and the straight line obtained rises slightly from 20° to 853°. However, this method yields an accuracy of only 1 part in 150.

3. Tellurium disulphide

Tellurium disulphide TeS_2, prepared as described by Taimni and Agarwal[12], decreases in weight from 180° to 370°. The level from 370° to 410° corresponds to elementary tellurium. A gain in weight from 410° to 620° reveals the formation of the oxide TeO_2.

4. Tellurium trisulphide

The trisulphide TeS_3, obtained by Salaria[13], loses weight from 180° to 380°; then, as above, the level from 380° to 430° corresponds to tellurium, and the oxide TeO_2 is formed from 430° to 630°. See Taimni and Tandon[14].

References p. 520

5. Lead tellurate

Lead tellurate is prepared from lead nitrate and sodium tellurate. It yields a horizontal between 103° and 397°. This is followed by a slight rise and then the curve begins to descend at 745°. The loss at 1000° amounts to 4.2 mg per 99.7 mg.

6. Luteochromic tellurate

Hexamminechromic tellurate (or luteochromic tellurate) is pale yellow. It was prepared by Bersin's procedure[15] with a reagent prepared in accord with Jörgensen's directions[16]. The product shows a horizontal level only as far as 53° where ammonia begins to be evolved. The fall is then rapid until the temperature reaches 311°. The final level is not strictly horizontal; it corresponds roughly to the composition $Cr_2O_3 + TeO_2$.

REFERENCES

[1] U. M. DOAN AND C. DUVAL, *Anal. Chim. Acta*, 5 (1951) 569.
[2] J. MEYER, *Z. anal. Chem.*, 53 (1914) 145.
[3] A. GUTBIER AND J. HUBER, *Z. anal. Chem.*, 53 (1914) 430.
[4] H. D. K. DREW AND C. R. PORTER, *J. Chem. Soc.*, (1929) 2091.
[5] K. MATSUURA, *J. Chem. Soc. Japan*, 52 (1931) 730.
[6] O. E. CLAUDER, *Z. anal. Chem.*, 89 (1932) 270.
[7] R. RIPAN AND C. G. MACAROVICI, *Bull. sect. sci. acad. roumaine*, 26 (1944) 283; *C.A.*, 41 (1947) 6493.
[8] P. BILEK, *Collection Czechoslov. Chem. Communs.*, 10 (1938) 430.
[9] A. JILEK AND J. KOT'A, *Collection Czechoslov. Chem. Communs.*, 6 (1934) 398.
[10] C. G. MACAROVICI, *Bull. sect. sci. acad. roumaine*, 26 (1944) 301; *C.A.*, 41 (1947) 6494.
[11] F. HECHT AND L. JOHN, *Z. anorg. Chem.*, 251 (1943) 14.
[12] I. K. TAIMNI AND R. P. AGARWAL, *Anal. Chim. Acta*, 9 (1953) 121.
[13] G. B. S. SALARIA, *Anal. Chim. Acta*, 15 (1956) 514.
[14] I. K. TAIMNI AND S. N. TANDON, *Anal. Chim. Acta*, 22 (1960) 554.
[15] T. BERSIN, *Z. anal. Chem.*, 91 (1932) 170.
[16] S. M. JÖRGENSEN, *J. prakt. Chem.*, 30 (1884) 1.

CHAPTER 51

Iodine

We will discuss in turn the I^- ion of iodides, the IO_3^- ion of iodates, and the IO_4^- and IO_6^{5-} ions of periodates.

1. Iodide ion I^-

(i) Palladium iodide Palladium iodide was prepared as directed by Winkler[1] who gave 132° as the drying temperature. This salt was heated by Dupuis and Duval[2] who found that the thermolysis curve was easy to interpret. Moisture is given off up to 84°. A good horizontal extends from 84° to 365°. This level corresponds to the iodide and may be used for the automatic determination of either iodine or palladium.Decomposition sets in at 365° and continues to 651°. The resulting palladium metal is not stable at this temperature; it rapidly combines with oxygen and the horizontal of the oxide PdO starts at 839°.

(ii) Silver iodide If protected from the light, silver iodide may be weighed between 60° and 900°. The corresponding straight line can be used for the automatic determination of either iodine or silver. A slight decomposition begins above 900° and becomes noticeable near 960°.

(iii) Lead iodide Lead iodide is seldom used as a weighing form because of the relatively high solubility of the salt. It is stable up to 370°; the corresponding straight line ascends very slowly, and the uptake of oxygen is only 1 mg per 237 mg of PbI_2. The decomposition then proceeds as far as 880° and yields the oxide PbO quantitatively. There is a transient inflection around 670°; it corresponds essentially to the basic iodide $PbI_2 \cdot 2PbO$.

(iv) Cuprous iodide The precipitation as cuprous iodide is discussed in Chapter Copper. It should be recalled that this salt is stable up to 296°. Matsuura[3] reported the corresponding level extending from 100° to 200°

(v) Thallium iodide See Chapter Thallium regarding the determination as thallium(I) iodide. Attention is called here to the fact that this salt gives a horizontal level from 70° to 473°.

2. Iodate ion IO_3^-

(i) Iodic acid Under our[4] method of working, iodic acid HIO_3, an anhydrous compound, does not change weight below 125°. An initial loss of water then occurs rather rapidly and results in the acid HI_3O_8. The latter is employed in the Unterzaucher method; its level extends from 151° to 229°. Although the acid $H_2I_4O_{11}$ is reported in various texts, the curve gives no indication of this compound. A new loss of water is completed near 248° and yields the oxide I_2O_5. This was the convenient procedure employed by Duval and Lecomte[5] for the preparation of this compound and the establishment of its structure. Decomposition sets in at higher temperatures and rapid sublimation commences at 410°.

(ii) Silver iodate Not much use is made of the precipitation as silver iodate $AgIO_3$ proposed by Gore[6]. In his text on analytical chemistry, Rose suggests that the salt be dried at 100°; actually the thermolysis curve shows that the compound is stable from 80° to 410°. This horizontal can be used for the automatic determination of iodates. The corresponding salt then begins to break down at 410° and yields silver iodide from 500° on. The latter, in turn, loses iodine around 900° as was pointed out above.

(iii) Benzidinium iodate Benzidinium iodate was prepared as directed by Kretsov[7]; it does not have a fixed composition. It continuously loses weight even at 20° and gives a curve that slants downward until 60° where a rather violent explosion occurs. We suggest that this method be discarded.

Discussions of other iodates will be found in Chapters Potassium, Calcium, Copper, Strontium, Cerium, Barium.

3. Periodate ion IO_4^- and IO_6^{5-}

Hauptman[8] has followed the dehydration of periodic acid HIO_4 which results in a compound with the formula $H_7I_3O_{14}$.

Information about several periodates is given in Chapters Lithium, Sodium, Potassium, Lead, and in a recent paper by Bestougeff[9].

We wished to learn whether the method of precipitation suggested by Kahane[10] would be suitable for a gravimetric procedure. The periodate is precipitated as the zinc salt. However, the case appears doubtful because the thermolysis curve slants downward continuously. Consequently, it is necessary to retain this method as a titrimetric procedure as Kahane advised.

REFERENCES

[1] L. W. WINKLER, Z. angew. Chem., 31 (1918) 101; J. Chem. Soc., 114 (1918) 237.
[2] T. DUPUIS AND C. DUVAL, Anal. Chim. Acta, 4 (1950) 615.
[3] K. MATSUURA, J. Chem. Soc. Japan, 52 (1931) 730.
[4] C. DUVAL, Anal. Chim. Acta, 16 (1957) 224.
[5] C. DUVAL AND J. LECOMTE. Rec. trav. chim., special number in honour of C. J. VAN NIEUWENBURG, 79 (1960) 519.
[6] G. GORE, Phil. Mag., 41 (1871) 310.
[7] A. KRETSOV, J. Russ. Phys. Chem. Soc., 60 (1928) 1427.
[8] Z. HAUPTMAN, Chem. Listy, 52 (1958) 1015.
[9] N. BESTOUGEFF, Diploma of Higher Studies, Paris, 1960.
[10] E. KAHANE et al., Bull. Soc. Chim. France, (1948) 70.

CHAPTER 52

Cesium

Cesium can be determined gravimetrically as the chloride, perchlorate, sulphate, hexanitrocobaltate (cobaltinitrite), hexachlorostannate (stannic chloride), hexachloroplatinate (platinum(IV) chloride), and as the tetraphenylboron compound.

Studies have been made also of the bisulphate, carbonate, ferrocyanide, and the hexachlorotitanate.

1. Cesium chloride

The product from the action of hydrochloric acid on cesium carbonate was heated by Duval and Duval[1] up to 877°. Constant weight was observed beginning at 110°. The loss of water and excess acid produced a distinct change in the slope of the graph at 32° as though a very unstable hydrate had been formed; in fact, the loss of weight is again found to be rapid from 38° on. The anhydrous chloride begins to sublime around 877°. Kitajima[2] suggested the region 530–640° as best for constant weight.

2. Cesium perchlorate

Cesium perchlorate $CsClO_4$ was prepared by treating the chloride or nitrate with 20% perchloric acid and evaporating the solution to dryness. The residue is washed with ethanol. Drying may be carried out up to 611°, where there is an abrupt decomposition with decrepitation. The loss of oxygen is complete at 677°. The suggestion that 230° be used as the drying temperature is correct but it is too exclusive.

3. Cesium sulphate

The sulphate Cs_2SO_4 was heated while slightly moist. The weight does not change from 105° to 876° and consequently the drying range is very extensive. Kitajima[2] gave a more restricted value, namely 480–870°. If the product contains sufficient water a horizontal can be observed between 45° and 67° but it does not correspond to a solid hydrate since the material in the crucible at this point is actually a saturated solution.

4. Cesium hydrogen sulphate

The acid sulphate $CsHSO_4$ was heated by Tomkova, Jirů and Rosický[3]. It is converted into the pyrosulphate $Cs_2S_2O_7$ at 270°, and the latter yields the sulphate Cs_2SO_4 at 470°.

5. Cesium nitrate

Cesium nitrate $CsNO_3$ is anhydrous and very stable. It maintains a constant weight up to 490° (see the cobaltinitrite below).

6. Cesium cobaltinitrite

In contrast to the analogous products obtained with potassium and rubidium, the material resulting from the action of sodium cobaltinitrite on cesium chloride contains no sodium and it satisfies the formula $Cs_3[Co(NO_2)_6]·H_2O$. See Dupuis and Duval[1]. All of the water of crystallization has been eliminated when the temperature reaches 110°. Decomposition then begins but there is no observable change in the slope of the curve; in other words the loss of the molecule of water and of the three nitro groups subsequently expelled do not produce any discernible difference. Therefore, the temperature of 110° recommended for drying is correct but the temperature range over which the anhydrous compound can exist is extremely narrow. For example, with an initial weight of 451.89 mg of precipitate, the loss at 110° is 10.80 mg (theoretical); at 127° it is already 13.00 mg. A good horizontal is found from 219° to 494°, proving the existence of a compound or a mixture of stable compounds. Qualitative analysis of the residue along this level reveals the presence of the oxide CoO and cesium nitrate $CsNO_3$ but not the slightest trace of nitrite. The explanation of this new reaction derives clearly from the fact that the cobalt oxide CoO accepts oxygen from the air and the gas thus made available becomes able, at the temperature in question, to convert the nitrite quantitatively into nitrate. The practical result with respect to analytical chemistry is a new method for determining cesium. Destruction of the cobaltinitrite between 219° and 494° yields a mixture CoO + 3 $CsNO_3$ which contains 60.44% of cesium and has a "molecular weight" of 659.7. The automatic method of determination would lend itself well to this case. If the contents of the crucible are then extracted with hot water and centrifuged, pure cesium nitrate can be recovered from the solution. The latter breaks down beyond 494° and so the above mixture will then yield a mixture of cesium oxides disseminated throughout the cobalt oxide.

7. Cesium carbonate

Reisman[4] made a study of the isobaric decomposition of cesium carbonate $2Cs_2CO_3 \cdot 7H_2O$ by thermogravimetry and differential thermal analysis.

8. Cesium hexachlorotitanate

The hexachlorotitanate has been heated by Morozov and Toptygin[5].

9. Cesium hexachlorostannate

The hexachlorostannate $Cs_2[SnCl_6]$ was prepared by the action of freshly distilled stannic chloride, dissolved in absolute alcohol, on a hydrochloric acid solution of cesium chloride. The white precipitate is washed with hydrochloric acid and alcohol. On heating, the product first undergoes a slight increase in weight much less than that of the corresponding rubidium derivative; constant weight is maintained between 130° and 344°, which is the region selected for drying. Decomposition sets in then and chlorine and a little cesium chloride are lost. A new constant weight level begins at 879°.

10. Cesium ferrocyanide

The water of hydration of the ferrocyanide $Cs_4[Fe(CN)_6] \cdot 4H_2O$ is given off between 45° and 105° according to Cappellina and Babani[6].

11. Cesium hexachloroplatinate

The chloroplatinate $Cs_2[PtCl_6]$, an orange complex, is readily prepared by treating cesium chloride with chloroplatinic acid. It does not become absolutely anhydrous below 200° and is stable up to 409°, where a very slow decomposition sets in. This decomposition becomes more rapid around 745° and the curve undergoes a change in shape at 947° with loss of cesium chloride. However, the decomposition is not yet complete at 1060°. Kitajima[2] gave the region 340–440° as proper for drying the complex.

12. Cesium tetraphenylboron compound

The tetraphenylboron compound $Cs[B(C_6H_5)_4]$, studied by Wendlandt[7], was prepared by the method of Raff and Brotz[8]. When heated at the rate of 4.5° per minute, and using a sample weighing 100–150 mg, the material was found to be stable up to 210° after being dried in the open air for 24 hours. Metaborate is formed at 825° as shown by the equation:

$$Cs[B(C_6H_5)_4] + 30\ O_2 \rightarrow CsBO_2 + 24\ CO_2 + 10\ H_2O$$

REFERENCES

[1] T. DUVAL AND C. DUVAL, *Anal. Chim. Acta*, 2 (1948) 205.
[2] I. KITAJIMA, *J. Chem. Soc. Japan*, 55 (1934) 199.
[3] D. TOMKOVA, P. JIRŮ AND J. ROSICKÝ, *Collection Czechoslov. Chem. Communs*, 25 (1960) 957.
[4] A. REISMAN, *Anal. Chem.*, 32 (1960) 1566.
[5] I. S. MOROZOV AND D. J. A. TOPTYGIN, *Zhur Neorg. Khim.*, 5 (1960) 88.
[6] F. CAPPELLINA AND B. BABANI, *Atti. accad. sci. ist. Bologna*, 11 (1957) 76.
[7] W. W. WENDLANDT, *Chemist Analyst*, 26 (1957) 38; *Anal. Chem.*, 28 (1956) 1001.
[8] P. RAFF AND W. BROTZ, *Z. anal. Chem.*, 133 (1951) 241.

CHAPTER 53

Barium

This chapter deals with the thermolysis of the hydroxide, chloride, iodate, sulphate and lithopone, the sulphamate, selenite, pyroselenite, nitrate, arsenates, antimonate, carbonate and witherite, the aluminate, chromate, ferrocyanide, formate, acetate, oxalate, and quinaldate.

1. Barium hydroxide

Duval[1] investigated several samples of commercial barium hydroxide, including the dry material furnished by Kahlbaum, a hydrated product from OSI*, and a rehydrated specimen prepared from the dry product. The latter obviously had taken up moisture in the ratio of a little more than one mole of water per mole of hydroxide. It starts to dehydrate at room temperatures, and the thermolysis curve shows a horizontal between 75° and 134° which almost agrees with $Ba(OH)_2 \cdot 1H_2O$. The compound $Ba(OH)_2 \cdot 0.97H_2O$ had been reported prior to Duval's study. The anhydrous compound $Ba(OH)_2$ was obtained from 166° to 180° by linearly heating an initial weight of 232 mg and increasing the temperature at the rate of 300° per hour. The thermogram remains horizontal up to 544°, and then a double change occurs: the hydroxide begins to dehydrate, and the resulting oxide BaO starts to take up oxygen. Half of the oxygen of the peroxide is slowly released between 664° and 772°; beyond this temperature there is obtained the horizontal of the oxide BaO which is free from carbonate as shown by its infra-red spectrum. When the OSI product was taken as the starting material, this allegedly hydrated product was found to contain around 5.8 H_2O instead of the normal 8 H_2O. The short preceding level for the monohydrate was not observed, but merely a simple inflection in the curve. Barium oxide was obtained above 728°. These two cases show that it is not feasible to accept the material just as it comes from the bottle, and in addition the

---

* Abbreviation for Omnium Scientifique et Industriel (Paris).

commercial labels are often deceptive. It is necessary to make a controlled dehydration starting from a carefully standardized solution. When the dry product was rehydrated at 15° in the absence of carbon dioxide, it yielded $Ba(OH)_2 \cdot 4.5H_2O$ at the start.

Burriel-Marti and Garcia-Clavel[2] have also followed the dehydration of barium hydroxide $Ba(OH)_2 \cdot 8H_2O$ continuously at the heating rate of 300° per hour, and then traced the isotherms at 40°, 50°, 80°, 90°, 110°, 120°, 125°, 130°. They found that one of the eight molecules of water is attached more firmly than the others. The transformation of the hydroxide into the oxide occurs between 450° and 650°. They then made a study of the solid state reaction between barium hydroxide and ammonium sulphate in the proportions of 1 : 1 and 2 : 1, which produced barium sulphate with evolution of ammonia.

2. Barium chloride

The dihydrate of barium chloride has been suggested on several occasions as a standard in aquametry. However, various workers have been disappointed when they used samples of this salt, which apparently were quite dry and well crystallized, for standardizing Karl Fischer reagent. This salt has likewise been suggested as a weighing form in a publication issued by the American Atomic Energy Commission (A.C.E.D. 2738). Duval[3] studied three commercial products, namely: Prolabo, new bottle; Prolabo, bottle opened several times; OSI, bottle opened several times. The anhydrous product was recrystallized from water at ordinary temperature. The thermolysis curves recorded with these materials appear very similar. They start horizontally and the salts begin to lose water around 60° when amounts from 180 to 250 mg are heated at the rate of 300° per hour. A short horizontal appears around 104–107° corresponding to the monohydrate, and then the anhydrous barium chloride yields a horizontal beginning at 150° and the weight holds constant up to 875° at least. Calculation showed that all of these specimens differ with regard to their content of water. They showed respectively: 1.98, 0.8, 1.8, and 1.95 H_2O. The dihydrate of barium chloride is not stable; it is a mixture of the monohydrate and the dihydrate. It cannot be employed as a reliable standard in aquametry or as a weighing form in gravimetry. It is preferable to dehydrate the salt of doubtful composition by heating it to near 160° before employing it for the preparation by direct weighing of standard solutions to be used in amperometry, the determination of sulphates, etc. This chloride has likewise been studied on other occasions[4] and by Orosco[5]; the latter reported 96° for obtaining the monohydrate, and 132° for obtaining the anhydrous compound.

3. Barium iodate

Barium iodate $Ba(IO_3)_2 \cdot H_2O$ has been investigated by two different modes of heating. The first rate was applied by Peltier and Duval[6] to the material as an analytical precipitate; the second heating was linear at the rate of 300° per hour and the material was treated as a standard by Duval[7] and Rocchiccioli[8]. The dehydration begins at 43°, and is barely discernible at 60° but accelerates near 79°. A change in the curve is observed when three-fourths of the water of crystallization has been removed. The loss of the rest of the water is much slower and is not really complete below 320°. The horizontal due to the anhydrous iodate extends from 320° to 476°, and then the Rammelsberg reaction takes place with violence, especially at 610°:

$$5\ Ba(IO_3)_2 \rightarrow Ba_5(IO_6)_2 + 9\ O_2 + 4\ I_2$$

The reaction is complete at 666°, and is followed by the horizontal of the white paraperiodate, which is stable up to at least 960°, the limit of the experiment. The determination of barium in the form of the iodate or the paraperiodate cannot be recommended.

4. Barium sulphate

The use of barium sulphate $BaSO_4$ has led to so many contradictory reports regarding the amounts of adsorbed substances that a particularly careful study was made of this mode of determination. Matsuura[9] reported merely that the salt should be dried below 900°.

(i) Precipitation while warm with stirring The precipitate was prepared by adding 0.05 N sulphuric acid, with vigorous stirring, to a boiling 2 N solution of barium chloride containing 2 N hydrochloric acid. The curve given by the precipitate, after a preliminary drying in the air at 25°, showed a horizontal up to 158°. From this temperature on and up to 950°, there is a total decrease in weight of 2.8% due to the loss of water and adsorbed hydrochloric acid. The extent of this loss at three temperatures was:

Temp. (°C)   300   500   900
Loss (%)   0.8   2   2.6

Strictly speaking, the ignition should be conducted only between 950°

and 1200°, but for practical purposes, the level corresponding to the sulphate becomes essentially horizontal at 780°. Several of the modern treatises on analytical chemistry suggest 800° as the ignition temperature.

(ii) Precipitation in the cold without stirring The precipitate was brought down in the cold and without stirring, but in a large volume (2 1 per 0.200 g). The curve has the same shape but the constant weight level begins at 180°. This finding is in agreement with the procedure given in the manuals used in a number of university laboratories.

(iii) Precipitation from boiling solution Barium sulphate was precipitated from a boiling solution containing ferric chloride and then allowed to ripen for 18 hours at 15°. Jeannin and Duval[10] found the thermolysis curve to be of the same form as that described in *(i)* when 188.1 mg of the sulphate, which had retained the maximum amount of 1.08% of iron, was heated at the rate of 300° per hour. The loss of water is again 2.8% between 158° and 950°. This water is not retained by the iron, which consequently is present as Fe_2O_3. No retention of Cl^- ions was detected.

Paulik and Erdey[11] employed differential thermal analysis and found that the impurities are removed in regular fashion from barium sulphate, and in the order: water, sulphuric anhydride, hydrochloric acid.

Lithopone Lithopone is a mixture of barium sulphate, zinc sulphide, and a trace of zinc oxide. Burriel-Marti, Barcia-Goyanes and Quiertant-Alvarez[12] showed that the presence of barium sulphate did not interfere with the thermal conversion of zinc sulphide into the oxide. These workers have discovered a rapid method of analysis which is accurate enough for the commercial product.

5. Barium thiosulphate

Duval[13] investigated barium thiosulphate which is used as a standard in iodometry. He found that its weight is constant up to 84°. Decomposition sets in then and a reoxidation starts at 447° leading to the production of barium sulphate.

This material was also prepared by the method of Gaspar y Arnal and Santos[14] which yields the monohydrate whose very short level goes from 68° to 75°. Then the curve descends constantly until the temperature reaches 410°; the decrease is due to the loss of one-half of the sulphur. The resulting sulphite oxidizes and the sulphate is obtained quantitatively at 900°.

6. Barium sulphamate

Barium sulphamate $Ba(NH_2-SO_3)_2$, which was studied by Capestan[15], crystallizes without water and begins to decompose in one step at 265°. The transformation into sulphate is complete at 405°.

7. Barium selenite

Barium selenite $BaSeO_3$ is stable up to 650°; it then oxidizes to the selenate. See Rocchiccioli[8].

8. Barium pyroselenite

The pyroselenite $BaSe_2O_5$ was found by Rocchiccioli[8] to decompose near 380° with loss of selenium dioxide. It changes into the neutral selenite around 500°.

9. Barium nitrate

Barium nitrate taken from various bottles was dry and anhydrous. It holds its weight when heated linearly at 300° per hour up to 592–600°. With six specimens the temperatures ranged between 590° and 605°, *i.e.* close to the melting point. Deep-seated decomposition sets in then and a constant weight due to the oxide BaO is observed at 980°. Some workers believe that the final product of these decompositions is a mixture of BaO and BaO_2, but our calculation gave 148 and 149 as the mean molecular weight of the residue, and this value lies between that of BaO and Ba. A physical study should be made to clarify the structure of this mixture.

10. Barium binarsenate

Bastick and Guérin[16,17] heated barium binarsenate $BaO.2As_2O_5$ in nitrogen at the rate of 150° per hour. They found that the metarsenate $BaO \cdot As_2O_5$, the first product, was succeeded by the pyroarsenate $2BaO \cdot As_2O_5$ with probable intermediate formation of the derivative $3BaO \cdot 2As_2O_5$.

$$BaO \cdot 2As_2O_5 \rightarrow BaO \cdot As_2O_5 + As_2O_3 + O_2$$
$$3(BaO \cdot As_2O_5) \rightarrow 3BaO \cdot 2As_2O_5 + As_2O_3 + O_2$$
$$2(3BaO \cdot 2As_2O_5) \rightarrow 3(2BaO \cdot As_2O_5) + As_2O_3 + O_2$$

The decomposition temperatures are greatly influenced by the rates of heating.

The decomposition of the binarsenate into metarsenate has been carried out successively also at 600°, 658°, and 691° on a sample weighing 200 mg, with the nitrogen flowing at the rate of 500 ml per minute.

The action of barium carbonate on these salts leads to transformations analogous to those resulting from the simple pyrolysis, but at distinctly lower temperatures.

11. Barium antimony compound

Dupuis[18] made a study of the product resulting from the double decomposition between "potassium pyroantimonate" and barium chloride. It should be written $Ba[Sb(OH)_6]_2$. Half of the water is removed up to 85° and the remainder departs slowly and regularly up to 700°. The total loss is precisely 6 H_2O per molecule. The horizontal obtained is due to the metantimonate $Ba(SbO_3)_2$, which was identified by chemical analysis and infra-red spectrography.

12. Barium niobate

The niobate $Ba_7Nb_{12}O_{37}\cdot26H_2O$ was heated, as were all the salts of this type, by Pehelkin and his associates[19]. This compound is converted into the stable pentahydrate from 120° to 175°; it yields the trihydrate at 220°, the dihydrate at 320°, the monohydrate at 480°. Finally, the metaniobate is formed at 550°.

13. Barium carbonate

Peltier and Duval[6] precipitated the carbonate $BaCO_3$ by means of ammonium carbonate (no excess); they found that the product retains water tenaciously. The dehydration is practically complete at 400°, but 481 mg of the true carbonate still lost 2 mg of water from 400° to 600°. If it is desired to weigh in this form, the product may be ignited in the temperature range 600–1000°.

Dry barium carbonate taken from a bottle with the intention of standardizing an appropriate acid solution does not change weight from 20° to 1000°. See Duval[1,4].

Witherite Caillère and Pobeguin[20] included witherite in their study of natural orthorhombic carbonates. The specimens were heated in air and in nitrogen as part of an excellent comparative study by differential thermal analysis, infra-red absorption, and X-ray absorption. In air, witherite begins to decompose at 1085° and in nitrogen at 975°. Accordingly, it is much more stable than cerussite, aragonite, and strontianite.

14. Barium aluminates

Maekawa[21] made a study of the three aluminates: $BaO \cdot Al_2O_3 \cdot 6H_2O$, $2BaO \cdot Al_2O_3 \cdot 5H_2O$, $3BaO \cdot Al_2O_3 \cdot 7H_2O$.

15. Barium chromate

Peltier and Duval[6] prepared barium chromate $BaCrO_4$ from barium chloride and sodium chromate. Because of the peculiarities of the curve obtained, they investigated the behaviour of an anhydrous product which had been dried for several days in the air at 20°. Some authors advise igniting at "white heat", whereas others caution against exceeding 180°. Strictly speaking, the compound retains the composition $BaCrO_4$ only as far as 60°. If the heating is continued up to 1015°, the result is a mixture of yellow chromate and green chromite. (We originally reported the oxide Cr_2O_3.) The curve includes two good horizontals that are quite reproducible but they do not correspond to any definite product. The extent of these variations is slight since the total loss of oxygen is much less than in the case of silver chromate, and does not reach 1% at 1015°. The first of these levels extends from 530° to 790°; the second begins at 930°. We believe that they correspond to equilibrium reactions.

16. Barium ferrocyanide

The ferrocyanide $Ba_2[Fe(CN)_6] \cdot 6H_2O$ was studied by Cappellina and Babani[22]. It shows a loss of water between 75° and 250°.

17. Barium formate

Because of its catalytic properties, barium formate has been decomposed by Zapletal et al.[23].

18. Barium acetate

Duval and Wadier[24] reported that barium acetate, taken from a previously opened bottle, corresponds to the monohydrate. It is not hygro-scopic and may be weighed as is. If the anhydrous salt is desired, the hydrate should be heated to around 105° in a thermostat. Normally the dehy-dration occurs between 72° and 96°; the anhydrous salt remains almost stable up to 240°. It slowly loses weight as far as 480° and acetone is evolved up to 542°. The residue, whose weight is constant, consists of barium carbonate. It is produced by the classic "Chancel" reaction, which seems to go back to Chenevix.

19. Barium oxalate

Barium oxalate BaC_2O_4, precipitated from a boiling solution by means of a saturated solution of ammonium oxalate, in the presence of alcohol, retains only a half-molecule and not one molecule of water of crystal-lization. Angelescu[25] advises drying at 70° whereas Diaz-Villamil[26] sug-gests 250°. These workers may be correct provided the formula of the material weighed is known precisely. The semi-hydrate is stable up to 76°. The water is eliminated from 76° to 110°. A perfectly horizontal level extends from 110° to 346°; it corresponds to the anhydrous oxalate. Carbon monoxide begins to come off then, with a sudden increase at 411°. The barium carbonate is completely formed at 476°, where a new horizontal begins and extends to beyond 1000°. Our numerical results have convinced us that weighing barium as the carbonate is more precise when the latter compound is prepared by destruction of the oxalate rather than by direct precipitation with ammonium carbonate.

20. Barium quinaldate

Thomas[27] prepared the white quinaldate $Ba(C_{10}H_6O_2N)_2.4H_2O$ by neutralizing a solution of quinaldic acid with barium carbonate. The salt begins to lose water at 70° if 300 mg are heated at the rate of 3° per minute, up to 700°, where the material is then kept for 2 hours. The level of the anhydrous salt extends from 135° to 365°, and the carbonate $BaCO_3$ starts to appear at 500°.

REFERENCES

[1] C. DUVAL, *Anal. Chim. Acta*, 13 (1955) 428.
[2] F. BURRIEL-MARTI AND M. E. GARCIA-CLAVEL, *16th Intern. Congr. Pure and Appl. Chem.*, Paris, 1957, p. 861.
[3] C. DUVAL, *Anal. Chim. Acta*, 13 (1955) 36.

[4] *1st Colloq. Thermography, Kazan*, 1953, p. 7, 110.
[5] E. OROSCO, *Ministério trabalho ind. e com., Inst. nac. tecnocol., (Rio de Janeiro)*, 1940; *C.A.*, 35 (1941) 3485.
[6] S. PELTIER AND C. DUVAL, *Anal. Chim. Acta*, 1 (1947) 360.
[7] C. DUVAL, *Anal. Chim. Acta*, 20 (1959) 263.
[8] C. ROCCHICCIOLI, *Thesis*, Paris, 1960.
[9] K. MATSUURA, *J. Chem. Soc. Japan*, 52 (1931) 730.
[10] Y. JEANNIN AND C. DUVAL, *Mikrochim. Acta*, (1959) 61.
[11] F. PAULIK AND L. ERDEY, *Acta Chim. Acad. Sci. Hung.*, 13 (1957) 117.
[12] F. BURRIEL-MARTI, C. BARCIA-GOYANES AND M. QUIERTANT-ALVAREZ, *Las Ciencias*, 24 (1959) 507.
[13] C. DUVAL, *Anal. Chim. Acta*, 16 (1957) 546.
[14] T. GASPAR Y ARNAL AND M. SANTOS, *Anal. soc. españ. fís. y quím.*, 40 (1944) 671.
[15] M. CAPESTAN, *Ann. Chim. (Paris)*, 5 (1960) 213.
[16] M. BASTICK, H. GUÉRIN AND L. DUBAILLY, *Bull. Soc. chim. France*, (1959) 626.
[17] M. BASTICK AND H. GUÉRIN, *La chimie des hautes températures. Colloq. nat. C.N.R.S.*, 1957, p. 139.
[18] T. DUPUIS, *Thesis*, Paris, 1954.
[19] V. A. PEHELKIN, Y. P. SIMANOV, A. V. LAPITSKII AND V. I. SPITSYN, *Zhur. Neorg. Khim.*, 1 (1956) 1784; *C.A.*, (1957) 2445.
[20] S. CAILLÈRE AND T. POBEGUIN, *Bull. soc. franç. minéral.*, 83 (1960) 36.
[21] G. MAEKAWA, *J. Soc. Chem. Ind. Japan*, 45 (1942) 130; 44 (1941) 912.
[22] F. CAPPELLINA AND B. BABANI, *Atti. accad. sci. ist. Bologna*, 11 (1957) 76.
[23] V. ZAPLETAL, J. JEDLIČKA AND V. RŮŽIČKA, *Chem. Listy*, 50 (1956) 1406; *C.A.*, (1957) 2438.
[24] C. DUVAL AND C. WADIER, *Anal. Chim. Acta*, 23 (1960) 257.
[25] B. ANGELESCU, *Bull. soc. Romania*, 5 (1923) 12.
[26] C. DIAZ-VILLAMIL, *Anal. soc. españ. fís. y quím.*, 34 (1936) 580.
[27] G. THOMAS, *Thesis*, Lyon, 1960.

CHAPTER 54

Lanthanum

Thermolysis curves have been traced for lanthanum hydroxide, fluoride, chloride, sulphate, sulphamate, nitrate, formate, acetate, oxalate, citrate, cupferronate, neocupferronate, oxinate, dichloro-, dibromo-, and diiodo-oxinates, 2-methyloxinate.

1. Lanthanum hydroxide

The hydroxide $La(OH)_3$ is prepared, in the air, by treating lanthanum nitrate with aqueous ammonia and briefly drying the precipitate. Under these conditions it takes up considerable carbon dioxide. The curve descends continuously as far as 944°, above which pure La_2O_3 is present. Other workers have reported 900° and 905°. Most of the retained water and ammonia have been eliminated when the temperature has reached 65°. The hydroxide stops losing water at 370° and oxide mixed with basic carbonate $La_2O_3 \cdot CO_2$ is obtained at this temperature. The latter is stable up to 467°, it slowly loses carbon dioxide up to 738°. The decomposition accelerates and is complete at 944°. See Duval and Duval[1].

2. Lanthanum fluoride

The fluoride, which was studied by Wendlandt and Love[2], was prepared by adding aqueous hydrofluoric acid to a solution of a lanthanum salt. The precipitate was heated at the rate of 5.4° per minute. The curve slants downward continually as far as 900°. It does not include a level of the anhydrous salt but an oxyfluoride was observed between 600° and 690°.

3. Lanthanum chloride

Wendlandt[3] studied the chloride $LaCl_3 \cdot 7H_2O$. It begins to lose water at 85°. The anhydrous salt gives a horizontal from 210° to 375°, and then becomes progressively contaminated with the oxychloride LaOCl, which is not completely decomposed even at 850°.

References p. 541

4. Lanthanum sulphate

After recrystallization, the sulphate has the formula $La_2(SO_4)_3 \cdot 9H_2O$ and not $8H_2O$. Wendlandt[3] reported that the first loss of water is at $50°$ and a break in the curve at $200°$ corresponds roughly to the dihydrate. The salt is completely dehydrated at $265°$ which is a much lower temperature than that given by Vickery[4] who reported $600°$. Duval and Wadier[5] repeated this experiment, and found $560°$ when 200 mg were heated at the rate of $300°$ per hour. Their initial product contained 8 molecules of water.

5. Lanthanum sulphamate

The amidosulphonate $La(NH_2-SO_3)_3 \cdot 2H_2O$ was prepared by Chrétien and Capestan[6] by dissolving the hydroxide $La(OH)_3$ in a water solution of sulphamic acid. It loses one molecule of water between $95°$ and $150°$. The level of the monohydrate is perfectly horizontal and extends almost to $300°$. Beyond this there is a rapid decrease in weight which continues to about $400°$ with formation of the double sulphate $(NH_4)_2SO_4 \cdot La_2(SO_4)_3$. The ammonium sulphate decomposes and anhydrous lanthanum sulphate is left at $520°$.

In connection with this study, these same workers investigated the thermolysis of another complex salt, namely $3(NH_4)_2SO_4 \cdot La_2(SO_4)_3$.

6. Lanthanum nitrate

The nitrate $La(NO_3)_3 \cdot 6H_2O$ was heated by Kato[7] who gave $745°$ as the minimum temperature for the formation of the oxide La_2O_3. Much later, Wendlandt[8] observed that the water of hydration begins to come off at $50°$ and that the curve shows breaks for the composition of the pentahydrate and the dihydrate. A level for the monohydrate begins at $170°$ and the anhydrous salt is formed at $240°$. The oxides of nitrogen begin to come off at $420°$ and the oxynitrate $LaONO_3$ in turn yields a level from $420°$ to $575°$. Finally, the oxide La_2O_3 appears at $780°$ when the heating rate is 4.5 per minute.

7. Lanthanum formate

Ambrozhii and Osipova[9] found that lanthanum formate $La(HCO_2)_3$ decomposes in two steps, giving the oxide. The two stages are $330-355°$, and $480-520°$.

8. Lanthanum acetate

Kato[7] reported that lanthanum acetate yielded the oxide La_2O_3 above 760°.

9. Lanthanum oxalate

.The action of oxalic acid on a neutral solution of lanthanum nitrate yields a precipitate of the oxalate $La_2(C_2O_4)_3 \cdot 11H_2O$, which is kept at 40° for two hours. Complete dehydration is observed in the vicinity of 300° where there is a change in slope. Carbon monoxide and carbon dioxide are lost up to about 700° and the resulting basic carbonate $La_2O_3 \cdot CO_2$ is stable to almost 800°. Hence it is evident that this carbonate does not exist over the same temperature range as the carbonate which is formed at the same time as the hydroxide. Strictly speaking, the basic carbonate weighed between 700° and 800° does not give a perfectly horizontal level, but the difference in the weight at the two extremities does not exceed 1/100. The oxide La_2O_3 is finally obtained at 876°, and accordingly it forms at a lower temperature than in the case of the hydroxide (carbonate). It may be weighed after ignition at 900° for instance.

Somina and Hirano[10] have followed this dehydration in a stream of carbon dioxide. Using the decahydrate, Wendlandt[11] found that the water starts to come off at 55°. The oblique level of the compound $La_2O_3 \cdot CO_2$ appears from 550° to 735°; the oxide La_2O_3 is formed from 800° on. The Stanton balance revealed a level due to the dihydrate when Padmanabhan, Saraiya and Sundaram[12] heated 500 mg of the decahydrate at the rate of 6° per minute. The carbonate $La_2O_3 \cdot CO_2$ starts to decompose at 770°.

Caro and Loriers[13] conducted the thermolysis in a vacuum; they were astonished not to find the same temperatures that we reported when working in the air. The only intermediate compound detected was the basic carbonate $La_2O_3 \cdot CO_2$. In the lanthanide series, the stability of this carbonate decreases as the atomic number of the metal rises.

10. Lanthanum citrate

Lanthanum citrate was studied by Shu-Chuan Liang and Li[14] and by Skramovsky[15].

11. Lanthanum cupferronate

The cupferronate, prepared by Wendlandt[16], appears to be stable up to around 280°. It yields the oxide La_2O_3 starting at 580°. Wendlandt does not state whether this cupferronate lends itself to analytical purposes.

12. Lanthanum neocupferronate

The neocupferronate of Wendlandt and Bryant[17] seems to be stable up to 260°. It is converted into the oxide La_2O_3 when heated above 605°.

13. Lanthanum oxinate

The procedure given by Pirtea[18] was used for the preparation and washing of the oxinate $La(C_9H_6ON)_3$. At the rate of heating employed by Duval, the curve slanted downward without interruption until the temperature reached 276°, where charring set in along with the evolution of a considerable amount of gas. The temperature reported by Pirtea for drying occurs on the curve only by chance. In another trial, Duval heated a specimen of lanthanum oxinate for 3 hours at 130° by keeping the thermobalance in a thermostat. A good horizontal was obtained thus demonstrating that it is possible to obtain a constant weight in this method of determination. It should be noted that this is not always true of various oxinates.

Wendlandt[19] repeated this experiment and reports that the excess of oxine is eliminated from 50° to 220°. From this temperature to 390°, the chelate is stable. The oxidation of the organic matter begins at 400° and is complete at 700° where the level of the trioxide La_2O_3 starts. Since it is extremely difficult to prevent the coprecipitation of oxine, the method cannot be recommended for determining lanthanum.

14. Lanthanum dichlorooxinate

The dichlorooxinate $La(C_9H_4Cl_2ON)_3$, prepared by Wendlandt[20], starts to lose weight at 95°. The decomposition rate increases rapidly above 210° and the level of the oxide La_2O_3 begins at 780°.

15. Lanthanum dibromooxinate

Wendlandt[20] also investigated the dibromooxinate $La(C_9H_4Br_2ON)_3$ which is slightly more stable than the dichlorooxinate; the first loss of weight occurs at 125°. The rate of decomposition rises rapidly beyond 340° and the level of the oxide La_2O_3 appears at 740°.

16. Lanthanum diiodooxinate

The diiodooxinate La(C$_9$H$_4$I$_2$ON)$_3$ was also first prepared by Wendlandt[21]. It begins to break down at 65°; the level of the oxide La$_2$O$_3$ appears at 585°.

17. Lanthanum 2-methyloxinate

The 2-methyloxinate La(C$_{10}$H$_8$ON)$_3$ is yellow. It was first reported by Wendlandt[16]. It is stable up to 65° and yields the oxide La$_2$O$_3$ at 645°. Whether it can be recommended for the determination of lanthanum is still to be decided.

REFERENCES

1 T. DUVAL AND C. DUVAL, *Anal. Chim. Acta*, 2 (1948) 218.
2 W. W. WENDLANDT AND B. LOVE, *Science*, 129 (1959) 842.
3 W. W. WENDLANDT, *J. Inorg. & Nuclear Chem.*, 5 (1957) 118.
4 R. C. VICKERY, *Chemistry of the Lanthanons*, Academic Press, New York, 1953, p. 235.
5 C. DUVAL AND C. WADIER, *Anal. Chim. Acta*, 23 (1960) 542.
6 A. CHRÉTIEN AND M. CAPESTAN, *Compt. rend.*, 248 (1959) 3177.
7 T. KATO, *J. Chem. Soc. Japan*, 52 (1931) 774.
8 W. W. WENDLANDT, *Anal. Chim. Acta*, 15 (1956) 435.
9 M. N. AMBROZHII AND Y. A. OSIPOVA, *Zhur. Neorg. Khim.*, 3 (1958) 2716.
10 T. SOMINA AND S. HIRANO, *J. Soc. Chem. Ind. Japan*, 11 (1931) 459.
11 W. W. WENDLANDT, *Anal. Chem.*, 30 (1958) 58; 31 (1959) 408.
12 V. M. PADMANABHAN, S. C. SARAIYA AND A. K. SUNDARAM, *J. Inorg. & Nuclear Chem.*, 12 (1960) 356.
13 P. CARO AND J. LORIERS, *J. Recherches C.N.R.S.*, 39 (1957) 107; *C.A.*, (1958) 11641.
14 SHU-CHUAN LIANG AND I. LI, *K'o Hsueh Chi Lu*, 2 (1958) 170; *C.A.*, (1959) 21122.
15 S. SKRAMOVSKY, *Silikáty*, 3 (1959) 74.
16 W. W. WENDLANDT, *Anal. Chem.*, 27 (1955) 1277.
17 W. W. WENDLANDT AND J. M. BRYANT, *Anal. Chim. Acta*, 13 (1955) 550.
18 T. I. PIRTEA, *Z. anal. Chem.*, 107 (1936) 191.
19 W. W. WENDLANDT, *Anal. Chim. Acta*, 15 (1956) 109.
20 W. W. WENDLANDT, *Anal. Chim. Acta*, 17 (1957) 428.
21 W. W. WENDLANDT, *Anal. Chim. Acta*, 15 (1956) 533.

CHAPTER 55

Cerium

The thermolysis studies of cerium compounds have dealt with the peroxide, chloride, iodate, sulphates, sulphamate, nitrate, ferrocyanide, formate, oxalate, cupferronate, neocupferronate, and the oxinate.

1. Cerium peroxide

The peroxide is obtained by treating a water solution of cerium(III) nitrate with aqueous ammonia and hydrogen peroxide (3%). The precipitate loses the greater part of its water up to 89°; the dehydration continues slowly until around 450°. Beyond this, the resulting CeO_2 does not change in weight up to 814°, the upper limit of the trials by Duval and Duval[1].

2. Cerium chloride

The chloride $CeCl_3 \cdot 6H_2O$ loses water above 55°; the anhydrous chloride is stable at 240° but becomes more and more contaminated with the peroxide CeO_2. The latter is stable beyond 550°. See Wendlandt[2]. Kato[3] gives 600° for this latter temperature.

3. Cerium potassium iodate

The iodate $2Ce(IO_3)_4 \cdot KIO_3 \cdot 8H_2O$ is determined particularly by titrimetric procedures. Chernikov and Uspenskaya[4] suggested that it be weighed after being washed with ether and dried at 40–45°. Actually, the pyrolysis curve descends continuously and between 410° and 630° shows a sudden fall corresponding to a loss of oxygen and iodine (analogous to the behaviour of the alkaline earth iodates). Between 650° and 746°, the material is a mixture of potassium iodide and cerium periodate; the latter has not been described previously. At 800° the residue contains cerium peroxide. Although the curve has no foreseeable use in gravimetric analysis, it may possibly be of value in studies in the inorganic field.

4. Cerium sulphates

(i) Cerium(III) Cerium(III) sulphate is obtained by treating cerium-(III) carbonate with sulphuric acid. Under the conditions employed by Duval and Duval[1], it retains a little more than 4 molecules of water, and consequently is a mixture of hydrates. It starts to lose weight around 90° and is anhydrous from 277° to 845°. These workers did not employ a temperature high enough to produce the peroxide, since this would have no analytical value nor would it be of interest from the standpoint of determining atomic weights. The two breaks at 150° and at 235° do not correspond to any simple formulas, doubtless because the product of the solution in sulphuric acid is a mixture of hydrates. Wendlandt[5], who has repeated this experiment with 85 to 100 mg of the pure pentahydrate heated at the rate of 5.4° per minute, found results which differed from those of Duval and Duval. The dihydrate appeared near 140° and the anhydrous salt, formed at 245°, began to lose sulphur trioxide at 650°.

(ii) Cerium(IV) Cerium(IV) sulphate, a commercial product used in cerimetry, was found not to be a definite material but rather it showed the behaviour of a mixture of the tetrahydrate and the anhydrous salt containing an average of 3.1 H_2O instead of 4 H_2O. A mixture of this kind holds its weight constant only as far as 47°. It loses water up to 190–197°. The anhydrous product is not stable for long; it starts to lose sulphuric anhydride slowly at 300° to form a basic yellowish white material $3CeO_2.4SO_3$. Therefore, this salt cannot be used directly for preparing a standard solution and all the more so because it is likely to contain other metals of the cerium earths. See Duval[6].

5. Cerium sulphamate

Cerium(III) amidosulphonate (sulphamate) was first isolated by Chrétien and Capestan[7]. The thermolysis curve can be interpreted like that of the lanthanum salt. The dihydrate $Ce(NH_2-SO_3)_3.2H_2O$ loses part of its water from 95° on and the level of the monohydrate extends from 150° to 300°. Beyond 300° there is a rapid loss of weight leading to the double sulphate $(NH_4)_2SO_4.Ce_2(SO_4)_3$ between 350° and about 460°. The ammonium sulphate then disappears and cerium(III) sulphate remains beyond 520°.

6. Cerium nitrate

Wendlandt[8] made a study of the nitrate $Ce(NO_3)_3.6H_2O$. He did not observe any intermediate level or any evidence of the oxynitrate. The loss of water begins at 80°, and a level commencing at 450° is due to the peroxide CeO_2. Kato[3] reported 830°, a temperature that is certainly too high.

7. Cerium potassium ferrocyanide

Spacu[9] reported $KCe[Fe(CN)_6] \cdot 4H_2O$ as the formula of the precipitate resulting from the action of cerium(III) nitrate and potassium ferrocyanide. This hydrate should be dried in vacuo at a low temperature. When heated, it undergoes a continuous loss of weight, and the evolution of cyanogen accelerates from 170° on. Duval[1] did not observe any level indicating the appearance of the trihydrate at 100°.

8. Cerium formate

Cerium(III) formate, like calcium formate, decomposes in two stages, the first from 250° to 310°, the second from 340° to 355°. The residue consists of the peroxide CeO_2. See Ambrozhii and Osipova[10].

9. Cerium oxalate

The product precipitated by the reaction between cerium(III) nitrate and oxalic acid and held at 60° for several hours contains 10–11 molecules of water; actually it is a mixture of hydrates. The curve shows no distinction between the departure of water, carbon monoxide, or carbon dioxide. The decrease continues until 450°. No level was observed at 100° corresponding to the trihydrate, which has been suggested as a weighing form. Calculation shows that the dehydration is complete at 229° but there is no possibility of weighing the anhydrous salt since it starts to decompose at this temperature. Somiva and Hirano[11] studied the decomposition of this oxalate in an atmosphere of carbon dioxide. Wendlandt[12], who repeated the thermolysis at the rate of 5.4° per minute, claims to have obtained the peroxide CeO_2 from 360° on. However, Duval found that the curve is still descending slightly, though almost imperceptibly, at this temperature.

10. Cerium cupferronate

Cerium(III) cupferronate was prepared by Wendlandt[13]. It is stable up to 150°. After the organic matter has been destroyed, the level of the oxide CeO_2 starts near 400°.

11. Cerium neocupferronate

Wendlandt and Bryant[14] report that the neocupferronate begins to break down near 80° and yields the oxide CeO_2 beyond 460°.

12. Cerium oxinate

The oxinate $Ce(C_9H_6ON)_4.2H_2O$ was prepared by the Pirtea[15] method. When the moist material is dehydrated, it yields a good horizontal extending from 128° to 233°, where a sudden decomposition sets in with release of vapours and charring. The temperature of 110°, suggested by Berg and Becker[16], thus strikes us as being a little low if constant weight is to be attained with certainty. Wendlandt[13] repeated this study and he gives 350° as the limit of the stability of this oxinate, and then 420° for obtaining the peroxide. We have found that in the case of oxine, this limit may vary quite a good deal from one experiment to another. It may range from 400° to 800° depending on the weight of the sample and the renewal of the air in the furnace.

REFERENCES

[1] T. Duval and C. Duval, *Anal. Chim. Acta*, 2 (1948) 222.
[2] W. W. Wendlandt, *J. Inorg. & Nuclear Chem.*, 5 (1957) 118.
[3] T. Kato, *J. Chem. Soc. Japan*, 52 (1931) 774.
[4] Y. A. Chernikov and T. A. Uspenkaya, *Zavodskaya Lab.*, 9 (1940) 276.
[5] W. W. Wendlandt, *J. Inorg. & Nuclear Chem.*, 7 (1958) 51.
[6] C. Duval, *Anal. Chim. Acta*, 15 (1956) 223.
[7] A. Chrétien and M. Capestan, *Compt. rend.*, 248 (1959) 3177; M. Capestan, *Ann. Chim. (Paris)*, 5 (1960) 222.
[8] W. W. Wendlandt, *Anal. Chim. Acta*, 15 (1956) 435.
[9] P. Spacu, *Z. anal. Chem.*, 104 (1936) 28.
[10] M. N. Ambrozhii and Y. A. Osipova, *Zhur. Neorg. Khim.*, 3 (1958) 2716.
[11] T. Somiva and S. Hirano, *J. Soc. Chem. Ind. Japan, Suppl. binding*, 34 (11) (1939) 459.
[12] W. W. Wendlandt, *Anal. Chem.*, 30 (1958) 58; 31 (1959) 408.
[13] W. W. Wendlandt, *Anal. Chem.*, 27 (1955) 1277.
[14] W. W. Wendlandt and J. M. Bryant, *Anal. Chim. Acta*, 13 (1955) 550
[15] T. I. Pirtea, *Bul. chim. soc. romîne chim.*, 39 (1937/38) 83.
[16] R. Berg and E. Becker, *Z. anal. Chem.*, 119 (1940) 1.

CHAPTER 56

Praseodymium

The thermolysis studies dealing with the salts of this element have included the hydroxide, chloride, sulphate, nitrate, formate, acetate, oxalate, cupferronate, neocupferronate, and oxinate.

1. Praseodymium hydroxide

Dupuis and Duval[1] precipitated the hydroxide by the method of Brinton and Pagel[2]. The curve reveals a rapid loss of water as far as 80°. The horizontal extending from 80° to 284° represents the existence zone of the hydroxide $Pr_2O_3 \cdot 3H_2O$, which contains a small amount of the carbonate. The mixture whose weight is constant and which is ordinarily written as Pr_6O_{11} is obtained from 676° on.

2. Praseodymium chloride

According to Wendlandt[3], the chloride $PrCl_3 \cdot 7H_2O$ begins to lose water beyond 65°. The hydrate $PrCl_3 \cdot H_2O$ is stable at 197° but it is contaminated with PrOCl. The latter is stable above 665°.

3. Praseodymium sulphate

The sulphate was prepared as described by Sarver and Brinton[4] by treating praseodymium nitrate with a slight excess of concentrated sulphuric acid. The thermolysis curve of the resulting product shows an elimination of retained water up to 110°. Sulphuric acid and sulphuric anhydride are then evolved up to 413° with a slight inflection at 280°. The anhydrous sulphate remains constant in weight beginning at 413°. The horizontal level has been followed as far as 950°. See Dupuis and Duval[1]. On the other hand, Wendlandt[5] found that the octohydrate starts to lose weight at 50°, and a break in the curve at 95° agrees approximately with the pentahydrate. A second break is seen at 150° but it does not correspond exactly to the dihydrate. The level of the an-hydrous sulphate commences at $240°$ (85–100 mg heated at the rate of $5.4°$ per minute).

4. Praseodymium nitrate

The nitrate $Pr(NO_3)_3 \cdot 6H_2O$ is not stable; it begins to lose water at $48°$ and does not yield intermediate hydrates. The anhydrous salt gives a level from $200°$ to $300°$ but it is slightly oblique. There is an inflection in the curve around $430°$. The oxide Pr_6O_{11} appears for the most part near $508°$ but it is not entirely pure. The horizontal really begins near $810-814°$ and continues as far as $950°$, the limit of the experiment. Kato[6] had re-ported $690°$ as the extreme limit of the oxide to which he assigned the formula Pr_2O_3. Wendlandt[7] repeated this study and gives $375°$ as the upper limit of the existence of the anhydrous salt, $470-480°$ for the ap-pearance of the oxynitrate $PrONO_3$, and $505°$ for that of the peroxide Pr_6O_{11}. These figures were obtained by heating 100–200 mg at the rate of $300°$ per hour.

5. Praseodymium formate

The formate $Pr(HCO_2)_3$ loses carbon monoxide directly between $450°$ and $475°$ according to Ambrozhii and Osipova[8].

6. Praseodymium acetate

Kato[6] reports that the acetate yields the oxide Pr_2O_3 above $800°$. He surely was dealing here with the peroxide Pr_6O_{11}.

7. Praseodymium oxalate

The decahydrate of the oxalate $Pr_2(C_2O_4)_3 \cdot 10H_2O$ was precipitated as directed by Sarver and Brinton[4] by treating the nitrate on the water bath for several hours with oxalic acid. The pyrolysis of this material shows first a loss of the excess water up to $78°$ where there is a change in di-rection of the curve corresponding to this decahydrate. The perfectly uniform dehydration is complete near $346°$ but the level of the anhydrous oxalate is very short and not perfectly horizontal. The oxalate then starts to decompose, there is a change in slope at about $543°$ and the decom-position is finished at $746°$ where the oxide Pr_6O_{11} may be weighed. The product which appears at 543° has no definite composition. Wendlandt[9] later heated 150 mg at the rate of 5.4° per minute. He obtained the anhydrous oxalate at 420° and the peroxide at 790°. Somiva and Hirano[10] decomposed the oxalate in an atmosphere of carbon dioxide.

8. Praseodymium cupferronate

Wendlandt[11] prepared a precipitate with cupferron. It appeared to be stable in the range 150–180°, and yielded a new level near 400°. The peroxide Pr_6O_{11} was formed beyond 650°.

9. Praseodymium neocupferronate

Wendlandt and Bryant[12] prepared the precipitate with neocupferron and heated it at the rate of 4.5° per minute. They found it to be stable up to 200°; the peroxide Pr_6O_{11} appeared above 650°.

10. Praseodymium oxinate

Dupuis and Duval[1] obtained a felted yellow precipitate with oxine in a solution containing acetic acid (pH 5). The product was readily filtered and washed. The precipitation is not quantitative and the thermolysis curve of the product reveals a continuous loss of oxine by sublimation along with a decomposition of the remaining oxine. A brown oxide fraction remains at 221°. In contrast to the findings of Dupuis and Duval, Wendlandt[13] reported that he obtained an oxinate which lost only around 1% of its weight between 250° and 280°. The organic matter then decomposed up to 520°, and the level of the peroxide began at 575°. Dupuis and Duval were not able to duplicate these results when they used linear heating at the rate of 5.4° per minute.

REFERENCES

[1] T. DUPUIS AND C. DUVAL, *Anal. Chim. Acta*, 3 (1949) 186.
[2] P. H. BRINTON AND H. A. PAGEL, *J. Am. Chem. Soc.*, 45 (1923) 1460.
[3] W. W. WENDLANDT, *J. Inorg. & Nuclear Chem.*, 5 (1957) 118.
[4] L. A. SARVER AND P. H. BRINTON, *J. Am. Chem. Soc.*, 49 (1927) 943.
[5] W. W. WENDLANDT, *J. Inorg. & Nuclear Chem.*, 7 (1958) 51.
[6] T. KATO, *J. Chem. Soc. Japan*, 52 (1931) 774.
[7] W. W. WENDLANDT, *Anal. Chim. Acta*, 15 (1956) 435.
[8] M. N. AMBROZHII AND Y. A. OSIPOVA, *Zhur. Neorg. Khim.*, 3 (1958) 2716.
[9] W. W. WENDLANDT, *Anal. Chem.*, 30 (1958) 58; 31 (1959) 408.
[10] T. SOMIVA AND S. HIRANO, *J. Soc. Chem. Ind. Japan, Suppl. binding*, 34 No. 11 (1931) 459.
[11] W. W. WENDLANDT, *Anal. Chem.*, 27 (1955) 1277.
[12] W. W. WENDLANDT AND J. M. BRYANT, *Anal. Chim. Acta*, 13 (1955) 550.
[13] W. W. WENDLANDT, *Anal. Chim. Acta*, 15 (1956) 109.

CHAPTER 57

Neodymium

Thermolysis studies have been made of the hydroxide, fluoride, chloride, sulphate, sulphamate, nitrate, formate, acetate, oxalate, cupferronate, neocupferronate, and the oxinate.

1. Neodymium oxide hydrate

The gelatinous hydrous oxide resulting from the action of aqueous ammonia on neodymium sulphate loses the excess water up to 73°. The good horizontal extending from 80° to 218° can be ascribed to the hydroxide $Nd(OH)_3$. A second level, parallel to the preceding one, but much shorter, goes from 270° to 317°; it corresponds to the existence region of the hydrate $Nd_2O_3 \cdot {}^3/_2 H_2O$, which was reported previously. The curve then descends slowly but gives no indication of the existence of the monohydrate $Nd_2O_3 \cdot H_2O$. The level corresponding to the oxide Nd_2O_3 starts at 608°. In contrast to lanthanum, no basic carbonate appears, and in contrast to praseodymium and cerium, there is no indication of a peroxide. See Duval and Duval[1].

2. Neodymium fluoride

The anhydrous neodymium fluoride NdF_3 was prepared by Wendlandt and Love[2] in accord with the directions of Popov and Knudsen[3]. It was heated at the rate of 5.4° per minute. Starting at 690° it is converted into the oxyfluoride but this change is not complete when the temperature reaches 900°.

3. Neodymium chloride

The chloride $NdCl_3 \cdot 6H_2O$ loses some water above 80° to give the monohydrate $NdCl_3 \cdot H_2O$ from 185° to 200°. The anhydrous salt is stable from 250° to 325°. The oxychloride $NdOCl$ is present above 550° according to Wendlandt[4].

References p. 551

4. Neodymium sulphate

The octahydrate of the sulphate $Nd_2(SO_4)_3 \cdot 8H_2O$ begins to decompose at 40°. Two breaks in the curve are observed, one at 85°, the other at 115°, and corresponding respectively to the pentahydrate and the dihydrate. Wendlandt[5] reports that the level of the anhydrous sulphate starts at 290°.

5. Neodymium sulphamate

The amidosulphonate (sulphamate) $Nd(NH_2-SO_3)_3 \cdot 2H_2O$ was discovered by Chrétien and Capestan[6]. It behaves similarly to the corresponding salts of lanthanum and cerium(III), *i.e.* it yields the monohydrate between 125° and approximately 300°, and then the double salt $(NH_4)_2SO_4 \cdot Nd_2(SO_4)_3$ between 400° and 450°. The anhydrous sulphate $Nd_2(SO_4)_3$ is obtained above 520°.

6. Neodymium nitrate

The nitrate $Nd(NO_3)_3 \cdot 6H_2O$ was converted into the oxide Nd_2O_3 by Kato[7] above 720°. The same result is obtained by rapid heating, but at 810° and with no evidence of an intermediate level. When slow heating is employed, the horizontal plateau between 219° and 260° agrees approximately with the anhydrous salt, which is progressively converted into the oxynitrate near 475°. Wendlandt[8] repeated this study with a heating rate of 4.5° per minute. He obtained the anhydrous salt between 290° and 380° but the oxide only appeared at 813°.

7. Neodymium formate

Ambrozhii and Osipova[9] report that the formate loses carbon monoxide between 445° and 455° leaving the oxide Nd_2O_3.

8. Neodymium acetate

Kato[7] heated the acetate and found that it yields the oxide Nd_2O_3 beyond 800°.

9. Neodymium oxalate

Neodymium oxalate is precipitated by adding a 10% solution of oxalic acid to a solution of the sulphate which is made just neutral by adding aqueous ammonia. The suspension is filtered after 12 hours and the precipitate is washed with a hot dilute solution of oxalic acid. The pyrolysis curve descends at first due to the loss of the wash liquid; the decahydrate is present at 69°. It undergoes a progressive loss of water up to about 290°; the trihydrate appears around 129° and the dihydrate near 196°. An essentially horizontal level extends from 290° to 350°; it represents the existence region of the anhydrous oxalate. Accordingly, neodymium can be weighed in this form and within these temperature limits. Carbon monoxide and carbon dioxide are evolved concurrently from 350° to 813°, but there is no indication of the intermediate production of the carbonate as in the case of calcium oxalate. The zone of the oxide Nd_2O_3 begins at 813°. See Duval and Duval[1]. This oxalate had been studied previously by Somiva and Hirano[10], and later by Wendlandt[11] who employed a heating rate of 5.4° per minute. He found the anhydrous salt at 445° and the level of the oxide at 735°.

10. Neodymium cupferronate

The cupferronate was isolated by Wendlandt[12]. It appears to be stable up to about 150°. The oxide Nd_2O_3 is obtained from 500° on.

11. Neodymium neocupferronate

Neodymium neocupferronate was prepared by Wendlandt and Bryant[13]. It apparently decomposes with explosive violence around 250°; the level of the oxide Nd_2O_3 starts at 750°.

12. Neodymium oxinate

Neodymium oxinate $Nd(C_9H_6ON)_3$, prepared by Wendlandt [14], appears to be stable up to 250°. The organic matter decomposes particularly between 330° and 510°, and the level due to the oxide begins at 730°. Apparently this oxinate has not been analyzed and there is no report with supporting data of its being used in analysis.

REFERENCES

[1] T. DUVAL AND C. DUVAL, *Anal. Chim. Acta*, 2 (1948) 226.
[2] W. W. WENDLANDT AND B. LOVE, *Science*, 129 (1959) 842.
[3] A. I. POPOV AND G. E. KNUDSEN, *J. Am. Chem. Soc.*, 76 (1954) 3921.
[4] W. W. WENDLANDT, *J. Inorg. & Nuclear Chem.*, 5 (1957) 118.

⁵ W. W. WENDLANDT, *J. Inorg. & Nuclear Chem.*, 7 (1958) 51.
⁶ A. CHRÉTIEN AND M. CAPESTAN, *Compt. rend.*, 248 (1959) 3177; M. CAPESTAN, *Ann. Chim. (Paris)*, 5 (1960) 222.
⁷ T. KATO, *J. Chem. Soc. Japan.*, 52 (1931) 774.
⁸ W. W. WENDLANDT, *Anal. Chim. Acta*, 15 (1956) 435.
⁹ M. N. AMBROZHII AND Y. A. OSIPOVA, *Zhur. Neorg. Khim.*, 3 (1958) 2716.
¹⁰ T. SOMIVA AND S. HIRANO, *J. Soc. Chem. Ind. Japan, suppl. binding*, 34 (11) (1931) 459.
¹¹ W. W. WENDLANDT, *Anal. Chem.*, 30 (1958) 58; 31 (1959) 408.
¹² W. W. WENDLANDT, *Anal. Chem.*, 27 (1955) 1277.
¹³ W. W. WENDLANDT AND J. M. BRYANT, *Anal. Chim. Acta*, 13 (1955) 550.
¹⁴ W. W. WENDLANDT, *Anal. Chim. Acta*, 15 (1956) 109.

CHAPTER 58

Samarium

This chapter will deal with the pyrolysis of the hydroxide, fluoride, chloride, sulphate, sulphamate, nitrate, formate, oxalate, cupferronate, neocupferronate, and the oxinate.

1. Samarium oxide hydrate

The gelatinous hydrous oxide precipitated by the action of aqueous ammonia on samarium sulphate yields a thermolysis curve which constantly descends until the temperature reaches 813° and then the oxide Sm_2O_3 attains constant weight. However, this curve shows two changes in slope. The first around 65° corresponds approximately to the composition $Sm_2O_3 \cdot 6H_2O$, and the second near 220° to $Sm_2O_3.3H_2O$. There is no indication of the formation of a basic carbonate. See Duval and Duval[1].

2. Samarium fluoride

Samarium fluoride SmF_3 was prepared by Wendlandt and Love[2] as described by Popov and Knudsen[3]. It is converted into the oxyfluoride between 600° and 690°, a change which continues up to around 900°.

3. Samarium chloride

Wendlandt[4] studied samarium chloride $SmCl_3.6H_2O$. It loses water above 90°. The chloride $SmCl_3 \cdot H_2O$ is obtained above 195°. The anhydrous salt is stable from 250° to 325° though contaminated with the oxychloride $SmOCl$. The latter is stable above 505°.

4. Samarium sulphate

The sulphate $Sm_2(SO_4)_3 \cdot 8H_2O$ starts to decompose at 100°. The level of the anhydrous sulphate is obtained at 295°. There are no breaks in the curve that indicate the formation of intermediate hydrates. See Wendlandt[5].

5. Samarium sulphamate

Chrétien and Capestan[6] were the first to report the sulphamate $Sm(NH_2-SO_3)_3 \cdot 2H_2O$. It loses one molecule of water from 130° to 180°, and then a level corresponding to the monohydrate extends from 180° to about 300°. A rapid loss of weight follows and continues to near 400°. The loss then becomes rapid and stops at about 520° with formation of the anhydrous sulphate $Sm_2(SO_4)_3$.

6. Samarium nitrate

The nitrate $Sm(NO_3)_3 \cdot 6H_2O$ is stable only up to 48° and the thermolysis curve descends continuously until the temperature reaches approximately 750° where the horizontal of the oxide Sm_2O_3 appears. Wendlandt[7] reported that he obtained a level between 450° and 490° corresponding to the oxynitrate $SmONO_3$. However, we were not able to record a level of this kind even though we used a very slow rate of heating.

7. Samarium formate

Samarium formate evolves carbon monoxide between 460° and 465° leaving the oxide Sm_2O_3. Note the difference as compared with lanthanum and cerium. See Ambrozhii and Osipova[8].

8. Samarium oxalate

The oxalate produced at 60° by the action of oxalic acid on samarium sulphate is the decahydrate. When the dry salt is heated it loses 4 molecules of water up to 90° and there is a very slight inflection in the curve. A new and more pronounced change of slope near 160° corresponds to the loss of two more molecules of water. The anhydrous oxalate is present only between 344° and 360°, and furthermore the corresponding level is not horizontal. Carbon monoxide and carbon dioxide are given off simultaneously above 360° and up to 800°. There is no indication of the neutral carbonate, but in the neighbourhood of 700° there is an abrupt change in slope which agrees rather well with the existence of the basic carbonate $2Sm_2O_3 \cdot CO_2$. The oxide Sm_2O_3 shows constant weight[1] above 800°. Somiva and Hirano[9] have heated this oxalate in an atmosphere of carbon dioxide. Wendlandt[10] has likewise repeated the investigation of the oxalate and published a thermolysis curve which is identical with that of neodymium oxalate but differing from that obtained by Duval[1]. The level of the anhydrous oxalate extends from 300° to 410°, and the resulting oxide begins to appear at 735° (heating rate: 5.4° per minute).

9. Samarium cupferronate

The cupferronate loses weight from the ordinary temperature on and yields the oxide Sm_2O_3 around 520° (read from the curve). See Wendlandt[11].

10. Samarium neocupferronate

The neocupferronate likewise produces a curve which descends continuously as far as 660° where the horizontal of the oxide begins. See Wendlandt and Bryant[12].

11. Samarium oxinate

Samarium oxinate was prepared by Wendlandt[13]. It is stable up 260°. Most of the combustion of the organic matter occurs between 410° and 500°. The oxide Sm_2O_3 is obtained at 700°.

REFERENCES

[1] T. DUVAL AND C. DUVAL, Anal. Chim. Acta, 2 (1948) 228.
[2] W. W. WENDLANDT AND B. LOVE, Science, 129 (1959) 842.
[3] A. I. POPOV AND G. E. KNUDSEN, J. Am. Chem. Soc., 76 (1954) 3921.
[4] W. W. WENDLANDT, J. Inorg. & Nuclear Chem., 5 (1957) 118.
[5] W. W. WENDLANDT, J. Inorg. & Nuclear Chem., 13 (1960) 51.
[6] A. CHRÉTIEN AND M. CAPESTAN, Compt. rend., 248 (1959) 3177; M. CAPESTAN, Ann. Chim. (Paris), 5 (1960) 222.
[7] W. W. WENDLANDT, Anal. Chim. Acta, 15 (1956) 435.
[8] M. N. AMBROZHII AND Y. A. OSIPOVA, Zhur. Neorg. Khim., 3 (1958) 2716.
[9] T. SOMIVA AND S. HIRANO, J. Soc. Chem. Ind. Japan, Suppl. binding, 34 (11) (1931) 459.
[10] W. W. WENDLANDT, Anal. Chem., 30 (1958) 58; 31 (1959) 408.
[11] W. W. WENDLANDT, Anal. Chem., 27 (1955) 1277.
[12] W. W. WENDLANDT AND J. M. BRYANT, Anal. Chim. Acta, 13 (1955) 530.
[13] W. W. WENDLANDT, Anal. Chim. Acta, 15 (1956) 109.

CHAPTER 59

Europium

The very pure europium oxide which we used for the preparation of the hydroxide, nitrate, and oxalate was kindly furnished by Mr. Felix Trombe, Director of Research at the C.N.R.S. It was part of the specimen which he had employed in 1938 for the isolation of metallic europium.

1. Europium hydroxide

Dupuis and Duval[1] prepared the hydroxide by the action of ammonium hydroxide at 60° on a solution of the nitrate $Eu(NO_3)_3 \cdot 4H_2O$. The suspension was allowed to stand for 2 hours. The curve reveals an initial loss of imbibed water up to 118°. The horizontal which starts at this temperature corresponds rather well to $Eu_2O_3 \cdot 3H_2O$ and extends up to 264°. The decomposition of the latter begins there and proceeds rapidly until the temperature reaches 347° where about 2/3rds of the combined water has disappeared, and then more slowly up to 650° which marks the beginning of the horizontal due to the oxide Eu_2O_3.

2. Europium chloride

The chloride $EuCl_3 \cdot 6H_2O$ was studied by Wendlandt[2]; it gives a level from 50° to 80°. A break in the curve corresponds approximately to $EuCl_3 \cdot 5H_2O$, and then from 225° to 265° a new level agrees almost with $EuOCl.2EuCl_3$. A new decomposition above 265° yields $EuOCl$ near 380°.

3. Europium nitrate

The nitrate $Eu(NO_3)_3 \cdot 4H_2O$ decomposes continuously up to 745° where the level of the oxide begins. A very short oblique level near 409° conforms roughly to the oxynitrate $EuONO_3$. The anhydrous nitrate did not appear on our curves. Wendlandt and Bear[3] repeated this study and give 730–760° as the temperature at which the oxide Eu_2O_3 is obtained.

4. Europium oxalate

The oxalate $Eu_2(C_2O_4)_3 \cdot 10H_2O$ was precipitated in a neutral solution of europium nitrate by adding a dilute aqueous solution of oxalic acid. The curve shows the loss of moisture and also of 5 molecules of water of crystallization by the time the temperature has reached 140° where there is a change in the curve but no horizontal. The rest of the water of crystallization is given off up to 280°. The existence of the anhydrous oxalate is indicated by the horizontal from 280° to 315°. The oxide Eu_2O_3 begins to appear at 695°. An intermediate horizontal between 320° and 400° corresponds to an anhydrous compound whose formula lies between $Eu_2O_3 \cdot CO_2$ and $Eu_2O_3 \cdot 2CO_2$. (This material yields carbonate bands when examined by infra-red spectrography.) The decahydrate was heated by Wendlandt[4] also. He found that the anhydrous oxalate was formed near 320° and the oxide only after 620° is reached. The heating rate was 5.4° per minute.

REFERENCES

[1] T. DUPUIS AND C. DUVAL, *Anal. Chim. Acta*, 3 (1949) 189.
[2] W. W. WENDLANDT, *J. Inorg. & Nuclear Chem.*, 9 (1959) 136.
[3] W. W. WENDLANDT AND J. L. BEAR, *J. Inorg. & Nuclear Chem.*, 12 (1960) 276.
[4] W. W. WENDLANDT, *Anal. Chem.*, 30 (1958) 58.

CHAPTER 60

Gadolinium

Thermolysis studies have been made of the hydroxide, fluoride, chloride, sulphate, nitrate, oxalate, cupferronate, neocupferronate, and the oxinate.

1. Gadolinium hydroxide

The hydroxide, which can be employed for the determination of this element, was precipitated at 70° by adding ammonium hydroxide to a solution of gadolinium nitrate. See Dupuis and Duval[1]. The thermolysis curve of the resulting gelatinous product reveals the departure of the moisture up to 100° followed by a horizontal from 100° to 160° which does not fully correspond to the formula $Gd_2O_3 \cdot 3H_2O$ but rather to a compound of higher molecular weight, which doubtless is a partially carbonated hydrous oxide. The horizontal related to the oxide Gd_2O_3 begins at ɕ50°.

2. Gadolinium fluoride

The anhydrous fluoride GdF_3 was prepared by Wendlandt and Love[2] as described by Popov and Knudsen[3] by precipitation with hydrofluoric acid. A sample weighing 100 mg was heated at the rate of 5.4° per minute. The conversion into the oxide commences between 600° and 690° but is not complete at 900°.

3. Gadolinium chloride

Gadolinium chloride $GdCl_3 \cdot 6H_2O$ loses the greater part of its water above 90°; the monohydrate is obtained at 210° and is stable from 270° to 330°. However, the latter is slightly contaminated with the oxychloride GdOCl, which is present in quantitative amount at 445°. See Wendlandt[4].

4. Gadolinium sulphate

The octahydrate of the sulphate $Gd_2(SO_4)_3 \cdot 8H_2O$ begins to lose water at 110°. The anhydrous salt is obtained around 260°, with no indication of any intermediate product according to Wendlandt[5].

Hall and Markin[6] observed that the octahydrate yields a very short level not going beyond 300°. The anhydrous salt $Gd_2(SO_4)_3$ gives a good level extending from 360° to 740°. Further heating results in the compound $Gd_2O(SO_4)_2$ at 830° and $Gd_2O_2SO_4$ at 880°.

5. Gadolinium nitrate

The thermolysis curve of gadolinium nitrate $Gd(NO_3)_3 \cdot 4H_2O$ descends from room temperature until 410° where the level of the oxide starts. An inflexed level between 160° and 200° agrees approximately with the existence of the anhydrous salt. On repeating this investigation, Wendlandt and Bear[7] announced that the tetrahydrate loses its water from 75° to 80°, and that the anhydrous nitrate produces the oxide Gd_2O_3 between 455° and 480°.

6. Gadolinium oxalate

The oxalate, precipitated from a hot solution for analytical purposes by the action of gadolinium nitrate with oxalic acid, is really a mixture, since the quantity of water retained by the product when dried in the air fluctuates between 9 and 10 molecules. The curve reveals a short level near 92° which corresponds to the hexahydrate. The existence of the tetrahydrate is signaled by an inflection at about 155°. The anhydrous salt is stable only between 300° and 350°; moreover weighing in this form cannot be recommended, since the portion of the curve corresponding to the anhydrous salt is not strictly horizontal. The carbon monoxide and dioxide are not evolved instantaneously; they are lost between 350° and 813° with an abrupt fall near 680°. Gadolinium oxide Gd_2O_3 is produced in the pure condition above 813°. See Dupuis and Duval[1]. Wendlandt[8], who repeated this study, used a decahydrate. He found the hexahydrate near 120°, and obtained the anhydrous oxalate at 315°, and the oxide at 700°, when he employed a heating rate of 5.4° per minute.

7. Gadolinium cupferronate

Gadolinium cupferronate is not stable even at ordinary temperatures. The level of the oxide starts between 500° and 600° so far as can be estimated from the curve published by Wendlandt[9].

References p. 560

8. Gadolinium neocupferronate

The neocupferronate is likewise unstable; it yields the oxide Gd_2O_3 above 490°. See Wendlandt and Bryant[10].

9. Gadolinium oxinate

Gadolinium oxinate is stable up to 250°. The organic matter burns from 400° to 525°. The level due to the oxide begins at 775°. See Wendlandt[11]. The precipitation of gadolinium is not quantitative (error: — 2.5%).

REFERENCES

[1] T. DUPUIS AND C. DUVAL, *Anal. Chim. Acta*, 3 (1949) 438.
[2] W. W. WENDLANDT AND B. LOVE, *Science*, 129 (1959) 842.
[3] A. I. POPOV AND G. E. KNUDSEN, *J. Am. Chem. Soc.*, 76 (1954) 3921.
[4] W. W. WENDLANDT, *J. Inorg. & Nuclear Chem.*, 5 (1957) 118.
[5] W. W. WENDLANDT, *J. Inorg. & Nuclear Chem.*, 7 (1958) 51.
[6] G. R. HALL AND T. L. MARKIN, *J. Inorg. & Nuclear Chem.*, 4 (1957) 137.
[7] W. W. WENDLANDT AND J. L. BEAR, *J. Inorg. & Nuclear Chem.*, 12 (1960) 276.
[8] W. W. WENDLANDT, *Anal. Chem.*, 30 (1958) 58; 31 (1959) 408.
[9] W. W. WENDLANDT. *Anal. Chem.*, 27 (1955) 1277.
[10] W. W. WENDLANDT AND J. M. BRYANT, *Anal. Chim. Acta*, 13 (1955) 550.
[11] W. W. WENDLANDT, *Anal. Chim. Acta*, 15 (1956) 109.

CHAPTER 61

Terbium

Wendlandt has studied the chloride and the oxalate.

1. Terbium chloride

The chloride $TbCl_3.6H_2O$ begins to lose water at 65°. A level extending from 250° to 300° accords with the formula $TbOCl·2TbCl_3$. Only the oxychloride $TbOCl$ remains above 300° and it yields a level at 425°. This oxychloride starts to decompose at 715° and yields the oxide Tb_4O_7. The complete conversion into oxide was not achieved since the operation was terminated at 850°. See Wendlandt[1].

2. Terbium oxalate

The oxalate $Tb_2(C_2O_4)_3·10H_2O$ yields the pentahydrate between 45° and 140°. The latter gives the monohydrate between 140° and 265°. The last molecule of water is lost between 265° and 435°. The oxide Tb_2O_3 begins to appear at 725°. See Wendlandt[2].

REFERENCES

[1] W. W. WENDLANDT, *Anal. Chem.*, 31 (1959) 408.
[2] W. W. WENDLANDT, *J. Inorg. & Nuclear Chem.*, 9 (1959) 136.

CHAPTER 62

Dysprosium

Wendlandt has investigated the fluoride, chloride, sulphate, nitrate, and the oxalate.

1. Dysprosium fluoride

The fluoride $DyF_3 \cdot H_2O$, prepared by Wendlandt and Love[1] in accord with the directions of Popov and Knudsen[2], loses its water between 40° and 60°. No level corresponding to the anhydrous fluoride was observed. The conversion into the oxyfluoride begins between 600° and 690° and is not complete at 900°.

2. Dysprosium chloride

The chloride $DyCl_3 \cdot 6H_2O$ begins to lose water at 90° according to Wendlandt[3]. There is a break in the curve at 190° and another at 235° corresponding to the composition $DyOCl \cdot 2DyCl_3$. The level of the oxychloride $DyOCl$ starts at 390°; pure oxide Dy_2O_3 begins to appear beyond 590°.

3. Dysprosium sulphate

The sulphate $Dy_2(SO_4)_3 \cdot 8H_2O$ decomposes directly with no manifestation of intermediate hydrates. It yields the anhydrous sulphate near 200° when 85–100 mg are heated at the rate of 5.4° per minute, and with a slow current of air passing through the furnace during the pyrolysis. See Wendlandt[4].

4. Dysprosium nitrate

When 80–90 mg of the nitrate $Dy(NO_3)_4 \cdot 4H_2O$ are heated at the rate of 300° per hour, water is given off between 75° and 80°. The anhydrous nitrate yields the oxide Dy_2O_3 between 455° and 480° (temperatures read from the curve). See Wendlandt[5].

5. Dysprosium oxalate

The oxalate $Dy_2(C_2O_4)_3 \cdot 10H_2O$ is converted to the tetrahydrate between 45° and 140°, and to the dihydrate between 140° and 220°. The anhydrous salt is obtained between 295° and 415°, and the oxide Dy_2O_3 between 435° and 725°. See Wendlandt[6].

REFERENCES

[1] W. W. WENDLANDT AND B. LOVE, *Science*, 129 (1959) 842.
[2] A. I. POPOV AND G. E. KNUDSEN, *J. Am. Chem. Soc.*, 76 (1954) 3921.
[3] W. W. WENDLANDT, *J. Inorg. & Nuclear Chem.*, 9 (1959) 138.
[4] W. W. WENDLANDT, *J. Inorg. & Nuclear Chem.*, 7 (1958) 51.
[5] W. W. WENDLANDT AND J. L. BEAR, *J. Inorg. & Nuclear Chem.*, 12 (1960) 276.
[6] W. W. WENDLANDT. *Anal. Chem.*, 31 (1959) 408.

CHAPTER 63

Holmium

Wendlandt has studied the chloride, sulphate, nitrate, and oxalate.

1. Holmium chloride

The chloride $HoCl_3 \cdot 6H_2O$ begins to lose water at 75°. Two breaks were observed in the curve, one at 175°, the other at 220°. The first corresponds to $HoCl_3$, the second to $HoOCl \cdot 2HoCl_3$. The oxychloride $HoOCl$ is produced completely at 305° and the oxide resulting from it appears above 510°. See Wendlandt[1].

2. Holmium sulphate

The octahydrate of the sulphate $Ho_2(SO_4)_3 \cdot 8H_2O$ decomposes above 105° but no intermediate hydrate was observed. The anhydrous product yields a level beyond 240°. See Wendlandt[2].

3. Holmium nitrate

Wendlandt and Bear[3] heated the nitrate $Ho(NO_3)_3 \cdot 4H_2O$ and found that it gives the oxide Ho_2O_3 directly near 600°.

4. Holmium oxalate

The oxalate has the formula $Ho_2(C_2O_4)_3 \cdot 10H_2O$. The water begins to be eliminated at 40°; a level between 200° and 240° corresponds to the dihydrate. The remainder of the water is given off above 240° and the anhydrous salt is present around 400°. Decomposition sets in and the oxalate yields the oxide Ho_2O_3 above 735°. See Wendlandt[4].

REFERENCES

[1] W. W. WENDLANDT, *J. Inorg. & Nuclear Chem.*, 9 (1959) 138.
[2] W. W. WENDLANDT, *J. Inorg. & Nuclear Chem.*, 7 (1958) 51.
[3] W. W. WENDLANDT AND J. L. BEAR, *J. Inorg. & Nuclear Chem.*, 12 (1960) 276.
[4] W. W. WENDLANDT, *Anal. Chem.*, 30 (1958) 58; 31 (1959) 408.

CHAPTER 64

Erbium

Wendlandt made a study of the pyrolysis of a number of erbium salts.

1. Erbium fluoride

The fluoride ErF_3, precipitated by means of hydrofluoric acid[1], loses its water from 315° and 405°. The formation of the oxyfluoride begins in the temperature range 600–690°.

2. Erbium chloride

The chloride $ErCl_3 \cdot 6H_2O$ is the most stable among the chlorides of the rare earths that have been studied. See Wendlandt[2]. The water does not begin to come off until the temperature has been raised to 95°, and breaks in the curve are observed at 175° and 220°. At the latter temperature, the composition is $ErOCl \cdot 2ErCl_3$. The level of the oxychloride begins at 380° and that of the oxide Er_2O_3 at 550°. The corresponding curve is entirely identical with that of the holmium salt.

3. Erbium sulphate

The octahydrate of the sulphate was studied by Wendlandt[3] on his thermobalance. There is a break in the curve at 110° corresponding to the pentahydrate. The level of the anhydrous sulphate starts near 500°.

4. Erbium nitrate

According to Kato[4], erbium nitrate is changed into the oxide Er_2O_3 above 680°. Wendlandt and Bear[5] report 600°.

5. Erbium oxalate

Wendlandt[6] studied the pyrolysis of the oxalate $Er_2(C_2O_4)_3 \cdot 6H_2O$ on his thermobalance. The water of hydration begins to disappear at 40°. The corresponding dihydrate yields a level which extends from 175° to 265°, and the dihydrate gives the anhydrous oxalate near 395°. The compound $Er_2(C_2O_4)_3$ in turn gives the oxide, whose level starts at 720° under linear heating at 5.4° per minute.

REFERENCES

1 W. W. WENDLANDT AND B. LOVE, *Science*, 129 (1959) 842.
2 W. W. WENDLANDT, *J. Inorg. & Nuclear Chem.*, 9 (1959) 138.
3 W. W. WENDLANDT, *J. Inorg. & Nuclear Chem.*, 7 (1958) 51.
4 T. KATO, *J. Chem. Soc. Japan*, 52 (1931) 774.
5 W. W. WENDLANDT AND J. L. BEAR, *J. Inorg. & Nuclear Chem.*, 12 (1960) 276.
6 W. W. WENDLANDT, *Anal. Chem.*, 30 (1958) 58; 31 (1959) 408.

CHAPTER 65

Thulium

Wendlandt has published studies of the chloride, nitrate, and oxalate.

1. Thulium chloride

The initial chloride begins to lose absorbed water at 40°; then a level from 55° to 90° agrees with the composition $TmCl_3 \cdot 6H_2O$. The hexahydrate loses water from 90° on and the curve shows two breaks. The first at 185° corresponds to $TmCl_3$, the other at 225° is due to $TmOCl \cdot 2TmCl_3$. The oxychloride $TmOCl$ is formed at 405° and the level of the subsequent oxide Tm_2O_3 starts at 535°. See Wendlandt[1].

2. Thulium nitrate

The nitrate $Tm(NO_3)_3 \cdot 4H_2O$ produces the oxide Tm_2O_3 near 600°. No intermediate compound was detected when 80–90 mg were heated at the rate of 300° per hour in a slow current of air. See Wendlandt and Bear[2].

3. Thulium oxalate

The oxalate $Tm_2(C_2O_4)_3 \cdot 5H_2O$ yields the dihydrate between 55° and 195°. The latter yields the oxide Tm_2O_3 between 335° and 730°. See Wendlandt[3].

REFERENCES

[1] W. W. WENDLANDT, *J. Inorg. & Nuclear Chem.*, 9 (1959) 138.
[2] W. W. WENDLANDT AND J. L. BEAR, *J. Inorg. & Nuclear Chem.*, 12 (1960) 276.
[3] W. W. WENDLANDT, *Anal. Chem.*, 31 (1959) 408.

CHAPTER 66

Ytterbium

Wendlandt has studied the thermolysis of several ytterbium salts.

1. Ytterbium chloride

Wendlandt[1] reports that the chloride $YbCl_3 \cdot 6H_2O$ begins to lose water at 90°. Two breaks in the curve are observed, one at 135°, and the other at 205° conforming to the composition $YbOCl \cdot 2YbCl_3$. The oxychloride is produced quantitatively at 395°. The decomposition into the oxide Yb_2O_3 begins at 585°.

2. Ytterbium sulphate

According to Wendlandt[2], the sulphate actually has the formula $Yb_2(SO_4)_3 \cdot 11H_2O$. This salt begins to lose water at 40°, and a break in the curve at 140° corresponds approximately to the dihydrate. The level of the anhydrous salt begins at 185°.

3. Ytterbium nitrate

The nitrate $Yb(NO_3)_3$ yields the oxide Yb_2O_3 above 590° according to Kato[3]. Wendlandt and Bear[4] found however that this nitrate first yields the oxynitrate $YbONO_3$ near 390°, and then the oxide Yb_2O_3 beyond 600° (80–90 mg were heated at the rate of 300° per hour).

4. Ytterbium oxalate

The oxalate $Yb_2(C_2O_4)_3 \cdot 5H_2O$ yields the dihydrate between 60° and 175°. This dihydrate yields the oxide Yb_2O_3 between 325° and 730°. See Wendlandt[5].

REFERENCES

[1] W. W. WENDLANDT, *J. Inorg. & Nuclear Chem.*, 9 (1959) 139.
[2] W. W. WENDLANDT, *J. Inorg. & Nuclear Chem.*, 7 (1958) 51.
[3] T. KATO, *J. Chem. Soc. Japan*, 52 (1931) 774.
[4] W. W. WENDLANDT AND J. L. BEAR, *J. Inorg. & Nuclear Chem.*, 12 (1960) 276.
[5] W. W. WENDLANDT, *Anal. Chem.*, 31 (1959) 408.

CHAPTER 67

Lutetium

Wendlandt has published the results of his studies of the chloride, nitrate, and oxalate.

1. Lutetium chloride

The chloride $LuCl_3 \cdot 6H_2O$ begins to lose water near 90°, and then the curve shows two breaks, one at 135° the other at 200°. The latter accords with the formula $LuOCl \cdot 2LuCl_3$. After a new loss of weight above 200°, the level of the oxychloride starts at 390°. The decomposition of the oxychloride to yield the oxide Lu_2O_3 begins at 505°. Complete conversion into this oxide appears not to have been reached. See Wendlandt[1].

2. Lutetium nitrate

The nitrate $Lu(NO_3)_3 \cdot 4H_2O$ produces the oxynitrate $LuONO_3$ at about 380°. The remainder of the nitrogen is then eliminated and the oxide Lu_2O_3 is obtained above 600° (temperatures taken from the curve; 80–90 mg heated at 300° per hour with a slow current of air). See Wendlandt and Bear[2].

3. Lutetium oxalate

The oxalate $Lu_2(C_2O_4)_3 \cdot 6H_2O$ yields the dihydrate between 55° and 190°. The latter produces the oxide between 315° and 715°. See Wendlandt[3].

REFERENCES

[1] W. W. WENDLANDT, *J. Inorg. & Nuclear Chem.*, 9 (1959) 139.
[2] W. W. WENDLANDT AND J. L. BEAR, *J. Inorg. & Nuclear Chem.*, 12 (1960) 376.
[3] W. W. WENDLANDT, *Anal. Chem.*, 31 (1959) 408.

CHAPTER 68

Hafnium

The Chapter Zirconium contains a discussion of the thermolysis curves of the oxychloride of unrefined zirconium and the oxychloride of hafnium-free zirconium. Dautel and Duval[1] have not reproduced all of these curves in the case of pure hafnium, but their study was limited to the six precipitates which they considered as characteristic and which are employed for the determination of hafnium, namely the formation of the hydroxide (by means of ammonium hydroxide and tannin), the selenite (basic and neutral), the mandelate, the *p*-hydroxyphenylarsonate, the phosphate, and the cupferronate. In each instance, Dautel[2] compared the thermolysis curve of the hafnium compound with that obtained with the zirconium compound, the precipitations being made under the same conditions.

The hafnium oxychloride $HfOCl_2 \cdot 8H_2O$ required for this research was obtained by the method of Street and Seaborg[3] employing Dowex-50 resin passed through a 100-mesh sieve.

1. Hafnium hydroxide

(i) Precipitation by aqueous ammonia The precipitate obtained by means of aqueous ammonia undergoes a rapid loss of water up to 199° when subjected to an increasing linear heating at the rate of 300° per hour and with a residual weight of 45.9 mg of hafnium oxide. The loss of weight becomes slower and the water of constitution is lost from around 270° on. The curve shows no discernible feature corresponding to the composition $HfO(OH)_2$ or $Hf(OH)_4$. The level of the oxide HfO_2 starts at 350°. (Under these same conditions it is necessary to carry the temperature up to 400° to obtain the anhydrous zirconium oxide.) Accordingly, this method permits the use of a relatively low temperature, but the lack of selectivity is a great disadvantage.

(ii) Precipitation by gaseous ammonia The precipitation was ac-

complished with a slow stream of air charged with ammonia gas as described by Trombe[4]. The thermolysis curve is very similar to the one just described. The dehydration proceeds slowly up to 350° where the level of the oxide is obtained, at least with the weight noted above. A determination of hafnium conducted by this method showed 5.3 ± 0.5 mg according to the measurements on the chart paper for a quantity of hafnium oxide amounting to 5.272 ± 0.001 mg as measured on the microbalance.

(iii) Precipitation by tannin We have used the Schoeller[5] procedure involving tannin. The curve is entirely identical with that of zirconium hydroxide and is attributed to reduction of the hafnium dioxide followed by a reoxidation.

2. Basic hafnium selenite

The method suggested by Claassen[6] was followed for the precipitation as the basic selenite. The observed level did not correspond to any definite compound, as shown by the calculation and as we had previously discovered for the homologous precipitation of zirconium.

3. Hafnium selenite

However, if the Claassen method is employed and the precipitate of hafnium selenite is digested at 80° for several hours, the crystalline precipitate corresponds closely to the composition $Hf(SeO_3)_2$. This hafnium selenite decomposes at a temperature distinctly lower than that applying to zirconium selenite, namely at 230° as compared with 537°. The method can be used for the determination of hafnium, and in two ways: weighing as the selenite, or as the oxide HfO_2. The following data demonstrate the efficacy of these two procedures:

Weight of hafnium oxide (weighed)   10.060 ± 0.001 mg
Weight of hafnium oxide (measured on the curve)   10.8   ± 0.5   mg
Weight of hafnium selenite (weighed)   22.750 ± 0.001 mg
Weight of hafnium selenite (measured on the curve) 23.8   ± 1   mg

The analytical factor is very favorable: $HfO_2/Hf(SeO_3)_2 = 0.4869$.

4. Hafnium mandelate

By generalizing the method of Kumins[7] and Hahn and Baginsky[8] we obtained a precipitate of hafnium tetramandelate whose composition corresponds to the formula $Hf(C_6H_5-CHOH-CO_2)_4$. It is stable up to 240°. The salt begins to decompose very slowly as far as 263° and then more rapidly, and the horizontal of the oxide commences at 497°. As in the case of zirconium, there is doubtless a slight reduction of this oxide because of the reducing atmosphere prevailing in the furnace. This reagent is very specific and in addition presents an advantageous analytical factor: $HfO_2/Hf(C_8H_7O_3)_4 = 0.2689$.

5. Hafnium p-hydroxyphenylarsonate

The precipitation as the p-hydroxyphenylarsonate was made by the Claassen[9] technique. The precipitate is white and flocculent and the action is highly selective. The thermolysis curves of the p-hydroxyphenylarsonates of hafnium and zirconium are entirely similar; the product does not appear to be a definite compound. In the case of hafnium, the curve slants downward continuously until the temperature reaches 600°; the level due to the oxide begins at 770°. Once more, and from 600° on, there is an increase in weight of the heated compound; this is ascribed to a reoxidation of the hafnium, subsequent to the reduction in the furnace by the arsenious oxide and the carbon monoxide. This is an excellent method for the determination of hafnium. One of our trials yielded the following data:

Weight of hafnium oxide (weighed)   11.85 ± 0.01 mg
Weight of hafnium oxide (measured on the curve) 11.9 ± 0.6 mg

6. Hafnium phosphate

By following the method proposed for zirconium by Stumpfer and Mettelock[10] which employs diammonium phosphate, we obtained a precipitate containing much water, most of which is lost by the time the temperature reaches 135°. The intervening level does not fit any simple formula. A phosphate corresponding to the formula $(HfO)HPO_4$ is obtained starting at 350°; it loses water slowly and then yields the pyrophosphate HfP_2O_7 near 700°. The findings are entirely in accord with those obtained for zirconium. The precision is good as evidenced by the figures we obtained: HfP_2O_7 calculated, 5.221 mg; measured, 5.30 mg. However, it is difficult to recover the hafnium combined with the phosphorus.

References p. 574

7. Hafnium cupferronate

The precipitate obtained with cupferron by the method of Schröder[11] and Ferrari[12] is very unstable, both with hafnium and zirconium. There is a constant decrease in weight which continues as far as 745° for zirconium and up to 670° for hafnium. However, this precipitation is not specific, and it is necessary to weigh the metal as the oxide. But it is possible to conduct the precipitation in a very acidic medium, or if it is desired to use this precipitant despite its shortcomings, it can be employed in an entrainment procedure.

In this same study, Dautel and Duval[1] described two methods of recovering hafnium.

Only the selenite and the mandelate methods were retained by these workers for the automatic determination of hafnium.

REFERENCES

[1] A. DAUTEL AND C. DUVAL, *Anal. Chim. Acta*, 20 (1959) 154.
[2] A. DAUTEL, *Diploma of Higher Studies*, Paris, 1957.
[3] K. STREET AND G. T. SEABORG, *J. Am. Chem. Soc.*, 70 (1948) 4628.
[4] F. TROMBE, *Compt. rend.*, 215 (1942) 539; 216 (1943) 888; 225 (1947) 1156.
[5] W. R. SCHOELLER, *Analyst*, 69 (1944) 259.
[6] A. CLAASSEN, *Z. anal. Chem.*, 117 (1939) 252.
[7] C. A. KUMINS, *Anal. Chem.*, 19 (1947) 376.
[8] R. B. HAHN AND E. S. BAGINSKY, *Anal. Chim. Acta*, 14 (1956) 45.
[9] A. CLAASSEN, *Rec. trav. chim.*, 61 (1942) 299.
[10] R. STUMPFER AND P. METTELOCK, *Bull. Soc. chim. France*, 8 (1947) 674.
[11] K. SCHRÖDER, *Z. anorg. Chem.*, 72 (1911) 89.
[12] F. FERRARI, *Ann. chim. appl.*, 2 (1914) 276.

CHAPTER 69

Tantalum

Tantalum may be determined by weighing the oxide Ta_2O_5 after precipitation with: tartaric acid, hexamethylenetetramine plus pyrogallol, cupferron, or phenylarsonic acid. Tannin seems to be used only for separations. See Doan and Duval[1]. Prior to our investigations, the minimum temperature for obtaining the oxide Ta_2O_5 had been fixed by Kiba and Ikeda[2] at 450°, and at 600° by Tashiro[3]. These temperatures strike us as being too low.

1. Tantalum tartrate

When tartaric acid is employed, the solution is prepared and the precipitation is carried out as directed by Schoeller and Webb[4], and Schoeller and Jahn[5]. The curve descends continuously as far as 894° which marks the beginning of the level of the oxide Ta_2O_5.

2. Tantalum hexamine

The procedure of Alimarin[6] with hexamethylenetetramine and pyrogallol leads to a mixture which decomposes progressively. The level of the oxide Ta_2O_5 starts at 769°.

3. Tantalum cupferronate

When cupferron is used as directed by Pied[7], the curve likewise slants downward. It is necessary to bring the crucible to at least 1000° to obtain complete destruction of the organic matter.

4. Tantalum phenylarsonate

The precipitation by means of phenylarsonic acid, which is due to Alimarin and Frid[8], seems even more unattractive than the preceding method. The curve descends continuously and arsenic is still coming off at 930°. The temperature must reach at least 1000° before constant weight is attained.

REFERENCES

[1] U. M. DOAN AND C. DUVAL, *Anal. Chim. Acta*, 6 (1952) 135.
[2] T. KIBA AND T. IKEDA, *J. Chem. Soc. Japan*, 60 (1939) 911.
[3] T. TASHIRO, *J. Chem. Soc. Japan*, 52 (1931) 727.
[4] W. R. SCHOELLER AND H. W. WEBB, *Analyst*, 54 (1929) 704.
[5] W. R. SCHOELLER AND C. JAHN, *Analyst*, 54 (1929) 320.
[6] I. P. ALIMARIN, *Zavodskaya Lab.*, 13 (1947) 547.
[7] H. PIED, *Compt. rend.*, 179 (1924) 897.
[8] I. P. ALIMARIN AND B. I. FRID, *Zavodskaya Lab.* 7 (1938) 913.

CHAPTER 70

Tungsten

The precipitates studied by De Clercq and Duval[1] with imposed thermal cycle and linear heating have given rise to the following comments:

(a) All of the proposed methods which terminate by weighing the oxide WO_3 are equivalent. They yield this oxide only at a quite high temperature, and consequently it is better to select the reagent which is most readily obtained at a reasonable price.

(b) These methods of precipitation have one feature in common, namely the material obtained is not a definite tungstate. Instead it consists of tungstic acid which is adsorptively retaining a variable amount of the organic precipitant. We are not in agreement with those who claim to have discovered definite compounds in the course of these precipitations with such materials as acridine, oxine, etc.

(c) A good practical method for determining tungsten has not yet been reported.

(d) None of the 23 procedures published thus far for the determination of tungsten is suitable for its automatic determination.

(e) These are the reasons why, since that date, we have focused attention on the determination in the form of the purpureocobaltic tungstate investigated by Dupuis[2].

This chapter will deal with the following determination forms: the blue oxide produced by zinc, stannous chloride, phenylhydrazine; calcium barium, cadmium, lead, mercury, and chloropurpureocobaltic tungstates, and the tungstates of the following amines: cumidine, α-naphthylamine, benzidine, acridine, β-naphthoquinoline, tetrabase, Michler's ketone, and Yoe's reagent (Wolfron), and the tungstates of the alkaloids: quinine, cinchonine, brucine, and totaquine; the complexes with nitron, α-benzoinoxime, Prontosil, tannin plus antipyrine, vanillidene–benzidine, and oxine.

1. Tungsten oxide

(i) Precipitation by zinc The precipitation with zinc in accord with

the directions of Dotreppe[3] yields an indefinite product which loses water rapidly. A level between 311° and 481° satisfies a variable composition of a mixture of blue and yellow oxides in such proportions that the ensemble looks green. Progressive heating of this mixture up to 1150° causes it to take up oxygen and the lower oxides are converted into the golden yellow oxide WO_3.

(ii) Precipitation by stannous chloride   The product obtained with stannous chloride as reducing agent, as suggested by Mdivani[4], yields a curve which is completely analogous to the one just described and its minimum is at the same temperature.

(iii) Precipitation by phenylhydrazine   The green precipitate resulting from the action of phenylhydrazine by the Dotreppe[5] method gives a curve similar to the preceding two. The product must be heated to at least 985° to obtain the yellow oxide WO_3.

2. Calcium tungstate

The precipitation as calcium tungstate is described in Chapter Calcium. It should be remembered that this salt becomes anhydrous beginning at 400°.

3. Barium tungstate

If no precautions are taken when sodium tungstate is treated with a boiling solution of barium chloride as described by Smith and Bradbury[6], the resulting tungstate retains a trace of barium carbonate which shows up on the thermolysis curve as a level extending from 93° to at least 423°. This finding demonstrates the necessity of working without access of air as was suggested by Buscarons-Ubeda and Herrera[7]. If the operation is conducted in the air, it is essential to carry the temperature to 1011° at least to destroy completely the small amount of barium carbonate (around 1/100) retained. The precipitate of the barium tungstate $BaWO_4$ is of course too heavy by an amount equivalent to the barium oxide.

4. Cadmium tungstate

The procedure of Smith and Bradbury[6] yields a precipitate of cadmium tungstate which becomes anhydrous from 290° on.

5. Lead tungstate

If tungsten is determined as lead tungstate as suggested by Bernouilli[8] the results are satisfactory if the precipitation is conducted at pH 6.0–6.5 as advised by Carrière and Berkem[9]. A perfectly horizontal level is found on the curve. It should be recalled that Dupuis[10] found that lead cannot be determined as the tungstate.

6. Mercurous tungstate

Mercurous tungstate (contaminated with mercurous nitrate, the precipitant) was prepared in accord with the directions of Gibbs[11] and Spitzin[12]. The salt decomposes progressively under the action of heat and does not give a level related to the anhydrous compound. Only the oxide WO_3 is left in the crucible at 880°. Yosida[13] reported 680° when a different heating rate was used.

7. Purpureocobaltic tungstate

The purpureocobaltic tungstate, prepared by Dupuis[2], is of the paratungstate type. When heated at the rate of 300° per hour it is stable up to 70°. The precipitate $5[CoCl(NH_3)_5] \cdot 12WO_3$ decomposes rapidly as far as 300° and then more slowly up to 500°. A good horizontal from 500° to 790° corresponds to a mixture of cobalt and tungsten oxides, which may be represented as $^5/_3 Co_3O_4 \cdot 12WO_3$. The curve relating to the metatungstate has precisely the same shape. A series of isotherms at 50°, 55°, 60°, 65°, 70°, and 80° showed that the precipitate may be dried at 70° for two hours without harm. If the compound is washed with ether, the isotherms are always perfectly horizontal straight lines from the beginning.

8. Cumidine tungstate

According to Kafka[14] the precipitate obtained by the action of cumidine is really a cumidine tungstate. However, it seems impossible to isolate it in a pure state. The curve descends continuously until the temperature reaches 570° where the horizontal of the oxide WO_3 begins.

9. α-Naphthylamine tungstate

The product obtained with α-naphthylamine by the procedure given by Tschilikin[15] yields a curve similar to the one just described. A constant weight from 771° on is due to the oxide WO_3.

10. Benzidine tungstate

The precipitate produced with benzidine appears to have little stability near 84° despite an inflection in the curve. It is necessary to heat the product to at least 600° to obtain constant weight of the oxide WO_3. See von Knorre[16].

11. Acridine tungstate

Using acridine and the method of Fidler[17], we obtained a tungstate whose thermolysis curve shows two successive horizontal levels, the first at 58–90°, the second from 200° to 280°. The calculated molecular weights of these levels do not conform to simple formulas and, in particular, not with the formula $5C_{13}H_9N \cdot 8H_2WO_4$ given by Fidler. Consequently it is necessary to decompose this precipitate by heat; the horizontal level of the golden-yellow oxide WO_3 is reached at 812°.

12. β-Naphthoquinoline tungstate

Platunov and Kirillova[18] suggested β-naphthoquinoline as precipitant. It yields a product whose thermolysis curve descends continuously until the temperature reaches 475°. This is the minimum temperature at which tungstic anhydride WO_3 is obtained.

13. Tetrabase tungstate

The precipitate which Papafil and Cernatesco[19] obtained with tetrabase also gives a curve of the same kind. The descent stops at 547°, the temperature which must be reached if the precipitate is to furnish a constant weight of the oxide WO_3.

14. Michler's ketone tungstate

The Kafka[14] procedure with Michler's ketone gives a precipitate which appears to us to be no more clearly defined than the others. Its thermolysis curve slants downward until 732° is reached, but the temperature must be carried to 1017° to reoxidize the blue oxide formed during the reduction.

15. Wolfron tungstate

Yoe and Jones[20] introduced the reagent known as Wolfron which yields a precipitate when added to a solution of sodium tungstate. The thermolysis curve is identical with the one just described; it descends until the temperature reaches 495°. The crucible then contains a mixture of the blue and yellow oxides, which oxidize when heated slowly as indicated by the upward course of the curve beyond this temperature. The heating should be continued up to 1020° for example, as in the preceding case.

16. Quinine tungstate

The method of Lefort[21] was used when quinine was taken as the precipitant. We found no definite tungstate in this case. The curve greatly resembles the preceding ones; it slants downward as far as 721°, and the material must be heated up to 1020° to bring about the complete disappearance of the blue oxide.

17. Cinchonine tungstate

The cinchonine method is regarded as the classic way of determining tungsten. We used the procedure of Fiorentino[22]. The curve obtained with a moist precipitate yielded the following findings. After the moisture has been eliminated, a good horizontal appears, extending from 50° to 105°, but is does not correspond to a definite compound. In fact, with an apparent molecular weight of 321 as shown this horizontal, 1 molecule of WO_3 is not holding even $1/3$ molecule of cinchonine. Therefore the precipitate must be an adsorption complex or it must be partially decomposed when it is washed with hydrochloric acid. Whatever the reason, this level cannot be used for the automatic determination of tungsten. When the heating is continued, a mixture of oxides is obtained at 475° and they reoxidize up to 1014°, the temperature necessary to obtain the yellow oxide WO_3.

18. Brucine tungstate

Grimaldi and Davidson[23] described the precipitation with brucine. The curve shows no level until the horizontal of the oxide WO_3 begins at 818°.

References p. 583

19. Totaquine tungstate

Philipp[24] suggested the use of Totaquine, which is a mixture of cinchona alkaloids. We, in agreement with his findings, observed that the precipitate must be heated to at least 750° to obtain the correct weight of the oxide WO_3.

20. Nitron tungstate

In contrast with the nitrate, perchlorate, and perrhenate, the precipitate obtained by Gutbier and Weise[25] with nitron and tungstic acid has no definite composition, even though it is stable up to 155° while showing a slight gain in weight. The horizontal corresponding to this adsorption product or decomposition product has an apparent molecular weight of 370, assuming 1 molecule of WO_3. Accordingly, as was reported by Gutbier and Weise, the precipitate must be heated to 800° to obtain constant weight of WO_3. We believe that the use of so expensive a reagent as nitron for the determination of tungsten is a needless waste.

21. Benzoinoxime tungstate

Yagoda and Fales[26] employed benzoinoxime but the precipitate is no more definite than those just considered. The curve descends gradually with a very flattened minimum between 425° and 450°. If the heating is continued to at least 961° the resulting mixture of oxides slowly reoxidizes to give WO_3.

22. Prontosil tungstate

The curve obtained with the product precipitated by Prontosil is identical with the preceding one. The procedure used was that of Shalyagin, Fomin and Starostina[27]. The precipitate, which has no definite formula, leaves a mixture of oxides whose weight is constant between 500° and 623°. This mixture slowly takes up oxygen as far as 1011°.

23. Tannin antipyrine tungstate

Moser and Blaustein[28] used tannin plus antipyrine as precipitant. The product likewise has no definite formula. The thermolysis curve reveals a loss of weight up to 540° which marks the minimum. The resulting mixture of oxides must be heated to at least 987° to obtain a correct weight.

24. Vanillidene-benzidine tungstate

The precipitate produced according to Hovorka[29] by means of the vanillidene–benzidine molecular association yields a curve containing a horizontal between 100° and 156°. Its apparent molecular weight of 345 does not agree with any definite compound. When heated to 610°, the organic matter is decomposed and the essentially horizontal level of the oxide WO_3 begins.

25. Oxine tungstate

The procedure employed for the precipitation by oxine is due to Jilek and Rysanek[30] who assign the incorrect formula $WO_2(C_9H_6ON)_2$ to the resulting product. They decomposed this precipitate and weighed the resulting oxide WO_3. The precipitate yields a good horizontal up to 218°. Ishimaru[31] reports this level as extending from 95° to 128° while Borrel and Pâris[32] give from 28° to 288°. It is easy to see that it does not conform to the announced formula; its composition changes from one experiment to another, and the product invariably contains less than one molecule of oxine per molecule of tungstic anhydride. For example, in one of our trials the apparent molecular weight was 435 instead of 503.9 required by the above formula. The oxide WO_3 begins to appear at 674° (Ishimaru gives 420°). The oxine method can therefore not be recommended for the determination of tungsten.

REFERENCES

[1] M. DE CLERCQ AND C. DUVAL, *Anal. Chim. Acta*, 5 (1951) 401.
[2] T. DUPUIS, *Mikrochim. Acta*, (1955) 851.
[3] G. DOTREPPE, *Chim. & Ind. (Paris)*, Special No. 173, March 1931; *C.A.*, 25 (1931) 3269.
[4] B. MDIVANI, *Bull. Soc. chim. France*, 9 (1911) 122.
[5] G. DOTREPPE, *Bull. Soc. chim. Belg.*, 38 (1929) 385.
[6] E. F. SMITH AND R. H. BRADBURY, *Ber.*, 24 (1891) 2931.
[7] F. BUSCARONS UBEDA, S. HERRERA AND E. LORIENTE GONZALES, *Anales soc. españ. fís. y quím.*, 42 (1946) 1139.
[8] F. BERNOUILLI, *Poggendorfs Ann.*, 111 (1860) 573.
[9] E. CARRIÈRE AND R. BERKEM, *Bull. Soc. chim. France*, 5 (1937) 1907.
[10] T. DUPUIS, *Anal. Chim. Acta*, 3 (1949) 663.
[11] W. GIBBS, *Z. anal. Chem.*, 21 (1882) 565.
[12] V. SPITZIN, *Z. anal. Chem.*, 75 (1928) 433.
[13] Y. YOSIDA, *J. Chem. Soc. Japan*, 61 (1940) 130.

[14] E. KAFKA, *Z. anal. Chem.*, 52 (1913) 601.
[15] M. TSCHILIKIN, *Ber.*, 42 (1909) 1302.
[16] G. VON KNORRE, *Ber.*, 38 (1905) 783.
[17] J. FIDLER, *Collection Czechoslov. Chem. Communs.*, 14 (1949) 648.
[18] B. A. PLATUNOV AND N. M. KIRILLOVA, *Khim. Referat. Zhur.*, 4 (1941) 73; *C.A.*, 37 (1943) 4983.
[19] M. PAPAFIL AND R. CERNATESCO, *Ann. sci. univ. Jassy*, 16 (1931) 526.
[20] J. YOE AND A. L. JONES, *Ind. Eng. Chem., Anal. Ed.*, 16 (1944) 45.
[21] J. LEFORT, *Compt. rend.*, 92 (1881) 1461.
[22] G. FIORENTINO, *Giorn. chim. ind. ed appl.*, 3 (1921) 56.
[23] F. S. GRIMALDI AND N. DAVIDSON, *U.S. Geol. Survey, Bull.* No. 950 (1946) 135; *C.A.*, 41 (1947) 49.
[24] P. PHILIPP, *Anais assoc. quím. Brasil*, 6 (1947) 161.
[25] A. GUTBIER AND L. WEISE, *Z. anal. Chem.*, 53 (1914) 426.
[26] H. YAGODA AND A. FALES, *J. Am. Chem. Soc.*, 60 (1938) 640.
[27] V. V. SHALYAGIN, V. V. FOMIN AND V. G. STAROSTINA, *Zavodskaya Lab.*, 13 (1947) 679.
[28] L. MOSER AND W. BLAUSTEIN, *Monatsh.*, 52 (1929) 351.
[29] V. HOVORKA, *Collection Czechoslov. Chem. Communs.*, 10 (1938) 518.
[30] A. JILEK AND A. RYSANEK, *Collection Czechoslov. Chem. Communs.*, 5 (1933) 136.
[31] S. ISHIMARU, *J. Chem. Soc. Japan*, 55 (1934) 201.
[32] M. BORREL AND R. PÂRIS, *Anal. Chim. Acta*, 4 (1950) 267.

CHAPTER 71

Rhenium

The following precipitation forms have been suggested for the gravi-metric determination of rhenium: sulphide, potassium perrhenate, thallium perrhenate, nitron perrhenate, tetraphenylarsonium perrhenate, tetron chlororhenate(IV). Tribalat and Duval[1] have studied the thermolysis of these precipitates.

1. Rhenium sulphide

The sulphide was precipitated in the manner suggested by Geilmann and Lange[2]. The curve reveals the loss of retained water up to 88-90°, where there is a mixture of dry sulphides in different valency states. It is impossible to determine by calculation either the composition of this mixture or of the products formed along the oblique level near 250°. Roasting occurs beyond this temperature and some sublimation, and the crucible is empty at 520°. We agree with a number of workers that this method cannot be recommended for the gravimetric determination of rhenium. Taimni and Tandon[3] reported that the sulphide Re_2S_7, which was prepared by Taimni and Salaria[4] and dried in the air, begins to lose weight at 80–90° and continues to do so up to 520°.

2. Potassium perrhenate

Potassium perrhenate $KReO_4$ is dry at 54°; it starts to decompose at 220°. Though it might be suitable for the determination of potassium, it is without interest for determining rhenium because of the too great solubility of the salt.

3. Thallium perrhenate

Thallium perrhenate $TlReO_4$ was precipitated as described by Krauss and Steinfeld[5]. After the water was removed, the salt proved to be stable and gave a horizontal extending from 90° to 569°. Unfortunately this level cannot be considered for the automatic determination of rhenium because again the solubility of the salt is too high. The rhenium and thallium oxides are driven off smoothly up to around 800°.

4. Nitron perrhenate

Nitron perrhenate was obtained by following the directions of Geilmann and Voigt[6,7] and Hecht and Donau[8]. The salt was found to be stable up to around 288–290°, but there was a slight increase in weight up to this temperature as in the case of nitron nitrate. The decomposition then proceeds in cascade fashion and the crucible is completely empty near 592°. Although this method has been widely employed for ordinary gravimetric purposes it is not suitable for the automatic determination of rhenium.

5. Tetraphenylarsonium perrhenate

Tetraphenylarsonium perrhenate was prepared by the method described by Willard and Smith[9]. When heated slowly (3rd speed) the moisture was lost up to 106°, or up to 126° when the heating was rapid (1st speed). The level of the tetraphenylarsonium perrhenate remains horizontal up to 183–185°, and then there is combination with oxygen up 397°. Decomposition occurs rapidly and the crucible is empty at 566°. We suggest the 126–183° level for the automatic determination of rhenium.

6. Tetron rhenium chloride

The reagent, commonly called tetron, was prepared in the laboratory by treating o-tolidine with methyl iodide. The rhenium was precipitated by this reagent as described by Geilmann and Hurd[10]. The thermolysis curve strikes us as being rather difficult of correct interpretation if the formula is accepted as that of a rhenium(IV) chloride. Although the weight remains quite constant as far as 169°, the corresponding level probably relates to a mixture. We suggest that this method be discarded in the absence of a known structure for the precipitate.

REFERENCES

[1] S. Tribalat and C. Duval, *Anal. Chim. Acta*, 6 (1952) 238.
[2] W. Geilmann and G. Lange, *Z. anal. Chem.*, 126 (1944) 321.

3 I. K. TAIMNI AND S. N. TANDON, *Anal. Chim. Acta*, 22 (1960) 35.
4 I. K. TAIMNI AND G. B. S. SALARIA, *Anal. Chim. Acta*, 12 (1955) 519.
5 F. KRAUSS AND H. STEINFELD, *Z. anorg. Chem.*, 197 (1931) 52.
6 W. GEILMANN AND A. VOIGT, *Z. anorg. Chem.*, 193 (1930) 311.
7 A. VOIGT, *Z. anorg. Chem.*, 249 (1942) 225.
8 F. HECHT AND J. DONAU, *Anorganische Mikrogewichtsanalyse*, Springer, Vienna, 1940, p. 170.
9 H. H. WILLARD AND G. M. SMITH, *Ind. Eng. Chem., Anal. Ed.*, 11 (1939) 305.
10 W. GEILMANN AND L. C. HURD, *Z. anorg. Chem.*, 213 (1933) 336.

CHAPTER 72

Osmium

The only published account dealing with the thermal behaviour of compounds of this element is that of Goto[1]. He reported that the oxide (osmic acid) OsO_4 should not be heated higher than 190° because of the danger of loss through volatilization. We have confirmed this finding.

REFERENCE

[1] H. Goto, *J. Chem. Soc. Japan*, 55 (1934) 326.

CHAPTER 73

Iridium

Iridium may be determined gravimetrically as the metal after precipitation by means of formic acid, or by 2-mercaptobenzothiazole, or as the sulphide. See Duval, Champ and Fauconnier[1].

1. Metallic iridium

The precipitation by means of formic acid is conducted as usual in the presence of ammonium acetate as buffer. Much difficulty is encountered when attempts are made to collect the precipitate, which readily goes into the colloidal state. This is why we have employed the following procedure, which we think is worthy of being recommended. We employ a glass filtering crucible of the type described in Chapter 4 and cover its bottom with quartz wool which retains the colloidal iridium effectively at its center. After careful washing of the quartz and the elementary iridium, the small pellet is removed cautiously and transferred to a crucible which fits the carrier ring of the thermobalance. Of course, a small silica vessel whose bottom has been perforated and lined with silica wool may also be used. Actually, after the heating has been concluded, the operator may treat this fibrous silica with hydrofluoric acid and thus recover the iridium; otherwise it is almost impossible to clean up a glass or a fritted silica crucible. There is no possibility of using a paper filter. Under these circumstance, the iridium precipitated in this manner yielded a perfectly horizontal line when heated to 879° on the thermobalance. Goto[2] reports 500° as suitable. In our opinion, there is no need to complicate this determination by heating and then cooling in hydrogen and nitrogen, as is still the practice in some laboratories.

2. Iridium 2-mercaptobenzothiazole

The voluminous orange precipitate, prepared by means of 2-mercapto-benzothiazole, as described by Barefoot, McDonnell and Beamish[3], does not show any level of constant weight although the curve shows a distinct change of slope between 135° and 250°. This corresponds essentially to 1 atom of iridium to 3.8 molecules of the mercaptobenzothiazole (calculated molecular weight: 837). The organic matter then decomposes and constant weight is obtained at 520° for iridium metal, which holds its weight constant up to 980° at least.

3. Iridium sulphide

Taimni and Salaria[4] used sodium sulphide and obtained a precipitate which was dried at 85°. They claimed its formula to be $Ir_2S_3 \cdot 10H_2O$. We repeated this experiment and found that the thermolysis curve showed, as was to be expected, no level in the vicinity of 85° where there is a sudden drop. The material loses weight from the ordinary temperature on. Slight reoxidation occurs above 410° and may be attributed to a conversion of the sulphide into the sulphate. Another loss of weight then occurs and iridium metal is obtained near 800°. Hence this method cannot be seriously considered.

REFERENCES

[1] C. DUVAL, P. CHAMP AND P. FAUCONNIER, *Anal. Chim. Acta*, 20 (1959) 152.
[2] H. GOTO, *J. Chem. Soc. Japan*, 55 (1934) 326.
[3] R. R. BAREFOOT, W. J. MCDONNELL AND F. E. BEAMISH, *Anal. Chem.*, 23 (1951) 514.
[4] I. K. TAIMNI AND G. B. S. SALARIA, *Anal. Chim. Acta*, 11 (1954) 329.

CHAPTER 74

Platinum

See the section on ammonium chloroplatinate in Chapter Ammonium regarding the detection of the oxidation of platinum in the spongy form.

Proposals have been made to determine platinum gravimetrically after precipitation by means of the following substances: magnesium, hydrogen sulphide and sodium sulphide, an ammonium salt, a potassium salt, a rubidium salt, a cesium salt, a thallium salt, thioformamide, thiophenol, thionalide, thiosemicarbazide, tetraphenylarsonium bromide, α-furildioxime. The corresponding thermolysis curves have been traced by Duval, Champ, and Fauconnier[1] with the photographing recording thermobalance using specimens averaging 200 mg. All of the heatings have been conducted in the air and not in hydrogen, along a linear pattern, and at a heating rate of 300° per hour.

Although we have not been able to duplicate the findings of some workers regarding the methods involving tetraphenylarsonium bromoplatinate and α-furildioxime, we nevertheless are in a position to propose a new gravimetric method employing thiophenol, by analogy with what we have noted, on the other hand, in connection with the determination of gold.

1. Metallic platinum

Among the numerous methods which have been recommended for precipitating platinum in the metallic state, we have followed the procedure described by Scheibler[2]. The neutral solution of the platinum salt is treated with a chip of magnesium. The system is heated and then the excess of magnesium is dissolved in dilute acetic acid. We have the feeling that this dissolution is not total and that a trace of magnesium remains mixed with the precipitated platinum. The thermolysis curve is essentially linear. A slight increase in weight starts at 646° and rises to 1/200th in the vicinity of 900°. It is very desirable that the platinum be dried between 100° and 600°.

References p. 594

Goto[3] points out that the precipitate must be heated to 550° to obtain pure platinum when it has been precipitated by means of mercurous chloride.

2. Platinum sulphide

The procedure of Jackson and Beamish[4] was used for the precipitation as sulphide from a warm solution of a platinum salt free of nitric acid; hydrogen sulphide was the precipitant. Taimni and Salaria[5] succeeded in preparing a definite pentahydrate by means of sodium sulphide, and Salaria[6] prepared a trihydrate. The thermolysis curve of all of these compounds slants downward continuously until the temperature reaches 421° and there is no discernible horizontal related to a definite sulphide. Taimni and Tandon[7] also were unable to discover such levels with the Stanton thermobalance. Above 421° and as far as 954° the curve is perfectly horizontal. The dry starting material no longer corresponds to the formula PtS_2 but contains an excess of sulphur. The method can be used for the automatic determination of platinum but only below 421°. Goto[3] gives 650°.

3. Chloroplatinates

It should be remembered that ammonium chloroplatinate yields platinum between 407° and 538°, and that potassium chloroplatinate is stable from 54° to 270°. Rubidium chloroplatinate is stable between 70° and 674°, cesium chloroplatinate between 200° and 409° and thallium chloroplatinate between 65° and 155°. In this last instance, Goto[3] indicates 250° as the upper limit and gives 700° for obtaining metallic platinum.

4. Bromoplatinic acid

Berg et al.[8] have made a study of the pyrolysis of bromoplatinic acid $H_2[PtBr_6] \cdot 9H_2O$.

5. Platinum thioformamide

The method of Gagliardi and Pietsch[9] was used for the precipitation by means of thioformamide, but no paper pulp was used in the filtration. The precipitate is a mixture which loses weight from the ordinary temperature on with a change of slope between 100° and 153°. A constant weight of platinum is obtained starting at 407°. There is no need to heat at 950° as recommended in the published report. As in the case of ammonium chloroplatinate, the resulting platinum black proved to be very oxidizable above 625°.

Goto[3] pointed out that if the precipitation is made with formic acid, it is necessary to heat the residual platinum above 100°.

6. Platinum thiophenolate

Fresh non-oxidized thiophenol was used as directed by Currah, Mc-Bryde, Cruikshank and Beamish[10]. The green-yellow precipitate behaves precisely like that obtained with gold. The thermolysis curve shows that the product maintains its weight almost constant between 230° and 300° and this level conforms with the formula $Pt(C_6H_5S)_2$ (therefore derived from bivalent platinum) at least within about 1/1000th. The residual platinum does not gain weight when it is heated as far as 960° (analogy with gold). The precipitate of platinum(II) thiophenolate is recommended for the automatic determination of platinum; the analytical factor is 0.4698. As in the case of gold, the homologues of thiophenol, which we tried in order to lower this factor still more, have proved to be less advantageous than thiophenol itself from the stability standpoint and also with respect to the precipitation.

7. Platinum thionalidate

According to Umemura[11], the complex containing 1 atom of platinum and 4 molecules of thionalide is stable up to 170°.

8. Platinum thiosemicarbazide

The material precipitated by means of thiosemicarbazide, as described by Naito, Kinoshita and Hayashi[12], does not appear to have a definite formula. It begins to lose weight at ordinary temperatures and the curve indicates that a constant weight of platinum is only obtained at 437°.

9. Platinum tetraphenylarsonium bromide

The orange precipitate obtained with tetraphenylarsonium bromide as directed by Bode[13] was prepared in amounts containing between 5 and 20 mg of platinum. These products yielded curves in which there was a level of constant weight extending up to around 210°, while the platinum resulting from the decomposition appears between 540° and 550° and it is not reoxidizable. Unfortunately, our trials did not yield any definite formula for the precipitate dried at 110°. Sometimes its weight is greater than that which corresponds to the expected formula $[(C_6H_5)_4As]_2[PtBr_6]$ and it is likely that some of the reagent is retained by adsorption. At other times the weight is below to that which corresponds even to $[(C_6H_5)_4As]_2PtBr$. Accordingly, the method does not impress us as being correct and it offers no advantage over the others if the final material weighed is metallic platinum. Until more information is available, we cannot regard this procedure as being more than a means of separation.

10. Platinum α-furildioxime

The precipitation with α-furildioxime is due to Ogburn[14]. The precipitate obtained does not conform to a definite formula and even though all of the metal is precipitated the product contains much less than 1 gram molecule of the dioxime per gram atom of platinum. The thermolysis curve of the precipitate includes however a horizontal stretch as far as 309°. Sudden decomposition then occurs and metallic platinum is obtained above 370°. The method is good for separation purposes.

REFERENCES

1 C. DUVAL, P. CHAMP AND P. FAUCONNIER, *Anal. Chim. Acta*, 10 (1954) 443.
2 C. SCHEIBLER, *Ber.*, 2 (1869) 295.
3 H. GOTO, *J. Chem. Soc. Japan*, 55 (1934) 326.
4 D. S. JACKSON AND F. E. BEAMISH, *Anal. Chem.*, 22 (1950) 813.
5 I. K. TAIMNI AND G. B. S. SALARIA, *Anal. Chim. Acta*, 11 (1954) 329.
6 G. B. S. SALARIA, *Anal. Chim. Acta*, 17 (1957) 395.
7 I. K. TAIMNI AND S. N. TANDON, *Anal. Chim. Acta*, 22 (1960) 554.
8 L. G. BERG, K. N. MOCHALOV, P. A. KURENKOVA AND N. P. ANOSHINA, *Izvest. Kazan Filiala Akad. Nauk. U.S.S.R.*, (1957) 127; *C.A.*, (1960) 5312.
9 E. GAGLIARDI AND R. PIETSCH, *Monatsh.*, 82 (1951) 656.
10 J. E. CURRAH, W. A. E. MCBRYDE, A. J. CRUIKSHANK AND F. E. BEAMISH, *Ind. Eng. Chem., Anal. Ed.*, 18 (1946) 120.
11 S. UMEMURA, *J. Chem. Soc. Japan*, 61 (1940) 25.
12 T. NAITO, Y. KINOSHITA AND J. HAYASHI, *J. Pharm. Soc. Japan*, 69 (1949) 361.
13 H. BODE, *Z. anal. Chem.*, 133 (1951) 95.
14 S. C. OGBURN JR., *J. Am. Chem. Soc.*, 48 (1926) 2493, 2507.

CHAPTER 75

Gold

A multitude of gravimetric methods have been proposed for determining gold, which is not surprising since reducing agents, including filter paper, decompose all of the gold salts with precipitation of the metal in a more or less finely divided state and of various colors.

Champ, Fauconnier and Duval[1] selected some of the precipitation methods for study and eliminated those which yield a product that does not filter well, or that is adsorbed by the filter paper, or that passes through it. Furthermore, another metal should not be introduced via the precipitant since the gold always retains traces of it.

These workers have traced about twenty thermolysis curves to demonstrate a fundamental property, namely that the gold retains oxygen depending on the state of division of the gold and the nature of the precipitating agent, and hence on the nature of the surrounding atmosphere at the time of the heating. When this uptake of oxygen occurs, the increase in weight may amount to a gain of 1/200th. It is reversible. If the cooling pattern is the same as the heating pattern, the gold releases precisely the quantity of oxygen that it has absorbed (or adsorbed) while the curve was ascending. This phenomenon is not shown by massive gold, in particular analytical equipment (crucibles, electrodes, wires, etc.) or by gold that has been deposited electrolytically. Consult in this connection the paper by Duval[2] dealing with the cleaning of gold-plated or gold objects that have become amalgamated.

All of the experiments have been conducted in the same aluminite crucible, and the gold precipitate was spread out as well as possible in such fashion that the apparent surface, exposed to the air, was constant and equalled 226 mm^2. Table 11 demonstrates the extent of this gain.

This rise in the thermolysis graph obviously precludes the automatic determination of gold by means of the thermobalance, but, as was pointed out above, the phenomenon is not general. From among all of the precipitants tried, we have in fact found that:

References p. 599

TABLE 11

INCREASE IN WEIGHT OF GOLD PRECIPITATES DURING THERMOLYSIS

Temperature t°	Precipitating reagent	Gain at t° (mg)	Weight of Au at 20° (mg)
959	Ferrous sulphate	3.9	400.00
957	Pyrogallol	6.5	256.78
958	Amidol	3.2	327.00
948	Dimethylglyoxime	0.5	427.22
957	Sulphur dioxide	4.0	415.02
957	Morpholine oxalate	4.0	425.32
956	Hydroxyhydroquinone	3.1	319.60
972	Pyrocatechol	0.9	344.50
972	Lead acetate	4.0	215.40
988	Oxalic acid	4.0	239.40

Citarin   Mercaptobenzothiazole
Thiophenol   Dimethylglyoxime
Hydroquinone   Ammonium sulphide
Resorcinol

yield gold which produces an excellent horizontal, at least over a wide range. According to Ishii[3], the sulphide yields gold above 350°. That which was prepared by Taimni and Agarwal[4] gives a level up to 220°, and then loses weight and yields the metal above 280°. See Taimni and Tandon[5]. We judge that the sulphide, the complex with mercaptobenzothiazole, and the complex with thionalide prepared by Umemura[6] are neither stable or definite. Thiophenol, alone among all of the precipitants for gold, yields a definite compound, which is white and stable up to 157°. It can be employed in microanalysis, and its formula C_6H_5SAu is that of aurous thiophenate; the analytical factor $Au/C_6H_5SAu = 0.6436$.

We recommend the use of citarin, thiophenol, and hydroquinone for the automatic determination of gold.

The precipitants used for gold have included:

Sulphur dioxide or alkali hydrogen sulphite	Treadwell[7]
Ferrous sulphate	Treadwell[7], Meineke[8]
Oxalic acid	Treadwell[7], Meineke[8], Magdalena[9]
Citarin	Vanino and Guyot[10]
Hydroquinone	Beamish, Russell and Seath[11]
Morpholine oxalate	Malowan[12]
Dimethylglyoxime	Thomson, Beamish and Scott[13]
Tetraethylammonium chloride	Maynard[14]

Thiophenol	Currah, McBryde, Cruikshank and Beamish[15]
Lead acetate+zinc	Wogrinz[16]
Mercaptobenzothiazole	Spacu and Kuraš[17]
Antimony trichloride	Levol[18]
Magnesium	Scheibler[19]
Saccharose	Leidler[20]
Alkaline hydrogen peroxide	Vanino and Seemann[21]
Palladium saturated with hydrogen	Kritschewsky[22]
Alkaline formaldehyde	Vanino[23]
Potassium nitrite	Jameson[24]
Copper and iron sulphides	Godshall[25]
Chlorine hydrate+sodium hydroxide	Dirvell[26]
Hypophosphorous acid	Hart[27]; Treubert[28]
Hydroxylammonium chloride	Lainer[29]
Acetylene	Makowka[30]

With hydroquinone as a model, Beamish, Russell and Seath[11] proposed the following as precipitants, but they gave no details of the procedures, and only a single analytical figure. We have not been able to filter off the precipitate obtained with amidol.

Rhodinol	Hydroxyhydroquinone	Phloroglucinol
Amidol	o-Phenylenediamine	o-Aminophenol
Resorcinol	Pyrocatechol	Phenol
Pyrogallol	Photol	

Pyrolysis of gold

We do not claim to reproduce all of the curves because they are too analogous; however we do wish to make a few remarks about those, which may serve as type examples (Fig.79).

Curve (1) precipitated by citarin The precipitation was made by means of citarin as suggested by Vanino and Guyot[10]. We observed a perfectly horizontal stretch between 45° and 965°.

Curve (2) precipitated by pyrogallol The precipitation was conducted with pyrogallol as proposed by Beamish, Russell and Seath[11]. The gold is dry immediately and weighed:

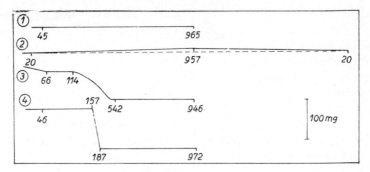

Fig. 79. Pyrolysis curves of gold precipitated by: (1) citarin, (2) pyrogallol, (3) resorcinol, (4) thiophenol.

257.63 mg at 20°
264.13 mg after heating at 957°
256.78 mg after cooling at 20°

The ensemble of the two curves traced in this way on the thermobalance shows the relatively greatest uptake of oxygen that we have recorded.

Curve (3) precipitated by resorcinol   The precipitant was resorcinol[11]. The precipitated gold occluded some of the reagent (56 mg per 434 mg in one of our experiments). The curve revealed the decomposition of the organic matter as far as 510°, with a horizontal level between 66° and 114°, which however did not correspond to a definite compound. The gold formed above 510° did not take up oxygen; the level is perfectly horizontal.

Curve (4) precipitated by thiophenol   Thiophenol was the precipitant as suggested by Currah, McBryde, Cruikshank and Beamish[15]. The white product became yellow after standing in the light for some time. It dries at once and maintains a constant weight up to 157°. Its composition agrees within 1/200th at least with C_6H_5AuS. The decomposition occurs between 157° and 187° and is accompanied by a rather disagreeable odour reminiscent of burning rubber. The residual gold does not take up oxygen. The two parallel levels may be used for the automatic determination of gold.

In one of their experiments, Kiba and Ikeda[31] claimed that the level of the precipitated gold was not obtained until the temperature reached 700°.

REFERENCES

[1] P. CHAMP, P. FAUCONNIER AND C. DUVAL, *Anal. Chim. Acta*, 5 (1951) 277.
[2] C. DUVAL, *Mikrochim. Acta*, (1956) 1433.
[3] K. ISHII, *J. Chem. Soc. Japan*, 52 (1931) 167.
[4] I. K. TAIMNI AND R. P. AGARWAL, *Anal. Chim. Acta*, 10 (1954) 312.
[5] I. K. TAIMNI AND S. N. TANDON, *Anal. Chim. Acta*, 22 (1960) 554.
[6] T. UMEMURA, *J. Chem. Soc. Japan*, 61 (1940) 25.
[7] F. P. TREADWELL, *Manuel de chimie analytique*, 2nd French Ed., Dunod, Paris, 1920, Vol. 2, p. 237.
[8] R. MEINEKE, *Lehrbuch der chemischen Analyse*, 2nd Ed., 1889, p. 197
[9] V. A. MAGDALENA, *Anales asoc quim. y farm. Uruguay*, 50 (1948) 27.
[10] L. VANINO AND O. GUYOT, *Arch. Pharm.*, 264 (1926) 98.
[11] F. E. BEAMISH, J. J. RUSSELL AND J. SEATH, *Ind. Eng. Chem.*, 9 (1937) 174.
[12] L. S. MALOWAN, *Mikrochemie*, 35 (1950) 104.
[13] S. O. THOMSON, F. E. BEAMISH AND M. SCOTT, *Ind. Eng. Chem.*, 9 (1937) 420.
[14] J. L. MAYNARD, *Ind. Eng. Chem.*, 8 (1936) 368.
[15] J. E. CURRAH, W. A. E. MCBRYDE, A. J. CRUIKSHANK AND F. E. BEAMISH, *Ind. Eng. Chem.*, 18 (1946) 120.
[16] A. WOGRINZ, *Z. anal. Chem.*, 108 (1937) 266.
[17] G. SPACU AND M. KURAŠ, *Z. anal. Chem.*, 104 (1936) 88.
[18] A. LEVOL, *Ann. Chim. et Phys.*, 30 (1850) 356.
[19] C. SCHEIBLER, *Ber.*, 2 (1869) 295.
[20] P. LEIDLER, *Chem. Zentr.* (1907 IV) 1867.
[21] L. VANINO AND L. SEEMANN, *Ber.*, 32 (1899) 1968.
[22] L. KRITCHEWSKY, *Z. anal. Chem.*, 25 (1886) 374; *Thesis*, Berne, 1885.
[23] L. VANINO, *Ber.*, 31 (1898) 1763.
[24] P. E. JAMESON, *J. Am. Chem. Soc.*, 27 (1905) 1444.
[25] GODSHALL, *Z. anal. Chem.*, 41 (1902) 307.
[26] P. J. DIRVELL, after SILVA, *Bull. Soc. chim. France*, 46 (1886) 806.
[27] HART, *Thesis*, 1906, after RÜDISÜLE, *Nachweis, Bestimmung und Trennung der chemischen Elemente*, Berne, 1913, Vol. 2.
[28] TREUBERT, *Thesis*, 1909, after Rüdisüle, ibid.
[29] A. LAINER, *Polytech. J.*, 284 (1892) 18.
[30] O. MAKOWKA, *Z. anal. Chem.*, 46 (1907) 149.
[31] T. KIBA AND T. IKEDA, *J. Chem. Soc. Japan*, 60 (1939) 911.

CHAPTER 76

Mercury

About twenty methods have been proposed for obtaining mercury in the metallic state and then condensing, collecting, and weighing it. Of these, we have selected only four procedures to serve as models, namely: electrolysis, reaction with potassium iodide and sulphuric acid, with hydrazine, and with hypophosphorous acid. See Duval and Dat Xuong[1].

The precipitation of mercuric chloride by benzidine and tolidine appears, according to Herzog[2], to be more suitable for determining these two bases. Consequently we will not include these reagents here with regard to mercury. Along this same line, Hovorka and Sykora[3] have investigated the precipitation of various metal salts by means of 5-methyl-3-isonitrosopyrazalone but it is difficult to discover from their paper whether they employed this reagent for the gravimetric determination of mercury. Finally, the microchemical determination of mercury vapours in air, as described by Moldawski[4], by absorption in bromine and weighing the resulting mercuric bromide after drying in the cold, has also been excluded from the present discussion. Therefore, only the following will be taken up here with respect to their thermolysis behaviour: mercury, mercuric nitrate and oxide, mercuric iodide, mercurous iodate, mercuric periodate, artificial mercuric sulphides and cinnabar, mercuric phosphate, mercuric arsenate, mercuric thiocyanate, mercurous chromate, dimercuriammonium chromate, mercuric dipyridine bichromate, the reineckate, vanadate, molybdate, mercurous tungstate, tetramminecadmium triiodomercurate, copper diethylenediamine iodomercurate(II), copper dipropylenediamine iodomercurate(II), mercurous oxalate, mercurous dimedone, mercuric anthranilate, mercuric pyridine chloride, the complex with dithiane, copper dibiguanide iodomercurate(II), the cupferronate, and the complexes with thionalide, 2-phenyloxine, 2-chloro-7-methoxy-5-thiolacridine.

In view of the volatility of mercury and its derivatives, and despite the low stability of the latter, it is rather remarkable that all of the temperatures given in this chapter are comparatively low as compared with those given by the other elements.

1. Metallic mercury

(i) Precipitation by electrolysis The method of Verdino[5] was used for depositing mercury electrolytically on a platinum cathode which had been previously plated with gold. The mercury begins to volatilize between 69° and 71°. A break in the curve is always observed at 165°; it corresponds to a slowing down of the evaporation. The gold is perfectly free of mercury at 261° and has regained its initial weight. Duval[6] has suggested that this finding be applied in the restoration of gold jewelry etc. which has accidentally been amalgamated.

(ii) Precipitation by zinc, potassium iodide and sulphuric acid François[7] used zinc, potassium iodide and sulphuric acid for the precipitation of mercury, and Douris[8] has also applied this procedure to the cyanide. The curve shows that the loss begins at 72°, and at 120° it amounts to 16 mg for an initial weight of 121 mg.

(iii) Precipitation by hydrazine Very finely divided black mercury was obtained when hydrazine was used as the reductant; this method is due to Duccini[9]. Water was lost in a smooth fashion up to 48° and the mercury collects into a shiny globule, which maintains its weight only up to 55°. This temperature should not be exceeded in this type of determination. We have also found the following losses by means of the curve: 5.6 mg at 94°, and 13.2 mg at 118°, for an initial weight of 126.3 mg.

(iv) Precipitation by hypophosphorous acid Winkler[10] proposed the use of hypophosphorous acid. The mercury is completely agglomerated at 71° and begins to volatilize. An initial weight of 127 mg showed a loss of 14 mg at 148°.

2. Mercuric nitrate

De Marignac[11] suggested that the oxide derived from the decomposition of mercuric nitrate be weighed. The method appears to have no present value for analysis, but we were interested in it from the inorganic point of view. When heated slowly, the nitrate is dry at 40° though it is very deliquescent. It is stable only up to 56° and the decomposition is essentially complete at 89°. The horizontal between 100° and 200° is due to oxide HgO. This oxide decomposes in accord with a reaction that has been well known for many centuries. The reaction stops abruptly at 540° since the crucible is empty.

3. Mercurous chloride

Mercurous chloride may be weighed after being dried up to 130°. Some authors give 100–105°, whereas others advise drying in the cold in a desiccator. The course of the sublimation is given in Chapter Chlorine. This chloride may be employed for determining chlorine, mercury, and in an indirect manner phosphites. We recommend the method of Winkler[10].

4. Mercuric iodide

Mercuric iodide was prepared by the procedure given by Liversedge[12] or by Pjankow[13]; the method was criticized by Wenger and Cimerman[14]. The salt should be dried at 40° in an oven according to these workers. The curve indicated to us that the precipitate is usually dry at 45°. The excellent horizontal which begins there extends to 88°. Sublimation sets in then and proceeds rapidly. We suggest that this level be used for the automatic determination of mercury.

5. Mercurous iodate

The method of Spacu and Spacu[15] was followed for the preparation of mercurous iodate $Hg_2(IO_3)_2$ by means of potassium iodate. They suggest that the salt be dried in vacuo at a low temperature. The curve shows that the material is stable up to 175° and gives a perfectly horizontal level which would be well suited to the automatic determination of mercury. The decomposition which follows takes place in four steps and is complete at 642° where the crucible is entirely empty. The inclined stretch between 230° and 449° probably corresponds to mercuric iodide; the latter sublimes and decomposes.

6. Mercuric periodate

The periodate, whose formula is alleged to be $Hg_5(IO_6)_2$, was prepared by the method of Willard and Thomson[16] who advocate drying for 3 hours at 100°. Actually, the curve descends continuously until the temperature reaches 650° where the crucible is empty. No horizontal was observed and consequently we recommend that the method be discarded.

7. Mercuric sulphide

(i) Precipitation by ammonium sulphide Various workers have studied the sulphide precipitated by means of ammonium sulphide; they recommend that the material be dried at $100°$, at $105°$, at $110°$, or at $115°$. Some have stated that there is no change in weight after the preparation has been heated for 2 hours at $130°$. The curve produced by this precipitate after it has been dried in the air contains a horizontal up to $109°$. Beyond this temperature the decomposition proceeds slowly as far as $210°$ and then abruptly accelerates to around $300–310°$; the crucible is empty at $478°$. According to Taimni and Tandon[17], the sulphide prepared by Taimni and Salaria[18] by means of sodium sulphide yields a horizontal as far as $300°$. Decomposition starts at $300–310°$ and the crucible is empty at $500°$.

(ii) Precipitation by sodium thiosulphate When the method described by Cattelain[19] was used, *i.e.* with sodium thiosulphate, a sulphide precipitates which he suggested drying at $100°$. However, the product is more stable than the sulphide prepared with ammonium sulphide since the decomposition does not begin until the temperature reaches $220°$. The horizontal extending from $75°$ and $230°$ may be used for the automatic determination of mercury.

(iii) Cinnabar Saito[20] heated a number of specimens of cinnabar in the air at the rate of $2°$ per minute. The mineral darkens at $200°$ and begins to volatilize and decompose. The formation of mercury starts at $388°$ when the grain-size is 0.10 mm and at $420°$ with grains whose size is 0.20 mm. The reaction producing mercury appears to be over near $450°$.

8. Mercuric phosphate

The phosphate $Hg_3(PO_4)_2$ was studied by Mehta and Patel[21]. It decomposes at $680°$ and produces the pyrophosphate $Hg_2P_2O_7$. The loss of oxygen is rapid at $630°$, which is the decomposition temperature of HgO. Some mercury is deposited on the cooler parts of the furnace.

9. Mercuric arsenate

The precipitate resulting from the reaction between mercuric nitrate and disodium arsenate should be dried between 95° and 100°, according to Pretzfeld[22], who suggested that this compound be used for the gravimetric determination of mercury; however the quantitative results have been strongly contested by Wenger and Cimerman[14]. The curve indicates that this arsenate $Hg_3(AsO_4)_2$ is stable up to 418°; but even though the weight is the same at 418° as at 45°, there is a variation in weight between these temperatures with a maximum near 151°. This variation is of the order of 0.8 mg per 171 mg of the precipitate. Beyond 418° there is a loss of mercury and then of arsenious anhydride, with an abrupt change in slope at 878°. The crucible is completely empty at 945°.

10. Cobalt mercurithiocyanate

As was pointed out in Chapter Cobalt, the precipitate of cobalt mercurithiocyanate, prepared as directed by Lamure[23], yields a horizontal extending from 50° to 200°. This level may be recommended for use in the automatic determination of cobalt or mercury.

11. Zinc mercurithiocyanate

Zinc mercurithiocyanate was precipitated as described by Jamieson[24]. It should be recalled that this complex salt is stable up to 270°, but since the level is not perfectly horizontal it is difficult to use it for determination purposes.

12. Mercurous chromate

Mercurous chromate Hg_2CrO_4, as was stated in Chapter Chromium, is perfectly stable from 52° to 256°. It loses mercury and oxygen from 256° to 671°; the residue consists of chromic oxide Cr_2O_3.

13. Dimercuriammonium chromate

The procedure followed for the preparation of dimercuriammonium chromate $(NHg_2)_2CrO_4 \cdot 2H_2O$ was that of Litterschied[25] who directed that the product be kept at 100° until constant weight is reached. Our study of the curve showed that this method of determining mercury cannot possibly be correct because the material suffers a continuous decrease in weight as far as 680°, where the level of the green oxide Cr_2O_3 begins. The curve undergoes two abrupt changes in direction, one at about 410°, the other near 529°. In our opinion, the method should be discarded.

14. Mercuric dipyridine bichromate

Mercuric dipyridine bichromate $[Hg(C_5H_5N)_2]Cr_2O_7$ was precipitated as directed by Spacu and Dick[26]. They recommend that the product be kept in a desiccator in the cold. The curve has a very short level from 56° to 66° which agrees well with the formula just given. Decomposition sets in immediately, and at 320° the crucible contains nothing except the green oxide Cr_2O_3. We suggest that this method be employed for the automatic determination of mercury; the precipitate washes and filters very well.

15. Mercuric Reineckate

The pink precipitate $Hg[Cr(SCN)_4(NH_3)_2]_2$ obtained with Reinecke's salt was prepared as described by Mahr[27] and repeated by Krupenio[28]. These workers prescribe that the product be dried at 105°-110° before it is weighed. Inspection of the curve confirms these directions; after the moisture is driven off, the curve shows a horizontal between 77° and 158°, and this level is suitable not only for the gravimetric determination of mercury but also for its automatic determination. Decomposition proceeds rapidly beyond 158°, particularly between 350° and 400°. At 749° the crucible contains only the green oxide Cr_2O_3, which may be weighed as a check.

16. Mercurous molybdate

As was pointed out in Chapter Molybdenum, the level extending from 75° to 135° does not relate to pure mercurous molybdate but to a mixture of the latter with basic mercuric nitrate.

17. Mercurous vanadate

Likewise, in Chapter Vanadium, it was noted that mercurous vanadate has no definite formula. It starts to lose water, mercury, and nitric acid at 60°. The method is very precise for determining vanadium in the form of the oxide V_2O_5, but it seems hardly suitable for the determination of mercury.

18. Mercurous tungstate

In Chapter Tungsten, it was pointed out that the thermolysis curve of mercurous tungstate descends without interruption until the temperature reaches 880°. From then on, the crucible contains only the oxide WO_3.

19. Tetramminecadmium triiodomercurate

It was stated in Chapter Cadmium that the complex tetramminecadmium triiodomercurate $[Cd(NH_3)_4][HgI_3]_2$ is stable up to 69°. It appears to be suitable only for the determination of cadmium.

20. Copper diethylenediamine iodomercurate

The complex salt copper diethylenediamine iodomercurate(II), whose formula may be written $[Cuen_2][HgI_4]$, was prepared as directed by Spacu and Suciu[29]. It should be dried in the cold in a desiccator under reduced pressure. The reasons have been given in Chapter Copper.

21. Copper propylenediamine iodomercurate

Likewise, the precipitate whose formula is $[Cupn_2][HgI_4]$, prepared by means of propylenediamine, was obtained as described by Spacu and Spacu[30]. It too has been discussed in the Chapter Copper. It should be noted once more that this product is stable up to 157° and that the horizontal corresponding to it may be employed for the automatic determination of either copper or mercury.

22. Mercurous oxalate

The method of Peters[31] and Spacu and Spacu[30] was employed for precipitating the oxalate $Hg_2C_2O_4$. After losing its moisture, it explodes with extreme violence at 104° where it is completely dry. Accordingly, it is necessary to dry it in the cold in a desiccator or at least below 100° with careful control of the temperature.

23. Mercurous dimedone

Dimedone, which is 5,5-dimethyl-1,3-cyclohexanedione, is a specific reagent for monovalent mercury and yields a characteristic blackish

brown precipitate. See Duval and Wadier[32]. When 200 mg of the precipitate was heated at the rate of 200° per hour, it maintained a constant
weight up to around 110°, where there is a loss of free mercury, which
comes off up to near 190°. The weight remains constant from 190° to 220°
and corresponds to the mercury dimedone compound which is white at
this point.

24. Mercuric anthranilate

The precipitation of the anthranilate $Hg(C_7H_6O_2N)_2$ was carried out
as described by Funk and Römer[33] who suggest that the product be
dried for 30 minutes at 105–110°. The result is in accord with the thermolysis curve, which shows that this anthranilate does not decompose below
113°. However, the initial level is not perfectly horizontal; the precipitate
begins by gaining weight as far as around 96° (which excludes it from
being used for the automatic determination). The uptake of oxygen is
slight and amounts to about 1 mg per 268 mg. Decomposition sets in
suddenly then, and mercury is given off from 113° to 159°. Kiba and
Sato[34] reported however that the salt is stable from 100° to 160°. Beyond
160°, the organic matter gradually is consumed and the crucible is found
to be completely empty at 639°.

25. Mercuric pyridine chloride

The precipitation and weighing of mercury as the mercuric pyridine
chloride $Hg(C_5H_5N)Cl$ cannot be advised. See Ryazanov and Pyshcheva[35]. Actually, the curve descends without interruption until the
temperature reaches 204°, where the crucible is completely empty.

26. Mercuric dithiane

The method of Schroyer and Jackman[36] was used for precipitating
divalent mercury with dithiane (diethylene disulphide). The reagent was
synthesized in our laboratory by the procedure of Bouknight and Smith[37].
The precipitate $Hg(C_4H_8S_2)Cl_2$ is stable only up to 103°; it then decomposes rapidly and the crucible is completely empty at 220°.

27. Copper biguanide iodomercurate

The pink precipitate of copper dibiguanide iodomercurate(II), prepared by the procedure of Bhadury and Rây[38], loses a slight amount of moisture and then yields a horizontal from 60° to 175°, which is excellent for the automatic determination of mercury. The reagent is readily obtained and the precipitate washes easily. Active decomposition sets in beyond 175°, and at 552° the crucible contains nothing but the black copper oxide CuO. However, we do not agree with the above workers, who claim that the precipitate loses one molecule of water at 105°.

28. Mercury cupferronate

When cupferron is employed as the precipitant, the method, as given by Pinkus and Katzenstein[39], terminates with an electrolytic determination. We have attempted to heat the cupferronate in order to obtain a level in the curve but the product decomposes before it is thoroughly dry.

29. Mercury thionalide

The method of Berg and Roebling[40] was used to obtain the thionalide compound $Hg(C_{12}H_{10}ONS)_2$, which filters very well. The thermolysis curve shows that this complex is stable between 90° and 169°; it yields a horizontal level which is suitable for the automatic determination of mercury. Therefore the temperature recommended by these workers is correct. The organic matter breaks down above 169°; the mercury escapes, and the crucible is empty at 749°.

30. Mercury phenyloxinate

The yellow phenyloxinate $Hg(C_6H_5-C_9H_5ON)_2 \cdot H_2O$ begins to lose water at 45°. See Bocquet and Pâris[41]. The anhydrous compound is orange; it is stable up to 275°.

31. Mercury 2-chloro-7-methoxy-5-thiolacridine

We suggest that the method employing 2-chloro-7-methoxy-5-thiolacridine be abandoned because the curve given by the resulting precipitate, as produced by Das-Gupta[42], descends continuously. The reagent is difficult to obtain, and we are not sure that the method recommended for its synthesis leads to the given composition.

REFERENCES

[1] C. Duval and Ng Dat Xuong, *Anal. Chim. Acta*, 5 (1951) 494.
[2] W. Herzog, *Chemiker Ztg.*, 50 (1926) 642.
[3] H. Hovorka and V. Sykora, *Collection Czechoslov. Chem. Communs.*, 11 (1939) 124.
[4] B. L. Moldawski, *Chem. Zentr.*, 102 (1931-I) 1644.
[5] A. Verdino, *Mikrochemie*, 6 (1928) 5.
[6] C. Duval, *Mikrochim. Acta*, (1956) 1433.
[7] M. François, *J. pharm. chim.*, 18 (1918) 129; 21 (1920) 85.
[8] R. G. Douris, *Ann. chim. anal.*, 23 (1941) 238.
[9] C. Duccini, *Gazz. chim. ital.*, 43 (1913) 693.
[10] L. W. Winkler, *Z. anal. Chem.*, 64 (1914) 262.
[11] C. de Marignac, *Ann.*, 72 (1849) 61.
[12] S. G. Liversedge, *Analyst*, 33 (1908) 217.
[13] W. A. Pjankow, *Chem. Zentr.*, 108 (1937-I) 3523.
[14] P.Wenger and C. Cimerman, *Helv. Chim. Acta*, 14 (1931) 718.
[15] G. Spacu and P. Spacu, *Z. anal. Chem.*, 96 (1934) 30.
[16] H. H. Willard and J. J. Thomson, *Ind. Eng. Chem., Anal. Ed.*, 3 (1931) 398.
[17] I. K. Taimni and S. N. Tandon, *Anal. Chim. Acta*, 22 (1960) 35.
[18] I. K. Taimni and G. B. S. Salaria, *Anal. Chim. Acta*, 12 (1955) 519.
[19] E. Cattelain, *J. pharm. chim.*, 22 (1935) 454.
[20] H. Saito, *Sci. Repts. Tôhoku Imp. Univ.*, 16 (1927) 137.
[21] S. M. Mehta and N. B. Patel, *J. Univ. Bombay*, A., 3 (1951) 82; *C.A.*, (1953) 992
[22] C. J. Pretzfeld, *J. Am. Chem. Soc.*, 25 (1903) 198.
[23] J. Lamure, *Bull. Soc. chim. France*, (1946) 661.
[24] G. S. Jamieson, *Ind. Eng. Chem., Anal. Ed.*, 11 (1919) 296.
[25] F. M. Litterschied, *Arch. Pharm.*, 241 (1903) 306.
[26] G. Spacu and J. Dick, *Z. anal. Chem.*, 76 (1929) 273.
[27] C. Mahr, *Z. anal. Chem.*, 104 (1936) 241.
[28] N. S. Krupenio, *Zavodskaya Lab.*, 7 (1938) 161.
[29] G. Spacu and G. Suciu, *Z. anal. Chem.*, 78 (1929) 329.
[30] G. Spacu and P. Spacu, *Z. anal. Chem.*, 89 (1932) 187.
[31] C. A. Peters, *Am. J. Sci.*, 9 (1900) 401; *Z. anorg. Chem.*, 24 (1900) 402.
[32] C. Duval and C. Wadier, *Compt. rend.*, 240 (1955) 433.
[33] H. Funk and F. Römer, *Z. anal. Chem.*, 101 (1935) 85.
[34] T. Kiba and S. Sato, *J. Chem. Soc. Japan*, 61 (1940) 133.
[35] I. P. Ryazanov and M. V. Pyshcheva, *C.A.*, 35 (1941) 6892.
[36] J. B. Schroyer and R. M. Jackman, *J. Chem. Educ.*, 24 (1947) 146.
[37] J. W. Bouknight and G. Mc P. Smith, *J. Am. Chem. Soc.*, 61 (1939) 28.
[38] A. Bhadury and P. Rây, *J. Indian Chem. Soc.*, 22 (1945) 229.
[39] A. Pinkus and M. Katzenstein, *Bull. Soc. chim. Belg.*, 39 (1930) 179.
[40] R. Berg and W. Roebling, *Angew. Chem.*, 48 (1935) 597.
[41] G. Bocquet and R. Pâris, *Anal. Chim. Acta*, 14 (1956) 201.
[42] S. J. Das-Gupta, *J. Indian Chem. Soc.*, 18 (1941) 43.

CHAPTER 77

Thallium

If we exclude the electrolytic determination of thallium collected at the cathode, or in its elementary form, or in the form of alloys, and the indirect determination through the weighing of gold, there remain the following gravimetric procedures, and even some of these were discarded long since: trioxide by the ferricyanide and by electrolysis, chloride, iodide, sulphate, sulphamate, thiostannate, cobaltinitrite, luteocobaltic chlorothallate(III), chromate, chloroplatinate, oxinate, 2-methyloxinate, 2-phenyloxinate, inner complexes with thionalide, mercaptobenzothiazole, tetraphenylboride, studied initially by Peltier and Duval[1] and by Takeno[2]. To these we have added some results given by various simple salts and complexes.

1. Thallium trioxide hydrate

(i) Precipitation by ferricyanide The product obtained when the hydrated trioxide is prepared by the ferricyanide method does not accord with the formula $Tl(OH)_3$ or $TlO(OH)$, in other words $Tl_2O_3 \cdot H_2O$. We invariably obtained a product containing less water than is demanded by the latter formula. Its weight remains strictly constant between 126° and 230°; this latter temperature is shown rather clearly on the curve but the former is much less distinct. The loss of water between 89° and 126° is insignificant (1 mg per 531 mg of oxide). When the curve was retraced, but with a very moist product, which intentionally was supplied with more than three molecules of water per each Tl_2O_3, a rapid descent is observed without break until the temperature reached 69°. Neither the trihydrate or the monohydrate was indicated and no difference was observed between the departure of the water that had been added and that which is normally retained by the product. A first loss of oxygen was found from 230° to 375°. The material remaining had the formula $3Tl_2O_3 \cdot Tl_2O$; it had previously been reported by Carnelley and Walker[3], but they assigned the range 440–565° to this molecular combination,

whereas our graphs, showed it only between 375° and 408°. These British workers operated in a discontinuous fashion and fixed the temperatures through the melting points of various salts. A rapid loss of weight occurs from 408° to 596–600°; this too had been found by Carnelley and Walker, who reported 585° and 815° as the limits of this part of the phenomenon, which is perfectly reproducible with different preparations. A certain portion of the thallous oxide of the system $3Tl_2O_3 \cdot Tl_2O$ disappears during this descent of the curve, but it should be remembered that we were working in an open crucible and under normal pressure. The curve ascends from 600° to 720°, and according to this ascending branch three-fourths of a molecule of Tl_2O takes up oxygen again so that pure Tl_2O_3 is found once more between 720° and 745°. Naturally, we do not regard the molecular associations $3Tl_2O_3 \cdot Tl_2O$ and $12Tl_2O_3 \cdot Tl_2O$ as definite compounds as long as an X-ray study of their structures has not been made. If the chemistry of thallium were not full of such singularities, there might be occasion for being astonished at finding that the same oxide has two existence ranges. The decrease in weight again manifests itself from 745°; it is due to decomposition and volatilization. The residue at 946° is a mixture of Tl and Tl_2O_3. In conclusion it should be pointed out that if thallium is determined by the ferricyanide method, it is necessary to dry the product between 100° and 230° (taking all possible precautions against exposure to the atmospheric carbon dioxide).

(ii) Precipitation by electrolysis Thallium can be determined by weighing the anode after electrolyzing a thallous salt in a medium containing sulphuric acid and in the presence of acetone. See Heiberg[4] and Besson[5]. The moist oxide obtained in this manner has been carefully compared with the preceding product; it retains about 0.8 molecule of water, a fact that was already pointed out, and consequently it is not the hydrate TlO(OH). The anhydrous oxide yields a level which is almost horizontal between 156° and 283°, and accordingly the drying which precedes the weighing must be conducted within this temperature range. Beyond 283°, a new level is observed between 411° and 677°; it corresponds to the double oxide $3Tl_2O_3 \cdot Tl_2O$ discussed above. However, it shows neither a maximum or minimum in the loss of weight, which is perfectly regular up to 950°, the limit of the experiment. The residue at 950° continues to be a mixture of Tl and Tl_2O_3 still in the course of formation. From this study it may be concluded that the electrolytic oxide retains water more firmly than the chemically-produced oxide, and that both on heating yield an intermediate product $3Tl_2O_3 \cdot Tl_2O$, but that

they differ in their behaviour above 677°. Consequently, we assume that there are two quite distinct forms of the oxide Tl_2O_3.

2. Thallous chloride

Thallous chloride is produced by treating the nitrate $TlNO_3$ with hydrochloric acid in the presence of ethanol. The curve obtained from this precipitate shows a loss of weight up to 56° due to the removal of the wash liquids; a horizontal extends from 56° to 425°. The salt begins to volatilize above 425° and at 475° it is found that 5.9 mg out of 264.0 mg have disappeared. Takeno[2] simply directed that the chloride should be kept below 120°.

3. Thallous iodide

The precipitation of thallous iodide as directed by Mach and Lepper[6] is a common method; it also constitutes a precise method of determining thallium. The curve shows a loss of water and acetone (used in the washing) up to 70°. The horizontal corresponding to the anhydrous iodide extends from 70° to 473°. Beyond this temperature, iodine is released and this loss becomes very rapid above 600°. The losses of iodine suffered by 421.32 mg of the dry iodide at various temperatures were:

Temp. (°C)	540	608	674	744	813
Loss (mg)	4.0	12.1	30.0	76.2	208.4

Takeno[2] specified that 100° should not be exceeded when this iodide is to be weighed.

4. Thallous sulphate

Ordinarily, the determination through the soluble sulphate Tl_2SO_4 consists in treating an insoluble thallium salt, such as the chloride $TlCl$, with sulphuric acid. The resulting sulphate is anhydrous between 92° and 355°. Previously, it loses its moisture and its excess acid with an abrupt change in speed at 64°; beyond 355° it breaks down and the conversion into the trioxide Tl_2O_3 is almost complete at 880°. Takeno[2] gives 400° and 700° as the limits of the level of the anhydrous salt.

5. Thallous sulphamate

The amidosulphonate (sulphamate) $Tl(NH_2-SO_3)$ was prepared by

Capestan[7]. Like the corresponding salts of the alkali metals, it is converted transitorily into the imidodisulphonate $NH(SO_3Tl)_2$ before the sulphate Tl_2SO_4 is obtained near 500°.

6. Thallous nitrate

The nitrate $TlNO_3$ chosen for the preparation of a standard solution is anhydrous and may be weighed directly. There is no change in its weight up to 410°; the decomposition is especially vigorous between 542° and 745°. The residue at 948° is not the oxide Tl_2O_3 but the mixture of Tl and Tl_2O_3 noted in (1). Wendlandt[8] has repeated the study of the thermolysis of this nitrate and found a decrease in weight between 265° and 370°.

7. Thallous carbonate

Thallium carbonate Tl_2CO_3 is anhydrous and stable up to 372° (a temperature higher than those reported by various workers). The thallium oxidizes beyond 372° and there is a loss of carbon dioxide accompanied by browning of the residue. The resulting compound $Tl_2CO_3.Tl_2O_3$ had already been described among the basic carbonates of thallium. This carbonate yields a horizontal from 418° to 615°. A rather complicated additional decomposition then begins; the remaining carbonate breaks down; the thallium oxide Tl_2O_3 is reduced and the resulting Tl_2O sublimes partially and also leaves metallic thallium. See Duval[9].

8. Thallous thiostannate

Hawley[10] has recommended the thiostannate Tl_4SnS_4 for the determination of thallium. It is our belief that this method should be discarded because the compound is hard to prepare, it filters slowly, and especially because the thermolysis curve descends continuously. There is a slight inflection between 281° and 343°, but this short range cannot be recommended with any certainty for the drying of the thiostannate which is already decomposing at the last temperature cited. Takeno[2] advises that the precipitate be dried below 105°.

9. Thallium sodium cobaltinitrite

The product resulting from the action of sodium cobaltinitrite on thallium nitrate does not accord with the composition $Na_2Tl[Co(NO_2)_6]$ or $NaTl_2[Co(NO_2)_6]$; it is closer to the first than to the second formula. Strecker and de la Peña[11] reported that the precipitate yields a level that is essentially horizontal after the water has disappeared, a level which may serve for the accurate determination of thallium, provided the composition of the nitrite is constant. Three NO_2 groups are eliminated from 110° to 200°. As in the analogous case of cesium, the resulting oxide of Co(II) takes up oxygen which then oxidizes the two residual nitrites, namely $TlNO_2$ and $NaNO_2$. A new horizontal extends from 283° to 345°; it corresponds to a mixture of CoO, $TlNO_3$, $NaNO_3$. Unfortunately this level cannot be used for gravimetric purposes since the ratio Tl:Na is not known. The resulting nitrates decompose in their turn above 345° and leave a mixture of oxides. Consequently, the weighing of thallium as the cobaltinitrite cannot be recommended in view of our present lack of precise knowledge about its mode of precipitation.

10. Luteocobaltic chlorothallate

The luteocobaltic chlorothallate(III), prepared as directed by Spacu and Pop[12], has the formula $[Co(NH_3)_6][TlCl_6]$. It becomes anhydrous in the region 50–210°, where it yields a perfectly horizontal level. It breaks down from 210° to 345° and loses the six molecules of ammonia, and the residue consists of the two chlorides $TlCl_3$ and $CoCl_3$. They sublime and decompose up to 813°. The cobaltic chloride, which was isolated by Hibert and Duval[13], is more stable than the thallic chloride at high temperatures. The residue at 880° consists of grey metallic cobalt and green undecomposed cobaltic chloride $CoCl_3$. We regard the method of Spacu and Pop as a good way to determine thallium.

11. Thallous chromate

Thallous chromate Tl_2CrO_4 is obtained by adding potassium bichromate to a boiling ammoniacal solution of thallous nitrate. The precipitate is allowed to stand for 12 hours and then is washed with a 1% aqueous solution of potassium chromate, followed by 50% ethanol. See Moser and Brukl[14]. The curve is extremely simple; a descending branch corresponding to the elimination of the retained water and alcohol is observed up to 97°. A perfectly horizontal stretch extends from 97° to 745° and relates to the anhydrous salt. Therefore, a broad temperature range is available for drying the compound in this classic method of determining thallium, though Takeno[2] recommends not carrying the temperature beyond 170°.

12. Thallium hexachlorotitanate

Morozov and Toptygin[15] have investigated the thermal stability of the hexachlorotitanate $Tl_2[TiCl_6]$.

13. Thallium hexachloroplatinate

The yellow precipitate of the chloroplatinate $Tl_2[PtCl_6]$ produced from thallous nitrate and hexachloroplatinic acid loses retained water and alcohol up to 65°. The salt is anhydrous and stable between 65° and 155°. Decomposition begins then, slowly at first up to 470° and then becomes more rapid. The decomposition is complete at 710°; the residue consists of pure platinum. Accordingly, this thallium salt is less sturdy than the corresponding salts of the alkali metals.

14. Thallium oxinate

The precipitation of the oxinate $Tl(C_9H_6ON)_3 \cdot H_2O$ is due to Feigl and Baumfeld[16]. The thermolysis was studied by Borrel and Pâris[17] and reveals complete dehydration from 44° to 150°. The anhydrous product is stable only within the narrow range 150–165°. The decomposition rate becomes much less near 300° and a level is obtained which corresponds essentially to a ratio of 3: 5 for thallium: organic matter. The decomposition becomes rapid above 380° and leads to finely divided metallic thallium, which begins to volatilize to an appreciable extent at 700°.

15. Thallium methyloxinate

Thallium methyloxinate $Tl(CH_3–C_9H_5ON)_3$, prepared by Borrel and Pâris[18], is yellowish. After a slight loss of moisture, it yields a level due to the anhydrous compound from 80° to 145°. Finally, it decomposes smoothly leaving a residue of metallic thallium, whose volatilization is appreciable even at 700°.

16. Thallium 2-phenyloxinate

The orange thallium 2-phenyloxinate $Tl[(C_6H_5–C_9H_5ON)_3] \cdot H_2O$, prepared by Borrel and Pâris[19], starts to lose water at 50°. The anhydrous compound gives a horizontal from 100° to 190° and then rapidly decomposes when subjected to linear heating which increases at the rate of 180° per hour.

17. Thallium thionalidate

The procedure of Berg and Fahrenkampf[20] was used for the preparation of the inner complex with thionalide. The product has the formula $C_{10}H_7$–NH–CO–CH_2–S–Tl. The thermolysis curve descends as the water and acetone used for washing are removed, and the anhydrous compound gives a level extending from 69° to 156°. Hence the temperature of 110° recommended by the above workers is suitable and lies within the above level. Decomposition begins beyond 156° with carbonation and evolution of gases. Umemura[21] gives 220° as the limit. In our opinion, the complex with thionalide is the best form for the gravimetric determination of thallium.

18. Thallium 2-mercaptobenzothiazole

The inner complex with 2-mercaptobenzothiazole $C_7H_4NS_2Tl$ was prepared as directed by Spacu and Kuraš[22]. The pyrolysis curve shows first a descending branch due to the loss of ammonia coming from the wash water. A horizontal extending from 69° to 217° represents the existence zone of the anhydrous complex. The latter then carbonizes and gives off gaseous products.

19. Thallium tetraphenylboride

Thallium tetraphenylboride $Tl[B(C_6H_5)_4]$ was discovered by Wittig and Raff[23] and studied by Wendlandt[24] with his thermobalance. The air-dried precipitate is stable up to 180°. It then slowly decomposes and yields a carbonaceous material at 325°. This residue oxidizes and a mixture of the metaborate $TlBO_2$ and the trioxide is obtained at 450°. Our findings show that the level can be employed for the automatic gravimetric determination of thallium; the analytical factor is 35.63% for metallic thallium.

20. Thallium amine complexes

Finally, for non-analytical purposes, Kochetkova and Tronev[25] investigated the decomposition of the complexes formed by thallium with ammonia or ethylenediamine.

REFERENCES

[1] S. Peltier and C. Duval, *Anal. Chim. Acta*, 2 (1948) 210.
[2] R. Takeno, *J. Chem. Soc. Japan*, 54 (1933) 741.
[3] T. Carnelley and J. Walker, *J. Chem. Soc.*, 53 (1888) 89.
[4] M. E. Heiberg, *Z. anorg. Chem.*, 35 (1903) 347.
[5] J. Besson, *Compt. rend.*, 224 (1947) 1226.
[6] F. Mach and W. Lepper, *Z. anal. Chem.*, 68 (1926) 40.
[7] M. Capestan, *Ann. Chim. (Paris)*, 5 (1960) 215.
[8] W. W. Wendlandt, *Texas J. Sci.*, 10 (1958) 392.
[9] C. Duval, *Anal. Chim. Acta*, 20 (1959) 266.
[10] L. F. Hawley, *J. Am. Chem. Soc.*, 29 (1907) 1011.
[11] W. Strecker and P. de la Peña, *Z. anal. Chem.*, 67 (1925/26) 257, 264.
[12] G. Spacu and A. Pop, *Z. anal. Chem.*, 120 (1940) 322.
[13] D. Hibert and C. Duval, *Compt. rend.*, 204 (1937) 780.
[14] L. Moser and A. Brukl, *Monatsh.*, 47 (1926) 709.
[15] I. S. Morozov and Toptygin, *Zhur. Neorg. Khim.*, 5 (1960) 88.
[16] F. Feigl and L. Baumfeld, *Anal. Chim. Acta*, 3 (1949) 83.
[17] M. Borrel and R. Pâris, *Anal. Chim. Acta*, 4 (1950) 267.
[18] M. Borrel and R. Pâris. *Anal. Chim. Acta*, 5 (1951) 573.
[19] G. Bocquet and R. Pâris, *Anal. Chim. Acta*, 14 (1956) 201.
[20] R. Berg and E. S. Fahrenkampf, *Z. anal. Chem.*, 109 (1937) 305.
[21] T. Umemura, *J. Chem. Soc. Japan*, 61 (1940) 25.
[22] G. Spacu and M. Kuraš, *Z. anal. Chem.*, 104 (1936) 90.
[23] G. Wittig and P. Raff, *Liebigs Ann.*, 573 (1951) 195.
[24] W. W. Wendlandt, *Anal. Chim. Acta*, 16 (1957) 216.
[25] A. P. Kochetkova and V. G. Tronev, *Zhur. Neorg. Khim.*, 2 (1957) 2043.

CHAPTER 78

Lead

There are numerous gravimetric methods for determining lead; inorganic and organic reagents are employed. A study of the treatment of these various precipitates and their thermal behaviour has been accompanied by a critical appraisal of these procedures. Not many of these methods are satisfactory; the high atomic weight of lead has resulted in the acceptance of methods which otherwise would better have been discarded. See Duval[1].

Most of these methods give a factor for lead between 50 and 100%; the most advantageous from the arithmetical point of view are those involving dibromooxine and picrolonic acid. Though lead tungstate has a very low solubility, it does not lend itself well to gravimetric operations. The cyanide method of Herz and Neukirch[2] is not quantitative and has been discarded. The precipitation by means of hematein, proposed by Moffath and Spiro[3], can be used for separation purposes but not for gravimetric determination. The method of Shredov, Goldshteyn and Seletkova[4] using the hydroxybromide need not be considered here since it concludes with a weighing of the chromate. The determination of lead chromate was discussed in Chapter Chromium, and the curves related to the arsenate, vanadate, selenate, molybdate, iodide, chlorite, and the fluorochloride are given in the respective Chapters Arsenic, Vanadium, Selenium, Molybdenum, Iodine, Chlorine and Fluorine.

The following gravimetric methods will be discussed: the oxide from the precipitated metal, the dioxide PbO_2, the oxide $Pb_5O_7.3H_2O$, the oxide Pb_3O_4 (minium), hydroxide, "peroxide", chloride, iodate, periodate, sulphide, sulphite, sulphate, potassium lead sulphate, pyrophosphate, carbonate, basic thiocyanate, phosphomolybdate, oxalate, phthalate, gallate, salicylate, anthranilate, complexes with dimethylglyoxime, salicylaldoxime, oxine, 5,7-dibromooxine, picrolonic acid, thionalide, mercaptobenzothiazole, mercaptobenzimidazole, 5-sulpho-7-nitrooxine, picolinic acid, quinaldic acid. To these we have now added studies of the sulphamate, selenite, selenate, nitrate, niobate, cerussite, acetate, and the 2-phenyloxinate.

1. Metallic lead

We have proved that the procedure of weighing lead in the metallic state after reduction of the sulphate with hydrogen (aluminium wire and hydrochloric acid as directed by Torossian[5]) is impossible. The spongy lead oxidizes too readily, and it is partially converted to litharge (PbO) during the washing. Consequently, the results are too high even though all of the operations are carried out at room temperature. Frankly, it would therefore be better to weigh as the oxide. Besides the Torossian method just noted, the reduction may also be accomplished as suggested by Stolba[6] with hydrochloric acid and zinc. However, the temperature 150–200°, advocated by Stolba, appears too low to us; actually the following uptake of oxygen was noted on the curve, which we recorded with an initial weight of 164.27 mg of lead:

Temp. (°C)	64	141	220	410	675	813	946
Gain (mg)	0	1.4	2.0	3.2	5.2	7.0	12.7 (theoretical)

The analytical factor for lead in this method is very high, and furthermore it is difficult to see how it can be of much use in view of the fact that the starting material is lead sulphate which can be weighed directly more advantageously.

2. Lead dioxide

(i) Preparation In the method devised by Rivot, Beudant and Daguin[7], lead dioxide is produced by passing chlorine into a suspension of lead sulphide in sodium hydroxide. If desired, this needlessly complicated procedure may be replaced by precipitating lead "peroxide" electrolytically as described by Hertelendi and Jovanovich[8]. These latter workers recommend heating this product to 650° to convert it into the monoxide. This finding is in remarkably close agreement with the thermolysis curve obtained with the dark brown oxide PbO_2 as control, which yielded identical results. Murayama[9] had reported 270° which obviously is too low. The minium (Pb_3O_4) revealed its existence by an essentially horizontal stretch extending from 500–510° to 619°. Then the oxide PbO forms above 649°. Murayama[9] noted also that the calcination of the monoxide should be conducted between 600° and 800°.

(ii) Orthorhombic and quadratic form Faivre and Weiss[10] studied

the thermal behaviour of lead dioxide in nitrogen below 600°. They worked with a non-stoichiometric variety with orthorhombic symmetry. The decomposition is complete and the monoxide is formed directly. The dark brown oxide PbO_2 yields minium (Pb_3O_4) under the same conditions. These workers compare on the same chart the 445° isotherms for the quadratic and the orthorhombic form of the dioxide.

The two varieties of the dioxide PbO_2 have been also investigated by Seigneurin and Brenet[11] who traced the two curves obtained continuously with the Eyraud balance. The two phases were prepared by the action of nitric acid on Pb_3O_4 (minium) and they were differentiated by X-rays. The non-quadratic phase yields the monoxide entirely without passing through the minium form.

(iii) Mixed oxides   Perrault and Brenet[12] have repeated the preparation of the quadratic dioxide and the disproportionation of mixed oxides such as Pb_3O_4. The thermolysis in the Chevenard apparatus did not reveal any loss of weight up to 300°, nor was there any absorption of gas or moisture on cooling. The decomposition proceeded via two intermediate stages: one, whose composition was approximately $PbO_{1.32}$, corresponds to a mixture of oxides, while the other, which is stable from 510° to 580°, consists of a mixture of Pb_3O_4 and quadratic PbO.

Isotherms between 540° and 575° yield β-PbO. After conducting the decomposition by slow heating up to 500°, these workers studied the isotherms for temperatures in the neighbourhood of 510°. The final product in that case was Pb_3O_4.

In an investigation, which included both thermal analysis and thermogravimetry, Spinedi and Gauzzi[13] prepared the oxide Pb_2O_3 by decomposing the carbonate $PbCO_3$ in the air at 300°, and then heating the different lead oxides. They accomplished the transformations:

$$PbO_2 \nrightarrow Pb_2O_3 \text{ between 375° and 460°}$$
$$Pb_2O_3 \nrightarrow Pb_3O_4 \text{ between 460° and 495°}$$
$$Pb_3O_4 \nrightarrow PbO \quad \text{ between 585° and 675°}$$

3. Lead "peroxide"

(i) Electrolytic peroxide   The electrolytic peroxide, prepared in the crucible-anode in accord with the familiar Classen[14] method, yields a curve analogous to that of the dioxide with a result ordinarily corresponding to $Pb/PbO_x = 0.8627$ instead of $Pb/PbO_2 = 0.8662$. The anode

should not be dried at a temperature above 340°. In contrast, Murayama[9] has given 270° as the proper drying temperature.

(ii) $Pb_5O_7 \cdot 3H_2O$ We followed the procedure of Das-Gupta, Roy and Sil[15] which begins with the peroxide $Pb_5O_7 \cdot 3H_2O$. This method is carried out most easily and it might even be said that it could lead to a good method for determining lead. However, we do not agree with these workers on several points. The thermolysis curve of the precipitate has a good level between 100° and 120° but it corresponds to the oxide PbO_2. It should be pointed out that the formula $Pb_5O_7 \cdot 3H_2O$ or $H_6Pb_5O_{10}$ differs from PbO_2 or Pb_5O_{10} only by the relatively insignificant weight of the hydrogen. Contrary to the statements of the above workers, the "peroxide" obtained is not stable up to 180° without decomposition, it breaks down slowly from 120° to 542° and produces the oxide PbO with no intermediate level for the Pb_3O_4.

But the product with which we worked differed from the dark brown oxide by its much lighter color and lesser stability. It could be used for the automatic determination of lead starting from a soluble salt, but a physical chemical study would be necessary to establish the nature of this "peroxide."

4. Minium

When minium Pb_3O_4 is the starting material, the curve shows that this oxide starts to take up oxygen at 47°. It is gradually converted into the dioxide PbO_2 and the maximum weight is reached at 220°. However, the conversion is not complete and involves the surface especially. For example, 2.1 mg of oxygen were taken up by 382.7 mg of Pb_3O_4. A horizontal is observed from 308° to 475° and corresponds to a compound having the appearance and the formula of minium; then it decomposes in turn and produces the monoxide PbO quantitatively from 610° to 619°. Perrault and Brenet[12] have also made a study of minium (see above).

4. Lead hydroxide

A number of workers have worked with lead hydroxide but nobody as yet has succeeded in preparing a material with the desired formula $Pb(OH)_2$. Genner[16] did prepare a truly hydroxylated product, which was identified by infra-red spectrography; its formula was $Pb_3O_2(OH)_2$ or

References p. 633

perhaps $Pb_2[PbO_2(OH)_2]$. When 250 mg of this material were heated at the rate of 300° per hour, it yielded litharge (PbO) from 175° on.

(i) Precipitation by ammonium hydroxide   With ammonium hydroxide, the precipitate gave a horizontal due to a basic nitrate, which decomposed between 100° and 350°. A second loss occurs between 400° and 550° (at 150° per hour) corresponding to the change from $PbO·H_2O$ to PbO. Robin and Théolier[17] deduced a method of preparing this hydrate in a pure state from the curve; they subjected the basic nitrate, which is first obtained by precipitation, to isothermal heating overnight at 280–300°. The general shape of the curve confirms the one traced by Duval[1].

(ii) Precipitation by sodium hydroxide   Nicol[18] treated lead nitrate with sodium hydroxide. With $N/30$ alkali, the very light white precipitate loses moisture from 18° to 134°, and the water of hydration is eliminated in two stages: 134° to 250° and 250° to 395°. The nitrate decomposes with transitory production of Pb_3O_4 from 395° to 552°, and decomposition of the minium with conversion to the monoxide from 552° to 660°. The quantitative study of these decreases in weight makes it possible to assign the formula $Pb(NO_3)_2·3Pb(OH)_2·2H_2O$ to the basic salt obtained. It is amorphous according to X-ray analysis.

When $N/10$ alkali is employed, the resulting voluminous white precipitate loses moisture in the desiccator. Water alone is lost in two equal parts at two stages: 134–200° and 200–370°. The formula arising from the analysis is $PbO·\frac{1}{2}H_2O$ or $Pb(PbO_2)·H_2O$. The material is crystalline.

6. Lead chloride

Lead chloride was precipitated by the action of concentrated hydrochloric acid on lead nitrate in the presence of alcohol-ether. The curve reveals a slight increase in weight from 55° to 526° where volatilization begins. Classen[19] for example recommends drying at 100°, but it is better to wash for a long time with an alcohol-ether mixture and then dry the precipitate in the desiccator.

In one experiment, the increase in weight was 2 mg per 328 mg of chloride. On the other hand, the losses due to the volatility of this salt and measured on a like weight of the chloride, were shown by the recording to be:

Temp. (°C)	528	675	788	826	859	915	928	946
Loss (mg)	0	4	30	51	80	180	220	292

At 946° the residue consists of lead chloride contaminated with a trace of lead monoxide produced by incipient decomposition.

7. Lead iodate

The iodate $Pb(IO_3)_2$ prepared as directed by Gentry and Sherrington[20] remains at almost constant weight up to 400° where the salt begins to decompose. The uptake of oxygen between 61° and 343° is slight though distinct. It amounts to nearly 1 mg per 335 mg of iodate and hence may be disregarded. The temperature of 140° indicated by the above workers is not exclusive. If the iodate is washed with acetone, it may even be better to dry the iodate in a desiccator. The iodate method is one of the better ways of determining lead. Along with the bromooxinate, it offers one of the most favourable analytical factors: $Pb/PbIO_3 = 0.3153$. We suggest that it be used for the automatic determination of lead. Vigorous decomposition sets in above 400° and there is loss of oxygen and iodine. The latter has entirely disappeared around 700° and there is a change in the slope of the curve but no distinct level is seen. Lead oxide PbO is obtained from 946° on. Murayama[9] specifies that the iodate be dried below 180°.

8. Lead periodate

A (meso)periodate was prepared by the procedure given by Willard and Thomson[21]; its curve contains no horizontal. They recommend 110° as the drying temperature but this is not suitable since it is situated on a descending portion of the curve. The weight remains practically constant between 141° and 280° and agrees with the approximate formula Pb_3 $(IO_5)_2$ which is intermediate between $Pb(IO_4)_2$ and $Pb_5(IO_6)_2$. The oxide PbO derived from the decomposition scarcely appears below 960°. Before that, from 560° to 745°, the same level is observed as in the case of the iodate and relates to a compound whose molecular weight is 250 for 1 gram atom of lead. Although determination as the periodate appears to have value from the point of view of the analytical factor, it is really not satisfactory if conducted entirely within the gravimetric technique. It is better to follow the advice of the above workers and conclude the procedure with a titration.

References p. 633

9. Lead sulphide

(i) Precipitation by hydrogen sulphide   Lead was precipitated as the sulphide in a medium containing tartaric acid by means of an acetone solution of hydrogen sulphide as proposed by Löwe[22]. The wash liquids are eliminated rapidly and then the sulphide appears within the short temperature interval 97.5–107.2°. Löwe therefore correctly recommended drying the sulphide at 105°. The curve ascends abruptly at 107° and constant weight is reached at 542°. The residual mixture cannot be considered for gravimetric purposes; although it consists mostly of lead sulphate it also contains lead oxide PbO, and even a small quantity of free lead. One of the phases of a well-known metallurgical procedure is recognized here. The distinctness of the sulphide level makes it very suitable for the automatic determination of lead. In a somewhat vaguer manner, Murayama[9] advised that the sulphide be dried below 150°.

(ii) Precipitation by sodium sulphide   Taimni and Salaria[23] prepared lead sulphide by means of sodium sulphide, and according to Taimni and Tandon[24] the product yields a horizontal up to 110°. Beyond this, it gains weight continuously; the gain becomes minimal beyond 500°.

(iii) Galena   In his general investigation of the sulphides, Saito[25] gave six groups of curves for galena. He estimated that the apparent oxidation point for finely ground galena is near 300°, and that this oxidation proceeds very slowly as far as 500°. Above 500°, the oxidation gradually becomes more intense. It may be represented by the expression:

$$a \text{ PbS} + b \text{ O}_2 \rightarrow c \text{ PbO}_2 + d \text{ PbSO}_4 + e \text{ SO}_2$$

Galena is difficult to oxidize and its oxidation products differ with the varieties and the conditions surrounding the oxidation. Various methods may be employed depending on the objectives of the roasting. Complicated interactions occur beyond 810–820° between PbS and PbO, and between PbS and PbSO$_4$. Coarse grains of galena have been seen to decrepitate.

10. Lead sulphite

Lead sulphite PbSO$_3$ has been precipitated by Gaspar y Arnal and Poggio-Mesorana[26] with sodium sulphite, and by Jamieson[27] with bisulphite, and by Hanuš and Hovorka[28] with metabisulphite. All of these precipitants yield the same product which is stable up to 440° (227 mg heated continuously at 300° per hour). Sulphur dioxide is evolved between 440° and about 500° with production of lead oxide PbO, while a more considerable portion oxidizes to lead sulphate. Near 971° the sulphate reacts with the lead oxide to give some metallic lead and there is a new disengagement of sulphur dioxide.

11. Lead sulphate

The precipitation of lead sulphate by adding sulphuric acid to a solution of lead nitrate in the presence of ethanol yields a product that is free of retained impurities from 271° on. The temperature of 600° recommended hitherto is not exclusive. Murayama[9] gave 600–800°. Actually, continuous weighing reveals that the level of the $PbSO_4$ continues unchanged up to 960°; no basic sulphate appears.

12. Lead potassium sulphate

The double sulphate $K_2SO_4.PbSO_4$ was prepared as directed by Tananaeff and Mizetzkaja[29]. It gives a horizontal stretch beginning at 40°, where the last traces of alcohol are removed, and extending up to 906° (limit of the experiment). The straight line obtained is suitable for the automatic determination of lead and the analytical factor is more favourable than that of the sulphate.

13. Lead sulphamate

The amidosulphonate, obtained by Capestan[30], decomposes:
$$Pb(SO_3-NH_2)_2 \rightarrow NH_3 + PbNH(SO_3)_2$$
the imidodisulphonate being formed between 230° and 250°.

14. Lead selenite

Lead selenite is stable up to 630° according to Rocchiccioli[31].

15. Lead selenate

Kato, Hosimija and Nakazima[32] report that lead selenate is stable only up to 280°.

References p. 633

16. Lead nitrate

Lead nitrate crystallizes in the anhydrous condition. Duval and Wadier[33] have heated it in a number of trials. Its thermolysis curve shows that it is stable up to 280°, and then decomposition proceeds slowly as far as 500°. It then accelerates and the horizontal of the oxide PbO starts at 648°. The nitrate has been studied likewise by Wendlandt[34] and Nicol[35]. The latter, through successive quenchings, guided by the thermolysis curve, has isolated the following compounds: up to 445° $Pb(NO_3)_2 \cdot PbO$, from 445° to 455° $Pb(NO_3)_2 \cdot 4PbO$, which is white and likewise amorphous, from 455° to 470° $Pb(NO_3)_2.5PbO$, which is brown-red and crystalline. When heated to 531°, the latter decomposes and produces orange minium Pb_3O_4.

17. Lead hydrogen phosphate

The precipitation of lead monoacid phosphate $PbHPO_4$ by means of diammonium phosphate can be carried out as described by Petrashen[36] or as suggested by Moser and Reif[37]. The thermolysis curve is very simple; the phosphate maintains its weight constant up to 310°, and it is permissible to weigh it after drying at 150°. However, the trilead phosphate is not obtained subsequently. Two molecules of the monoacid phosphate lose exactly one molecule of water between them from 310° to 355° and the resulting pyrophosphate $Pb_2P_2O_7$ remains constant in weight up to 933°. Trau[38] pointed out that the pyrophosphate attacks glass and porcelain, and that its weight remains constant from 410° to 1050°.

18. Lead niobate

According to Pehelkin, Lapitskii, Spitsyn and Simanov[39], the niobate $Pb_7Nb_{12}O_{37} \cdot 26H_2O$ loses water and yields the hydrates containing 8.5 H_2O at 140°, with 6 H_2O at 220°, and with 4 H_2O at 240°. The metaniobate $Pb(NbO_3)_2$ appears at 460°.

19. Lead carbonate

(i) Precipitated The carbonate precipitated as directed by Jilek and Kot'a[40] has the formula $PbCO_3$ which it retains up to 142°. Accordingly, drying at 100° is indicated, and the method is suitable for the automatic determination of lead. Decomposition then begins and continues as far as 745°, and produces the oxide PbO. When about one-half of the carbon dioxide has been given off, there is an abrupt change in the direction of the thermolysis curve signaling the appearance of the basic carbonate $PbCO_3.PbO$.

(ii) In storage batteries Takagaki[41] has employed the thermobalance to follow the absorption of carbon dioxide by the paste used in storage batteries. The basic carbonate $PbCO_3 \cdot Pb(OH)_2$ is formed.

(iii) Cerussite Cerussite $PbCO_3$ was subjected to a lengthy study by Caillère and Pobeguin[42]. When heated in nitrogen, it begins to decompose at 260° (Chevenard thermobalance). This was the lowest decomposition temperature obtained with the four natural orthorhombic carbonates.

(iv) Artificial carbonate Lamure[43], on the other hand, heated the artificial carbonate in the air or in nitrogen. He employed the Chevenard thermobalance and heating rates of 25° per hour and 300° per hour, and also isothermal heating. Only the basic carbonate $PbCO_3.2PbO$ was found. By working in an atmosphere of carbon dioxide, the author produced $2PbO \cdot CO_2$ and $3PbO.CO_2$; the latter is stable up to around 430°.

(v) Ceruse Burriel-Marti and Barcia-Goyanes[44] heated four specimens of commercial Spanish ceruse (white lead) on the Chevenard thermobalance. The water of constitution was lost between 120° and 220°, and about half of this water is eliminated from 160–165°. Carbon dioxide is given off around 215–220°. The decomposition is complete in the vicinity of 400–450°. The final product is the oxide PbO. Consequently, ceruse is not a definite compound; analysis shows its composition to be approximately $2PbCO_3.Pb(OH)_2$.

(vi) Carbon dioxide absorption The absorption of carbon dioxide by lead oxide PbO, freshly prepared by heating the carbonate in nitrogen at 420°, begins slowly at 230°; it accelerates and is rapid at 390°. After 24 hours of contact, the total quantity of CO_2 taken up becomes closer to that corresponding to $3PbO \cdot CO_2$ the more nearly the temperature approaches the decomposition temperature of the latter.

20. Basic lead thiocyanate

The directions of Spacu and Dick[45] were followed for the preparation of the basic thiocyanate Pb(OH)SCN. It should be well dried in the desiccator. When heated slowly, its curve reveals a gain of weight from 42° to 93°. Decomposition begins then with release of cyanogen. The precipitate filters well and is easily washed.

21. Lead phosphomolybdate

The phosphomolybdate, prepared by the method of Beuf[46], likewise is easy to filter. The lead content is only 54.8%. As this worker reported, the water is eliminated with difficulty. Constant weight, in fact, is only obtained above 436°. The curve does not differ much from a straight line.

22. Lead tungstate

Dupuis[47] points out that weighing lead as the tungstate is not practical because the precipitate is hard to manipulate. After the moisture is removed, the compound $PbWO_4$ maintains constant weight up to 945° at least (perfectly horizontal thermolysis line).

23. Lead acetate

Lead acetate, studied as a standard by Duval and Wadier[33], is usually stated for analytical purposes to contain 3 molecules of water. We found a little less (around 2.3 H_2O) if account is taken of the weight of the residue. In any case, this salt could scarcely be used for a direct weighing.

24. Lead oxalate

The precipitation as the oxalate as described by Böttger[48] yields a product that is quite dry at 50° and the level of the anhydrous compound extends up to 300°. There is a very slight rise in the curve between 154° and 283° (hardly 1 mg gain per 230 mg). Murayama[9] reported 150–290° as the limits of the level. The decomposition is sudden and leads directly to the oxide PbO from 378° on. The curve does not reveal the formation of either the sub-oxide Pb_2O (in which no one believes any longer) or the basic carbonate, or minium, etc. From the standpoint of the automatic determination by means of anhydrous lead oxalate, it would be best to select the portion of the curve near 100°.

25. Lead o-phthalate

Zombory[49] reported that lead o-phthalate precipitates readily in a neutral medium and in the presence of a sufficient quantity of alcohol. The pure anhydrous salt is shown by the thermolysis curve to be present only between 228° and 320°; the analytical factor for lead is 0.5580. The oxide produced by the decomposition begins to appear at 542°. The phthalate still contains two molecules of water at 110°; these are slowly lost.

26. Lead gallate

The method described by Mayr[50] was used for the preparation of the gallate. The decomposition curve shows that the anhydrous salt may be weighed up to 152°. Metallic lead (due to reduction by the organic matter) is obtained at 345° and is stable in this state up to 410° where it begins to reoxidize slowly to the oxide PbO which is produced completely only at 850°. The anhydrous gallate may be employed as the basis of an automatic method of determining lead. It should be filtered on a No. 4 glass frit; the precipitate does not adhere to the container. It should be noted that lead occupies the position of the hydrogen of the carboxyl group as well as that of the phenolic group since the precipitate has the formula $Pb_2(C_6H_2O_3CO_2)$.

27. Lead salicylate

The precipitation as salicylate, due to Murgulescu and Dobrescu[51], leads to a curve which includes no level except that of the oxide PbO above 906°. The curve rises until the temperature reaches 240°, and the decomposition, as in the preceding case, yields metallic lead between 425° and 555° but there is no distinct level. Weighing as the salicylate cannot be carried on in a satisfactory manner. The above workers, in their paper, direct that the salt be converted into sulphate.

28. Lead dithizonate

Lead dithizonate was prepared by Pariaud and Archinard[52]. Two curves were traced with the Chevenard thermobalance at the rates of 150° and 300° per hour, respectively. The complex is absolutely stable up to 80°. The decomposition becomes violent and is almost complete at

215°. The two final phases of the decomposition occur as in the case of silver dithizonate but are much more accentuated. The weight does not change more than 1.5% between 215° and 340°. A level corresponding to close to 4% of PbS begins at 555°; this sulphide is converted into pure PbO from 1020°.

29. Lead thionalidate

The excellent thionalidate method has been repeated as given by Berg and Fahrenkampf[53]. It and the sulphide method are the most quantitative of all the proposed procedures. The heating curve fixes the limits of the horizontal due to the complex as 71° and 134°. Umemura[54] had given 100–145°. The compound then decomposes, and lead sulphide is formed beginning at 675°. Drying at 100–105° is highly recommended, since these temperatures lie within the horizontal level. The monohydrate may be weighed after drying at the ordinary temperature in a desiccator.

30. Lead mercaptobenzothiazole

Spacu and Kuraš[55] recommended that the precipitate with mercaptobenzothiazole be dried at 110°. The curve given by this complex has a good horizontal up to 120° and is well suited to the automatic technique. An abrupt explosion occurs at 188° and the crucible from then on contains only lead sulphide which gradually is oxidized to the sulphate.

31. Lead mercaptobenzimidazole

The procedure described by Kuraš[56] yields a complex with mercaptobenzimidazole whose formula agrees well with $PbOH(C_6H_5N_2CS)$. The curve includes a horizontal extending from 97° to 172° that corresponds to this formula and is the weighing range for the automatic determination. The complex then decomposes and metallic lead appears near 450°. Its reoxidation is complete at 618°. This is a good method for determining lead but the analytical factor for lead 0.5549 is still rather large.

32. Lead anthranilate

Funk and Römer[57] gave directions for the precipitation of lead anthranilate $Pb(C_7H_6O_2N)_2$. It is stable as far as 198°. Kiba and Sato[58] report 230°. When it decomposes, this salt yields metallic lead at 792°;

the metal oxidizes immediately. The transformation into the oxide PbO
is complete at 940°.The level due to the anthranilate up to 198° is suitable
for the automatic determination of lead.

33. Lead dimethylglyoxime

The complex with dimethylglyoxime $Pb_2(C_4H_6O_2N_2)$ is produced
quantitatively but its existence level is rather short and not strictly linear.
When prepared as described by Funahashi and Ishibashi[59] it seems to be
stable only between 60° and 87–88°. The decomposition is complete at
610°, where the excellent horizontal of the lead oxide PbO begins. From
the curve we have calculated that the molecular weight is 546, which
corresponds to the formula

$$Pb(OH)_2 \left[\begin{array}{c} CH_3\text{-}C \underline{\hspace{2em}} C\text{-}CH_3 \\ \overset{\|}{} \quad \overset{\|}{} \\ NO\text{-}Pb\text{-}NO \end{array} \right] = 545.536$$

34. Lead salicylaldoxime

The complex with salicylaldoxime, prepared as directed by Ishibashi
and Kishi[60], is much more stable than the dimethylglyoxime complex;
its excellent horizontal extends from 45° to 180°. Explosive decomposition
occurs there and a clinging and unpleasant odour makes itself evident
near 220°. Metallic lead appears at 675°; it then oxidizes as usual and is
converted into lead oxide PbO beginning at 900°. The method is good for
the automatic determination, but care must be taken not to heat the
complex beyond 177°.

35. Lead oxinate

Hovorka[61], whose procedure was used for the preparation of lead
oxinate, stated that it should be dried at 105°, but actually the curve
shows that this compound suffers a continuous loss of weight from 46°
to 820°. Strictly speaking, it should be dried in the cold in a desiccator
if a correct weight is to be obtained. The metallic lead produced during
the decomposition is slightly contaminated with carbon; the metal starts
to oxidize at 820°. The oxide PbO is formed by 919°. See also Sekido[62].

36. Lead 5,7-dibromooxinate

The dibromooxinate method, which provides the lowest analytical factor for lead, is however not entirely satisfactory from the experimental standpoint. It was proposed by Zan'ko and Bursuk[63]. The thermolysis curve ascends from the ordinary temperature, with a maximum at 93°. The increase in weight amounts to 3 mg per 170 mg. Lead oxide PbO is obtained at 320° and maintains its weight up to 449°. A new oxidation starts there and gives rise to a product close to minium, which in turn loses oxygen almost imperceptibly up to 950° where litharge is again formed. The dibromooxinate method consequently requires that the precipitate be dried in a desiccator in the cold.

37. Lead 5-sulpho-7-nitrooxinate

The complex with 5-sulpho-7-nitrooxine, prepared as directed by Molland[64], should not be seriously considered. The curve descends continuously until the temperature reaches 410° where the level of the oxide PbO begins. No level for the complex is observed. Consequently, we regard this method as useless for determination purposes since no use is derived from the relatively low analytical factor 0.2778. The complex should be dried in the cold in a desiccator, at any event not above 48°. The preparation of the reagent is no simple matter and it seems to us that the method given by Molland does not produce a homogeneous material.

38. Lead 2-phenyloxinate

Bocquet and Pâris[65] tried the 2-phenyloxinate. The product is yellow; its formula is $Pb(C_6H_5-C_9H_6ON)_2$. It contains two molecules of water which it begins to lose at 50°. The chelate is stable up to 295° and yields a horizontal which may be suitable for the automatic determination of lead.

39. Lead picrolonate

The procedure used with picrolonic acid was that given by Hecht and Donau[66]. The curve shows a constant weight from 58° to 112°, followed by a slight loss up to 220° where the material explodes suddenly with production of extremely fine carbon. Lead picrolonate filters remarkably well and does not stick to the walls of the container. Our results do not agree with those reported by Kiba and Ikeda[67] who stated that the sesquihydrate is stable below 140° and that the hemihydrate is stable between 220° and 260°. These figures depend greatly on the rate of heating in the case of explosive materials.

40. Lead picolinate

Lead picolinate $Pb(C_6H_4O_2N)_2$ is white. It retains two molecules of water, which it begins to lose at $50°$. See Thomas[68].

41. Lead quinaldate

Lead quinaldate $Pb(C_{10}H_6O_2N)_2$ is white and anhydrous. It is stable up to $320°$. The organic matter burns near $600°$. See Thomas[68] and Lumme[69].

REFERENCES

1 C. DUVAL, *Anal. Chim. Acta*, 4 (1950) 159.
2 W. HERZ AND E. NEUKIRCH, *Z. anorg. Chem.*, 130 (1923) 343.
3 M. R. MOFFATH AND H. S. SPIRO, *Chemiker-Ztg.*, 31 (1907) 639.
4 V. P. SHREDOV, E. O. GOLDSHTEYN AND N. I. SELETKOVA, *Zhur. Anal. Khim.*, 3 (1948) 109.
5 G. TOROSSIAN, *Ind. Eng. Chem.*, 8 (1916) 331.
6 F. STOLBA, *J. prakt. Chem.*, 101 (1867) 150.
7 RIVOT, BEUDANT AND DAGUIN, *Compt. rend.*, 37 (1853) 126.
8 L. HERTELENDI AND J. JOVANOVICH, *Z. anal. Chem.*, 128 (1948) 151.
9 K. MURAYAMA, *J. Chem. Soc. Japan*, 51 (1930) 786.
10 R. FAIVRE AND R. WEISS, *Compt. rend.*, 245 (1957) 2513.
11 L. SEIGNEURIN AND J. BRENET, *Compt. rend.*, 245 (1957) 1427 .
12 G. PERRAULT AND J. BRENET, *Compt. rend.*, 250 (1960) 325.
13 P. SPINEDI AND F. GAUZZI, *Ann. chim.*, (*Roma*), 47 (1957) 1297.
14 A. CLASSEN, *Ber.*, 27 (1894) 163.
15 P. N. DAS GUPTA, G. C. ROY AND K. M. SIL, *J. Indian Chem. Soc.*, 5 (1928) 657.
16 R. GENNER, *Diploma of Higher Studies*, Paris, 1960.
17 J. ROBIN AND A. THÉOLIER, *Bull. Soc. chim. France*, (1956) 680.
18 A. NICOL, *Compt. rend.*, 226 (1948) 810.
19 A. CLASSEN, *Ausgewählte Methoden der analytischen Chemie*, 1901, p. 18.
20 C. H. R. GENTRY AND L. G. SHERRINGTON, *Analyst*, 71 (1946) 31.
21 H. H. WILLARD AND J. J. THOMSON, *Ind. Eng. Chem., Anal. Ed.*, 6 (1934) 425.
22 J. LÖWE, *J. prakt. Chem.*, 77 (1859) 73.
23 I. K. TAIMNI AND G. B. S. SALARIA, *Anal. Chim. Acta*, 11 (1954) 54.
24 I. K. TAIMNI AND S. N. TANDON, *Anal. Chim. Acta*, 22 (1960) 554.
25 H. SAITO, *Sci. Repts. Tôhoku Imp. Univ.*, 16 (1927) 115.
26 T. GASPAR Y ARNAL AND J. M. POGGIO-MESORANA, *Anal. soc. españ. fís. y quím.*, 43 (1947) 439.
27 G. S. JAMIESON, *Am. J. Sci.*, 40 (1915) 157.
28 J. HANUŠ AND V. HOVORKA, *Chem. Listy*, 31 (1937) 489.
29 I. V. TANANAEFF AND I. B. MIZETZKAJA, *Zavodskaya Lab.*, 12 (1946) 529.
30 M. CAPESTAN, *Ann. Chim. (Paris)*, 5 (1960) 222.
31 C. ROCCHICCIOLI, *Thesis*, Paris, 1960.
32 H. KATO, H. HOSIMIJA AND S. NAKAZIMA, *J. Chem. Soc. Japan*, 60 (1939) 1115.
33 C. DUVAL AND C. WADIER, *Anal. Chim. Acta*, 23 (1960) 257.
34 W. W. WENDLANDT, *Texas J. Sci.*, 10 (1958) 392.
35 A. NICOL, *Compt. rend.*, 226 (1948) 253.
36 V. I. PETRASHEN, *C.A.*, 35 (1941) 1346.
37 L. MOSER AND W. REIF, *Mikrochemie, Emich Festschr.*, 1930, p. 215.

[38] J. TRAU, *Chem. Anal. (Warsaw)*, 4 (1959) 557.
[39] A. PEHELKIN, A. V. LAPITSKII, V. I. SPITSYN AND Y. P. SIMANOV, *Zhur. Neorg. Khim.*, 1 (1956) 1784; *C.A.*, (1957) 2445.
[40] A. JILEK AND J. KOT'A, *Collection Czechoslov. Chem. Communs.*, 5 (1933) 396.
[41] L. TAKAGAKI, *J. Electrochem. Soc. Japan*, 23 (1955) 399; *C.A.*, 50 (1956) 10566.
[42] S. CAILLÈRE AND T. POBEGUIN, *Bull. soc. franç. minéral.*, 83 (1960) 36.
[43] J. LAMURE, *Compt. rend.*, 236 (1953) 926.
[44] F. BURRIEL-MARTI AND C. BARCIA-GOYANES, *28th Congr. chim. ind. Liège*, 1958, p. 1008.
[45] G. SPACU AND J. DICK, *Z. anal. Chem.*, 72 (1927) 289; *Bul. soc. stiinte Cluj*, 4 (1928) 75.
[46] H. BEUF, *Bull. Soc. chim. France*, 3 (1890) 852.
[47] T. DUPUIS, *Anal. Chim. Acta*, 3 (1949) 663.
[48] W. BÖTTGER, *Pharm. Post*, 40 (1907) 679.
[49] L. ZOMBORY, *Magyar Chem. Folyoirat*, 44 (1938) 160; *C.A.*, 33 (1939) 4157.
[50] C. MAYR, *Monatsh.*, 77 (1947) 65.
[51] I. G. MURGULESCU AND F. DOBRESCU, *Z. anal. Chem.*, 128 (1948) 203.
[52] J. C. PARIAUD AND P. ARCHINARD, *Rec. trav. chim.*, 71 (1952) 634.
[53] R. BERG AND E. S. FAHRENKAMPF, *Z. anal. Chem.*, 112 (1938) 161.
[54] T. UMEMURA, *J. Chem. Soc. Japan*, 61 (1940) 25.
[55] G. SPACU AND M. KURAŠ, *Z. anal. Chem.*, 104 (1936) 88.
[56] M. KURAŠ, *Collection Czechoslov. Chem. Communs.*, 11 (1939) 313, 367.
[57] T. KIBA AND T. IKEDA, *J. Chem. Soc. Japan*, 60 (1939) 911.
[58] T. KIBA AND S. SATO, *J. Chem. Soc. Japan*, 61 (1940) 133.
[59] H. FUNAHASHI AND M. ISHIBASHI, *J. Chem. Soc. Japan*, 59 (1938) 503.
[60] M. ISHIBASHI AND H. KISHI, *Bull. Chem. Soc. Japan*, 10 (1935) 362.
[61] V. HOVORKA, *Chem. Listy*, 31 (1937) 273.
[62] E. SEKIDO, *Nippon Kagaku Zasshi*, 80 (1959) 80; *C.A.*, (1960) 13950.
[63] A. M. ZAN'KO AND A. Y. BURSUK, *J. Appl. Chem. (U.S.S.R.)*, 9 (1936) 2297.
[64] J. MOLLAND, *Tids. Kjemi Bergvesen*, 19 (1939) 119; *C.A.*, 34 (1940) 1932.
[65] G. BOCQUET AND R. PÂRIS, *Anal. Chim. Acta*, 14 (1956) 201.
[66] F. HECHT, W. REICH-ROHRWIG AND H. BRANTNER *Z. anal. Chem.*, 95 (1933) 152; F. HECHT AND J. DONAU, *Anorganische Mikrogewichtsanalyse*, Springer, Vienna, 1940, p. 151.
[67] H. FUNK AND J. RÖMER, *Z. anal. Chem.*, 101 (1935) 85.
[68] G. THOMAS, *Thesis*, Lyon, 1960.
[69] P. O. LUMME, *Suomen Kemistilehti*, B, 32 (1959) 198.

CHAPTER 79

Bismuth

According to Panchout and Duval[1] and Panchout[2], bismuth may be determined gravimetrically in various forms with the aid of the following reagents: as metal by formaldehyde, glucose, hypophosphorous acid, lead, as basic carbonate, hydroxide, as oxychloride by ammonium chloride, as oxyiodide by potassium iodide, as iodate, sulphide by hydrogen sulphide, sodium sulphide, thiosulphate, as sulphite, sulphate and basic sulphate, selenite, basic nitrate, orthophosphate, arsenate, molybdate, chromithiocyanate, iodobismuthate of cobaltiethylenediamine, basic formate, benzoate, iodobismuthate of hexamethylenetetramine, cupferronate, pyrogallate and gallate, inner complex with salicylaldoxime, oxinate, inner complexes with 2-mercaptobenzothiazole, thionalide, antipyrine–methyleneamine, phenyldithiodiazolonethiol, phenylarsonate, and inner complex with α-naphthoquinoline.

The precipitation by means of manganous nitrate–potassium permanganate which yields the basic nitrate that is transformed into bismuth oxychloride, as suggested by Kallmann and Pristera[3], has not been included.

1. Metallic bismuth

(i) Precipitation by formaldehyde We have repeated the well-known method of Vanino and Treubert[4] with formaldehyde. They advise drying the precipitated bismuth at 105°. The curve shows a sudden release of the retained alcohol up to 73°, and then a horizontal as far as 150°, and then a slow ascent. The bismuth oxidizes and a mixture of the metal and its oxides is present at 959°. Therefore the temperature of 105°, which has been recommended, is suitable. The method is suggested for the automatic determination of bismuth.

(ii) Precipitation by glucose When glucose is the precipitant, a large excess of this reagent and sodium hydroxide is used as suggested by

References p. 645

Cousin[5] who started from bismuth oxychloride. Actually, the method does not appear to be applicable to all bismuth compounds. The black precipitate is not pure bismuth and remains contaminated even after being kept in a desiccator. The curve gives no worthwhile result; there is an initial rapid loss of weight but no level is observed for the elementary bismuth. The final product is a mixture in which the oxide Bi_2O_3 predominates.

(iii) Precipitation by hypophosphorous acid If, in accord with the directions of Mawrow and Muthmann[6], bismuth oxychloride and 50% hypophosphorous acid (sp. gr. 1.27) are brought together, the resulting black precipitate filters poorly and has a marked tendency to assume the colloidal condition. It is therefore necessary to heat the filtrate to coagulate the metal. The curve includes a horizontal from 48° to 230° which corresponds to a product whose formula approximates Bi_2O. In fact, as we learned from the Debye and Scherrer spectrum, we are not dealing here with a definite compound but with metallic bismuth which has adsorbed oxygen in purely fortuitous amounts. The mixture oxidizes rapidly above 230° and the oxide Bi_2O_3 is obtained quantitatively beyond 840°. We suggest that the method be discarded.

(iv) Precipitation by lead Metallic bismuth precipitated by lead as directed by Meineke[7] and which he recommended be dried at 150°, yielded a thermolysis curve showing a horizontal from 65° to 113°. A slight indication of the oxide is already visible at 150°. A mixture in which the yellow oxide Bi_2O_3 predominates is present at 946°.

2. Basic bismuth carbonate

The Meineke[7] procedure for the precipitation as the basic carbonate by the reaction of bismuth nitrate and ammonium carbonate yields a white product which washes easily. Its thermolysis curve reveals a loss of water up to 68° followed by a good horizontal related to a basic carbonate $(BiO)_2CO_3$, which has not been previously reported. It is stable up to 308°. The mixture obtained near 417° contains the yellow oxide Bi_2O_3, which remains as such up to 732° and then the curve ascends slightly because of an uptake of oxygen. At 960° there still remain some black grains that have escaped total oxidation. Instead of weighing as the oxide, as recommended by Meineke, it is better to use the basic carbonate and to employ its level for the automatic determination.

3. Bismuth hydroxide

The procedure described by Ostroumov[8] for precipitating the hydroxide by means of pyridine appears to be excellent for obtaining a product which filters nicely, but the curve did not yield results of interest to us. The product must be heated to at least 960° to obtain pure bismuth trioxide. In fact, a level which rises slightly is observed from 688° to 960°. Kho Wha Lee[9] reported 550° as the minimum temperature for obtaining the oxide when the hydroxide is the starting material.

4. Bismuth oxychloride

The method of Luff[10] for preparing the oxychloride by means of ammonium chloride yields a white precipitate. It is stable and its formula may be written as $BiOCl + H_2O$ or $BiCl(OH)_2$. This material was already known, and it retains this composition up to 285°. Beyond 285°, it rapidly loses the elements of one molecule of water and the resulting oxychloride $BiOCl$ yields a new level extending from 328° to 805°. At higher temperatures, there is a rapid loss of chlorine, and the oxide Bi_2O_3 is obtained above 950°. The two parallel levels of the curve are suitable for the automatic determination of bismuth.

5. Bismuth oxyiodide

The directions given by Hecht and Reissner[11] were used for precipitating the oxyiodide by means of potassium iodide. The product has a coppery lustre; it filters and washes easily. Its thermolysis curve has a horizontal stretch up to 365°, but there is a slight rise between 231° and 307° (1 mg per 126 mg). Therefore the compound should preferably be dried below 231°. Beyond 360° there is simultaneous loss of iodine and gain of oxygen; however, the product obtained above 841° does not correspond to pure Bi_2O_3 (molecular weight 466) but rather to a greenish mixture containing dispersed iodine; the ensemble has a mean "molecular weight" of 480. The precipitated oxyiodide contains some iodine and hence the level noted above for this material cannot be recommended for the automatic determination of bismuth.

6. Bismuth iodate

Kho Wha Lee[9] reported that the curve of the anhydrous iodate

$Bi(IO_3)_2$ contains a level from $100°$ to $450°$ but we have not been able to reproduce this curve. We employed the method recommended by Buisson and Ferray[12] for the volumetric determination of bismuth. The curve inclines downward as far as $883°$ where a mixture is obtained consisting of the yellow oxide Bi_2O_3 tinted greyish green by admixed iodine, as in the case of the oxyiodide.

7. Bismuth sulphide

(i) Precipitation by hydrogen sulphide The bismuth sulphide precipitated with hydrogen sulphide and freed of excess sulphur by treatment with carbon disulphide as suggested by Moser[13] yields an extremely capricious curve. Between $100°$ and $165°$ there is a slightly ascending level corresponding to the probable region of the sulphide; moreover Kho Wha Lee[9] reported $100–170°$ as the limits. Ordinarily it is recommended that the material be dried at $100–110°$ after the removal of the last traces of sulphur. The final product is a mixture containing some bismuth sulphate.

(ii) Precipitation by sodium sulphide The precipitate produced with sodium sulphide was reported by Taimni and Salaria[14] to be pure. It yielded a horizontal up to $160°$ according to Taimni and Tandon[15].

(iii) Precipitation by sodium thiosulphate When the procedure described by Faktor[16] using sodium thiosulphate was employed, a very impure product resulted and the curve showed such pronounced irregularities that it is very difficult to make use of this recording. It provides no horizontal in any case, and the residue at $959°$ is a mixture of the white sulphate $(BiO)_2SO_4$ and the yellow oxide Bi_2O_3.

8. Bismuth sulphite

The precipitate obtained when a bismuth salt is treated with sodium sulphite as directed by Gaspar y Arnal and Poggio-Mesorana[17] is very impure. The curve showed a loss of water and sulphur dioxide accompanied by an oxidation. The result at $475°$ is bismuthyl sulphate $(BiO)_2SO_4$ whose weight stays constant up to $946°$ at least.

9. Bismuth sulphate

The bismuth sulphate obtained from the carbonate as directed by

Meineke[7] releases water and retained sulphuric acid up to 236°. The horizontal due to the definite stable sulphate $Bi_2(SO_4)_3$ continues then up to 405°, a finding which agrees with the results reported by Bailey[18] but not with those of Kho Wha Lee[9] who found the sulphate to be intact up to 210° at least. This level is well suited to the automatic determination of bismuth. Decomposition sets in above 405° and bismuthyl sulphate $(BiO)_2SO_4$ is obtained at 810°. The latter is more stable than was formerly reported; in fact its horizontal extends as far as 975°.

10. Basic bismuth sulphate

The precipitate of the basic sulphate obtained by the method given by Luff[10] seems to yield a material whose formula is $(BiO)_2SO_4 \cdot 2H_2O$. The corresponding thermolysis curve contains no level before 854° where that of bismuthyl sulphate begins in analogy to that in the preceding case. If the product is ignited, the material weighed between 854° and 962° will therefore be $(BiO)_2SO_4$ and not Bi_2O_3, unless a still higher ignition temperature is used and there is no point in doing this.

11. Bismuth selenite

The precipitation conducted as directed by Funakoshi[19] yields a mixture of selenites which produce a horizontal between 86° and 221°, which however does not correspond to $Bi_2(SeO_3)_3 \cdot H_2O$. Ignition leaves the oxide Bi_2O_3 above 948°. The method offers little of interest.

12. Bismuth nitrate

The nitrate $Bi(NO_3)_3 \cdot 5H_2O$ was heated by Duval[20] as a standard. This salt is very hygroscopic and commences to lose water of crystallization at around 49°. The thermolysis curve descends very slowly and the oxide Bi_2O_3 is not obtained until the temperature has reached the neighbourhood of 600°. Kho Wha Lee[9] reported 550° for the conversion into the oxide Bi_2O_3.

By application of the method of Löwe[21] as modified by Luff[10], we obtained a product close to $2Bi_2O_3 \cdot N_2O_5$. The precipitation is not quantitative if the Löwe procedure is used. The curve is essentially the same for both products; it descends continuously until 650° is reached where the horizontal of the oxide Bi_2O_3 starts. Wendlandt[22] likewise heated this product and found that it decomposes at a temperature lower than that reported by us.

13. Bismuth phosphate

Various workers have precipitated bismuth phosphate with diammonium phosphate. See Moser[13], Stähler and Scharfenberg[23], Schöller and Waterhouse[24], Bladsdale and Parle[25], and Stähler[26]. The precipitate filters well and gives a curve which shows the loss of retained water up to 379°. The excellent horizontal which begins there corresponds approximately to the phosphate $BiPO_4$, which is still stable at 1000°. Kho Wha Lee[9] has reported the region 480–800° for this compound.

14. Bismuth arsenate

Salkowski[27] precipitated bismuth arsenate by means of arsenic acid in an ammoniacal medium. The product $BiAsO_4$ is white and filters well. Its weighing has given rise to considerable discussion. Water is lost up to 47° and then the curve shows a level extending as far as 400°. The material may be heated as high as 607° without much error, but at a higher temperature beginning at 854° one atom of oxygen is lost between two molecules and the arsenate arsenite $BiAsO_4 \cdot BiAsO_3$ results. We doubtless were the first to report this white salt. On the other hand, Yosida[28] called attention to the existence of a stable semihydrate below 150° which yielded the anhydrous arsenate, which then produced a horizontal between 450° and 550°. Kho Wha Lee[9] found the semihydrate to be present from 100° to 110° while the anhydrous product yielded its horizontal from 450° to 800°.

15. Bismuth chromate

The orange precipitate of the chromate, prepared as directed by Löwe[21], gives rise to a curve which is difficult to interpret since the precipitate is actually a mixture which releases oxygen in stages. The bright green residue contains a chromite which is not attacked by acids. We suggest that this method be discarded.

Kho Wha Lee[9] assumed the formation of the phases:

$Bi_2O_3 \cdot 2CrO_3 \cdot H_2O$ below 250°
$Bi_2O_3 \cdot 2CrO_3$ between 350° and 450°.

16. Bismuth molybdate

The precipitate, produced as directed by Miller and Cruser[29] by the action of ammonium molybdate in ammoniacal medium, yields a curve which descends without interruption until the temperature reaches 410°. A horizontal begins at 410° and continues as far as 840°; it is related to the mixture $Bi_2O_3 \cdot 4MoO_3$. The latter oxide sublimes rapidly above 840°. Kho Wha Lee[9] gives the limits 500–700° for this level.

17. Bismuth chromithiocyanate

The reagent potassium chromithiocyanate $K_3[Cr(SCN)_6]$ and the precipitate $Bi[Cr(SCN)_6]$ were prepared by the procedure of Mahr[30] who suggested that the product be weighed after drying at 120–130°. The curve contradicts this proposal; it shows no horizontal before 876°, where a green mixture is produced consisting of the oxides of chromium and bismuth. The level observed between 81° and 173° is a straight but inclined line; its start corresponds almost to the formula given above. Hence the complex is necessarily beginning to decompose at 130°.

18. Cobaltic ethylenediamine iodobismuthate

The iodobismuthate of cobaltic ethylenediamine was prepared by the procedure of Spacu and Suciu[31] which yields a complex whose formula is $[Coen_2]I(BiI_4)_2$. The product is very stable up to 188° and in this range gives a perfectly horizontal stretch. Hence it is not necessary to dry the complex in a vacuum, but an oven at 100° is satisfactory. On heating, most of the bismuth sublimes as iodide. A horizontal which conforms to a mixture of Co_3O_4 and Bi_2O_3 starts at 758°. The level extending as far as 188° may be used for the automatic determination of bismuth; the analytical factor is 23.24%.

19. Basic bismuth formate

The basic formate $BiO \cdot HCO_2$, also known as bismuthyl formate, was precipitated as described by Kallmann[32]. It decomposes at once when heated. The curve includes a good horizontal between 45° and 155° which agrees with a material whose molecular weight is 251.5. The calculated value for $(BiO)_2CO_3$ is 255. The latter decomposes in turn and the oxide Bi_2O_3 is obtained from 370° on. The carbonate may be used for the automatic determination.

References p. 645

20. Bismuth benzoate

In the opinion of its authors, Jewsbury and Osborn[33], the quantitative precipitation as the benzoate serves especially as a separation method. The curve does not include any level indicating constant weight at 110°. There is no reduction if filter paper is not used; a horizontal conforming to the oxide Bi_2O_3 begins at 623°.

21. Hexamine iodobismuthate

The directions given by Solodovnikov[34] were followed with regard to determining bismuth as the iodobismuthate of hexamethylenetetramine. He claimed to have obtained a precipitate whose formula he gave as $BiI_3[(CH_2)_6N_4]_3$, whereas Kakita[35] found that the precipitate is merely bismuth oxyiodide BiOI, which he suggested be dried between 120° and 200°. We have confirmed the finding that the precipitate contains no hexamethylenetetramine. The curve descends until the temperature reaches 600° and then it has practically the same shape as that of the oxyiodide. We advocate the abandonment of this method.

22. Bismuth cupferronate

The precipitation by means of cupferron is conveniently carried out as directed by Pinkus and Dernies[36]. The white product loses weight as far as 400°. However, it is not necessary to dry the material at 400° because the yellow oxide Bi_2O_3 is contaminated with the lower black oxide. The latter disappears very slowly during the subsequent heating and notably between 758° and 946°. It is correct to assume that pure Bi_2O_3 is present at this latter temperature.

23. Bismuth pyrogallate

With pyrogallol we repeated the procedure given by Feigl and Ordelt[37] who claim to have obtained the yellow compound $BiC_6H_3O_3$, which may be dried at 110°. The curve descends until the temperature reaches 140° where sudden decomposition occurs with evolution of carbon monoxide. The residue between 191° and 409° consists of a mixture of bismuth, bismuth oxide, and carbon. The curve then ascends slightly and the oxide Bi_2O_3 is obtained at 841°.

24. Bismuth gallate

We, like Kieft and Chandlee[38], obtained a sulphur-yellow unstable precipitate with gallic acid. The formula is close to that of dermatol $C_6H_2(OH)_3 \cdot COOBi(OH)_2$ or $C_6H_2(OH)O_2CO_2Bi.2H_2O$, which is also known as bismuth subgallate. However, this product cannot be seriously recommended as a weighing form. The thermolysis curve shows a steady loss of weight with an oblique level between 60° and 140°. The oxide Bi_2O_3 is obtained at 948° but it is slightly contaminated with bismuth.

25. Bismuth salicylaldoxime

The inner complex, produced in neutral surroundings, with salicylaldoxime as described by Flagg and Furman[39], does not decompose in simple fashion. The curve includes a slightly inclined level between 90° and 190° but it cannot be considered for determining bismuth. The destruction of the organic matter (clinging and disagreeable odour) proceeds up to 310° and then the carbon burns. The bismuth takes up oxygen and the oxide Bi_2O_3 is obtained from 877° on. This is the weighing form.

26. Bismuth oxinate

The procedure used with oxine was that of Hecht and Reissner[40] who suggested that the inner complex $Bi(C_9H_6ON)_3$ be weighed after drying at 140°. However, the curve shows no level for this temperature, in fact it is descending rapidly. The oxide Bi_2O_3 is obtained at 798°, though slightly contaminated with bismuth. We therefore are forced to reject this method.

27. Bismuth 2-mercaptobenzothiazole

When the precipitation is made by means of 2-mercaptobenzothiazole, as described by Spacu and Kuraš[41], the product carries an excess of bismuth hydroxide because it is necessary to work in a slightly ammoniacal medium. The precipitate produces a horizontal extending as far as 173°. The residue at 497° consists of a mixture of bismuth and the oxide Bi_2O_3. Slow oxidation proceeds beyond this temperature but it is essential to maintain the crucible at 927° to obtain constant weight of the Bi_2O_3.

References p. 645

28. Bismuth thionalidate

The thionalidate $Bi(C_{12}H_{10}ONS)_3 \cdot H_2O$ was prepared as directed by Berg and Roebling[42] who recommended drying the precipitate at 100°. The precipitation is quantitative (test with thiourea); it is necessary to employ an excess of alcohol in the solution of the reagent. The curve includes a horizontal extending from 45° to 134°; it is suitable for the automatic determination and agrees with the anhydrous compound. The ignition residue above 944° is a mixture of bismuth oxide Bi_2O_3 and the sulphate $(BiO)_2SO_4$ and it is not usable for analysis. Umemura[43] made a study of this method and found that the compound can be weighed as the monohydrate after it has been kept at the ordinary temperature, and as the anhydrous compound after heating to 140–160°.

29. Antipyrine methyleneamine iodobismuthate

Since the antipyrine–methyleneamine reagent is not on the market, we prepared it in our laboratory. The method used was that of Takaki and Takasi[44] and yielded an orange precipitate of iodobismuthate $[C_{11}H_{11}N_2$ $(OCH_2)_3N]_2 \cdot 3HBiI_4$, which produces a horizontal up to around 160°. The authors of the method suggest drying the product at 110°. Two very short levels are observed at 307° and 572° before reaching the oxide Bi_2O_3 at 839°. The oxide contains iodine as in the analogous cases involving a iodobismuthate.

30. Bismuth phenyldithiodiazolonethiol

The brick-red precipitate $Bi(C_8H_5N_2S_3)_3 \cdot \frac{1}{2}H_2O$ was obtained with phenyldithiodiazolonethiol as described by the originator of the method Majumdar[45] who kindly supplied us with the reagent. The curve includes a perfectly horizontal stretch from 40° to 150°. Majumdar suggested 105° as the drying temperature. This level may be used for the automatic determination (with $\frac{1}{2}$ H_2O). The precipitation occurs readily and the product is washed and dried with no difficulty. It is unfortunate that the reagent is so uncommon. The residue above 600° is the oxide Bi_2O_3 which likewise gives rise to an equally good horizontal level.

31. Bismuth phenylarsonate

Majumdar[46] used phenylarsonic acid to obtain the precipitate C_6H_5-

$AsO_3.BiOH$. Its thermolysis curve shows a good horizontal extending from 60° to 300–310° which may serve for the automatic determination of bismuth. The yellowish residue is a mixture of bismuth arsenate and bismuth oxide.

32. α-Naphthoquinoline iodobismuthate

α-Naphthoquinoline iodobismuthate, obtained by the method of Hecht and Reissner[47], yields thermolysis curves that were not reproducible. They contained a level at temperatures below 300° but this level did not always correspond to a compound of the same apparent molecular weight. We found 1959 and 1425 successively instead of 897. In this second instance, washing the precipitate with benzene revealed that the product contained occluded amine. The final level above 900° corresponds to the oxide Bi_2O_3. The method cannot be retained for the automatic determination of bismuth. Furthermore, it does not seem proper to recommend it for the determination along the ordinary lines with Bi_2O_3 as the weighing form since the decomposition of the organic matter produces toxic materials.

With the above information at her disposal, Panchout[2] was in a position to make a critical and comparative study of the best methods for determining bismuth.

REFERENCES

[1] S. PANCHOUT AND C. DUVAL, *Anal. Chim. Acta*, 5 (1951) 170.
[2] S. PANCHOUT, *Diploma of Higher Studies*, Paris, 1952.
[3] S. KALLMANN AND F. PRISTERA, *Ind. Eng. Chem., Anal. Ed.*, 13 (1941) 8.
[4] L. VANINO AND F. TREUBERT, *Ber.*, 31 (1898) 1303.
[5] J. COUSIN, *J. pharm. chim.*, 28 (1923) 179.
[6] W. MAWROW AND F. MUTHMANN, *Z. anorg. Chem.*, 13 (1897) 209.
[7] R. MEINEKE, *Lehrbuch der Chemischen Analyse*, 1904, Vol. 2, p. 45.
[8] E. A. OSTROUMOV, *Zavodskaya Lab.*, 8 (1939) 1226.
[9] KHO WHA LEE, *J. Chem. Soc. Japan*, 52 (1931) 167.
[10] G. LUFF, *Z. anal. Chem.*, 63 (1923) 330.
[11] F. HECHT AND R. REISSNER, *Z. anal. Chem.*, 103 (1935) 283.
[12] BUISSON AND FERRAY, *Monit. sci.*, 3 (1873) 903.
[13] L. MOSER, *Z. anal. Chem.*, 45 (1906) 19.
[14] I. K. TAIMNI AND G. B. S. SALARIA, *Anal. Chim. Acta*, 11 (1954) 54.
[15] I. K. TAIMNI AND S. N. TANDON, *Anal. Chim. Acta*, 22 (1960) 555.
[16] F. FAKTOR, *Pharm. Post*, 33 (1900) 301, 317.
[17] T. GASPAR Y ARNAL AND J. M. POGGIO-MESORANA, *Anales soc. españ. fís y quím.*, 43 (1947) 439.
[18] G. H. BAILEY, *J. Chem. Soc.*, 51 (1887) 679.
[19] O. FUNAKOSHI, *Bull. Chem. Soc. Japan*, 10 (1935) 359.
[20] C. DUVAL, *Anal. Chim. Acta*, 16 (1957) 223.

21 J. Löwe, *J. prakt. Chem.*, 67 (1856) 464.
22 W. W. Wendlandt, *Texas J. Sci.*, 10 (1958) 392.
23 A. Stähler and W. Scharfenberg, *Ber.*, 38 (1905) 3865.
24 W. R. Schöller and E. F. Waterhouse, *Analyst*, 45 (1920) 435.
25 W. C. Bladsdale and W. C. Parle, *Ind. Eng. Chem., Anal. Ed.*, 8 (1936) 352.
26 A. Stähler, *Chemiker-Ztg.*, 31 (1907) 615.
27 H. Salkowski, *J. prakt. Chem.*, 104 (1868) 170.
28 Y. Yosida, *J. Chem. Soc. Japan*, 60 (1940) 915.
29 E. H. Miller and F. Cruser van Dyke, *J. Am. Chem. Soc.*, 27 (1905) 116.
30 C. Mahr, *Z. anal. Chem.*, 120 (1940) 6; *Z. anorg. Chem.*, 208 (1932) 313.
31 G. Spacu and G. Suciu, *Z. anal. Chem.*, 79 (1929) 196.
32 S. Kallmann, *Ind. Eng. Chem., Anal. Ed.*, 13 (1941) 897.
33 A. Jewsbury and G. H. Osborn, *Anal. Chim. Acta*, 3 (1949) 652.
34 P. P. Solodovnikov, *Trans. Kirov. Inst. Chem. Tech. Kazan*, 8 (1940) 60; *C.A.*, 35 (1941) 2438.
35 Y. Kakita, *J. Chem. Soc. Japan*, 65 (1944) 435.
36 A. Pinkus and J. Dernies, *Bull. soc. chim. Belges*, 37 (1928) 267.
37 F. Feigl and H. Ordelt, *Z. anal. Chem.*, 65 (1925) 448.
38 L. Kieft and G. C. Chandlee, *Ind. Eng. Chem., Anal. Ed.*, 8 (1936) 392.
39 J. F. Flagg and N. H. Furman, *Ind. Eng. Chem., Anal. Ed.*, 12 (1940) 529.
40 F. Hecht and R. Reissner, *Z. anal. Chem.*, 103 (1935) 261.
41 G. Spacu and M. Kuraš, *Z. anal. Chem.*, 104 (1936) 88.
42 R. Berg and W. Roebling, *Angew. Chem.*, 48 (1935) 597.
43 T. Umemura, *J. Chem. Soc. Japan*, 61 (1940) 25.
44 S. Takaki and Y. Takasi, *J. Pharm. Soc. Japan*, 56 (1936) 405.
45 A. K. Majumdar, *J. Indian Chem. Soc.*, 21 (1944) 240, 347.
46 A. K. Majumdar, *J. Indian Chem. Soc.*, 21 (1944) 119, 187, 188.
47 F. Hecht and R. Reissner, *Z. anal. Chem.*, 103 (1935) 88.

CHAPTER 80

Thorium

The chemistry of thorium has profited during the last years from the discovery of precipitates with a large number of organic acids, so that the hydroxide, iodate, selenite, thiosulphate, and pyrophosphate have been demoted to a lower level. However, these organic precipitates prove to have very unequal values. We do not include here the precipitation by means of sodium hypophosphate, suggested by Koss[1], since it is not quantitative, and furthermore Rosenheim[2] and then Hecht[3] had already condemned this procedure. Beck[4] reported that sodium alizarin-3-sulphonate yields a precipitate with thorium but apparently he did not investigate its quantitative aspects. The same is true of quinaldic acid recommended by Erämetsä[5], of gallic acid by Neish[6], and phenylarsonic acid by Rice, Fogg and James[7].

1. Thorium oxide hydrate

(i) *Precipitation by aqueous ammonia*   The colloidal hydrous oxide $ThO_2 \cdot nH_2O$ is obtained by keeping a mixture of thorium nitrate and ammonium hydroxide at 70° for several hours, as directed, for example, by Kroupa and Hecht[8]. The curves show an inflection point near 150° according to Ott[9,10], and some of the preparations show a portion of the curve with a little less slope up to near 300°. These findings are not sufficient to tell whether the compound $ThO_2 \cdot 2H_2O$ exists. The final level of the anhydrous oxide is generally attained with difficulty around 950° by keeping the material thermostatted at this temperature. The horizontal starting at 500° continues up to 1000° if the precipitate has been prepared by adding thorium chloride to potassium hydroxide. Kato[11] gives 780° as the minimal temperature of the level related to ThO_2. A great number of factors are involved here.

(ii) *Precipitation by gaseous ammonia*   Trombe[12] devised the procedure in which a slow stream of air charged with ammonia is passed into the aqueous solution of thorium nitrate being determined. A quantitative precipitation is obtained, beginning at pH 3.0, and the product agglomerates quickly and can be filtered immediately when the pH reaches 6.0, and the material appears to be purer than the one discussed in (i). The curves given by this non-gelatinous product show an abrupt change of direction near 130° in the course of the dehydration and agree rather well with $ThO_2 \cdot 2H_2O$. See Dupuis and Duval[13,14]. However, Ott[9] has shown by infra-red spectrography that the material is not the hydroxide $Th(OH)_4$ but rather a less condensed hydrate than the preceding one. The dehydration is complete at 470°; the level of the oxide ThO_2 begins there.

(iii) Precipitation by hydrogen peroxide   Using the directions of Denis and Kortright[15] and the further information supplied by Wyrouboff and Verneuil[16] and Benz[17], we treated thorium nitrate with 2.5% hydrogen peroxide. The white precipitate was colloidal but it could be filtered easily. On heating, it loses moisture, and then there is a sudden change in direction near 100° where a level begins that is essentially horizontal over about 20 degrees. Calculation shows that this level relates to $ThO_2 \cdot 3H_2O_2 \cdot 3H_2O = 419\text{–}420$. No level was found corresponding to such compounds as $Th(O_2H)_4$ or $Th(O_2H)(OH)_2$, or $Th(O_2H)(OH)_3$, which had previously been reported in the literature. The infra-red spectra reveal only bands due to liquid water. More deep-seated dehydration of the peroxide compound just mentioned leads, between 296° and 450°, to a compound with the approximate formula $ThO_2 \cdot 2H_2O$ which yields an almost horizontal level. The latter product loses its water and the anhydrous oxide ThO_2 appears from 650° on.

(iv) Precipitation by hexamine   The hexamethylenetetramine method of Ismail and Harwood[18] serves principally for the separation of thorium from cerium; it yields a white precipitate containing the organic compound[13]. The curve descends until the temperature is near 900° and gives no indication of an intermediate transitory material. The organic matter retained by the precipitate is eliminated in a discontinuous fashion between 180° and 200°.

(v) Precipitation by tannin   When tannin was the precipitant, the procedure used was identical with that employed for the precipitation of zirconium. See Neish[6] and Schoeller[19]. The resulting colloidal mixture gives a thermolysis curve which descends in irregular fashion until the temperature reaches 475°; the horizontal due to the thorium oxide is then observed. See also Section 20 which deals with thorium tannate.

2. Thorium fluoride

Although thorium fluoride precipitates quantitatively, it is very difficult to filter. It cannot be recommended for gravimetric purposes. Matsuura[20] apparently found a level due to $ThF_4 \cdot H_2O$. The conversion into thoria ThO_2 is complete at 700°.

3. Thorium iodate

The procedure of Meyer and Speter[21] and Meyer[22] was employed for the precipitation (in very acid surroundings) with iodic acid. Very few chemists employ this iodate gravimetrically; they prefer to determine the iodine titrimetrically or to heat the iodate with ammonia and then weigh the resulting thorium dioxide. Chernikhov and Uspenskaya[23] suggested that the precipitate has the complicated formula $4Th(IO_3)_4 \cdot KIO_3 \cdot 18H_2O$; however, our analysis is in closer agreement with the formula published more recently by Moeller and Fritz[24]. We are dealing here with the normal iodate $Th(IO_3)_4$, which furthermore is unstable and readily hydrolyzes while it is being washed. The curve shows a rapid loss of water up to approximately 100°. The anhydrous iodate maintains a constant weight between 200° and 300°; iodine and oxygen are lost as the temperature mounts. From 674° on, the residue consists of iodine-free thorium oxide ThO_2. Consequently, the material may be weighed after it has been heated between 674° and 947°. In our opinion the method has little to offer with regard to the gravimetric determination of thorium since no advantage is taken of the relatively high atomic weight of iodine. However, the method should be kept in mind because of its value for separation purposes.

4. Thorium thiosulphate

The precipitation by means of a thiosulphate was conducted as described by Chydenius[25] and modified by Hintz and Weber[26], Drossbach[27], and Hauser and Wirth[28]. Even when a No. 4 glass frit is used, the wash liquid is turbid; the precipitate retains free sulphur. The curve descends until 900° is reached, which is the beginning of the region of the thorium dioxide. The curve is difficult to interpret because of the variable quantity of impurities contained in the precipitate. The method should be regarded merely as a means of separating thorium and the rare earth metals, and even here better procedures are known.

5. Thorium selenite

The directions given by Kot'a[29] were followed when selenious acid was used as the precipitant. The white precipitate loses weight up to 946°, where only thorium oxide is present. No level was observed at 150°, the temperature recommended by Kot'a for drying the selenite.

6. Thorium nitrate

Commercial thorium nitrate is a mixture of the tetrahydrate and hexahydrate; the loss of water begins at 50°. The tetrahydrate $Th(NO_3)_4 \cdot 4H_2O$ appears near 110° but there is no clear-cut level. Oxides of nitrogen are suddenly released between 216° and 270°, and the level of the thorium oxide starts between 600° and 630°. See also Wendlandt[30].

7. Thorium pyrophosphate

Thorium pyrophosphate was prepared by the procedure of Carney and Campbell[31]. Its thermal desiccation curve is extremely simple. Moisture is given off up to 130°; then there is an abrupt change in the course of the curve and the retained water is slowly eliminated up to 540°. The horizontal of the anhydrous pyrophosphate then appears. No loss of phosphorus was observed up to 950°, even though the authors of this method believed such a loss did occur.

8. Thorium oxalate

The directions given by Hecht and Krafft-Ebing[32] were followed for the precipitation of the oxalate; the procedure used in microanalysis is given by Hecht and Donau[33]. Precipitated thorium oxalate has a variable water content. The pyrolysis curve has a break near 90° which corresponds to the dihydrate $Th(C_2O_4)_2 \cdot 2H_2O$. A short level near 153° conforms to the composition $Th(C_2O_4)_2 \cdot 1.5H_2O$. It is difficult to obtain (and hence to weigh) the anhydrous product at 350°. Almost as soon as it is formed it loses carbon monoxide and carbon dioxide, and the horizontal due to the thorium oxide ThO_2 starts at 610°. The temperature of 950° given in the paper is too restrictive.

Padmanabhan, Saraiya and Sundaram[34], who apparently are not aware of our work, heated this oxalate on a Stanton balance at the rate of 180° per hour.

9. Thorium stearate

The stearate, which was studied by Wendlandt[35], has the composition of a basic salt $Th(OH)_2(C_{17}H_{35}CO_2)_2$. It starts to lose weight near 50°; above 205° the decomposition rapidly accelerates and the level of the thorium oxide is reached at 450°. This method of determining thorium is more accurate than that employing the *m*-nitrobenzoate (Section 17).

10. Thorium fumarate

The Metzger[36] procedure was employed for the precipitation of the fumarate, but the milieu must not be too acidic. The pyrolysis curve shows a peculiarity. After the anhydrous salt, which yields no level, is decomposed, a horizontal appears at 405° whose ordinate agrees well with the oxide ThO_2.

11. Thorium sebacate

Thorium sebacate $Th[CO_2-(CH_2)_8-CO_2]_2$ was prepared as described by Kaufman[37]; it filters very rapidly. The anhydrous salt is present between 70° and 125°. It then decomposes by a seemingly complicated mechanism and thorium oxide is completely formed above 650°. This is an excellent method for determining thorium. With regard to the automatic determination, we particularly suggest the use of the level given by the anhydrous salt. The analytical factor for thorium is 0.3670 in this case.

12. Thorium benzoate

The precipitate produced by benzoic acid was studied by Wendlandt[35]; its formula is $Th(C_6H_5CO_2)_4$. It is stable up to 285°. There is a change in slope at 435° during the destruction of the organic matter but no level is observed. Pure thorium oxide ThO_2 appears from 510° on.

13. Thorium phenylacetate

Contrary to the report by Datta and Banerjee[38], the phenylacetate, prepared by Wendlandt[35], corresponds to a normal salt; its formula is $Th(C_6H_5-CH_2-CO_2)_4$. It begins to lose weight at 100°, slowly at first, and then more rapidly beyond 275°. The curve has a break at 355°. Finally, the level of the thorium oxide ThO_2 starts at 500°.

14. Thorium 2,4-dichlorophenoxyacetate

Wendlandt[35] prepared the 2,4-dichlorophenoxyacetate $Th(OH)(C_6H_3Cl_2-OCH_2-CO_2)_3$. It does not begin to lose weight below 140°. Above 300° the decomposition becomes very rapid and leads to a level near 475°. This precipitate is well suited to the separation of thorium from numerous other ions.

15. Thorium 2,4,5-trichlorophenoxyacetate

The 2,4,5-trichlorophenoxyacetate has been prepared by Datta[39] and Wendlandt[40]; its formula is $Th(OH)(C_6H_2Cl_3-OCH_2-CO_2)_3$. It is stable only up to 55°. The compound decomposes rapidly beyond 205°; the curve shows a break at 325°. The carbonaceous matter burns slowly; the level of the thorium oxide starts at 620°.

16. Thorium m-hydroxybenzoate

The m-hydroxybenzoate, prepared by Wendlandt[35], likewise has a basic structure $Th(OH)_2(HO-C_6H_4-CO_2)_2$. Its initial loss of weight occurs at 50°; the decrease remains very slow up to 350°; the rapid decomposition which then ensues leads to the level of the thorium oxide beginning at 550°. This precipitate serves for the separation of thorium and cerium (III).

17. Thorium m-nitrobenzoate

The method of Neish[6] was used for the formation of thorium m-nitrobenzoate $Th(NO_2-C_6H_4-CO_2)_4$. The curve reveals that the anhydrous salt can exist from 70° to 153°, and this temperature range may serve for the automatic determination of thorium. When heated further, the salt decomposes explosively, with spattering, at 300–305°. A new horizontal begins at 413°; it corresponds to the thorium oxide ThO_2.

18. Thorium anthranilate

Wendlandt[35] found that the o-aminobenzoate (anthranilate) appears to correspond to a formula between $Th(OH)_2(NH_2-C_6H_4-CO_2)_2$ and $Th(OH)_3(NH_2-C_6H_4-CO_2)$. When heated at the rate of 4.5° per minute, the first loss of weight is at 75°; the decrease rapidly accelerates beyond 325° and the level of the thorium oxide starts at 450°.

19. Thorium pyrogallate

Wendlandt[35] prepared the precipitate with pyrogallic acid; it has a basic structure and its formula is $Th(OH)_2(C_6H_3O_3)_2$. It starts to lose weight slowly at 50°. The decomposition speeds up abruptly above 150° and the level of the oxide ThO_2 begins at 675°.

20. Thorium tannate

Wendlandt[40] studied the thermolysis of the tannate whose approximate formula is $Th(OH)(C_{14}H_9O_9)_3$. The decomposition becomes very rapid beyond 215° and the level of the oxide starts at 565°.

21. Thorium m-cresoxyacetate

According to Wendlandt[35], the m-cresoxyacetate corresponds closely to the formula $Th(OH)_2(CH_3-C_6H_4-OCH_2-CO_2)_2$. Following a slight decrease in weight at 85°, there is a constant weight level which extends to about 210°. The salt starts to decompose rapidly beyond 210° and yields thorium oxide from 500° on.

22. Thorium phenoxyacetate

Wendlandt[40] reports that the phenoxyacetate $Th(OH)_2 (C_6H_5O-CH_2-CO_2)_2$ begins to break down at 50° and a break in the curve appears at 105°. No level is observed before that of the thorium oxide ThO_2 which begins at 575°.

23. Thorium phenylpropionate

When the phenylpropionate $Th(OH)_2(C_6H_5-CH_2-CH_2-CO_2)_2$ is heated the first decrease in weight is observed at 40°. The decomposition be-

comes rapid at 180° and the thorium oxide is formed completely at 475°.
See Wendlandt[40].

24. Thorium cinnamate

Wendlandt[35] reports that the cinnamate $Th(OH)(C_6H_5-CH=CH-CO_2)_3$ has good stability up to 180°. The speed of decomposition increases near 300°. The level of the oxide starts at 490°. This precipitate may be used for separating thorium from the rare earths present in monazite sands.

25. Thorium diphenate

The thermolysis curve of the 2,2'-diphenate was followed by Wendlandt[40]. The compound has the formula $Th(OH)_3(CO_2H-C_6H_4-C_6H_4-CO_2)$. The first loss of weight is observed at 50°. The decomposition is slow at first becomes faster above 375°. The level of the thorium oxide starts at 575°.

26. Thorium o-acetylsalicylate

According to Wendlandt[40], the o-acetylsalicylate $Th(OH)_2(CH_3CO-O-C_6H_4-CO_2)_2$ is stable only up to 40°. The decomposition accelerates beyond 140° and the level of the thorium oxide ThO_2 begins at 665°.

27. Thorium 5-bromosalicylate

The 5-bromosalicylate $Th(HO-C_6H_3Br-CO_2)_4$ corresponds closely to a normal salt and hence, as observed by Wendlandt[40], it does not begin to decompose until the temperature reaches 200° since its stability is greater than that of a basic salt. However, the salt decomposes rapidly beyond 200° and the level of the thorium oxide begins at 635°.

28. Thorium 5-nitrosalicylate

The 5-nitrosalicylate explodes suddenly with loss of thorium at about 300°. The level of the oxide begins at 595°. It is not possible to determine the formula of the initial precipitate by means of a thermobalance. See Wendlandt[40].

29. Thorium 1-hydroxy-2-naphthoate

The 1-hydroxy-2-naphthoate or approximately $Th(OH)_2(HO-C_{10}H_6-CO_2)_2$ is not stable and begins to lose weight at 35°. The decomposition rate increases markedly around 310°, and the curve shows a break at 335°. The level of the oxide begins at 485°.

30. Thorium phenylglycine-o-carboxylate

The phenylglycine-o-carboxylate corresponding to the formula $Th(OH)_3(CO_2H-CH_2-NH-C_6H_4-CO_2)$ is not much more stable. The first loss of weight is observed at 45°. The decomposition becomes rapid beyond 240°. The oxide residue is obtained near 530°. See Wendlandt[40].

31. Thorium cupferronate

Thornton[41] made studies which led to the precipitation of thorium cupferronate. If not too much sulphuric acid is used, the product filters well and does not adhere to the walls of the vessels. The anhydrous compound is not stable; it is difficult to obtain with it the theoretical formula. A product is invariably obtained whose molecular weight is too low, even if the drying is conducted in the open air. The inflection observed near 91° does not correspond to the anhydrous compound, since an explosion has already occurred at around 60° with evolution of vapours. A horizontal starts at 408° and extends as far as 946°; it corresponds to the thorium oxide ThO_2.

32. Thorium benzenesulphinate

The preparation of thorium benzenesulphinate was carried out in accord with the unpublished studies of Feigl, Hecht and Korrisch[42]. The curve slants downward as far as 540° where the level of the thorium oxide begins. No level is observed for the anhydrous compound, but only an abrupt change in direction near 85° when most of the moisture has been eliminated.

The curves recorded with thorium p-chlorobenzenesulphinate proved to be unsatisfactory. The reagent is too fragile to lead to quantitative results.

33. Thorium oxinate

The curve traced with thorium oxinate, prepared as described by Hecht and Ehrmann[43], showed no horizontal anywhere between 20° and 945° (difference from uranium). The curve descends continuously. Consequently, this method can only be accepted with much reservation.

Borrel and Pâris[44] repeated the pyrolysis on the assumption that the compound after being dried at room temperature corresponds essentially to the formula $Th(C_9H_6ON)_4 \cdot C_9H_7ON$ with no water of crystallization. It is stable up to 80°, but especially above 136° it probably loses 1 molecule of water, originating from 2 molecules of the oxinate:

$$2\ Th(C_9H_6ON)_4 \cdot C_9H_7ON \rightarrow H_2O + 2\ Th(C_9H_6ON)_4 \cdot (C_9H_6N)_2O$$

This latter compound corresponds to the inclined level extending from 142° to 175°, and by the loss of 1 molecule of the dioxyquinolyl ether it yields the compound $4Th(C_9H_6ON)_4 \cdot (C_9H_6N)_2O$, which is revealed by the level 232–245°. Finally, it was only after starting at 275° and going up to 345° that these workers obtained the stability level of the normal oxinate $Th(C_9H_6ON)_4$, *i.e.* at a temperature distinctly higher than that reported by Berg (105–110°). However, Borrel and Pâris point out that prolonged heating below 275° will certainly result in the production of the normal oxinate.

Wendlandt[45], who doubtless heated 100 mg at the rate of 5.4° per minute, observed the level of the anhydrous oxinate around 250° if it had previously been dried for 24 hours in the air. .

34. Thorium 5,7-dihalogenooxinates

The 5,7-dichlorooxinate, prepared by Wendlandt[45], showed no horizontal stretch before that of the thorium oxide, *i.e.* near 600°. The same can also be said regarding the 5,7-dibromooxinate. The 5,7-diiodooxinate of Wendlandt[46] proved to be stable below 90° when 100–150 mg were heated at the rate of 300° per hour. The thorium dioxide appeared above 520°.

35. Thorium ferronate

If the directions given by Ryan, Donnell and Beamish[43] are followed precisely, thorium may be separated and determined by means of ferron which is also known as loretine (7-iodo-8-hydroxy-quinoline-5-sulphonic acid). The curve shows that the anhydrous precipitate is stable from 110°

to 216°; beyond 216°, the organic matter disappears and thorium dioxide is obtained quantitatively at 570°. Despite the relative difficulty of carrying out this method, we recommend it for the automatic determination of thorium.

36. Thorium methyloxinate

The cream-colored methyloxinate $Th(C_{10}H_8ON)_4$ was studied by Wendlandt[48]. It was found to be relatively stable. Decomposition does not set in below 185° and then becomes rapid at 325°. The level of the thorium oxide starts at 435°.

37. Thorium picrolonate

Thorium picrolonate was prepared by the method of Hecht and Ehrmann[43] as given in the manual by Hecht and Donau[33]. The anhydrous product is obtained between 60° and 200°. It decomposes explosively and suddenly at 266° if heated slowly (rate 3). The portion of the residue not ejected from the crucible consists of a mixture of carbon and thorium oxide ThO_2. The method could be suggested for the automatic determination of thorium if it were not for the rarity of the reagent. However, the procedure employing sebacic acid is better.

38. Thorium mercaptobenzothiazole

Spacu and Pirtea[49] developed the method employing mercaptobenzothiazole as the precipitant. The product $Th(C_7H_4NS_2)_4$ begins to lose weight at 35°, slowly at first and rapidly above 180°. The curve shows breaks at 375° and 675°. The curve slopes downward continuously as far as 910°.

REFERENCES

1 M. Koss, *Chemiker-Ztg.*, 36 (1912) 686.
2 A. Rosenheim, *Chemiker-Ztg.*, 36 (1912) 821.
3 F. Hecht, *Z. anal. Chem.*, 75 (1929) 28.
4 G. Beck, *Mikrochemie*, 27 (1939) 47.
5 O. Erämetsä, *Suomen Kemistilehti*, B, 17 (1944) 30.
6 A. C. Neish, *J. Am. Chem. Soc.*, 26 (1904) 780.
7 A. C. Rice, H. L. Fogg and C. James, *J. Am. Chem. Soc.*, 48 (1926) 895.
8 E. Kroupa and F. Hecht, in F. Hecht and J. Donau, *Anorganische Mikrogewichts-analyse*, Springer, Vienna, 1940.
9 C. Ott, *Compt. rend.*, 240 (1955) 68.

[10] C. CABANNES-OTT, *Thesis*, Paris, 1958.

[11] T. KATO, *J. Chem. Soc. Japan*, 52 (1931) 774.

[12] F. TROMBE, *Compt. rend.*, 215 (1942) 539; 216 (1943) 888; 225 (1947) 1156.

[13] T. DUPUIS AND C. DUVAL, *Anal. Chim. Acta*, 3 (1949) 589.

[14] C. DUVAL AND T. DUPUIS, *Compt. rend.*, 228 (1949) 401.

[15] L. M. DENIS AND F. L. KORTRIGHT, *Z. anorg. Chem.*, 6 (1894) 35; 13 (1897) 412.

[16] G. N. WYROUBOFF AND A. VERNEUIL, *Compt. rend.*, 126 (1898) 340.

[17] E. BENZ, *Chemiker-Ztg.*, 15 (1902) 297.

[18] A. M. ISMAIL AND N. F. HARWOOD, *Analyst*, 62 (1937) 185.

[19] W. R. SCHOELLER AND H. W. WEBB, *Analyst*, 54 (1929) 704.

[20] K. MATSUURA, *J. Chem. Soc. Japan*, 52 (1931) 730.

[21] R. J. MEYER AND M. SPETER, *Chemiker-Ztg.*, 35 (1910) 306.

[22] R. J. MEYER, *Z. anorg. Chem.*, 71 (1911) 65.

[23] Y. A. CHERNIKHOV AND T. A. USPENKAYA, *Zavodskaya Lab.*, 9 (1940) 276.

[24] T. MOELLER AND N. D. FRITZ, *Anal. Chem.*, 20 (1948) 1055.

[25] J. J. CHYDENIUS, *Poggendorfs Ann.*, 119 (1863) 46.

[26] E. HINTZ AND H. WEBER, *Z. anal. Chem.*, 36 (1897) 27.

[27] G. P. DROSSBACH, *Z. angew. Chem.*, 14 (1901) 655.

[28] O. HAUSER AND F. WIRTH, *Z. angew. Chem.*, 22 (1909) 484.

[29] J. KOT'A, *Chem. Listy*, 27 (1933) 79, 100, 128, 130, 194.

[30] W. W. WENDLANDT, *Anal. Chim. Acta*, 15 (1956) 435.

[31] R. J. CARNEY AND E. D. CAMPBELL, *J. Am. Chem. Soc.*, 36 (1914) 1136.

[32] F. HECHT AND H. KRAFFT-EBING, *Mikrochemie*, 15 (1934) 43.

[33] F. HECHT AND J. DONAU, *Anorganische Mikrogewichtsanalyse*, Springer, Vienna, 1940, p. 212.

[34] V. M. PADMANABHAN, S. C. SARAIYA AND A. K. SUNDARAM, *J. Inorg. & Nuclear Chem.*, 12 (1960) 356.

[35] W. W. WENDLANDT, *Anal. Chem.*, 29 (1957) 800.

[36] F. J. METZGER, *J. Am. Chem. Soc.*, 24 (1902) 275, 916.

[37] L. E. KAUFMAN, *J. Appl. Chem. (U.S.S.R.)*, 8 (1935) 1520.

[38] S. K. DATTA AND G. BANERJEE, *J. Indian Chem. Soc.*, 32 (1955) 231.

[39] S. K. DATTA, *Anal. Chim. Acta*, 14 (1956) 39.

[40] W. W. WENDLANDT, *Anal. Chim. Acta*, 18 (1958) 316.

[41] W. M. THORNTON JR., *Am. J. Sci.*, 42 (1916) 151.

[42] F. FEIGL, F. HECHT ÁND F. KORRISCH, unpublished results cited in F. HECHT AND J. DONAU, *Anorganische Mikrogewichtsanalyse*, Springer, Vienna, 1940, p. 214.

[43] F. HECHT AND W. EHRMANN, *Z. anal. Chem.*, 100 (1935) 98.

[44] M. BORREL AND R. PÂRIS, *Anal. Chim. Acta*, 4 (1950) 267.

[45] W. W. WENDLANDT, *Anal. Chem.*, 28 (1956) 499.

[46] W. W. WENDLANDT, *Anal. Chim. Acta*, 15 (1956) 533.

[47] D. E. RYAN, W. J. MCDONNELL AND F. E. BEAMISH, *Anal. Chem.*, 19 (1947) 416.

[48] W. W. WENDLANDT, *Anal. Chim. Acta*, 17 (1957) 274.

[49] G. SPACU AND T. PIRTEA, *Acad. rep. populare Romîne*, 2 (1950) 669.

CHAPTER 81

Uranium

Gravimetrically, uranium has been determined by precipitation as ammonium pyrouranate (by ammonium hydroxide, pyridine, ethylenediamine, mercuric oxide, ammonium benzoate, hexamethylenetetramine, tannin), as uranium(VI) "peroxide", oxyfluoride, sulphide, phosphate, as uranium(IV) oxalate, cupferronate, and as inner complexes with β-isatoxime, oxine, 5,7-dichlorooxine, 5,7-dibromooxine, 5,7-diiodooxine, quinaldic acid. With the exception of the phosphate, all of these materials produce the oxide U_3O_8 on heating to a suitable temperature. The precipitation with ethylenediamine offers no advantage as compared with pyridine, or even ammonium hydroxide, but only entails the use of a more expensive reagent. See Duval[1]. The use of mercuric oxide as precipitant as directed by Alibegoff[2] has been discarded, first because of the difficulty of working with it, and then because the procedure does not lead to quantitative results, as shown by Schwarz[3]. Weighing as $V_2O_5 \cdot 2UO_3$, which was recommended by Blair[4], is discussed in Chapter Vanadium.

To this list we have added: the oxide UO_2, the tetrafluoride, the perchlorate, the sulphate, the sulphamate, the selenite, the nitrate, the arsenates, the complex carbonates, wyartite, the acetate, and uranium(IV) oxalate.

1. Ammonium pyrouranate

(i) Precipitation by aqueous ammonia The yellow precipitate of ammonium pyrouranate $(NH_4)_2U_2O_7$ was prepared by the method of Kern[5] with avoidance of the use of ammonium chloride so that the volatile oxychloride will not be formed. The curve confirms the findings of Lebeau[6] and of Tammann and Rosenthal[8] but does not substantiate the work of Fischer[8]. The trioxide UO_3 yields an essentially horizontal level between 480° and 610°. Its decomposition proceeds and is finished at 745°, and the oxide U_3O_8 is present from 745° to 946°. This is

the weighing form. At still higher temperatures, the curve descends and the product becomes blacker and blacker. No matter what the starting material, the decomposition which results in the formation of the oxide UO_2 appears to begin at 946° on all of the recordings. This temperature should not be exceeded if the heating is conducted in the air or in oxygen.

More recently, Gasco Sanchez and Fernandez-Cellini[9] heated 0.7 g of this ammonium pyrouranate at the rate of 300° per hour on a Chevenard thermobalance. The curve descends from 20° to 500°. The slightly inclined level observed between 500° and 600° corresponds closely to $UO_{2.90}$; the level of the oxide U_3O_8 starts at 900°.

(ii) Precipitation by gaseous ammonia By employing the technique outlined previously for aluminium and gallium, Duval[1,10] precipitated uranium from a solution of uranium(VI) sulphate which had not been previously treated with an ammonium salt. The precipitant was a slow stream of air charged with ammonia gas in accord with the technique recommended by Trombe[11] for the rare earths. The precipitate of ammonium pyrouranate begins to appear at pH 3.8; it may be filtered off, washed, and weighed at once. The thermolysis curve reveals a very distinct discontinuity at 283° where there is a sudden loss of ammonia gas. The resulting oxide UO_3 gradually loses weight as far as 675°. The oxide U_3O_8 yields a horizontal extending from 675° to 946°. The equilibrium between these two oxides is only slowly established. From 475° to 675°, the variation in weight (loss of oxygen) does not amount to 1 mg per 118 mg of the oxide U_3O_8, so that the automatic determination of uranium may be carried out after the precipitation has been made by the very practical Trombe method. Any point on the curve between 475° and 946° may be selected for the calculation.

(iii) Precipitation by pyridine The differences observed when the precipitation is made with ammonia, pyridine, ethylenediamine, tannin, hexamethylenetetramine, or even ammonium benzoate are not so marked with uranium as in the case of aluminium. All of these precipitants yield products which filter nicely. Consequently, the use of pyridine, as directed by Ostroumov[12], is not indispensable. The pyrolysis curve of the golden-yellow precipitate reveals a rapid loss of water up to 80°. The pyrouranate then slowly decomposes and suffers a sudden removal of pyridine at 325°. The resulting oxide UO_3 is gradually produced up to 745°. The oxide U_3O_8 yields a good horizontal from 745° to 946°.

(iv) Precipitation by ammonium benzoate The technique employed with ammonium benzoate was that of Shemyakin, Adamovich and Pavlova[13]. The curve is identical at all points with the one given by pyridine, but the oxide U_3O_8 shows constant weight beginning at 691°.

(v) Precipitation by hexamine The precipitation by means of hexamethylenetetramine is due to Rây[14]. The product adheres rather tenaciously to glass and to paper. It begins to lose weight at ordinary temperatures, and the organic matter stops coming off at 300° where the curve shows a distinct break. The oxide UO_3 provides a horizontal from 540° to 650°. The region of the oxide U_3O_8 is again found to extend from 745° to 946°.

(vi) Precipitation by tannin The tannin method of Das-Gupta[15] produces a voluminous precipitate which filters well only 15 minutes after it has been formed. The curve corresponding to this material reveals that weight is lost from room temperature on and water is given off until the temperature is near 140°. The "tannate" slowly breaks down and at 400° the residue consists of a mixture of UO_3 and unburned carbon. The oxide U_3O_8 begins to form at 539° but it starts to break down even at 878°, doubtless because the slight trace of carbon adhering to the oxide exercises a reducing action. The tannin method was the only one observed in which the oxide U_3O_8 produced from the precipitate decomposed below 946°.

Kato, Hosimija and Nakazima[16] and also Gasco Sanchez and Fernandez-Cellini[9] likewise investigated the thermal behaviour of the oxide U_3O_8.

2. Uranium oxides

(i) The oxide UO_2 Peakall and Antill[17] prepared the oxide UO_2 (sp. gr. 10.5) stoichiometrically and heated it on a thermobalance from 350° to 1000° in a current of air flowing at the rate of 500 ml per minute. The reaction was finished at 900° and yielded U_3O_8. Frequently, a protective coating forms on the pellet of UO_2. In general, the oxidation follows a linear course after the induction period which extends from 350° to 600°.

(ii) The oxide U_3O_8 and schoepite The oxide U_3O_8 was heated by

Kato, Hosimija and Nakazima[16], Gasco Sanchez and Fernandez-Cellini[9], and Protas[18]. The latter, in particular, made a study of the thermolysis of schoepite from Katanga, namely $UO_3 \cdot 2H_2O$. An initial loss of water between 60° and 135° is followed by a second loss (water of constitution) from 135° to 450°; he obtained the oxide UO_3 up to 510°. The artificial schoepite usually contains more water than the natural product.

(iii) Minerals Hydrolysis of uranyl acetate at 120° yields the hydrates $UO_2(OH)_2$ and $UO_3 \cdot 0.9H_2O$. Protas has also traced the dehydration of the following materials: becquerelite, the double oxide $2SrO \cdot 7UO_3 \cdot 10H_2O$, also $SrO \cdot 6UO_3 \cdot 10H_2O$, $BaO \cdot 3UO_3 \cdot 5H_2O$, billietite $BaO \cdot 6UO_3 \cdot 11H_2O$, also $3PbO \cdot 8UO_3 \cdot 10H_2O$, artificial curite, also four-marierite, vandendriesscheite, also artificial uranospherite. All of these materials have been heated at the rate of 100° per hour, and 350 mg specimens were used. They furnished perfectly distinct thermolysis curves, which were complemented by differential thermal analysis curves.

3. Uranium "peroxide"

The so-called "peroxide", which is ordinarily given the formula $UO_4 \cdot 2H_2O$, is precipitated in acidic surroundings by treating uranium(VI) nitrate with 12 volume hydrogen peroxide. The method was condemned by Schwarz[3] because he found it to be inexact, but our experience with it has been the opposite. Actually, the method is very precise and it lends itself also very well to the separation of uranium from other metals. Wunder and Wenger[19] employed this method for separating the uranium and beryllium occurring in Madagascar pegmatites. The pyrolysis curve can be interpreted only by abandoning the "peroxide" formula $UO_4 \cdot 2H_2O$, and by turning to infra-red absorption as a means of studying the structure. The formula as written above is correct from the standpoint of percentage composition but the atoms are not grouped properly. Moisture is lost smoothly up to 90° and only at this temperature does the compound correspond to the empirical formula just given. Exactly one molecule of hydrogen peroxide is lost over the interval 90–180°. It is rather astounding to find such relative stability exhibited near 100° by a compound containing a molecule of combined hydrogen peroxide. The residue, whose formula is $UO_3 \cdot H_2O$, slowly gives up one molecule of water and produces a horizontal due to the oxide UO_3 between 560° and 672°. In turn, the oxide U_3O_8 is stable between 800° and 946°, the interval used by the analyst. Accordingly, it appears that the formula $UO_3 \cdot H_2O_2 \cdot H_2O$ conforms more closely to the findings, and also it appears correct in view of the facts that the infra-red spectrum of the powder is identical with that of UO_3 and that no bands characteristic of an XO_4 group were distinguished. Similarly, the transient hydrate $UO_3 \cdot H_2O$ cannot be formulated as H_2UO_4 as has been incorrectly written by some writers.

Gasco Sanchez and Fernandez-Cellini[9] heated this "peroxide" at the rate of 40° per hour and traced the isotherms at 70°, 96°, 110°, 135°, 160°, 180°, 190°, 220°. Subsequently, they worked in oxygen, and then in nitrogen. In the latter case, they reached $U_3O_{7.908}$ at 1037°.

4. Uranium fluoride

Uranium(IV) fluoride UF_4 was heated by Gasco Sanchez and Fernandez-Cellini[9] at the rate of 300° per hour. It is essentially stable up to 350° and then gradually produces the oxide U_3O_8 up to 830° but the reaction is not really completed until the temperature has reached 950°. This same paper[9] contains pyrolysis curves of $UF_4 + UO_2$ mixtures, which present a maximum weight between 555° and 613°. Curves are also included for the ammonium salt $NH_4F \cdot UF_4$, obtained either by the wet or the dry method.

5. Uranium oxyfluoride

The oxyfluoride method of determination applies only to compounds containing tetravalent uranium. The initial solution is obtained from the reduction of uranium(VI) sulphate by pure zinc and sulphuric acid. Its U(IV) content was determined by titrating with standard permanganate solution. The precipitate $UOF_2 \cdot 2H_2O$ is produced in a platinum dish; it is readily washed by decantation. It can be weighed very precisely if care is taken to maintain the platinum crucible between 812° and 954° as indicated by the curve. Hydrofluoric acid is given off very smoothly, beginning at ordinary temperatures and continuing up to 812°. Weighing the anhydrous salt UOF_2 is out of the question. It has been reported sometimes that this method of determination is difficult to carry out because of the oxidation of uranium(IV) salts to uranium(VI) salts but this oxidation is insignificant if care is taken to maintain a certain degree of acidity in the solution. Under proper conditions, a uranium(IV) –uranium(VI) mixture keeps its strength unchanged for at least four hours in a vessel open to the air, and this stability is amply sufficient to permit the precipitation and the determination.

6. Potassium uranyl periodate

Burriel-Marti and Barcia-Goyanes[20] have traced the pyrolysis curve of a new precipitate, namely potassium uranyl periodate which they wrote $K_2(UO_2)_2I_2O_{10}\cdot 5H_2O$. This compound loses its five molecules of water from 20° to 100°. The anhydrous product then slowly loses oxygen up to around 350° where a horizontal begins and extends as far as 600°. This level corresponds to the compound $K_2(UO_2)_2I_2O_5$. Iodine comes off in turn, and the pyrouranate $K_2U_2O_7$, which is very stable, is left near 800°.

7. Uranium oxysulphide

By applying the method of Rose[21] to the preparation of the oxysulphide, a precipitate is obtained which forms spontaneously in the cold. It loses water and sulphur up to around 285°. Even at this temperature, the resulting UO_2 takes up oxygen and the net effect of two phenomena is observed. The uptake of oxygen is finished at 412° and the UO_3 formed is then stable up to 635°. Since the corresponding level is horizontal, the uranium may be weighed in this form. The oxide U_3O_8 is not pure until the temperature has reached 860° and it remains so up to 948°. This method is precise but the precipitation is really a delicate matter; sometimes it is necessary to wait for 24 hours before the filtrate is clear enough, and there is danger of introducing carbon dioxide into the solution under analysis which would then give lower results, since the expected oxysulphide forms complexes with ammonium carbonate.

We have not been able to verify the assertion by Pierlé[22] that the final oxide contains sulphate ion.

8. Uranyl sulphate

The sulphate $UO_2SO_4\cdot 3H_2O$ has been heated by Gordon and Denisov[23]. We likewise studied this salt a number of times, and also various commercial specimens. Although most of them contain some trihydrate, they also include a certain amount of the monohydrate, which we have never been able to detect through the formation of a corresponding level. Hence when 200 mg of the dry material were heated at the rate of 150° per hour the compound was found to be stable up to 190° $\pm$ 5°. The loss of water is regular after that, and the level of the anhydrous salt is reached at 400° $\pm$ 10°. Our experiments did not go beyond 637°.

9. Uranyl sulphamate

The amidosulphonate (sulphamate) $UO_2(NH_2-SO_3)_2\cdot4.5H_2O$ was first prepared by Capestan[24]. It loses 1.5 molecules of water from $70°$ to $100°$ and another molecule from $100°$ to $185°$. The intermediate very unstable hydrate loses water as soon as it is formed and is revealed on the thermolysis curve only by a change in slope. From $185°$ to $300°$, this curve includes a level which corresponds to $UO_2(NH_2-SO_3)_2.2H_2O$. However, the latter is not a dihydrate; in reality we are dealing here with its isomer, namely the double ammonium uranyl sulphate $(NH_4)_2UO_2(SO_4)_2$, which has been identified by chemical analysis, X-ray analysis, and infra-red spectrometry. The transformation occurs below $185°$. Beyond $300°$, the double ammonium uranyl sulphate loses ammonium sulphate in two steps. The intermediate product $(NH_4)_2SO_4\cdot2UO_2SO_4$ is not very stable. It does not yield a level but is revealed by a distinctly inclined section of the thermogram.

10. Uranyl selenites

(i) *Normal selenite*   The normal uranyl selenite UO_2SeO_3 was studied by Claude[25]. It is stable up to $550°$. From this temperature on it decomposes in accord with the reaction:
$$6\ UO_2SeO_3 \rightarrow 6\ SeO_2 + 2U_3O_8 + O_2 \tag{1}$$
when heated at the rate of $150°$ per hour.

(ii) *Acid selenite*   The acid selenite $UO_3\cdot2SeO_2\cdot2H_2O$ is dehydrated (or more precisely loses water of constitution) between $150°$ and $200°$:
$$UO_3\cdot2SeO_2\cdot2H_2O \rightarrow UO_2Se_2O_5 + 2\ H_2O \tag{2}$$
and then from $290°$ to $350°$:
$$UO_2Se_2O_5 \rightarrow UO_2SeO_3 + SeO_2 \tag{3}$$
and reaction (1) occurs from $550°$ on.

(iii) *Sodium uranyl selenite*   Claude also studied the thermal decomposition of sodium uranyl selenite $Na_2UO_2(SeO_3)_2$, which is stable up to about $450°$. It then decomposes:
$$2\ Na_2UO_2(SeO_3)_2 \rightarrow Na_2SeO_3 + Na_2U_2O_7 + 3\ SeO_2$$
as is the case with disodium diuranyl triselenite $Na_2(UO_2)_2(SeO_3)_3$, and monosodium hydrogen uranyl diselenite $NaH(UO_2)(SeO_3)_2$, and likewise the corresponding lithium, ammonium, potassium, and magnesium salts.

11. Uranyl nitrate

The nitrate $UO_2(NO_3)_2 \cdot 6H_2O$ is quite dry but not very stable. It begins to lose water at 46°, and it is impossible to distinguish between the loss of water and disintegration. See Duval and Wadier[26]. However, the decomposition rate decreases near 300° when the residue UO_3 is being approached, and it in turn loses oxygen, and the oxide U_3O_8 is obtained from 760° on. A standard solution of uranyl nitrate can be prepared by direct weighing if the material is taken from a fresh bottle.

Other workers have investigated this nitrate, especially Protas[18] who obtained the level of the oxide UO_3 near 490° and at 400° in a thermostat. Wendlandt[27], Bridge, Melton, Schwartz and Vaughan[28], and Vaughan, Bridge and Schwartz[29] have likewise studied uranyl nitrate.

12. Uranyl hydrogen phosphate

The determination of uranium through the monoacid phosphate, which was recommended by Coomans[30] in his critical investigation, has been repeated as described by Kern[5]. The operation is rather messy since the precipitate adheres tenaciously to paper, clogs the filter, etc. Its thermolysis curve reveals the loss of moisture up to 80° where the hydrate $UO_2HPO_4 \cdot H_2O$ appears. The latter is only slightly stable at this temperature and begins to break down at 130°. The anhydrous salt is produced at 272° but it too has little stability, and is converted into the pyrophosphate $(UO_2)_2P_2O_7$ beginning at 281°. The latter is stable beyond 673° and does not appear to break down even in the neighbourhood of 1000°. The level corresponding to this pyrophosphate is perfectly horizontal. The hygroscopicity of the product is no hindrance since it should be remembered that the crucible is not removed from the furnace during the weighing.

13. Ammonium uranyl arsenates

Yosida[31] reports that ammonium uranyl arsenate $NH_4UO_2AsO_4 \cdot 3.5 H_2O$ begins to lose water at 60°. It yields the stable monohydrate between 200° and 310°, and then produces the pyroarsenate $(UO_2)_2As_2O_7$ between 620° and 680°.

14. Mixed uranyl carbonates

Chernyaev et al.[32] have made a similar study of ammonium uranyl

tricarbonate, barium uranyl carbonate, ammonium diuranyl diaquopen-
tacarbonate monohydrate, and also the barium octohydrate, ammonium
uranyl diaquodicarbonate, the barium dihydrate, and ammonium diuranyl hydroxopentaaquotricarbonate.

Stonhill[33] employed differential thermal analysis and thermogravimetry to determine the fate of the simple uranyl carbonate UO_2CO_3,
which decomposes into UO_3 and CO_2. He also studied sodium uranyl
carbonate $Na_4UO_2(CO_3)_3$ which breaks down:

$$Na_4UO_2(CO_3)_3 \rightarrow Na_2UO_4 + Na_2CO_3 + 2\ CO_2$$

with intermediate stage:

$$2\ Na_4(UO_2)(CO_3)_3 \rightarrow Na_2U_2O_7 + 3\ Na_2CO_3 + 3\ CO_2$$

The other uranyl carbonates investigated decompose as follows:

$$2\ Na_2UO_2(CO_3)_2 \rightarrow Na_2U_2O_7 + Na_2CO_3 + 3\ CO_2 \rightarrow 2\ Na_2UO_4 + CO_2$$
$$Na_6(UO_2)_2(CO_3)_5 \rightarrow Na_2U_2O_7 + 2\ Na_2CO_3 + 3\ CO_2 \rightarrow 2\ Na_2UO_4 +$$
$$Na_2CO_3 + CO_2$$
$$Na_6(UO_2)_2(UO_3)_3(CO_3)_5 \rightarrow 2\ Na_2U_2O_7 + Na_2CO_3 + UO_3 + 4\ CO_2$$
$$\rightarrow Na_2UO_4 + 2\ Na_2U_2O_7 + CO_2$$

Wyartite   Wyartite, a new mineral from Shinkolobwe (Katanga) has
the formula $3CaO \cdot UO_2 \cdot 6UO_3 \cdot 2CO_2 \cdot 12\text{–}14H_2O$. It has been studied carefully by Guillemin and Protas[34] who found that the water is given off in
two stages, near 80° and around 160°. These findings were checked by
differential thermal analysis. Carbon dioxide is released and the UO_3 is
converted into U_3O_8 near 630°.

15. Uranyl acetate

Uranyl acetate $UO_2(CH_3CO_2)_2 \cdot 2H_2O$ is stable up to 100°. Duval[35]
found that the water of crystallization is released up to 159°. The level
of the anhydrous salt extends from 159° to 247°. The acetate then begins
to decompose very slowly and continues up to 388°, and then in an
explosive manner up to 412°. The residue consists of the oxide U_3O_8 which
doubtless is mixed with a little UO_2 since the thermolysis curve shows
a slight bend and a moderate reoxidation.

16. Uranium(IV) oxalate

The determination of uranium as uranium(IV) oxalate [in the presence
of uranium(VI)] appears to be as accurate as the fluoride method; it
has the advantage over the latter that it is not essential to work in plat-

inum vessels. The reduction was carried out with Bertrand zinc, as described
for the fluoride, and the procedure used was that of Rossi[36]. It is necessary
to use oxalic acid and not an alkali oxalate. The pale green precipitate
$U(C_2O_4)_2 \cdot 6H_2O$ loses four molecules of water between 50° and 78°; the
dihydrate (not reported hitherto) yields a short horizontal from 78° to
93°. Then the anhydrous oxalate gives an excellent horizontal extending
from 126° to 188°; this level is well suited to the automatic determination
of uranium. Above 188°, and particularly beyond 250°, the oxalate
suddenly decomposes and loses, as is usual with oxalates, carbon monoxide and carbon dioxide, while the residue takes up oxygen. In fact, it
is very astonishing to note that the residual oxide does not have the
composition UO_2, as would be expected, but instead it is the oxide U_3O_8.
The latter is not stable for long; it absorbs the quantity of oxygen needed
for the quantitative conversion into the oxide UO_3; finally the latter
again produces the oxide U_3O_8 above 700°. Because of the small amounts
of oxygen involved, these transformations have hitherto escaped direct
observation through weighing, but the continuous recording reveals them
very clearly even though as little as 100 mg of U_3O_8 is used.

Cappellina, Carassiti and Fabbri[37] repeated the thermal study of the
oxalate (with 5 H_2O) which they heated at the rate of 150° per hour.
They confirmed the finding that the compound begins to lose water at
50°. The recording was not continued beyond 250°.

Potassium uranium(IV) oxalate $K_4[U(C_2O_4)_4] \cdot 4H_2O$ was heated in the
same manner. It loses water between 50° and 180°.

17. Uranium(VI) oxalate

Uranium(VI) oxalate or $UO_2C_2O_4 \cdot 3H_2O$ was studied by Padmanathan,
Saraiya and Sundaram[38] who prepared this compound by the procedure
given by Wendlandt[39]. The monohydrate is produced at 100° and the
latter loses its water at 210°. Carbon monoxide and carbon dioxide are
evolved simultaneously at 365° and uranium oxides are left.

18. Uranium cupferronate

The excellent method of determining uranium(IV) with cupferron
yields results of great precision when conducted as described by Holladay
and Cunningham[40]. The thermolysis curve discloses the instability of the
precipitate which explodes at 92°. The solid residue does not possess a
definite composition even though it maintains a constant weight up to

219°. The oxide UO_3 yields an almost horizontal level extending from 609° to 744° and the oxide U_3O_8 is present from 800° to 946°. Holladay and Cunningham advise heating in a stream of oxygen but this precaution is superfluous. Good results are obtained by heating in the air, employing an electric furnace kept at 880°, for instance. It should be pointed out that this method of determining uranium is very rapid; the yellow light brown precipitate may be filtered off immediately and it can be satisfactorily washed in several minutes.

19. Uranium β-isatoxime

The method of precipitation employed for obtaining the inner complex with β-isatoxime is due to Hovorka, Sykora and Vorisek[41]; the French version of their paper, however, did not specify the kind of wash liquid to be used. The oxime loses moisture up to 68°; it then holds its weight almost constant until the temperature reaches 110°. However, weighing the anhydrous complex salt in this form is difficult. The organic matter decomposes slowly up to 218° but proceeds more rapidly from 281° to 343°. It is indeed astonishing to discover that the oxide U_3O_8 is reached at 408°; its level continues in horizontal fashion up to 945°.

20. Uranyl oxinate

The precipitation of uranium(VI) by oxine provides a precise and rapid method which can serve for the automatic determination in less than an hour. The technique adopted was that of Hecht and Donau[42] or that of Claassen and Visser[43]. They obtained a precipitate containing one molecule of oxine of crystallization; the thermogravimetric recording made with this precipitate disclosed a new fact. The complex ordinarily used in the analysis has the formula $UO_2(C_9H_6ON)_2.C_9H_7ON$. It is stable up to 157°. Furthermore, it is so slightly wetted by water that it dries almost instantly, especially if the final washing is with alcohol and ether. Hence the temperatures between 110° and 140° recommended hitherto for drying the precipitate are correct. If now the heating is carried above 157°, the supplementary molecule of oxine sublimes and a new horizontal starts at 252° and extends to 346° where sudden decomposition occurs. Wendlandt[44] reported 380° but he did not state the heating rate used. Complete carbonization is rather difficult and for a weight of approximately 250 mg, the oxide U_3O_8 remains contaminated with carbon up to at least 1000°. We have not been able to confirm the statement by

Wendlandt that the level of the oxide U_3O_8 is reached at 450°. Analysis of the residue leaves no doubt that it contains some carbon. This is true of most of the residues yielded by oxinates.

21. Uranyl 5,7-dichlorooxinate

The 5,7-dichlorooxinate is less attractive than the oxinate because it is more difficult to eliminate the retained reagent by sublimation. The curve given by Wendlandt[44] is hard to read. It appears that the level of the oxides is reached near 570°.

22. Uranyl 5,7-dibromooxinate

The 5,7-dibromooxinate likewise begins to yield the chelate UO_2 $(C_9H_4Br_2ON)_2 \cdot C_9H_5Br_2ON$ near 200°. The level due to the oxides is reached at around 580°. We have found that the residue still contains a sticky tar at this temperature.

23. Uranyl 5,7-diiodooxinate

The alleged diiodooxinate $UO_2(C_9H_4I_2ON)_2 \cdot C_9H_5I_2ON$ reported by Wendlandt[45] does not lose the supplementary molecule of the reagent when heated. The chelate begins to decompose at 75° and the decomposition continues rapidly above 350°. The level of the oxide U_3O_8 appears at 668°.

We have repeated this experiment, which was made in 1938 but not published because the results were so poor. Contrary to Wendlandt's findings, we believe that : (1) the precipitate does not have the reported formula; moreover, he did not publish his analytical data; (2) the residue at 668° is not the oxide U_3O_8; instead it consists of the oxide UO_3 containing carbon and finely divided iodine, which gives it almost the same colour as that of U_3O_8. Consequently, we believe this method should be discarded as a means of determining uranium.

24. Uranyl 2-methyloxinate

The 2-methyloxinate $UO_2(C_{10}H_8ON)_2 \cdot C_{10}H_9ON$, prepared by Wendlandt[46], is stable up to 55° when heated at the rate of 4.5° per minute. A break in the curve at 258° corresponds to the normal chelate; the latter is stable up to 290°. It then starts to decompose and yields the oxide U_3O_8 beginning at 480°.

25. Uranium quinaldate

Rây and Bose[47] report that quinaldic acid yields a precipitate which suffers a continuous loss of weight up to 475°. A good horizontal appears at this temperature and extends as far as 946°; it corresponds to the oxide U_3O_8. The precipitate does not have a constant composition; the curve shows nothing which indicates the presence of the anhydrous compound. Consequently, it is a shame to waste a reagent as expensive as quinaldic acid to prepare eventually only the oxide U_3O_8. Ammonium hydroxide can produce the same effect and at a much lower price.

26. Uranium acetylacetonate

In conclusion, reference should be made to a study of the acetylacetonate conducted by Wendlandt, Bear and Horton[48]. They employed calorimetry, thermogravimetry, and differential thermal analysis.

REFERENCES

[1] C. DUVAL, *Anal. Chim. Acta*, 3 (1949) 335.
[2] G. ALIBEGOFF, *Liebigs Ann.*, 233 (1886) 143.
[3] R. SCHWARZ, *Helv. Chim. Acta*, 3 (1920) 330.
[4] A. A. BLAIR, *Proc. Am. Phil. Soc.*, 52 (1913) 201.
[5] E. F. KERN, *Chem. News*, 84 (1901) 224, 236, 251, 260, 271, 283; *J. Am. Chem. Soc.*, 23 (1901) 685.
[6] P. LEBEAU, *Compt. rend.*, 174 (1922) 388.
[7] G. TAMMANN AND W. ROSENTHAL, *Z. anorg. Chem.*, 156 (1926) 20.
[8] A. FISCHER, *Z. anorg. Chem.*, 81 (1913) 202.
[9] L. GASCO SANCHEZ AND R. FERNANDEZ-CELLINI, *Anales real. soc. españ. fís. y quím.*, B, 54 (1958) 181.
[10] C. DUVAL, *Compt. rend.*, 227 (1948) 679.
[11] F. TROMBE, *Compt. rend.*, 215 (1942) 539.
[12] E. A. OSTROUMOV, *Z. anal. Chem.*, 106 (1936) 244.
[13] F. M. SHEMYAKIN, V. V. ADAMOVICH AND N. P. PAVLOVA, *Zavodskaya Lab.*, 5 (1936) 1129.
[14] P. RÂY, *Z. anal. Chem.*, 86 (1931) 20.
[15] P. N. DAS-GUPTA, *J. Indian Chem. Soc.*, 6 (1929) 777; *Analyst*, 55 (1930) 154.
[16] H. KATO, H. HOSIMIJA AND S. NAKAZIMA, *J. Chem. Soc. Japan*, 60 (1939) 1115.
[17] K. A. PEAKALL AND J. E. ANTILL, *J. Mater. Nucl.*, 2 (1960) 194.
[18] J. PROTAS, *Bull. soc. franç. minéral.*, 82 (1959) 239.
[19] M. WUNDER AND P. WENGER, *Z. anal. Chem.*, 53 (1914) 371.
[20] F. BURRIEL-MARTI AND J. BARCIA-GOYANES, *Anales soc. españ. fís y quím.*, B, 50 (1954) 285.
[21] H. ROSE, *Poggendorfs Ann.*, 116 (1862) 352.
[22] C. A. PIERLÉ, *J. Ind. Eng. Chem.*, 12 (1920) 60.
[23] B. E. GORDON AND A. M. DENISOV, *Ukrain. Khim. Zhur.*, 19 (1953) 368; *C.A.*, (1955) 5062.
[24] M. CAPESTAN, *Thesis*, Paris, 1960; *Ann. Chim. (Paris)*, 5 (1960) 222.

25 R. CLAUDE, *La chimie des hautes températures, 2e Colloq. natl. C.N.R.S.*, 1957, p. 147; *Thesis*, Caen, 1960; *Ann. Chim. (Paris)*, 5 (1960) 186.

26 C. DUVAL AND C. WADIER, *Anal. Chim. Acta*, 23 (1960) 259.

27 W. W. WENDLANDT, *Anal. Chim. Acta*, 15 (1956) 435.

28 R. BRIDGE, C. W. MELTON, C. M. SCHWARTZ AND D. A. VAUGHAN, *U.S. Atomic Energy Comm.*, BMI-110, 1956.

29 A. VAUGHAN, J. R. BRIDGE AND C. M. SCHWARTZ, *U.S. Atomic Energy Comm.*, BMI-1205 1957.

30 R. COOMANS, *Ing. chim.*, 10 (1926) 213.

31 Y. YOSIDA, *J. Chem. Soc. Japan*, 60 (1940) 915.

32 I. I. CHERNYAEV, V. A. GOLOVNYA, G. V. ELBERT, R.N. SHCHELOKOV AND V.P. MARKOV, *Proc. U.N. 2nd. Intern. Conf. Peaceful Uses Atomic Energy*, Geneva, 28 (1958) 235.

33 L. G. STONHILL, *Anal. Chim. Acta*, 23 (1960) 423.

34 C. GUILLEMIN AND J. PROTAS, *Bull. soc. franç. minéral.*, 82 (1959) 80.

35 C. DUVAL, *Anal. Chim. Acta*, 20 (1959) 266.

36 G. ROSSI, *Thesis*, Munich, 1902.

37 F. CAPPELLINA, V. CARASSITI AND G. FABBRI, *Ann. chim. (Rome)*, 50 (1960) 615.

38 V. M. PADMANATHAN, S. C. SARAIYA AND A. K. SUNDARAM, *J. Inorg. & Nuclear Chem.*, 12 (1960) 356.

39 W. W. WENDLANDT, *Anal. Chem.*, 30 (1958) 58.

40 J. A. HOLLADAY AND T. R. CUNNINGHAM, *Trans. Am. Electrochem. Soc.*, 6 (1929) 329.

41 V. HOVORKA, V. SYKORA AND J. VORISEK, *Ann. chim. anal.*, 29 (1947) 268; *Collection Czechoslov. chem. Communs.*, 10 (1938) 83.

42 F. HECHT AND J. DONAU, *Anorganische Mikrogewichtsanalyse*, Springer, Vienna, 1940, p. 205.

43 A. CLAASSEN AND J. VISSER, *Rec. trav. chim.*, 65 (1946) 211.

44 W. W. WENDLANDT, *Anal. Chem.*, 28 (1956) 499.

45 W. W. WENDLANDT, *Anal. Chim. Acta*, 15 (1956) 533.

46 W. W. WENDLANDT, *Anal. Chim. Acta*, 17 (1957) 274.

47 P. RÂY AND M. K. BOSE, *Z. anal. Chem.*, 95 (1933) 400.

48 W. W. WENDLANDT, J. L. BEAR AND G. R. HORTON, *J. phys. Chem.*, 64 (1960) 1289.

CHAPTER 82

Plutonium

Faugeras, Anselin and Grison[1] studied the thermal behaviour of plutonium(III) oxalate. Skljarenko and Chubukova[2] determined the stability regions of: the sulphate $Pu(SO_4)_2$, 280° to 450°; the iodate $Pu(IO_3)_4$, 200° to 250°; the oxalate $Pu(C_2O_4)_2$, 120° to 150°; the oxinate $Pu(C_9H_6ON)_4$, 50° to 230°. All of these compounds yield the oxide PuO_2 above 800°. See also Dawson and Elliott[3].

REFERENCES

[1] P. FAUGERAS, F. ANSELIN AND E. GRISON, *C.A.*, (1956) 11149.
[2] I. S. SKLJARENKO AND T. M. CHUBUKOVA, *Zhur. Anal. Khim.*, 15 (1960) 706.
[3] J. K. DAWSON AND R. M. ELLIOTT, *Atomic Energy Research Establishment (Great Britain)*, C/R 1207 (1957).

CHAPTER 83

Americium

1. Americium sulphate

The findings of Hall and Markin[1] regarding the behaviour of the hydrated sulphate have been confirmed. Levels are observed between 300° and 350°. The monohydrate $Am_2(SO_4)_3 \cdot H_2O$ yields a horizontal extending from 360° to 480°, and that of the anhydrous sulphate extends from 500° to 720°. The oxide AmO_2 is obtained near 850°.

2. Americium nitrate

The nitrate leaves this same oxide at 800°.

3. Americium oxalate

Markin[2] heated americium(III) oxalate at the rate of 3° per minute in air and in vacuo. He did not state the make of thermobalance used. In the air, the starting material $Am_2(C_2O_4)_3 \cdot 7H_2O$ yields in succession the hydrates carrying 4, 3, and $\frac{1}{2}$ H_2O. The anhydrous compound gives a horizontal from 240° to 320°. The dioxide AmO_2 is obtained from 470° on.

The hydrates with 3 and $\frac{1}{2}$ H_2O do not appear in vacuo. The anhydrous oxalate yields its horizontal from 240° to 300°. In this case, the brownish-red sesquioxide Am_2O_3 is obtained at 620°.

REFERENCES

[1] G. R. HALL AND T. L. MARKIN, *J. Inorg. Nuclear Chem.*, 4 (1957) 139.
[2] T. L. MARKIN, *J. Inorg. Nuclear Chem.*, 7 (1958) 290.

CHAPTER 84

Studies in organic chemistry

Even though the title of this book bears the adjective "inorganic", we believe it advisable to devote this closing chapter to a discussion of the results obtained with a number of purely organic compounds closely linked with inorganic analysis.

1. Carbohydrates

We have previously reported[1] on the burning of a filter paper. The curve obtained is fundamentally the same as those given by cellulose, glucose, starch, inulin, lactose, saccharose, light-colored dextrin, absorbent cotton, and cereal flours[2]. Starch has likewise been studied by Waters[3]. We invariably obtained the same curve with saccharose[4] no matter if the specimen was cane, beet, or maple sugar, or powdered or crystallized. When 200 mg were heated at the rate of 300° per hour, we found the weight to hold constant up to 231°, even though the product had already browned at this temperature. Evolution of gases then set in and the sample began to burn; the combustion was in full force at 327° and the reactions were over near 619°, where the crucible was practically empty. Only a trace of inorganic material (lime?) remains at 900° plus a little unburned carbon.

(i) Flour　The problem of determining the moisture and ash content of flours is difficult. In the first place, these materials are hygroscopic and it is almost impossible to obtain them in the anhydrous condition because the starch in them begins to lose water of constitution before all of the moisture can be eliminated. Furthermore, the residual ashes are themselves hygroscopic and may retain fine particles of carbon which burn only with difficulty. Consequently, the analyst may be deceived in one or both instances if he employs an ordinary balance. With wheat flour in particular, we constructed two isotherms, the first at 105° for 4 hours, the other at 130° for 105 minutes, and these have been adopted

References p. 683

by several other countries. The isotherms descend continuously; it is apparent that a constant weight cannot be obtained, and determination of the moisture content by means of an oven set at these temperatures leads only to self-deception on the part of the analyst.

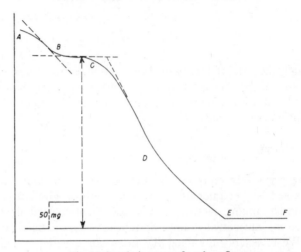

Fig. 80. Pyrolysis curve of a wheat-flour.

We first attempted to secure a counter-check by a chemical method, namely through a direct measurement of the moisture content by the Karl Fischer procedure. Seven kinds of cereal grains: wheat, oats, rye, barley, buckwheat, maize, rice, were heated linearly at the rate of 300° per hour. The samples weighed 300 mg. The moisture was given off from A to B (Fig. 80); an almost horizontal level BC is obtained, save in the case of rye, where it is sharply inclined. Water of constitution (plus nitrogen) is released from C to D. The change of direction at D shows that only carbon remains in the crucible; this carbon is found to be amorphous when studied by X-rays. It burns smoothly so that the loss of weight is linear. Ordinarily, we terminated the heating near 820°. From E to F and beyond, there is a horizontal stretch showing the weight of the perfectly dry colorless residue. The ordinate of this level EF readily gives the weight of the ash which can be read directly from the chart paper. Obviously, this is not constant for a given cereal; it depends of course on the variety, the soil in which it was grown, and especially on the bolting. The Table 12 gives some moisture and ash values.

TABLE 12

MOISTURE AND ASH CONTENTS OF FLOUR

Cereal	Water (%)	Ash (%)
Wheat	15.40	1.86
Oats	8.10	2.80
Rye	11.12	2.32
Barley	11.37	0.63
Maize	10.64	2.40
Buckwheat	11.41	2.82
Rice	13.12	2.25

The question of extraneous water (moisture) is more difficult to resolve. Since the level BC is oblique, it is possible, as in the case of a polarogram or a potentiometric determination curve, etc. to take the ordinate of $\frac{1}{2}$ the level extending from AB and DC prolonged into a straight line. There is no difficulty with regard to CD since it is a straight line. As to AB, it is necessary to be content with taking a tangent of inflection. The ordinate of $\frac{1}{2}$ the level gives a weight which *we believe to be suitable to accept as that of the anhydrous form.* From this it has been possible to deduce the loss due to moisture and obtain values which agree within almost 1% of the figure given by the Karl Fischer method. Some of these values are included in Table 12; they are dependent on the hygrometric state of the air and on the way in which the flour has been handled, also on the plant, the harvesting season, and the year in which the grain was harvested. This study has often been cited as an example of automatic determination.

(ii) Cellulose Bouttemy[5] studied the thermal decomposition of a cellulose and numerous carboxymethylcelluloses, using 200 mg samples which were heated in platinum at the rate of 330° per hour. The various carboxymethylcelluloses (substitution degree = 0.50) were prepared from bleached chestnut cellulose (polymerisation degree = 700).

Two kinds of thermograms were obtained.

(a) In the case of cellulose and carboxymethylcellulose, and the copper, zinc, and silver salts (*i.e.* those which yield an oxide or a metal), the water of hydration of the cellulose is found to depart up to 200–230°. Then the structure breaks down with loss of water and carbon dioxide near 300°.

(b) In the case of the sodium, potassium, and barium salts, whose decomposition yields a carbonate, the thermolysis curves differ from the preceding ones in that there is a very distinct level which stops abruptly at a point not far from 650° in the case of sodium and potassium and near 400° in the case of the barium salt at the moment at which the carbonaceous skeleton starts to burn.

2. Proteins

A study along other lines was made in association with Robert[6], namely of the hydration of various proteins. The investigations dealt with casein and gelatin kept for several days at 16–20° under a bell jar in an atmosphere saturated with water vapour. Two kinds of measurements were made: (1) at constant temperature (80°, 110°, 150°) attained with the speed of 100° per hour, and (2) under continuous heating from 20° to about 250° by raising the temperature of the furnace linearly at the rate of 50°, 100° or 200° per hour. The curves obtained in this way had the following characteristics. They were S-shaped and resembled the curves found among the adsorption isotherms. A horizontal level is not always observed; frequently the curve descends continuously, and occasionally the slope is very slight, which is in accord with the findings reported by H. B. Bull in *J. Am. Chem. Soc.*, 66 (1944) 1499. However, certain proteins yield isothermal levels of greater or less length, along which there is no change in weight, but the quantity of water lost depends intimately on the temperature of the isotherm. Kinetic study of the loss of weight as a function of the heating rate is capable of providing information about the activation energy of the dehydration of the proteins whereas the hysteresis phenomenon occasionally introduces an ambiguity into the thermodynamic interpretation of adsorption isotherms.

3. Organic acids

A number of organic acids have received attention because they can serve as standard materials in analytical chemistry.

(i) Oxalic acid Oxalic acid[7] retains its two molecules of water up to 38°. The dehydration continues up to about 132° where a change in direction of the curve indicates that the anhydrous compound, which is not stable under these conditions, is beginning to sublime. According to Martens[8], who heated 350 mg at the rate of 150° per hour, the crucible is empty at 270°.

(ii) Malonic acid   Malonic acid[9] is stable and anhydrous up to 144°; it then decomposes rapidly up to 227° and the last traces of carbon are gone from the crucible near 420°.

(iii) Succinic acid   Succinic acid[9] is likewise anhydrous and stable up to 195°; decomposition sets in suddenly and continues to around 313°, and the crucible is empty from 449° on.

(iv) Tartaric acid   No change of weight is shown by *d*-tartaric acid until its melting point (170°) is reached. The decomposition begins imperceptibly near 173° and speeds up suddenly around 214°; consequently this acid may be weighed after it has been dried for several minutes in the solid state[10].

(v) Citric acid   Citric acid[11] retained its molecule of water of crystallization up to 56° when 132 mg were heated at the rate of 150° per hour. Water is lost in a smooth fashion from 56° to 82°, and the anhydrous acid may be kept between 82° and 131°. Slow decomposition begins then and continues up to 161°, where it accelerates, and at 192° the loss of weight amounts to 61 mg per 200 mg. Therefore, when citric acid is to be employed for the preparation of buffer solutions, it may be recrystallized and dried below 56.°

(vi) Benzoic acid   Benzoic acid is usually dry and may be weighed just as it is when taken from the bottle[12]. The temperature should not exceed 94° if drying is deemed necessary; the acid sublimes rapidly beyond this temperature.

(vii) Phthalic acid   Dry anhydrous phthalic anhydride[9] is stable up to 123°; it then sublimes rapidly and the crucible is empty at 260°. The curve obtained with *o*-phthalic acid[9] is identical with that of the anhydride. This acid is dry and stable up to 187°. It then loses water and sublimes; the crucible is empty at 279°.

4. Guanidine and hydrazine

Duval and Wadier[9] heated 200 mg of guanidinium carbonate at the rate of 300° per hour. This compound is anhydrous and stable up to 150° at least; beyond this, the decomposition proceeds in three stages. Fauth[13] employed the thermobalance described by Gallant[14] for studying the thermal behaviour of a number of unstable derivatives of guanidine and hydrazine. He used 200 mg specimens and heated them at the rate of 480° per hour. The temperatures were read from the chart; the figures in the second column of Table 13 show the limit of the starting horizontal level.

TABLE 13

THERMOLYSIS OF GUANIDINE AND HYDRAZINE DERIVATIVES

Compound	Max. temp. (°C)	
N-Ethylguanidinium sulphate	110	loss of 12% up to 260°
Guanidinium sulphate	270	loss of 2 % up to 295°
Nitroguanidine	220	decomposes at 240°
Guanylurea styphnate	130	detonates at 182°
Aminoguanidinium styphnate	205	detonates at 210°
Hydrazinium styphnate	105	detonates at 105°
N-Ethylguanidinium styphnate	120	detonates at 125°
N-Methylguanidinium styphnate	100	detonates at 105°
Guanidinium styphnate	130	detonates at 180°
Hydrazinium picrate	100	decomposes at 125°
N-Methylguanidinium picrate	150	decomposes at 190°
N-Ethylguanidinium picrate	195	decomposes at 210°
Guanidinium picrate	220	decomposes at 222°
Aminoguanidinium picrate	160	decomposes at 175°
Guanylurea picrate	150	loss of 72 % up to 280°
Guanidinium nitrate	100	loss of 95 % up to 250°

5. Reagents for aldehydes and ketones

Duval and Dat Xuong[15] made a thermogravimetric study of the precipitates obtained with the functional reagents for aldehydes and ketones.

(i) 2,4-Dinitrophenylhydrazine First of all, 2,4-dinitrophenylhydrazine when applied to 14 aldehydes and 19 ketones yielded precipitates which, with two exceptions (*p*-methoxybenzophenone and *p*-ethylbenzophenone), yielded horizontal straight lines up to the vicinity of the melting points. The trials were conducted with a thermobalance provided with a pen recorder; 80–200 mg were heated at the rate of 150° per hour.

(ii) Dimedone The same holds for the precipitates obtained by the action of dimedone on 26 aldehydes belonging to very diverse series and

whose purity was checked by infra-red absorption tests. When dimedone itself (Eastman-Kodak, m.p. 148–149°) was heated, it proved to be far from stable up to its melting point. It starts to lose weight near 107° at the heating rate indicated above, and at 126.2 mg of this reagent[16] showed a loss of 16.0 mg at 153°.

(iii) Thiocarbohydrazide It was also found that thiocarbohydrazide yields crystalline precipitates with most of the aldehydes and some ketones; the products have sharp melting points. These products likewise yield straight thermolysis horizontals up to these melting points. The reagent itself decomposes, before melting at 94°, when it is heated linearly at the rate of 150° per hour[17].

6. Reagents for metals

(i) Oxine Oxine (8-hydroxyquinoline) is a well-known analytical reagent; it ordinarily is dry and may be weighed directly for the preparation of standard solutions[12]. It does not lose weight below 78° where it starts to sublime. This compound can be completely burned only with difficulty, and the total conversion of an oxinate into a metal oxide is greatly dependent on the weight of the specimen.

(ii) 5,7-Diiodooxine On the other hand, Wendlandt[18] found that 5,7-diiodooxine is not very stable. If 100–150 mg are heated at the rate of 240–300° per hour, slight sublimation occurs at 135° and then rapid decomposition ensues. Nothing but carbon remains at 310° and the crucible is empty at 700°. We, however, are somewhat skeptical about the existence of this diiodo-compound in view of the findings by Dupuis[19]; its composition can be only approximate.

(iii) o-Hydroxyphenyl-2-benzoxazole Wendlandt[20] has also investigated the behaviour of o-hydroxyphenyl-2-benzoxazole, a reagent introduced by Walter and Freiser[21], which forms chelates with various metals. It begins to lose weight at 100° (m.p. 124–126°) and does not sublime. It leaves a carbonaceous material at 218°.

7. Shale oil

Orosco[22], among others, has made a dry-distillation of shale oil with a new thermobalance. He observed levels related to the departure of free and combined water.

References p. 683

8. Complexones

Finally, in the field of the complexones, we, since 1947, have conducted a variety of thermolysis on samples kindly provided by Prof. Schwarzenbach.

(i) Cyanoacetic acid   Cyanoacetic acid proved to be stable up to 286° when heated at the rate of 300° per hour[23]. Wendlandt[24] reports 240° for the specimen he examined.

(ii) EDTA   When heated at this same rate, we found 230° to be the stability limit of ethylenediaminetetraacetic acid (EDTA) as measured on a specimen supplied by Prolabo and heated at 300° per hour. Wendlandt[24] heated a specimen supplied by J. T. Baker at the rate of 180° per hour and reported 230°. He also found 265° for the brand known as sequestrene AA put out by Geigy Chemical Company. The disodium salt $Na_2EDTA \cdot 2H_2O$ yielded various stability temperatures when investigated by Wendlandt: 125° (Na_2 sequestrene), 114° (Eastman), 105° (Baker). The commercial products available to us in Paris did not contain even one molecule of water; all of them were mixtures. Consequently, Duval[12] has suggested that they be completely dehydrated between 100° and 120° prior to weighing when it is desired to prepare a standard solution of reliable strength. The ignition residue invariably consists of sodium carbonate Na_2CO_3.

9. Antibiotics

Aside from their biological and pharmacological significance, certain antibiotics have been suggested for analytical purposes. Wendlandt and Zief[25] have heated some of these, namely:

Tetracycline hydrochloride	decomposes at 145°
Terramycin	decomposes at 165°
Aureomycin	decomposes at 175°
Aureomycin hydrochloride	decomposes at 170°
Erythromycin	level from 95° to 165° after loss of water
Leucomycin base	level from 90° to 155° after loss of water
Oleandomycin phosphate	level from 90° to 145° after loss of water
Carbomycin	continuous loss from 40° to 570°
Sulphocydin	stable up to 45°, level from 70° to 110°, 5% residue at 535°

Sodium penicillin G	loss above 50°, rapid decomposition above 165°, formation of sodium carbonate at 755°
Chloromycetin	stable up to 160°
Polymyxin sulphate	loss from 40° on, level from 120° to 160°
Viomycin sulphate	loss from 55° on, rapid loss at 190°
Pulvomycin	slow loss from 50° to 490°

Wendlandt and Zief's findings confirm our observations that certain organic materials lose some weight below their melting points. Therefore, it may be asked in all seriousness of what fundamental value are these melting points which are so highly regarded as criteria in organic chemistry?!

REFERENCES

[1] C. DUVAL, *Anal. Chim. Acta*, 2 (1948) 92.
[2] C. DUVAL, *Chim. anal.*, 36 (1954) 61.
[3] P. WATERS, *Nature*. 178 (1956) 324; *Coke and Gas*, 30 (1958) 341.
[4] C. DUVAL, *Anal. Chim. Acta*, 20 (1959) 263.
[5] M. BOUTTEMY, *Compt. rend.*, 240 (1955) 305.
[6] C. DUVAL AND L. ROBERT, *Compt. rend.*, 238 (1954) 282.
[7] C. DUVAL, *Anal. Chim. Acta*, 16 (1957) 221.
[8] P. MARTENS, *Bull. inst. agron. et stas. recherches Gembloux*, 21 (1953) 131.
[9] C. DUVAL AND C. WADIER, *Anal. Chim. Acta*, 23 (1960) 541.
[10] C. DUVAL. *Anal. Chim. Acta*, 15 (1956) 223.
[11] C. DUVAL, *Anal. Chim. Acta*, 13 (1955) 427.
[12] C. DUVAL, *Anal. Chim. Acta*, 16 (1957) 545.
[13] M. I. FAUTH, *Anal. Chem.*, 32 (1960) 654.
[14] W. K. A. GALLANT, *133rd Meeting A.C.S., San Francisco*, 1958.
[15] C. DUVAL AND NG DAT XUONG, *Anal. Chim. Acta*, 10 (1954) 520.
[16] C. DUVAL AND NG DAT XUONG, *Anal. Chim. Acta*, 12 (1955) 47.
[17] C. DUVAL AND NG DAT XUONG, *Mikrochemie*, (1956) 747.
[18] W. W. WENDLANDT, *Anal. Chim. Acta*, 15 (1956) 533.
[19] T. DUPUIS, *Mikrochemie*, 35 (1950) 449.
[20] W. W. WENDLANDT, *Anal. Chim. Acta*. 18 (1958) 638.
[21] J. L. WALTER AND H. FREISER, *Anal. Chem.*, 24 (1952) 984.
[22] E. OROSCO, *Ministério trabalho ind. e com., (Inst. nac. tecnol. Rio de Janeiro)* 1940
[23] C. DUVAL. *Anal. Chim. Acta*, 20 (1959) 20.
[24] W. W. WENDLANDT, *Anal. Chem.*, 32 (1960) 848.
[25] W. W. WENDLANDT AND M. ZIEF, *Nature*, 181 (1958) 1207; 182 (1958) 665; *Naturwissenschaften*, 45 (1958) 467.

AUTHOR INDEX

SUBJECT INDEX

Uranium vanadate, 659
Uranospherite, 662
Uranyl acetate, 154, 159, 662, 667
Uranyl amidosulphonate, 665
Uranyl ammonium sulphate, 121
Uranyl carbonate, 666
Uranyl 5, 7-dibromooxinate, 670
Uranyl 5, 7-dichlorooxinate, 670
Uranyl 5, 7-diiodooxinate, 670
Uranyl hydrogen phosphate, 666
Uranyl 2-methyloxinate, 670
Uranyl nitrate, 666
Uranyl oxinate, 669
Uranyl pyroarsenate, 666
Uranyl pyrovanadate, 297
Uranyl selenites, 665
Uranyl sulphamate, 121, 665
Uranyl sulphate, 664
Urea, 227, 417, 475

Vanadic anhydride, 112, 117, 188, 196, 295, 605
Vanadium, 161, 294
Vanadyl cupferronate, 299
Vanadyl dicyanodiamidine, 299
Vanadyl hydroxide, 294
Vanadyl 2-methyloxinate, 299
Vanadyl α-nitroso β-naphthol, 298
Vanadyl oxinate, 299
Vanadyl sulphate, 294, 295, 518, 519
Vanadyl sulphite, 294
Vandendriesscheite, 662
Vanillidene benzidine molybdate, 468
Vanillidene benzidine tungstate, 583
Vaterite, 130
Vermiculite, 234
Viomycin sulphate, 683
Vitamin C, 484
Vittel water, 325
Vulkacite-774, 333

Xylidine, 447

Wagnerite, 276
Wheat flour, 677
White lead, 627
Witherite, 533, 534
Wolfron tungstate, 581
Wyartite, 667

Yatren, 283
Yellow mercuric oxide, 150, 314, 337
Ytterbium, 161, 568
Ytterbium chloride, 568
Ytterbium nitrate, 568

Ytterbium oxalate, 568
Ytterbium oxychloride, 568
Ytterbium oxynitrate, 568
Ytterbium sulphate, 568
Yttrium, 161, 442
Yttrium chloride, 442
Yttrium cupferronate, 443
Yttrium fluoride, 442
Yttrium hydroxide, 442
Yttrium methyloxinate, 444
Yttrium neocupferronate, 443
Yttrium nitrate, 443
Yttrium oxalate, 443
Yttrium oxinate, 444
Yttrium oxychloride, 442
Yttrium oxyfluoride, 442
Yttrium oxynitrate, 443
Yttrium sulphate, 443

Zettlitz kaolin, 233
Zinc, 162, 405
Zinc, 336, 355, 374, 577, 597, 601
Zinc acetate, 154, 159
Zinc amidosulphonate, 408
Zinc ammonium phosphate, 408
Zinc anthranilate, 411
Zinc basic carbonate, 406
Zinc basic sulphate, 407
Zinc blende, 407
Zinc borneolglucuronate, 405
Zinc bromate, 406
Zinc bromide, 406
Zinc 5-bromoanthranilate, 411
Zinc carbonate, 96
Zinc chloride, 305
Zinc cyanamide, 410
Zinc 2, 7-diaminofluorene, 413
Zinc dithizonate, 413
Zinc formate, 410
Zinc hydroxide, 98, 99, 405
Zinc lithium pyrouranate, 168
Zinc mercurithiocyanate, 410, 604
Zinc methyloxinate, 412
Zinc monosalicylaldoxime, 412
Zinc nitrate, 408
Zinc oxalate, 411
Zinc oxide, 405, 531
Zinc oxinate, 411
Zinc phenyloxinate, 412
Zinc periodate, 523
Zinc picolate, 413
Zinc picrolonate, 414
Zinc pyridine 2-carboxylate, 413
Zinc pyridine thiocyanate, 410
Zinc pyrophosphate, 243, 408, 409

PRINTED IN THE NETHERLANDS BY A. W. SYTHOFF, N.V., LEYDEN